New Energy

Caineng Zou

New Energy

石油工业出版社
PETROLEUM INDUSTRY PRESS

Springer

Caineng Zou
Research Institute of Petroleum Exploration & Development
Beijing, China

ISBN 978-981-15-2727-2 ISBN 978-981-15-2728-9 (eBook)
https://doi.org/10.1007/978-981-15-2728-9

Jointly published with Petroleum Industry Press
The print edition is not for sale in China. Customers from China please order the print book from: Petroleum Industry Press.

This Springer imprint is published by the registered company Springer Nature Singapore Pte Ltd.
The registered company address is: 152 Beach Road, #21-01/04 Gateway East, Singapore 189721, Singapore

Preface

The earth was formed 4.6 billion years ago, and human beings have been born for 6 million years. Since primitive humans picked up the fire for the first time, energy, water and food constitute the three main elements of human survival.

Energy refers to the power resources that can be provided in nature for human survival and social progress. Energy comes mainly from in three ways, the sun, the earth and the moon. Energy types include primary energy and secondary energy. Energy is usually divided into two categories, traditional energy and new energy. Traditional energy mainly includes wood, coal, oil and natural gas, etc. and the "new energy" discussed in this book mainly refers to non-fossil new energy sources, focusing on solar energy, wind energy, geothermal energy, electricity, hydrogen energy and energy storage by using new materials and so on.

Scientific and technological progress and social civilization are the two major drivers of energy development. The world energy structure has undergone two transitions, of which the first transition has achieved the energy revolution from fuelwood to coal and the second transition has achieved the energy revolution from coal to petroleum. The current energy development of human beings is entering the third major transition period from traditional fossil energy to new energy.

Energy forms will be converted from solid (wood + coal), liquid (oil) to gaseous (natural gas), and carbon containing in energy will extend from high carbon (wood and coal), low carbon (oil and natural gas) to carbon-free (new energy). In the future, energy will develop along three major trends, such as utilizing resource types with reducing carbon, intensive production technology and diversification of utilization methods. Network big data system based on artificial intelligence, intelligent energy network system based on internet plus energy network, nano-materials, graphene, battery energy storage and other technologies are changing with each passing day, cost of new energy power generation is reducing and battery energy storage technology is breaking through, which will strongly promote the arrival of the new energy era.

Global energy is forming three new trends of oil and gas, conventional and unconventional, fossil and non-fossil collaborative development. Stabilizing oil and increase gas is the development trend, therefore, natural gas will develop rapidly, the production rate will surpass oil, and the twenty-first century will enter the era of natural gas development. To focus on both conventional/unconventional oil and gas has been the development strategy of major oil companies, insisting the conventional oil and gas as the main part of exploration, extending new areas of unconventional oil and gas strategy. According to the estimation of oil and gas resources, the life cycle of oil industry development will be more than 300 years. Traditional fossil energy is not renewable and renewable non-fossil new energy is bound to complete the ultimate revolution of traditional energy. Electric power, hydrogen energy and so on are the important direction of energy utilization in the future. Hydrogen energy and electrical energy are easy to exchange, which has a relatively better energy storage effect, and hydrogen energy can be used as the third energy or multiple energy for energy conversion, so the exchanges between hydrogen energy and power, and multiple energy conversion will be a new direction of energy utilization.

World energy production is forming the world energy structure mainly composing of four members, e.g. oil, natural gas, coal and new energy. In 2017, global energy production was 13.32 billion tons of oil equivalent, of which oil accounted for 32.93%, natural gas accounted for 23.76%, coal accounted for 28.29% and new energy accounted for 15.02%. The world's energy consumption is entering an era dominated by oil and gas (about 57%), however, due to abundant coal, shortage of oil and gas and delayed development of new energy, China is still in the era of energy consumption mainly depending on coal (accounting for 61%).

Global oil industry has entered the "stable period" of development, the world's proved remaining recoverable reserves of oil is 239.3 billion tons, the ratio of reserves over production is 54.5, and annual production rate is 4.39 billion tons. Natural gas is entering the stage of "climax", the world's proved remaining recoverable reserves of natural gas is 193.5 trillion cubic meters, the reserve production ratio is 52.3, and annual output of 3.7 trillion cubic meters. Coal is entering the the "transition period" of development, the world's proved remaining recoverable reserves of coal is 1.17 trillion tons, the average reserve production ratio reaches 134, and the annual output of coal 3.77 billion tons of oil equivalent. New energy is gradually entering the "golden period" of development, the global new energy available can be used for tens of millions of years, which is inexhaustible, theoretically the solar radiant energy reaching the earth's surface every year is 130 trillion tons of standard coal equivalent. The world's total production of new energy including nuclear energy, hydropower and renewables, reached 2.001 billion tons of oil equivalent in 2017.

As for energy, China is rich in coal but short of oil and gas, energy production faces three challenges. One is the worries about the increasing external dependence on oil and gas. Preliminary forecast between 2030 and 2035, China's annual oil production will be 160 million to 180 million tons, annual production of natural gas will be 180 billion to 200 billion cubic meters, however, annual oil consumption will be 650 to 700 million tons, annual consumption of natural gas will be 550 billion to 600 billion cubic meters, so it is necessary to establish early warning system of China's oil and gas "safe consumption peak" to avoid the safety risk of oil and gas supplies. Second, the composition of energy production is unreasonable. China's energy structure is in the form of "one major component and three non-major components" (in 2017, coal accounted for 70% of the country's energy production, oil accounted for 8%, natural gas accounted for 5% and new energy accounted for 17%). Third, the future pillar industries in the domain of new energy are still not clear. The development of wind, light, hydropower in China has a certain scale already and the production capacity of nuclear power is at the stage of massive construction. Geothermal, biomass energy, ocean energy, hydrate and other domains are still at the stage of starting exploration and development or learning.

In recent years, China's new energy industry has undergone a strong momentum of development, the proportion in the structure of primary energy production expands to 17%, and becomes an important part of energy. In China, hydropower resources are rich, nuclear power generation continues to grow and the proportion accounting for national electricity generation is also increasing. Hydrogen can be considered as one of the most promising secondary green energy resources in the twenty-first century, and the United States, Japan, Germany and other developed countries have successively promoted the hydrogen energy industry to the national energy strategy. Hydrogen and electricity are obtained through the conversion of primary energy (solar, wind, ocean energy, thermal energy, etc.). Hydrogen can generate electricity through fuel cell technology, and electricity can also be used to make hydrogen from water and therefore to connect and transfer energy among the entire energy network can be realized. Within the range of power grid extension, it is advisable to have the convenience of electricity and to use hydrogen for energy storage and supply outside the power grid. Hydrogen can be a strategic new energy resource in the future.

Energy consumption forecasted in China will reach a peak value of 4.4 billion tons of oil equivalent in 2030. Facing the new period of global climate change and vigorously developing low-carbon energy, China needs to speed up the production and industrialization pilot of conventional-unconventional oil and gas, coalbed methane, hydrate and hydrogen, speed up the realization of "two scale production" of clean utilization of coal and new energy. At present, it is necessary to speed up the realization of "production revolution" of conventional-unconventional oil and gas, the "clean utilization revolution" of coal development, the "speed revolution" of new energy development, and strive to realize the revolution of energy structure from "one major component and three non-major components" to "three major components" period of coal, oil and gas, new energy with the same importance before and after 2050. At that time, coal will account for about 40% of energy consumption, oil and gas will account for 30%, new energy will account for 30%. Around 2100, national "energy independence" is expected by relying on new energy development, fossil energy will account for about 30% of the energy consumption, non-fossil energy will account for 70%, achieving the transition of the historical status of fossil and non-fossil energy.

Mankind's eagerness to reduce carbon dioxide emissions and desire for high-carbon fossil energy to transition to non-carbon energy, so the size and speed of new energy taking the place of traditional energy sources will likely be beyond expectations. Energy is also the cornerstone of a country's strong power and security. Each country has proposed their own energy strategies, such as America's "energy independence", China's "energy revolution", and Japan's "Hydrogen Society". With the improvement of human demand for green ecological environment and the arrival of low-carbon society, the transition from traditional fossil energy to non-fossil new energy is an inevitable trend and choice for energy development. Energy resources have been moving and developing along the main route of clean, low-carbon, ultimately to the carbon-free energy continuously. Global energy production and utilization is moving towards cleaner, more efficient, more convenient, safer and more sustainable direction. Maybe, the replacement of conventional energy by new energy will come before the depletion of fossil energy.

The purpose of this book is to reveal the basic laws of world energy development, analyze the current situation and trends of energy production and consumption in the world, evaluate science and technology revolution and technological milestones of energy, propose the strategy of energy development, and the plan of oil and gas and other energy companies' strategic development based on the energy resources of China, and speed up the coming of low-carbon natural gas and the new energy era.

The book is striving to be popular, practical, strategic and forward-looking. It can be used as a textbook or reference book for energy-related institutions, and it can also be used as a reference and evidence for making relevant policies by energy companies and relevant national departments.

The book is divided into four parts and 12 chapters. The first part talks about energy trends, including 3 chapters which are the laws of energy development, the world energy map, energy development trends, respectively. The second part focuses on new energy revolution, including 2 chapters which are revolutionary energy technology, energy internet technology, respectively. The third part presents new strategy of traditional energy companies, including 2 chapters which are the new plan of energy of oil companies and the new plan of coal power company respectively. The fourth part is regarding theories of new energy, including 5 chapters which are hydrogen energy, energy storage and new materials, geothermal energy, nuclear energy, wind, light, tides and other new energy, respectively.

The preface was written by Zou Caineng. The introduction was written by Zou Caineng, Zhao Qun and Xiong Bo. The first part was mainly written by Zou Caineng and Zhao Qun, Zheng Dewen, Zhang Jinhua was involved in the preparation. The second part was mainly written by Zhang Qian, Sun Fenjin, Zheng Dewen, Chen Yanpeng and Chen Shanshan, Zhang Fudong and Zhang Jinhua were involved in the preparation. The third part was mainly written by Zhang Fudong, Zhang Qian and Gu Jiangrui, Zheng Dewen and Zhang Jinhua were involved in the preparation. Chapter 8 in the fourth part was mainly written by Zheng Dewen and Zhang Qian, Zhang Fudong participated in the preparation; Chap. 9 was mainly written by Zheng Dewen, Ge Zhixin, Dong Zhen, Xue Huaqing and Miao Sheng,

Zhang Jinhua, Zhang Fudong and Peng Yong participated in the preparation; Chap. 10 was mainly written by Zeng Bo, Cao Qian and Fang Chaohe, Zheng Dewen and Zhang Qian participated in the preparation; Chap. 11 was mainly written by Liu Renhe, Zhang Mengyuan, Xiao Hongping and Liu Weihong, Zhang Jinhua and Zhang Xi participated in the preparation; Chap. 12 was mainly prepared by Fang Chaohe, Zhang Fudong and Zheng Dewen, Zhang Jinhua and Zhang Qian participated in the preparation. Liao Qing was responsible for the preparation of the drawings and charts of this book. The book was finally compiled and revised by Zou Caineng, Zhang Fudong, Zheng Dewen, Zhao Qun, Zhang Qian and Zhang Jinhua.

Special thanks go to colleagues of the new energy team of CNPC Research Institute of Petroleum Exploration and Development for their strong support and assistance, and Ma Xinfu, Ran Yifeng and so on from Petroleum Industry Press for their careful editing and proofreading in the process of writing the book. Due to limited knowledge of the main authors of this book, lack of practice and experience engaged in new energy domain and tight schedule, the book may inevitably demonstrate inappropriate understanding or opinions. It is also possible that participants in the writing of the book or references are not all listed here. The book is definitely not perfect and your understanding and feedback will be highly appreciated with which perfection will hopefully be achieved when it's reprinted.

Beijing, China
September 2019

Caineng Zou

Executive Summary

Based on the analysis of global energy development trends and the laws of energy transition, this book introduces the principles, key technologies and main types of new energy utilization comprehensively and systematically. Starting from the basic law of energy development, this book points out the inevitable trend from fossil energy to non-fossil new energy resources, explains the significance of developing new energy to comply with the law of energy development and safeguarding national energy security scientifically and sincerely, introduces various technologies of new energy in detail, summarizes the new strategies of traditional energy companies, and the development status and application prospect of all kinds of new energy resources are also discussed respectively. The book is divided into four parts, the first part is "energy trend", including the law of energy development, the world's energy map and energy development trend. The second part is "new energy revolution", including revolutionary energy technologies and energy internet technologies. The third part is "new strategy of traditional energy companies", including the new plan of energy of oil companies and the new plan of coal power company, respectively. The fourth part is "theories of new energy", including hydrogen energy, energy storage and new materials, geothermal, nuclear energy, wind, light, tides and other new energy.

This book can be used as a textbook for teachers and students majoring in new energy in colleges and universities, and can also be used as a reference by researchers, managers, investors and national energy-related decision-making departments engaged in new energy domains.

Introduction

Basic Connotation of Energy

Energy refers to the energy resources that can be provided in nature for human survival and social progress. Energy comes mainly from three ways, i.e. the sun, the earth and the moon. Energy is divided into two types, i.e. primary energy and secondary energy resources. Primary energy is divided into renewable sources (hydro energy, wind, solar, geothermal, nuclear, etc.) and non-renewable energy (firewood, coal, oil, natural gas, etc.); secondary sources include electric power, hydrogen energy, gasoline, diesel, coke lasers and so on. Energy is usually divided into two other categories, i.e. traditional energy and new energy. Traditional energy mainly includes wood, coal, oil and natural gas, etc. and the "new energy" discussed in this book mainly refers to non-fossil energy resources, mainly including solar energy, wind energy, geothermal energy, electric energy, hydrogen energy, energy storage by new materials and so on. Energy generally has three attributes, i.e. natural, social and national properties. With the support of revolutionary, disruptive multidisciplinary subjects and new energy science and technology innovation, the era of artificial intelligence is imminent. One of the urgent desires of human beings is reducing carbon dioxide emission, and changing from high-carbon fossil energy to non-carbon energy which will make the speed and scale of changing from traditional energy to new energy beyond our expectations. Energy is also the power of a country's strength and the cornerstone of security. Each country has proposed energy strategies, such as the U.S.'s "energy independence", China's "energy revolution", and Japan's "hydrogen society". With the impetus of scientific and technological innovation and human civilization, the big step from traditional fossil energy to non-fossil new energy is an inevitable trend and this road must be chosen for energy development. The replacement of conventional energy with new energy will possibly happen early before the depletion of fossil fuel resources.

The Third Transition of Energy

The earth was formed 4.6 billion years ago and human beings first walked the earth 6 million years ago. Since primitive humans began to use fire for the first time, energy, water and food have constituted the three main elements for human survival. The world energy structure has undergone two transitions, of which the first transition is the energy revolution from firewood to coal and the second transition is the energy revolution from coal to petroleum. At present, the energy development of human beings is in the third important energy transition from traditional fossil energy to new energy. Easy access to firewood can meet the basic survival requirements of heating, cooking and so on at the early stage of human beings. With the progress of coal mining technology, coal with high energy density has been widely used. The invention of steam engine by Watt in 1769 followed by the construction of coal-fired power plants resulted in the rapid development of the coal industry. The proportion of coal in primary energy consumption exceeded firewood in the 1780s, and has become the largest role in primary energy so far. The first major transition from firewood to coal was completed. The demand for oil and gas as an efficient energy resource increased significantly in 1886 due to the invention and application of internal combustion engines. Progress of oil and gas geological development theory, drilling and completion engineering and refining technologies have resulted in a significant increase in oil and gas production and a rapid rise in the proportion of primary energy consumption structures. Oil and gas accounted for more than 50% in 1965, took the place of coal as the largest energy source, and the second major transition from coal to oil and gas was achieved. With the progress of social civilization, the ecological environment problems caused by the utilization of high carbon energy such as coal and petroleum are becoming more and more serious, which arose people's high alertness and profound reflection. With the increasing demand for a green ecological environment and the arrival of a low-carbon society, the third major transition from traditional fossil energy to non-fossil new energy will inevitably happen. Energy resources have been making progress along the main road of clean, low-carbon, towards carbon-free continuously. The trend of global energy production and utilization is to be cleaner, more efficient, more convenient, safer, smarter and more sustainable.

Two Driving Forces of Energy Development

The two major driving forces of scientific and technological progress and social civilization have driven energy development. From firewood to coal, coal and oil and gas to renewable energy, the replacement of each energy type and each transition are not only the result of scientific and technological innovation, but also the result of meeting the needs of the civilized development of human society. The continuous progress of the development and utilization of science and technology runs through the whole life cycle from the appearance to the prosperity of the same kind of energy. The development of social civilization drives energy demand. The energy of primitive society mainly satisfies the living demand; The feudal society has improved the quality of human life, the primary industrial production has greatly increased the energy demand, the social civilization has accelerated the development since the industrial revolution, and the human demand for transportation, information and cultural entertainment has greatly increased, and the modern industrial demand for energy has reached an unprecedented level. The development and utilization of high carbon energy has resulted in wastewater, waste gas, waste residue and other ecological environment problems, so ecological demand for more environmentally-friendly energy production and consumption has been a new requirement. Carbon reduction as well as green, intelligent and secure energy technologies have become new trends.

Three New Structures of Energy Development

With the progress of social civilization and the improvement of scientific and technological level, global energy is forming a new pattern of coordinated development of oil and gas, conventional and unconventional, fossil and non-fossil. To stable oil and increase gas production is the development trend, during which natural gas will play a more important role than oil and the twenty-first century will enter the era of natural gas development. The development strategy of major oil companies has already focused on both conventional and unconventional oil and gas at the same time, persisting in the exploration of conventional oil and gas as the main part, starting up a new field of unconventional oil and gas strategy, realizing the development of "artificial reservoirs" through the horizontal well, large platform and factory operation. According to resource estimation, the life cycle of development of petroleum industry will last for more than 300 years, of which 160 years have passed since the commence of the world petroleum industry in 1859, meaning there will be at least 140 years for further development left. Traditional fossil energy is not renewable, renewable non-fossil new energy is bound to complete the ultimate revolution of traditional energy. Wind energy, solar energy, geothermal energy and today's rapid development of energy storage, hydrogen energy, nuclear energy, etc., all of which show broad prospects.

Three Trends in Energy Development

The states of energy will be converted from solid (wood and coal), liquid (oil) to gaseous (natural gas) and carbon containing in energy will extend from high carbon (wood and coal), low carbon (oil and natural gas) to carbon-free (new energy). In the future, energy will develop along three major trends, such as utilizing resource types with reducing carbon, intensive production technology and diversification of utilization methods. Primary energy will develop from high carbon to carbon-free resource, that is, from fossil energy to non-fossil energy, and the world will eventually enter the era of clean energy. The mode of production of energy resources is developed from simple production to technological production. Primitive humans directly obtained firewood from nature as energy, and engineering technology will play a more important role in oil field development than coal mining. The developments of nuclear energy, wind energy, solar energy and other new energy resources are all technology-intensive industries. The energy utilization is advancing from primary energy to secondary energy or multiple energy, electricity, hydrogen energy and so on are important directions of power utilization in the future. It is easy to convert between hydrogen energy and electrical energy, and it has relatively better energy storage effect, which can be used as tertiary energy or multiple sources of energy for conversion, so the multiple energy conversion by a collaborative exchange between hydrogen energy and electricity will be a new direction of energy utilization.

Ten Rules of Oil and Gas Energy

Since the Neoproterozoic, two important plate tectonic division and merge cycles between supercontinents of Rodinia and Pangea have controlled four tectonic domains of Tethys, Laura, Gondwana and the Pacific, as well as the formation of six types of sedimentary basins such as craton, passive continental margin, rift, foreland, pre-arc and arc back. Evolution formed six sets of main source rocks, two types of reservoirs of carbonate rocks and clastic rocks, two sets of regional seal, e.g. shale and gypsum rock. Under the control of the above-mentioned factors, the distribution of global oil and gas has ten laws listed below: (1) symbiotic orderly accumulation of conventional-unconventional oil and gas resources; (2) the formation and distribution of the global oil and gas enrichment zone controlled by Tethys domain; (3) the foreland thrust zone controls the distribution of oil and gas field group; (4) the craton inner uplift controls the distribution of extra-large oil and gas fields; (5) the edge of the platform controls the belt like distribution of the large reef beach oil and gas field group; (6) passive continental margin controls the formation and distribution of the large offshore oil and gas fields; (7) the occurrence of large-scale heavy oil bitumen controlled by the foredeep slope of foreland; (8) the sedimentary slope of basin controls the accumulation of tight oil and gas and coalbed methane; (9) the rich organic deposits under deep water in the basin control the retention of shale oil and gas; (10) low temperature and high pressure seafloor deposition controls the distribution of hydrate. The ratio of conventional to unconventional oil and gas resources is 2:8. Conventional oil and gas resources are mainly distributed in the Middle East, Russia, North America and Latin America, and unconventional oil and gas resources are mainly distributed in four major regions, i.e. North America, Asia Pacific, Latin America and Russia. It has been found that 68% of oil and gas reserves come from Tethys Domain, and the passive continental margin basin accounts for 49% of the global oil and gas resources to be discovered. The future exploration of oil and gas will mainly focus on three major fields—"two deep and one unconventional", e.g. the offshore deep-water area, onshore deep formations and unconventional. Oil and gas are non-renewable, but can be reused.

Four Major Types of Energy Production

The world energy enters a new stage of the third transition of coal, oil and gas to new energy, and is forming a world energy structure of "Four Major Types in the World", e.g. oil, natural gas, coal and new energy. The production of global energy is 13.32 billion tons of oil equivalent in 2017, with oil accounting for 32.93%, natural gas accounting for 23.76% and coal accounting for 28.29%, and new energy accounting for 15.02%. The global oil reserves are generally plentiful, and the ratio of reserves over production has been maintained at more than 50, especially in Central and South America and the Middle East, where the ratio of reserves over production is as high as 120 and 70, respectively. Global oil production continued to grow steadily, with an average growth rate of 8% in the past 10 years, which has entered a stable period of development. Resources of natural gas are very huge (471 trillion cubic meters), and are mainly distributed in four major regions, i.e. Middle East, Russia, North America and South America. Natural gas production is growing at an average annual rate of about 4%, and will enter the peak of development with this rapid growth. Coal, as the cheapest fossil energy resource, will continue to play an important role in the world's energy structure. With the increasing care of humans for protecting ecological environment, coal utilization will be transitioned into the direction of efficient and clean, and the development of coal is entering the transition period. New energy refers to all kinds of energy, such as solar energy, geothermal energy, wind energy, hydrogen energy, ocean energy, biomass energy, nuclear fusion energy, etc., which are at the starting stage of development or are under study in addition to traditional energy resources. When the new energy revolution is underway, new energy will gradually enter the golden period of development with the continuous progress of internet plus, artificial intelligence, new materials and other technologies.

Coal Entering the Transition Period

In 2017, the proved remaining recoverable reserves of coal in the world are 1.17 trillion tons, the average reserve production ratio reached 134, and the production of coal is 3.77 billion tons of oil equivalent. In 2017, China's proved remaining recoverable reserves of coal are 138.8 billion tons, the ratio of reserves over production reached 39, and production of coal is 1.75 billion tons of oil equivalent. Coal utilization will be transitioned into a centralized, efficient and clean direction, and its share in the world's primary energy consumption structure will be further reduced. As the cheapest fossil energy resource at present, coal will continue to play an important role in the world's energy structure. Centralized, high efficiency and clean power generation by using coal are the main directions of coal resource utilization, leading to reduce the combustion of bulk coal, with more than half of coal being consumed in centralized power generation. The thermal efficiency of coal generator sets can be increased by about 50% through the technology of large capacity and high parameter coal power generation, as well as large circulating fluidized bed power generation, integrated coal gasification combined cycle power generation and so on. Coal chemical energy conversion has become a new direction of clean coal development, such as producing gas and oil from coal. Taking the gas production from coal at surface or underground as an example, its efficiency of energy conversion can exceed 50%, the emissions of sulfur, nitrogen and dust and other harmful substances can be well controlled in the process of production, and commercial development can be achieved under the current technical and economic conditions.

Petroleum Entering a Stable Stage

Since its discovery in 1859, the history of petroleum has been more than 160 years, and is known as "the blood of industry", and has played an irreplaceable role in the world economy and human development. In 2017, the proved remaining recoverable oil reserves are 239.3 billion tons, the ratio of reserves to production rate is 54.5, production rate is 4.39 billion tons, and the overall trend of oil production showed a steady growth, accounting for 33% in primary energy production. As conventional oil exploration extends into deep water, and deep formations and the Arctic, the world's new proved oil reserves continue to grow. Conceptual innovation and technological breakthroughs that drive the oil industry from conventional to unconventional is a big step forward, and unconventional oil has become a new field of oil development in the future. The global oil resources are much abundant and have entered a stable period of development, and oil production is expected to approach the peak of 5 billion tons in 2040.

Natural Gas Entering the Peak Period

Natural gas, which contains low carbon and is environmentally friendly, is the cleanest fossil energy and has the property of cleaning; natural gas can meet most of the basic living needs of humans, with the property of livelihood; natural gas is the bridge of fossil energy transitioning to new energy, which has the property of transition. Natural gas will become an insurmountable bridge between fossil energy and new energy, or a co-existing and co-prosperous partner of new energy resources. In 2017, the world's proved natural gas remaining recoverable reserves are 193.5 trillion cubic meters, the ratio of reserves over production rate is 52.3 and production rate is 3.7 trillion cubic meters, accounting for 23.8% in primary energy production. It is the fastest growing fossil energy and has entered the peak period of development. The revolution of unconventional natural gas, represented by shale gas in the United States, has significantly increased the size of the world's natural gas production. Global conventional natural gas and three types of unconventional natural gas resources (tight gas, shale gas, coalbed methane) are rich. Natural gas hydrate resources reach 20,000 trillion cubic meters, and are 20 times that of the conventional oil and gas resources, and the remaining natural gas resources are huge. Natural gas consumption is expected to surpass coal in 2030, overtake oil in 2040 and enter the era dominated by natural gas, expecting a peak of 5 trillion cubic meters of production in 2060.

New Energy Gradually Entering the Golden Period

The global utilization of new energy resources can last for tens of millions of years and the new energy will be inexhaustible. Theoretically, the solar radiant energy that reaches the surface of earth each year is 130 trillion tons of standard coal equivalent, and the energy of nuclear fusion is 3 trillion tons of oil equivalent each year. The accelerated pace of development and utilization of new energy has become a new driving force for global energy growth, and the development of new energy resources is gradually entering the golden period. According to the data of the International Energy Agency (IEA), the world's total production of new energy resources, such as nuclear energy, hydropower and renewable energy, reached 2.001 billion tons of oil equivalent in 2017, accounting for 14.98% of the total energy consumption structure. The investment in new energy is $333.5 billion, close to the investment of $408 billion in upstream of oil and gas. With the progress of technology, the development and utilization cost of new energy has been declining, and the cost is more competitive than fossil energy. According to the data of the International Renewable Energy Agency (IRENA), the cost of onshore wind power around the world in 2017 was US $0.06/kwh, the cost of solar photovoltaic power was US $0.10/kwh, and the cost of coal power was US $0.066–0.105/kwh. Renewable energy (including hydropower) has become the major segment in the global development of new energy and accelerated development is ongoing. The fast changing of network big data system based on artificial intelligence, intelligent energy network system with internet plus energy network, nanomaterials, graphene, battery energy storage and other technologies, the reducing cost of new energy power generation and the technology breakthrough of battery energy storage, will strongly promote the arrival of the new energy era.

Oil and Gas of the World and Coal of China

There is a great imbalance between production and consumption of global energy. World energy consumption is entering the era in which it will be dominated by oil and gas (57%), however, China is in the energy era dominated by coal (61%), due to abundant coal, shortage of oil and gas, and the delayed development of new energy in China. In 2017, primary energy consumption around the world was 13.511 billion tons of oil equivalent, of which coal was 3.732 billion tons of oil equivalent (accounting for 27.62%), petroleum was 4.622 billion tons of oil equivalent (34.21%), natural gas was 3.156 billion tons of oil equivalent (23.36%) and new energy was 2.001 billion tons of oil equivalent (14.82%). In 2017, primary energy consumption in China was 3.132 billion tons of oil equivalent, of which coal was 1.893 billion tons of oil equivalent (60.43%), petroleum was 608 million tons of oil (19.42%), natural gas was 207 million tons of oil equivalent (6.6%) and new energy was 424 million tons of oil equivalent (13.55%).

World Fossil Energy Map

The global fossil energy is mainly composed of oil, natural gas and coal, and with the deepening of theoretical understanding and the significant improvements of the level of exploration technology, the new map of fossil energy resources in the world has been reshaped. Global oil production in 2017 was 4.39 billion tons (14% of which were unconventional) and natural gas production was 3.69 trillion cubic meters (25% of which were unconventional). The production technology progress of horizontal well and platform-type industrialization operation has caused the revolution of unconventional oil and gas, which is pushing oil and gas production pattern of the world to be profoundly adjusted. Over the past 10 years, world oil production has grown steadily and natural gas production has climbed faster. Affected by the release of coal production capacity in emerging economies such as China, the imbalance in world coal production has increased, and the Asia-Pacific coal production has been strengthened. In 2017, global coal production was 7.73 billion tons, Asia Pacific coal production was 5.36 billion tons (69.3%) and China's coal production was 3.52 billion tons (accounting for 45.5%). With the strong growth of energy demand in emerging economies and the approaching limit of ecological environment, human beings must make a choice between different energy products. This choice has a direct and profound impact on the world's new map of fossil energy consumption and reshape it. Global energy consumption is related to the level of social and economic development and the ease of access to resources. Energy demand in developed countries such as the United States and Europe has remained stable, while energy demand of emerging economies in Asia-Pacific has grown rapidly, and the fossil energy consumption map is changing from the “three major components” situation consisting of North America, Europe and Asia-Pacific to the "two poles" status consisting of the eastern and western hemispheres. Total global energy consumption reached 13.51 billion tons of oil equivalent in 2017, an increase of 43.9% in comparison with the year 2000. Among them, energy consumption in the Asia-Pacific region reached 5.74 billion tons of oil equivalent in 2017, an increase of 116.7% in comparison with the year 2000, making it the main driving force of global energy consumption growth. Energy consumption in Europe and North America in 2017 was 2.95 billion tons of oil equivalent and 2.77 billion tons of oil equivalent, respectively, an increase of 4.8% and 0.6%, respectively, in comparison with the year 2000, and generally stable.

World's New Energy Map

Influenced by the level of science and technology and economic development and existing resources, the development and utilization of new energy resources mainly formed in three major centers: Europe, Asia Pacific and North America. Affected by the nuclear power accident, the global nuclear power development shows a differentiated trend whereby the United States, Japan and South Korea are gradually slowing down the development of nuclear power and China and other emerging countries are actively developing nuclear power due to strong demand of energy. Nuclear power consumption in Europe, North America and Asia and the Pacific in 2017 was 258 million tons of oil equivalent (43.3%), 216 million tons of oil equivalent (36.2%) and 112 million tons of oil equivalent (18.7%). World hydropower technology tends to mature, industry development is mainly controlled by the distribution of hydro energy resources, and four regions of Asia Pacific, Europe, North America and Central and South America are formed. Hydropower consumption in Asia and the Pacific, Europe, North America and Central and South America in 2017 was 372 million tons of oil equivalent (40.4%), 187 million tons of oil equivalent (20.4%), 164 million tons of oil equivalent (17.9%) and 162 million tons of oil equivalent (17.7%) respectively. Other renewable energy consumption, such as solar and wind energy, are growing rapidly and the proportion of it is keep on increasing. Renewable power generation has become the main mode of energy utilization. With the continuous progress of science and technology, other renewable energy sources are forming the renewable energy map consisting of three areas of Asia Pacific, Europe and North America. Other renewable energy consumption in Asia Pacific, Europe and North America in 2017 was 175 million tons of oil equivalent (36%), 163 million tons of oil equivalent (33.4%) and 110 million tons of oil equivalent (22.5%) respectively.

China's Fossil Energy Map

The existing status of abundant coal resources but relatively insufficient oil and gas in China determines its special ratio of both energy production and energy consumption, which constitutes China's energy structure of "one major component and three non-major components". In 2017, as for China's energy production, coal accounts for 70%, oil accounts for 8%, natural gas accounts for 5% and new energy accounts for 17%. Fossil energy production has been growing steadily, and coal production has been a dominator. In 2017, the total fossil energy production of China reached 2.07 billion tons of oil equivalent, of which coal accounted for 84.5%, oil accounted for 9.3% and natural gas accounted for 6.2%. After coal production peaked in 2013, coal production shows a downward trend due to the declined demand for coal. China's oil production has been stable overall in the past years, with production reaching 200 million tons in 2010, a peak of 215 million tons in 2015 and production decreasing to 192 million toes in 2017. China's conventional natural gas is entering a period of sustainable growth and unconventional natural gas is entering a leap-forward development period. For a long time, the proportion of coal in China's energy consumption structure is too high and the proportion of oil and natural gas is low. From the point of view of coal consumption, China's energy structure is in the era of high carbon. It is predicted that by 2050, China's coal consumption will account for 40%, reaching the world's ratio of 41% of coal in 1965, which is nearly 100 years later than the world.

China's New Energy Map

In recent years, China's new energy industry has developed strongly, and its proportion in a primary energy production structure had continued to expand to 17%, becoming an important part of energy. China's hydropower resources are abundant, and the pace of hydropower development is accelerating, accounting for 10.5%. Since the twenty-first century, with the Three Gorges Dam put into operation as a symbol, hydropower construction technology has continuously set a world record. Driven by the national policy, China's wind power industry is booming and newly installed capacity as well as accumulated installed capacity of wind power is firmly ranked first in the world. In 2017, China's newly installed capacity was 19.66 million kilowatts, the cumulative installed capacity reached 188 million kilowatts, an increase of 11.7% over a year earlier. The installed capacity of photovoltaic projects is likely to fall due to the government setting two upper limits of the development of the two areas. In 2017, China's photovoltaic installed capacity was more than 96 gigawatts, the expected solar photovoltaic installed capacity in 2018 will still reach 105 gigawatts, an increase of 11% over a year earlier. China's geothermal resources are relatively rich, the total geothermal resources account for 7.9% of the world and recoverable reserves are equivalent to 462.65 billion tons of standard coal. Geothermal energy is mainly used directly, and there are 5 geothermal power stations with a total of 27.78 MW. The utilization ratio of nuclear power equipment in China is declining, and the challenge of nuclear power consumption and acceptance still exists. China's nuclear power generation will continue to grow, as does its share of the country's electricity generation. China's total nuclear power is small, but in recent years, the pace of nuclear power construction is accelerating, the scale of nuclear power under construction ranks first in the world. From 2010 to 2017, the average utilization hours for nuclear power equipment in China are generally decreasing, and the average utilization hours for nuclear power equipment in China were 7,108 h in 2017. Nuclear power generation in 2017 was 248.3 billion kWh, an increase of 16.7% over 2016.

Green Energy Strategy

The green energy strategy refers to the goal of producing cleaner energy, replacing high-carbon fossil energy with cleaner low-carbon or carbon free energy. The existing energy structure is highly dependent on coal and oil, and it is necessary to gradually change the energy structure and increase the proportion of green and clean energy in the energy structure. Renewable energy is growing rapidly, but it is not currently able to meet global energy demand. Natural gas, as a transitional energy source, will play an insurmountable bridge or coexistent role. Hydrogen energy is a kind of high-quality new energy and can play a synergistic effect, working together to build an industry chain system containing wind, light, electricity and hydrogen energy, achieve the balanced development of distributed energy production and centralized consumption, create efficient, intelligent energy Internet and realize the development of "intelligent energy" based on the Internet of things and blockchain technology.

Hydrogen Energy Technology Revolution

Hydrogen has many advantages, such as the lightest quality, the standard state density of 0.09 g/L, liquefaction at −252.76 °C; the best thermal conductivity, 10 times higher than that of most gas high calorific value, 3 times higher than that of gasoline except for nuclear fuel; good combustion, wide combustible range, high ignition point, fast combustion; various forms, appearing as gaseous, liquid or solid metal hydride; wide range of resources, widely present in water in the form of compounds; non-toxic and environmentally friendly, clean reaction products, reducing greenhouse effect. Hydrogen can be regarded as one of the most promising secondary green energy resources in the twenty-first century, and the United States, Japan, Germany and other developed countries have consecutively upgraded the hydrogen energy industry to the national energy strategy. The storage of high electrical energy is difficult, and power generated from wind, light and geothermal has an impact on the power grid. The problem of power storage can be solved through the mutual conversion between electricity and hydrogen, which is currently the ideal solution for new energy. Hydrogen energy and electrical energy can be exchanged to achieve cross-border energy connectivity. Hydrogen and electricity can be obtained through the transition of primary energy (solar, wind, ocean energy, thermal energy, etc.). Hydrogen can generate electricity through fuel cell technology, and electricity can also be used to make hydrogen from water to connect the entire energy network. Within the range of power grid extension, it is advisable to take the convenience of electricity and to use hydrogen to store energy and supply energy outside the power grid. Hydrogen could become a strategically new resource of energy in the future.

Energy Internet Technology

Energy Internet refers to the product of the combination of Internet information technology and renewable energy. The development of energy Internet will fundamentally change the dependence on traditional energy utilization modes and promote the transition of traditional industries into new industries based on renewable energy and information networks, which is a fundamental revolution of the way of life of human society. We should promote the deep integration of new technologies in the fields of energy and information, coordinate the construction of infrastructure networks such as energy and communications and transportation, and build a coordinated development of “resource–internet–load-storage”, integrated and complementary energy Internet. New energy technologies such as renewable energy, distributed power generation, energy storage, electric vehicles, etc., and Internet technologies such as the Internet of things, big data, cloud computing and mobile Internet are deeply integrated. Energy Internet has become another strategic direction in the field of energy after smart grid.

Energy Storage Technology Revolution

The direct utilization of new energy resources such as wind, light and nuclear power is restricted by the locations, and secondary or tertiary transition is the main direction of their development. The storage of electrical energy and hydrogen energy and so on is critical to the development of new energy. Energy storage is divided into four types in the form of energy conversion: cold and hot energy storage, mechanical energy storage, chemical energy storage and electromagnetic energy storage. As a new type of cheap and efficient battery system, liquid metal battery has both low cost for energy storage and long life, and is expected to have better application in energy storage domain. Since hydrogen is the cleanest energy resource, fuel cells are the cleanest energy conversion devices, fuel cell technology has made significant progress in battery life and cost, clean technology for hydrogen production has also made breakthroughs, so the combination of fuel cells and clean hydrogen production technology will become the most competitive energy supply technology in the future.

King of New Materials—Graphene

Petroleum is well known as the "black gold" of the twentieth century and graphene is known as the "black gold" of the twenty-first century. The graphene, known as the second "black gold", has excellent characteristics such as fast electron migration rate, high strength, good conductivity of heat and electricity, high light transmittance and light mass, and is known as the "king of new materials". Superior material properties of graphene are expected to trigger key technological changes in areas such as new energy, petrochemicals, electronic information, composite materials, biomedicine and energy conservation and environmental protection. Graphene is expected to be a strategic new material which will lead the next generation of industrial technology revolution. The application of graphene in the field of electrochemical energy storage, especially in the field of supercapacitors and batteries, will speed up the process of new energy revolution. In recent years, with the development of graphene battery technology, graphene technology has had a significant impact on new energy development.

Nuclear Power Technology

Nuclear energy is the energy released when the structure of the nucleus changes, and the release of nuclear energy includes nuclear fission and nuclear fusion. Uranium, the raw material used for nuclear fission, can use only one gram in nuclear fission and can release the energy equivalent to 30 tons of coal, while only 560 tons of deuterium used in nuclear fusion can provide energy for the world for one year. The reserves of deuterium in the ocean, which can be used by human beings for billions of years, are inexhaustible new energy resources. At present, the utilization of nuclear energy is mainly based on nuclear fission, and the development of nuclear energy will gradually change from nuclear fission to nuclear fusion. The nuclear fission resources on the earth belong to mineral resources and reserves are limited. After nuclear energy is released, it cannot be restored to nuclear fuel itself in the short term, that is, the material used for nuclear fission is non-renewable energy. Although the life cycle of nuclear fission resources is relatively long, the scarcity of resources makes it difficult to become the world's main resource of energy. Nuclear fusion can be regarded as renewable energy to a certain extent, because the raw material used for nuclear fusion, i.e., the isotope of hydrogen, is abundant in the ocean, but it is more difficult to obtain than hydrogen, the application technology is not mature, so it is difficult to form large-scale industry in a short time.

Geothermal Technology

Geothermal energy is the second largest clean energy resource after solar energy. In recent years, the global geothermal industry has flourished, and by 2020, the proportion of geothermal energy utilization in China's total energy consumption will rise from the current 0.5 to 1.5%. China Petroleum & Chemical Corporation, or Sinopec, has basically completed the plan of geothermal industry in medium and deep formations. Geothermal and waste heat resources owned by China National Petroleum Corporation, or CNPC, are abundant. With geothermal exploration and development technology and market advantages, CNPC is optimizing the direction of geothermal utilization, expanding the scale of geothermal utilization, building a demonstration project with social impact on geothermal utilization, promoting China's oil energy conservation and emission reduction, reducing cost and increasing efficiency, and realizing cleaner production in the oilfield.

Renewable Energy

Renewable energy mainly includes solar energy, wind energy, hydro energy, biomass energy, ocean energy, tidal energy, geothermal energy, etc., mainly from the conversion of solar energy, which can be recycled and regenerated from nature in human history. Fossil energy such as coal and petroleum, which has been developed and utilized by human society at a large scale, cannot be recycled and regenerated in human history and is a non-renewable energy source. Considering the resource utilization and environmental problems brought forward by the development of fossil energy in the world, people begin to realize the importance of renewable energy utilization.

Three Major Challenges of Energy in China

Since China's energy is rich in coal but short of oil and gas, energy production faces three challenges. First, continuously rising external dependence of oil and gas is concerned. Domestic oil and gas production cannot meet national demand, e.g. external dependence of oil in 2017 is 67.79%, and external dependence of natural gas is 39.4%. According to this trend, the external dependence will reach 80% and it is urgent to control the fast growing of external dependence and keep the dependence level within a reasonably safe and controllable range. The contradiction between supply and demand is becoming more and more obvious, the autonomy of supply-side is weakening, and the unstable factors of safe supply are increasing. China's oil and gas production capacity by itself is insufficient, and it can only implement the global multi-channel supply strategy. Second, the structure of energy production is unreasonable. China's energy structure can be described as "one major component and three non-major components", i.e. in 2017, coal accounted for 70% of the country's energy production (one major component) and oil accounted for 8%, natural gas accounted for 5% and new energy accounted for 17% (three non-major components). In the view of carbon from the evolution of the energy structure, China has been lagging behind the world for nearly 100 years, and the clean development and utilization of coal is imminent. "Outstanding" coal production has caused a weaker complementary linkage with oil, natural gas and new energy, which leads to the weak ability of national energy supply to resist risk. Third, the future pillar industries in the new energy domain are still foggy. China's wind, light and hydropower development has reached a certain scale, nuclear power production capacity is under concentrated construction, and geothermal, biomass, ocean energy, hydrate and other energy domains are still in the initial stage of exploration or learning stage. From the scientific point of view, "restriction mode" is not a long-term idea to solve the fundamental problem. It is an inevitable choice to promote the utilization of clean transition by "dredging mode".

The Safe Consumption Peak of Oil and Gas

By 2017, China consumes 0.43 tons of oil and 170 cubic meters of natural gas per capita, far below the level of developed countries, and the demand for oil and gas consumption is continuing to rise. The early warning system of "peak safe consumption" of oil and gas in China needs to be established. Preliminary forecast indicates that, by 2030, China's oil production will be 160 million to 180 million tons, natural gas production will be 180 billion to 2000 billion cubic meters, oil consumption will be 650 million to 700 million tons, natural gas consumption will be 5500 billion to 6000 billion cubic meters, external dependence of oil and gas will exceed 70%. The "shortage" of natural gas supply in the winter of 2017 fully reflected the safety risk of China's natural gas imports. With the increasing dependence of natural gas on the outside world, it is necessary to strengthen the construction of natural gas infrastructure. It is estimated that the import by onshore gas pipeline will be 160 billion cubic meters, imports by LNG will reach 200 billion to 250 billion cubic meters, underground gas storage capacity will be 80 billion to 100 billion cubic meters. Therefore, on the basis of artificial intelligence and big data analysis, it is necessary to closely track the oil and gas production, consumption, climate, transportation path, inventory and politics at home and abroad, establish the early warning system of "safe consumption peak" of China's oil and gas, suggest that the external dependence degree of oil and gas should be controlled within 70%, and avoid oil and gas supply safety risks systematically.

Three Leapfrog Development Periods of Oil Company

With the improvement of oil and gas exploration and development in China, oil companies need to coordinate the three overall situations of conventional and unconventional, domestic and foreign, oil and gas and new energy, and promote the "three greater leaps". The first leap is the greater leap from conventional oil and gas to unconventional oil and gas. We should make efforts to enhance the production rate of domestic conventional oil and gas, try our best to improve the production capability and reserves of unconventional oil and gas, maintain long-term and fundamentally stable production of oil, and accelerate the production of natural gas. The second leap is the greater leap from home to abroad which means to participate deeply in the construction of the "one belt and one road" initiative, create upgraded version of the "one belt and one road" oil and gas cooperation based on the existing overseas cooperation areas, and speed up "four production capacity" construction of domestic oil and gas, pipeline oil and gas, oil and gas storage, and LNG in advance. The third leap is the greater leap from oil and gas to new energy. Solar and wind power generation is growing fastest, hydropower and nuclear power accounted for the highest portion in renewable energy generation, hydrogen energy, energy storage, new materials, renewable energy is the most revolutionary energy, the clean utilization of coal and early arrival of "two scale" of new energy should be accelerated, and the time span and safety pressure of oil and gas in China's energy utilization should be reduced.

China's Energy Strategy

The ultimate goal of accelerating transition revolution of China's energy structure is to build a modern energy system in new era, the core essence of which is to optimize and improve two aspects of the energy system, i.e. going clean and containing low carbon, and at the same time being safe and efficient. It is necessary to explore a way of low energy consumption and sustainable development from the objective reality of China's characteristics of underground resources. China's energy consumption is forecast to reach a peak value of 4.4 billion tons of oil equivalent in 2030. In front of the new period of global climate change and fast development of low-carbon energy, China needs to speed up the production of conventional-unconventional oil and gas, coalbed methane, hydrate and hydrogen, speed up the earlier arrival of "two scale" of clean utilization of coal and new energy. At present, it is necessary to speed up the realization of "production revolution" of conventional-unconventional oil and gas, the "clean revolution" of coal development, the "speed revolution" of new energy development, and strive to realize the energy structure revolution of from status of "one major component and three non-major components" to "three major components" consisting of coal, oil and gas, new energy around 2050. At that time, coal will account for about 40% of the primary energy consumption, oil and gas will account for 30% and new energy will account for 30%; Around 2100, national "energy independence" by relying on new energy is expected, fossil energy will account for about 30% of the primary energy consumption, non-fossil energy will account for 70%, and the transition of the historical status of the two roles can be achieved.

Contents

About the Author

Dr. Caineng Zou is the founder of unconventional petroleum geology and the scientist in energy strategy in China. He serves as the academician in Chinese Academy of Sciences, and is interested in petroleum exploration and development.

He devotes himself to the establishment and application of unconventional petroleum geology, promoting the exploring strategy and the industrial breakthrough. He analyzes the current strategic situation for energy needs and supplies in the world, proposing the "Hydrogen China" and "Energy Independence" strategies in China.

He has published 257 journal papers and 8 books including Unconventional Petroleum Geology worldwide. He has been selected in the list of Chinese Most Cited Researchers by Elsevier in 2019. He gained 3 approvals for the national standards, and several awards in the National Prize for Progress in Science and Technology.

Part I
Energy Trend

Chapter 1
Laws of Energy Development

Since the first time the primitive human beings started using fire, energy, water and food together constitute the three main elements of human survival. In the long history of energy development of human beings, energy resources have continued to evolve along the clean, low-carbon, and carbon-free main road. Over the past hundred years, following the trend towards low carbon, the world's energy structure has undergone two huge conversions, of which the first conversion achieved the energy revolution from firewood to coal and the second conversion achieved the energy revolution from coal to oil and gas. At present, energy development is entering the third major conversion period from traditional fossil energy to new energy and will eventually realize the revolution of replacing fossil energy with new energy.

In the 21st century, under the background of the ever-increasing demands of the quality of environment by human beings and with the continuous progress of science and technology such as Internet and human civilization, the world energy pattern is undergoing the third major change quietly, and the global energy production and utilization is developing towards cleaner, more efficient, more convenient, safer and sustainable development. In the near future, people will clearly feel the profound impact of renewable energy as the core of the new energy on daily lives. At present, it is on the eve of the third major energy transition. In the future, it will enter a clean, efficient, carbon-free green energy era (solar energy, wind energy, hydrogen energy, etc.).

1.1 Transition of Energy Development

1.1.1 From Firewood to Coal

The first transition: Coal surpasses firewood. Since the first use of fire by primitive humans, energy has become an essential resource for human survival. The wood that is easily obtained on the spot satisfies the basic survival needs of human heating,

C. Zou, *New Energy*, https://doi.org/10.1007/978-981-15-2728-9_1

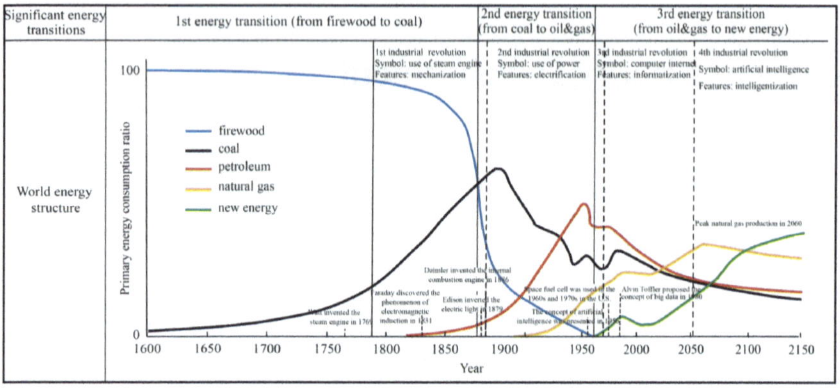

Fig. 1.1 Road map of world science and technology development and energy revolution

cooking and so on. With the advancement of coal mining technology, coal with higher energy density has been widely used. In the Tang and Song Dynasties, coal was called stone charcoal, which has been applied in large scale in society. During the reign of Su Dongpo in Xuzhou, a coal mine with considerable reserves and good quality was discovered in Baidi Town, southwest of Xuzhou, in the first year of Yuanfeng (AD 1078). In the middle of the 17th century, the book named "Tian Gong Kai Wu" by Song Yingxing systematically recorded the mining techniques of ancient Chinese coal, including geology, development, coal mining, support, ventilation, artificial lifting, and gas emission, which indicated that coal mining had developed to a certain scale at that time.

Before the industrial revolution, however, firewood has always dominated the main position of energy consumption to meet the basic needs of human society for cooking, heating and simple industrial production. In 1769, Watt invented the steam engine. As the first industrial revolution led to the development of productivity, firewood could no longer meet people's demand for energy (Fig. 1.1). Coal has become a power resource for driving steam engines because of its advantages of easy mining, high heat of combustion, and easy storage and transportation. In 1880, it surpassed firewood in the proportion of primary energy consumption and became the largest energy resource for consumption (Zou et al. 2016). The era of coal energy is the first stage in the era of human society entering the fossil energy era (Fig. 1.1).

1.1.2 From Coal to Petroleum

The second transition: Oil surpasses coal. The name of petroleum comes from the Greek Petra and Oleum. The earliest text was recorded in Shen Kuo's "Meng Xi Bi Tan, Volume 1": "There is oil in the territory of Yan Yan, called Gao Nu County in the past, the produced greasy water is actually petroleum." In 1848, Russian engineer

F.N. Semyenov drilled the first modern oil well in the northeast of Baku on the Aspheron peninsula. The petroleum distillation process was invented in 1853. Polish scientist, Ignacy Lukasiewicz obtained kerosene from petroleum by distillation. The first petroleum reservoir was discovered in Bobrka near Krosno, southern Poland in the following year. Meerzoeff built the first Russian refinery on an existing oilfield in Baku in 1861.

In 1876, Otto invented the internal combustion engine, which officially took to the historical stage as the power of the internal combustion engine (Fig. 1.1). The invention of internal combustion engine promotes the transformation of transportation tools, makes the demand of oil and natural gas grow rapidly and drives the development of petrochemical and natural gas industry (Zou et al. 2016). Oil and natural gas have gradually become the most important energy varieties and chemical raw materials after coal and human society has entered the second stage of fossil energy era-oil and gas era.

1.1.3 From Oil and Gas to New Energy

The third transition: New energy takes the place of oil and gas. The way humans utilize energy will experience the third major conversion from oil and gas to new energy after the conversion from firewood to coal and the transition from coal to oil and gas have been basically completed (Zou et al. 2018). With the sustained growth of energy demand and civilized progress in economic development, the third major transition from traditional fossil energy to non-fossil new energy sources will become inevitable.

Natural gas is a bridge or co-existing resource with the transition from fossil energy to new energy resources. Driven by the development of social civilization and science and technologies, energy development transitioned from solid (firewood + coal), liquid (oil) to gaseous (natural gas), natural gas has become an insurmountable bridge or an associated period from fossil energy to new energy, and will ultimately promote the harmonious development of energy consumption and ecological environment of human society (Fig. 1.1). In recent years, the ecological environment problems caused by the utilization of high carbon energy such as coal and petroleum have become more and more serious, i.e. the formation of "city of fog" of London, England in the early 20th century and the current extensive haze weather in China that results from the large-scale utilization of high-carbon fossil energy such as coal. With the increasing demand of humans for a green ecological environment, the proportion of natural gas and new energy as clean energy in primary energy structure will gradually increase. According to BP's world energy statistics in 2018, oil accounted for 34%, natural gas accounted for 23%, coal accounted for 28%, and other renewable energy sources, such as nuclear power and hydropower, accounted for 15% of the global energy consumption structure in 2017 (Fig. 1.1). A global energy source is entering the "four major types in the world" structure consisting of oil, gas, coal and new energy (Zou et al. 2018).

Each energy transition has been a long historical process, and the change from old energy type to new energy type often takes at least half a century or even centuries. World energy development has undergone two major changes which are coal took the place of firewood and petroleum took the place of coal, and is undergoing changes from fossil to renewable resources of energy. From the perspective of energy transitions, the fuelwood era has lasted for nearly 250 years, the energy era dominated by coal lasted for about 85 years, and the energy era dominated by petroleum lasted for about 80 years. The era of natural gas we are experiencing is expected to last for 50 years, with gas production peaking in 2060. It is expected that the era of green and clean energy will come by the end of 21st century (Fig. 1.1).

1.2 Energy Development Structure

With the progress of social civilization and the improvement of scientific and technological level, global energy is forming a new structure of mutual development of oil and gas, conventional and unconventional, fossil and non-fossil. From the perspective of international energy development situation and the exploration and development trends of oil companies, it is the general movement that natural gas will take the place of oil first and enter the era of natural gas development. The combination of conventional and unconventional resources has been incorporated into the development strategy of oil companies. The strategy is to persist in conventional oil and gas as the main part of exploration and further improve the development of it, to fully understand the key technology theories of and gradually achieve effective development of unconventional resources (Zou et al. 2016, 2018). Traditional fossil energy is not renewable, so renewable non-fossil new energy is bound to fulfill the ultimate revolution of traditional energy. Wind energy, solar energy, geothermal energy amd the current rapid development of energy storage technology and hydrogen energy, all of which have show broad prospects for development, perhaps the new energy revolution will come early before the depletion of fossil energy.

1.2.1 Petroleum and Natural Gas

From the view of the development status of international energy and the exploration and development trend of oil companies, the global natural gas resources are abundant, energy development is entering the natural gas era, and the current global energy development has formed a new structure characterized by "four major types of global energy", consisting of oil, natural gas, coal and new energy. According to data of BP (2018), global energy production was 13.32 billion tons of oil equivalent in 2017, with oil accounting for 32.93%, natural gas accounting for 23.75%, coal accounting for 28.29% and new energy accounting for 15.02%. The global energy consumption in 2017 was 13.51 billion tons of oil equivalent, of which oil accounted for 34.2 1%, natural gas accounted for 23.36%, coal accounted for 27.62% and new energy accounted for 14.81% (Table 1.1).

The proportion of natural gas in the primary energy structure has gradually increased, and together with oil, it forms the main part of the world's primary energy production and consumption. According to data of BP (2018), the total global oil production in 2017 was 4.387 billion tons, and the total consumption was 4.469 billion tons (Figs. 1.2 and 1.3). The total global natural gas production in 2017 was 3.164 billion tons of oil equivalent, with a total consumption of 3.156 billion tons of oil equivalent (Figs. 1.2 and 1.3). Compared with the year of 2000, global oil production and consumption increased by 21% and 24% respectively in 2016 and global natural gas production and consumption increased by 46% and 45% respectively in 2016 (Figs. 1.2 and 1.3).

Natural gas, as the cleanest fossil energy, is the best choice for human beings to meet the needs of ecological environment at present. With the demand of ecological environment and the progress of science and technology, the world energy will enter a new era of natural gas in the 21st century. The energy consumption structure has further accelerated its transition to clean energy with low carbon. In 2017, the proportion of coal consumption decreased by 0.31% compared with 2016, and the proportion of natural gas and non-hydropower renewable energy consumption increased by 0.14% and 0.16%, respectively. Natural gas is expected to surpass coal around 2030 and surpass oil in 2040, becoming the most important resource of energy for non-fossil energy development and entering the natural gas era. In the end, with the continuous growth of economic and social demand for energy and the arrival of a low-carbon society, the third major conversion of traditional fossil energy to non-fossil new energy resources will become inevitable.

1.2.2 *Conventional and Unconventional*

The combination of conventional and unconventional resources has been integrated into the development strategy of oil companies, insisting on the conventional oil and gas as the main part of exploration, further improving the development of conventional oil and gas, better understanding the key technologies of unconventional resources, and gradually achieving effective development. Driven by the theory and technology of oil and gas, the global oil and gas exploration and development are changing from simple to complex, from structure to lithology, from shallow to deep, from onshore to offshore, from conventional to unconventional. Since 2000, due to the innovation in understanding unconventional shale oil and gas accumulation theory and the breakthroughs of key engineering techniques of multi-stage fracturing in horizontal wells, the potential of unconventional oil and gas resources has been re-recognized. The global amount of recoverable unconventional oil resources is 620 billion tons, roughly the same as the amount of conventional oil resources, and the amount of recoverable non-conventional natural gas resources is about 4,000 trillion

Table 1.1 Comparison of energy production/consumption of the world, the United States and China in 2017

Energy type		World		U.S.		China	
		Production/Consumption (100 million tons of oil equivalent)	Ratio (%)	Production/Consumption (100 million tons of oil equivalent)	Ratio (%)	Production/Consumption (100 million tons of oil equivalent)	Ratio (%)
Petroleum	Production	43.871	32.93	5.71	29.62	1.915	7.69
	Consumption	46.219	34.21	9.133	40.87	6.084	19.42
Natural gas	Production	31.646	23.75	6.316	32.77	1.283	5.15
	Consumption	31.56	23.36	6.358	28.45	2.067	6.60
Coal	Production	37.686	28.29	3.713	19.26	17.472	70.13
	Consumption	37.315	27.62	3.321	14.86	18.926	60.43
Nuclear power	Production	5.964	4.48	1.917	9.95	0.562	2.26
	Consumption	5.964	4.41	1.917	8.58	0.562	1.79
Hydropower	Production	9.186	6.90	0.671	3.48	2.615	10.50
	Consumption	9.186	6.80	0.671	3.00	2.615	8.35
Renewable energy	Production	4.868	3.65	0.948	4.92	1.067	4.28
	Consumption	4.868	3.60	0.948	4.24	1.067	3.41
Total	Production	133.221	100.00	19.275	100.00	24.914	100.00
	Consumption	135.112	100.00	22.35	100.00	31.321	100.00

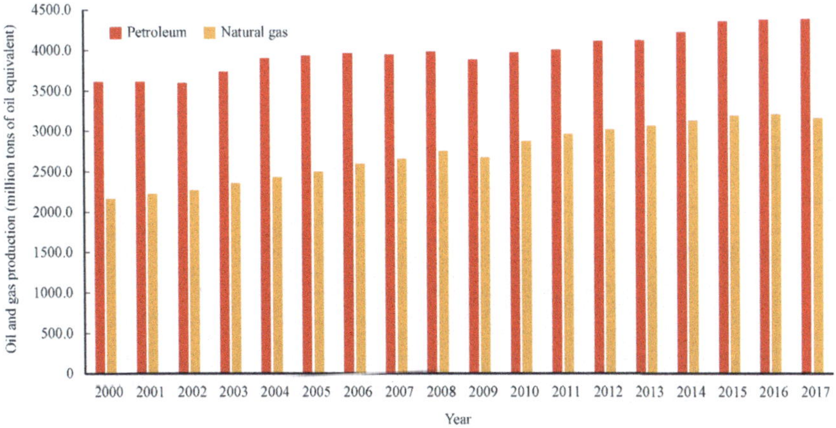

Fig. 1.2 Comparison of global oil and natural gas production from 2000 to 2017

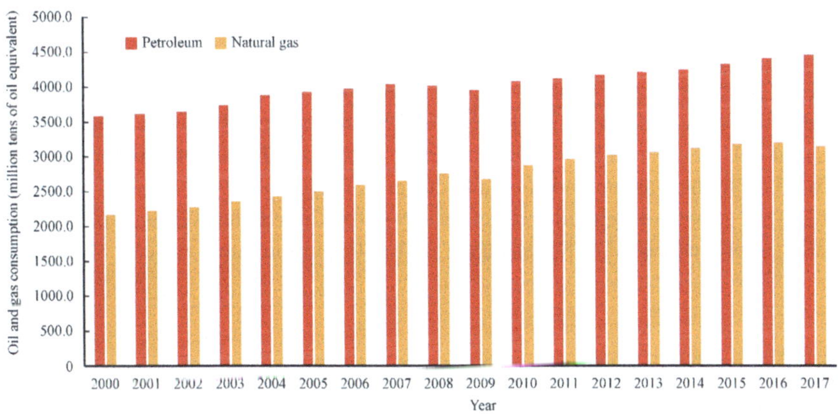

Fig. 1.3 Comparison of global oil and natural gas consumption from 2000 to 2017

cubic meters, roughly 8 times of the amount of conventional natural gas resources (Zou et al. 2015).

The United States is a pioneer of unconventional oil and gas in theory and technology, and a leader in unconventional oil and gas development. Unconventional oil and gas production has grown rapidly and has become the main part of American oil and gas production (Zou et al. 2014, 2018). Total oil production of U.S. in 2017 was 571 million tons, of which 220 million tons was tight oil, with unconventional oil accounting for 39% of total production, and the ratio of importing was 34% (Fig. 1.4). Natural gas production of U.S. in 2017 was 739.5 billion cubic meters, of which shale gas was 477.2 billion cubic meters, tight gas 120 billion cubic meters, coalbed methane was 28 billion cubic meters, unconventional accounted for 85% and

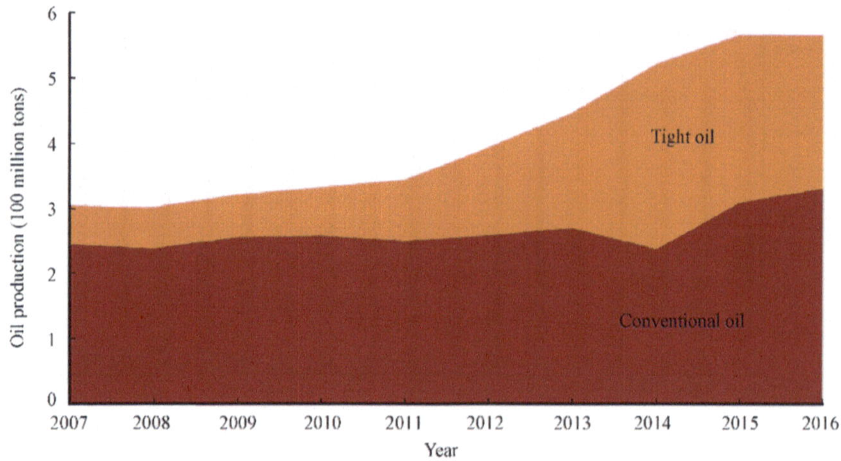

Fig. 1.4 Composition of oil production in the United States (according to EIA)

external dependence was 0.4% (Fig. 1.5). The growth of shale oil and gas production has built a strategic foundation for the "energy independence". According to the prediction of the U.S. Energy Information Administration (2016), the tight oil production of U.S. will reach 350 million tons in 2040, with the largest growth potential of tight oil in the Bakken and Austin Cretaceous formations, which will account for 80% of the growth of tight oil production across the United States (Fig. 1.6). In 2040, shale gas production of U.S. can reach up to 800 billion cubic meters. Marcellus,

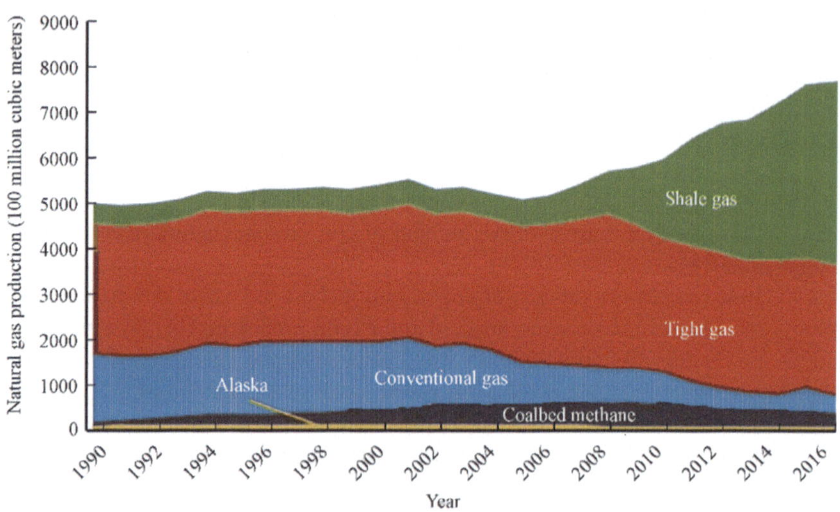

Fig. 1.5 Composition of natural gas production in the United States (according to EIA)

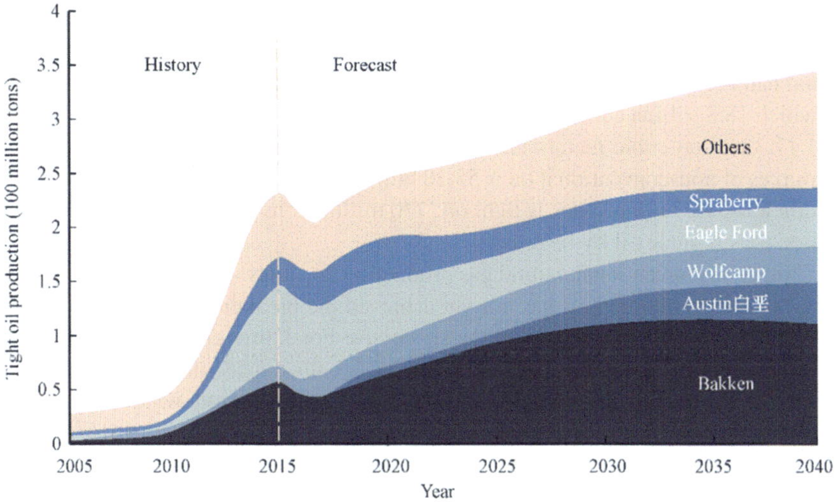

Fig. 1.6 Forecast of tight oil production in the United States

Utica and Haynesville shale gas have the greatest growth potential, accounting for 75% of shale gas production growth across the United States (Fig. 1.7).

China has become the world's third largest producer of unconventional oil and gas resources after the United States and Canada. The large amount of unconventional oil and gas resources in China is the basis of future increasing reserves and production

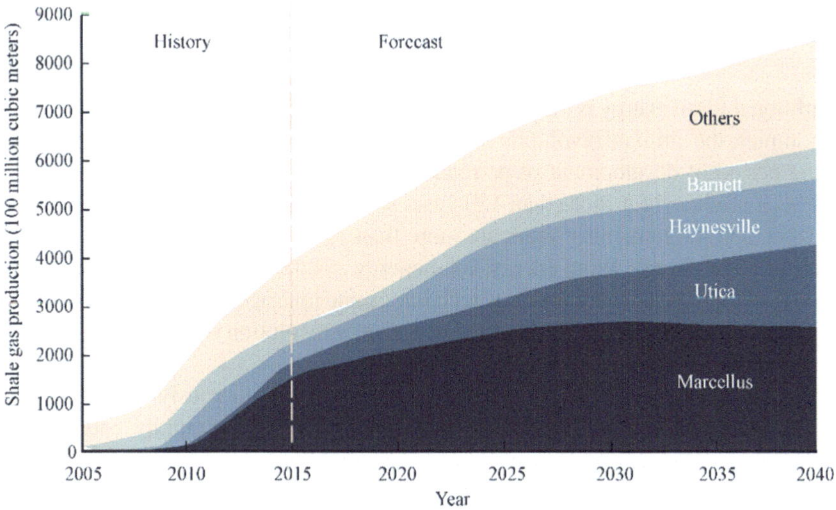

Fig. 1.7 Prediction of shale gas production in the United States

of oil and gas. According to the evaluation of the Research Institute of Petroleum Exploration and Development, the amount of recoverable resources of unconventional natural gas in China is 46.6 trillion cubic meters from the view of technology, of which 18.8 trillion cubic meters is shale gas, 16.5 trillion cubic meters is tight gas and 11.3 trillion cubic meters is coalbed methane, and the amount of recoverable resources of non-conventional oil is 55.18 billion tons, of which oil shale oil is 13.18 billion tons, 1.23 billion tons is tight oil, 770 million tons is oil sands, more than 40 billion tons is shale oil by forecast.

China's unconventional natural gas production accounts for an important proportion, but unconventional oil has not yet achieved the breakthrough of practical and large-scale development yet. With the successive breakthrough in the development of unconventional natural gas, such as tight gas, coalbed methane and shale gas, the production rate is growing rapidly and entering a leapfrog period, which has become the main force for the growth of natural gas production. Compared with 2016, China's unconventional natural gas production increased by 14% to 39.8 billion cubic meters in 2017, accounting for 28.6% of total natural gas production. China's tight gas production made a breakthrough in 2005, more than 30 billion cubic meters was produced in 2012. In 2015, production reached 35 billion cubic meters and tended to stable. In 2017, the production was 34.3 billion cubic meters, accounting for 23.7% of total natural gas production. After a breakthrough in the development of coalbed methane in China in 2006, production growth was relatively slow due to various reasons, the production was 4.5 billion cubic meters in 2017, accounting for 3% of total natural gas production. Production grew rapidly after the breakthrough in shale gas development in China in 2013. In 2017, the production rate reached 9 billion cubic meters, accounting for 6% of total natural gas production.

1.2.3 Fossils and Non-fossils

Traditional fossil energy is not renewable, renewable non-fossil new energy is bound to complete the ultimate revolution of traditional fossil energy (Zou et al. 2016, 2018). If the period of development of petroleum industry is recognized for 300 years, it has been produced for more than 150 years since the world oil industry was started in 1859, and the remaining years are more than 150 years, which could be the life cycle of fossil energy. Wind energy, solar energy, geothermal energy and the current rapid development of energy storage technology and hydrogen energy, all show broad prospects for development, perhaps the new energy revolution will come early before the depletion of fossil energy.

The innovations and breakthroughs are happening continuously in new energy, production and consumption of non-fossil energy are growing rapidly, and the proportion in the primary energy structures is rising rapidly. According to BP (2018), the total global fossil energy production was 11.32 billion tons of oil equivalent in 2017, the total global non-fossil energy production was 2.001 billion tons of oil equivalent in 2017, accounting for 85% and 15%, respectively. Compared with 2000, global fossil energy production increased by 39.42% in 2017, non-fossil energy production increased by 62.16%, and non-fossil energy production grew much faster than fossil energy. In particular, renewable energy excluding hydroelectricity increased by 8.9 times, from 49 million tons of oil equivalent in 2000 to 487 million tons of oil equivalent to 2017 (Figs. 1.8 and 1.9).

It is estimated that in 2050, the world's demand of primary energy will reach 17.5 billion tons of oil equivalent, an increase of about 27% compared with 2016, with an average annual growth rate of 0.65%. Among them, the average growth rate between 2017 and 2030 will be 0.95%, and the average growth rate during the period from 2031 to 2050 will be 0.45%, and the growth rate showed a slowing trend. Renewable energy will grow rapidly at an average annual rate of 6% during the period, with average annual growth rates of natural gas and oil in fossil energy will be 1.3% and

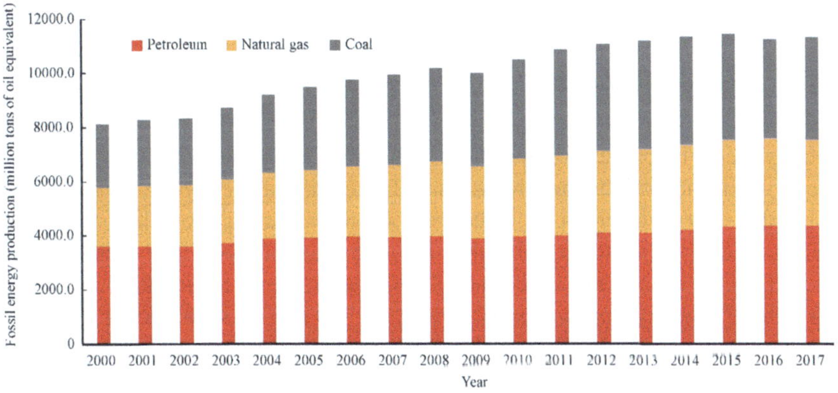

Fig. 1.8 Global fossil energy production from 2000 to 2017

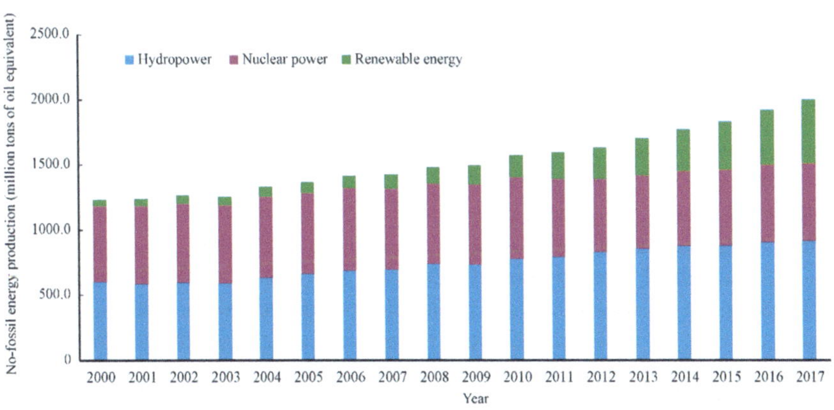

Fig. 1.9 Global non-fossil energy production from 2000 to 2017

0.3%, respectively, at relatively low growing rate. The demand of coal will reduce with an average annual decline rate of 0.8%, showing a negative trend.

1.3 Energy Development Trends

From the standpoint of energy resources type, mode of production and utilization, there will be three major trends in energy development in the future: the development of resource types from high carbon to low carbon and carbon free; the methods of development are changing from producing simply to producing with advanced technologies; and the utilization mode is evolving from direct and primary usage to multiple conversions (Zou et al. 2016, 2018).

1.3.1 Carbon Reduction of Resource Types

Primary energy types are developed from high carbon to low carbon. That is, from fossil energy to non-fossil energy, the world will enter the era of clean energy. The overall trend of energy development is the move from fossil energy to non-fossil energy sources. Carbon content per unit calorific value is as follows: 26.37 tons/terajoule for coal, 20.1 tons/terajoule for crude oil, 15.3 tons/terajoule natural gas (1 terajoule $= 1 \times 10^{12}$ J), and hydropower, wind power, nuclear energy, solar energy and so on do not contain carbon. In the process of the development from coal to oil and gas and from oil and gas to new energy, the amount of pollutants and carbon emissions produced by various types of energy will be lower and lower, adapting to and meeting the needs of green development of the ecological environment.

In the future, energy will be further developed to high energy density, greening and diversification. Energy evolution basically follows the process from low energy density to high energy density. In the age of burning firewood, the energy density was 10–20 MJ/kg (1 MJ $= 1 \times 10^6$ J) level; the industrial revolution marks the energy density gradually increases to 30 MJ/kg or more after entering the coal era. After the second industrial revolution, entering the era of oil, the energy density was further increased to 40 MJ/kg (Fig. 1.10).

Renewable energy is almost carbon-free, and the share of primary energy consumption has increased significantly in recent years. The global clean energy market has maintained a rapid growth trend. By the end of 2016, the cumulative installed capacity of wind power and photovoltaics reached 467 and 296 Gigawatts (1 Gigawatts $= 1 \times 10^9$ watts), respectively, maintaining a high growth rate. Total global investment in clean energy was $66.9 billion in the third quarter of 2017, increased by nearly 40% year-over-year. Among them, the proportion of clean energy power has steadily increased year by year, from January to July 2017, nuclear power generated by OECD (Organization for Economic Cooperation and Development) countries accounted for 17.6% of total electricity generation, hydropower accounted for 14.5%,

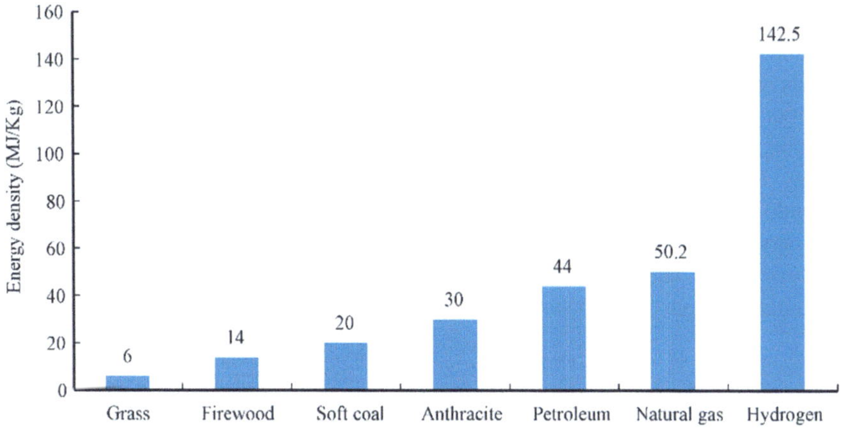

Fig. 1.10 Energy density of various energy sources

geothermal, wind power, solar energy and so on accounted for 9.7%. Among them, geothermal, wind power, solar energy and other non-nuclear new energy development is faster, power generation was 594.2 TKWH from January to July 2017, compared with the same period in 2016, the electricity generation increased by 12.5%, and the proportion increased by 1%. The proportion of renewable energy in China is rising rapidly. In 2016, China's wind power, photovoltaic, nuclear power generation accounted for 4%, 1.1% and 3.6%, respectively. In the first three quarters of 2017, the proportion reached 4.5, 1.8, and 3.9%, showing a trend of increasing year by year.

1.3.2 Technology Intensive Production

The mode of production of energy resources is developed from simple production to technology intensive production. From the view of the general trend of energy development, primitive human beings directly obtained firewood from the natural world as energy. From coal mining to oilfield development, the importance of engineering technology is increasingly obvious. The development nuclear energy, wind energy, solar energy and other new energy resources are technology-intensive industries. From the development history of a certain type of energy, it also reflects the importance of technology. Taking oil and gas exploitation as an example, oil exploitation at the early stage is dominated by vertical wells, the application of horizontal well technology and hydraulic fracturing technology made many wells with low productivity to produce possible and economically. The application of multi-stage fracturing technology in horizontal wells has resulted in a revolution of shale oil and gas in the domain of energy.

The development and utilization of new energy reflects the highest level of scientific and technological development today. Taking new energy power generation to the grid as an example, compared with the traditional centralized power generation, the distributed energy system of new energy has the characteristics of diversification, dynamics and complexity. The system contains many participants and multi-level energy and information flow, capital flow not only can have a better control over the sudden power interruption, but also have more flexible arrangements and plans for energy consumption. This distributed energy network needs to give full play to the current big data and artificial intelligence technology to solve the problem of discontinuous, unstable power production and consumption.

1.3.3 Diversification of Utilization Methods

Energy utilization is developed from primary energy to secondary energy resources or multiple energy resources, and electricity and hydrogen energy are the direction of power utilization in the future. Before the first industrial revolution, firewood and coal as energy were mainly used for direct heat utilization; with the invention of the steam engine in the 1769, energy utilization expanded toward power; and in 1831, when Faraday discovered electromagnetic induction, the energy utilization mode developed towards electricity, starting the era of electrification of energy utilization. However, with the matter of power storage technology, it has been difficult to make a major breakthrough, and it has greater limitations as a power source. Hydrogen energy and electrical energy are easily interchangeable, and has a relatively better energy storage effect. Hydrogen can be converted as secondary energy or multiple resources of energy conversion (Wang et al. 2017). Therefore, the mutual exchange between hydrogen energy and electric energy and multiple energy conversion will be a new direction of energy utilization in the future.

Electric energy has the characteristics of convenient transmission and convenient utilization, and fossil, nuclear energy and renewable energy can be converted into electric energy for transmission or utilization, which is one of the main energy utilization methods at present. Global energy Internet with UHV power grid as the frame of the grid (channel), which is led by the transmission of clean energy. It is a national smart grid of transcontinental and international frame network, covering power grids of different countries and voltage level (transmission grids, distribution networks). The power grid is connected to "one pole and one equator" and the large-scale energy base in various continents to adapt to a variety of needs of distributed power access. It can deliver renewable energy resources such as wind, solar and ocean energy to all types of users. It is a global energy allocation platform with a wide range of services, strong optimizing capabilities, high safety and reliability, and green and low carbon. It is estimated that by 2020, countries will accelerate the development of clean energy and interconnection of domestic power grids, and greatly increase the power grid allocation capacity, intelligence level and proportion of clean energy of each country. From 2020 to 2030, the large-scale, wide-range and efficient optimization of clean energy is expected to be realized by promoting the development of large-scale energy bases and the cross-border interconnection of power grids of clean energy in the continent. From 2030 to 2050, the global energy network will be basically built to achieve the goal of clean energy as the dominator on a global scale, comprehensively addressing issues such as world energy security, environmental pollution and greenhouse gas emissions.

It is difficult to store power energy, and energy power generation such as wind, light and geothermal has a great impact on the power grid. Therefore, the mutual conversion of electricity and hydrogen can solve the problem of power storage, is currently the most ideal solution of new energy. Hydrogen energy and electrical energy can be exchanged to achieve cross-border energy connectivity. Hydrogen and electricity are obtained through the conversion of primary energy (solar, wind, ocean energy, thermal energy, etc.), while hydrogen can generate electricity through fuel cell technology, which can also be used to make hydrogen from water, thus enabling the connectivity of the entire energy network (Yi et al. 2018). The convenience of electricity can be utilized within the extended range of power grid, and hydrogen energy storage and energy supply can be used outside the power grid. The hydrogen energy market continues to grow rapidly, and its role in the energy sector is rising rapidly. China's hydrogen production exceeded 10,000 tons for the first time in 2009 and continued to grow rapidly. In 2017, hydrogen production in China reached 19.1 million tons, which is equivalent to 213.9 billion cubic meters of gas on the condition of standard state. In 2017, China's natural gas production was only 147 billion cubic meters (Table 1.2). Hydrogen has a market share in China over natural gas and is mainly used in agricultural synthetic fertilizers and other products.

Hydrogen energy and hydrogen fuel cells have emerged in the early stages of commercialization in the field of transportation, stationary power generation, communications base station backup power supply and material handling (Yi et al. 2018), forming the German model and the Japanese model. The German hydrogen energy development model focuses on the development of electricity to gas model, providing a convenient infrastructure for downstream hydrogen energy applications, and then

Table 1.2 Hydrogen production in China and forecast production in 2020

Production	2016		2017		Forecast of 2020	
	10,000 tons	100 million m^3	10,000 tons	100 million m^3	10,000 tons	100 million m^3
Global hydrogen production	6923	7754	7637	855	10378	11623
China's hydrogen production	1797	2012	1910	2139	2542	2847
China's natural gas production		1371		1470		1800

activating downstream application scenarios. The government and industrial capital also actively promote the construction of hydrogen energy infrastructure, focusing on the core city and relying on the natural gas pipeline system to expand gradually to the outside. Japan hydrogen energy development model is the popularization of downstream applications of hydrogen energy and hydrogen fuel cell, and continue to expand the size of downstream market. In 2014, the government of Japan has clearly promoted the popularization of hydrogen fuel cells for household and industrial utilization. In 2015, it began to rapidly popularize hydrogen fuel cell vehicles. It plans to invest 5.3 million household hydrogen fuel cells in the market in 2030, therefore, there will be a household fuel cell in every 10 Japanese families.

1.4 The Driving Force of Energy Development

From the history of the change of energy industry, from firewood to coal and from coal to oil and gas, the replacement or conversion of each energy is driven not only by the result of science and technology, but also the result of meeting the energy demand of the development of human society civilization. Therefore, scientific and technological innovation and the development of social civilization are the two driving forces of energy development.

1.4.1 Driven by Scientific and Technological Progress

Scientific and technological progress drives energy change. From the three major transitions of energy development, the change of every energy is the result of energy science and technology innovation. For the same kind of energy, the whole life cycle from the appearance to the prosperity is related closely with the continuous innovation and progress of the development and utilization of science and technology.

In 1769, Watt invented the steam engine, which started the first industrial revolution. Human beings entered the era of machine power. Coal entered a new development period as a fuel with high energy density. Faraday discovered the phenomenon of electromagnetic induction in 1831, Edison invented the electric light in 1880, so the second industrial revolution was started thereafter and mankind entered the era of the power industry. Coal was the main raw material for thermal power, and its demand was further increased (Table 1.3). At the same time, coal mining technology continued to improve and entered the era of mechanization, and coal production has increased greatly.

In the 1880s, coal overtook firewood as the largest energy source, completing the first energy conversion. As a result of the invention of internal combustion engine, oil as the main raw material of internal combustion engine, its demand has greatly increased, and energy development of human beings gradually entered the oil and gas era. In the middle of the 20th century, oil surpassed coal as the largest energy resource,

Table 1.3 Summary of key factors of the three technological revolutions and energy transitions

Parameter	1st industrial revolution (industrial revolution)	2nd industrial revolution (electrical revolution)	3rd industrial revolution (information revolution)
Start time	1860s	1970s	2040s–2050s
End time	First half of 19th century	End of 19th century and start of 20th century	Undergoing
Major symbols	Invention and widespread use of steam engine	Invention and wide application of internal combustion engine and electric power	Major breakthroughs in the fields of atomic energy, computer, aerospace engineering, biotechnology, etc.
Theoretical fundamentals	Newton's mechanics	Faraday's electromagnetic	Cybernetics and Einstein's theory of relativity
Emerging industries	Metal smelting, machine manufacturing	Electric power, chemical industry, automobile manufacturing, shipbuilding industry	Electronics, nuclear, aerospace, laser, information industry, etc.
Transportation	Steamboat(ship), train	Tram, car, airship, airplane	Spaceship
Power energy	Improved steam engine (coal)	Generators and motors (electricity), internal combustion engines (petroleum)	Atomic or nuclear energy, solar energy
Era	Steam era	Electrical era	Information era
Production mode	Mechanized production	Electrification production	Automatic production
Summary	From manual labor to machine production	From steam era to electrical era	From electrification to automation and intelligence

completing the second energy conversion. According to BP statistics, the proportion of oil in world's primary energy consumption peaked in 1973 (accounting for 48.7%), and then decreased year by year. By 2017, the proportion of oil was 33.46%, and that of natural gas continued to rise from 15.8% in 1965 to 23.62% in 2017, an increase of about 8%; the proportion of coal recovered slightly after falling to its lowest point in 1999 (about 25%), with a small rebound to 27.93% in 2017.

The firewood can be obtained directly in nature, the energy acquisition method is simple, the energy density is low, and the simple utilization mode of heating and cooking is dominant; the acquisition of coal energy has two ways of open-pit mining and underground mining, which need mining technology, safety technology and delivery technology, etc. during producing. Coal can be used as the main fuel for steam engine and thermal power. Compared with coal, the exploration and development technology of oil and gas resources is more complex, and the discovery of oil and gas resources needs the integration of multidiscipline and multi work types such as geology, geophysics, drilling engineering, reservoir engineering, surface engineering, especially the current development of oil and gas exploration and development to the deep sea, deep formations and unconventional domains, which reflects the highest level of technology today and is a typical technology-intensive industry.

Taking the development of oil and gas as an example, the history of petroleum industry is a history of science and technology, and the continuous innovation of oil and gas geology, theory and technology provides inexhaustible impetus for the sustainable development of economy and society (Zou et al. 2016). In the development history of oil and gas industry for 150 years, there are two major innovations in oil and gas theory (Zou et al. 2014). The first innovation is to find conventional trap oil and gas reservoirs, the second innovation is to find unconventional "sweet area." The leapfrogging revolution of petroleum science and technology from conventional to unconventional oil and gas includes conventional oil and gas trap accumulation theory, unconventional oil and gas continuous occurrence theory, conventional oil and gas vertical well drilling technology, nanomaterial and gas flooding to improve oil and gas recovery technology (Zou et al. 2015). Theoretical technology improves the continuous development of the oil industry, and promotes steady growth of world oil and gas reserves and production. In 2017, the global proved remaining recoverable oil and gas reserves was 239.3 billion tons, 193.5 trillion cubic meters respectively, and the production was about 7.55 billion tons of oil equivalent. Scientific and technological progress promotes the discovery and utilization of oil and gas resources and satisfies the demand for oil and gas in the development of human society.

1.4.2 Driven by Social Civilization

The development of social civilization drives energy demand. The energy of primitive society mainly satisfies the needs of survival. The quality of human life in the feudal society has improved, the primary industrial production has greatly increased the energy demand, the social civilization has accelerated the development since the industrial revolution, the human demand for transportation, information and cultural entertainment has greatly increased, and the modern industrial demand for energy has reached an unprecedented height (Zou et al. 2016). In recent years, waste water, waste gas and waste residue caused by high-carbon energy in the process of development and utilization resulted in a series of ecological environment problems, ecological demand of energy production and consumption have entered the process of energy development.

Productivity levels of primitive society are low, and human survival is facing challenges. Firewood was used as an energy resource for basic survival needs such as cooking, heating and lighting. The productivity level of feudal society has improved, the quality of human life has improved, and the primary industrial production has greatly increased the energy demand, and simple cooking and heating cannot meet the needs of life. In feudal society, the demand for iron, porcelain and other products increased significantly, and firewood could not meet the industrial demand because of its low energy density. Coal has high energy density and large amount of heat, and gradually enters human life and production, and is widely used in smelting, porcelain and other industries. Since the industrial revolution, the development of social civilization has accelerated, internal combustion engine has been used as a high-efficiency power tool in many fields, which has greatly increased the demand for oil and gas. The emergence of means of transportation, such as automobiles, ships and airplanes, has shifted the demand for energy in human society to oil and gas and promoted the second major conversion of energy.

Since the industrial revolution, the population and living standards have also increased significantly as a result of the dramatic increase in productivity, resulting in an unprecedented increase in human demand for energy. The environmental problems caused by the combustion of high carbon energy are becoming more and more serious, and environmental pollution and global warming have become two major environmental challenges facing mankind today. China is affected by its existing conditions of energy resources, coal-based energy consumption structure makes the problem of environmental pollution particularly serious. In the past 20 years, Beijing and Tianjin have increased the consumption of natural gas and new energy, and the proportion of coal has declined significantly (Hou et al. 2015). However, the surrounding area is still dominated by coal, resulting in serious environmental pollution. Coal combustion produces soot, sulfur dioxide, nitrogen oxides and carbon dioxide, etc., the pollution of the atmosphere is very serious. Carbon monoxide, carbon-oxygen compounds, nitrogen oxides and sulphides emitted from automobile exhaust cause serious harm to the atmosphere. Beijing and central and eastern China were covered and influenced by haze in early 2013, similar to London 60 years ago, but the air quality have improved significantly in recent years.

Fossil fuels (such as coal, oil, etc.) emit large amounts of greenhouse gases, such as carbon dioxide, which makes the global climate tend to be warmer. The average global temperature between 1981 and 1990 rose by 0.48 °C from 100 years ago. In the 20th century, the world average temperature climbed about 0.6 °C. The spring ice and snow melting period of Northern Hemisphere is 9 days earlier than 150 years ago, while the autumn frost begins about 10 days later. The 1990s was the warmest decade since the start of temperature recording in the mid-19th century, and the hottest years in the records were 1998, 2002, 2003, 2001 and 1997. The warming of the world has led to a significant increase of global mega-typhoons, hurricanes, tsunamis and other disasters, and the melting of the Earth's glaciers has caused significant changes in the ecological environment.

In order to solve the problem of environmental pollution and climate warming caused by high carbon energy, the demand of human development on renewable energy such as wind, solar and geothermal energy has greatly increased. Before the advent of the new energy era, natural gas has become an insurmountable bridge between fossil energy and new energy, which will finally promote the harmonious development of human energy consumption and ecological environment. According to forecast data of BP, the proportion of oil and coal in primary energy consumption will be about 28% and 24%, respectively, which will decrease by nearly 5% by 2035 in comparison with 2015. The share of natural gas will increase by 1% to about 25% in comparison with 2015. The share of nuclear energy in primary energy consumption remains still low, increasing by about 1.5%. The share of renewable energy has changed the most and will rise by nearly 7.5% over the next 20 years.

References

Hou, Qijun, Xingshan Zhu, and Wu Wang. 2015. Improving natural gas competitiveness is a key issue in optimizing China's energy structure. *International Petroleum Economy*, 6: 20–22.

Wang, Geng, Jinyang Zheng, Lijun Jiang, et al. 2017. The development of hydrogen energy in China. *Science and Technology Review* 35 (22): 105–110.

Yi, Wenjing, Qi Liang, and Qingbing Zhao. 2018. Enhance the hydrogen application in China's energy system to accelerate the energy transition: Status and progress. *Environmental Protection* 2: 30–34.

Zou, Caineng. 2018. Energy revolution and oil company transformation strategy of new era. *Journal of Beijing Petroleum Managers Training Institute* 25 (4): 3–15.

Zou, Caineng, Shizhen Tao, Lianhua Hou, et al. 2014. *Unconventional oil and gas geology*. Beijing: Geological Publishing House.

Zou, Caineng, Zhi Yang, Rukai Zhu, et al. 2015. Progress in China's unconventional oil & gas exploration and development and theoretical technologies. *Acta Geologica Sinica* 89 (6): 979–1007.

Zou, Caineng, Qun Zhao, Guosheng Zhang, et al. 2016. Energy revolution: From a fossil energy era to a new energy era. *Natural Gas Industry* 36 (1): 1–10.

Chapter 2
Map of World Energy

Because of the difference of crustal formation and evolution, the global fossil energy distribution has strong regional characteristics, i.e., exploration and development, production and consumption are largely imbalanced. With the continuous progress of social civilization, demand of human beings for renewable energy, hydropower, nuclear power, biomass fuel and other new energy is increasing. In recent years, the rapid development of unconventional oil and gas, and the energy demand of developing countries such as China, India and other developing countries has grown rapidly. They both have had a major impact on the traditional energy structure from the two sides of supply and demand respectively. Oil and gas have formed the map consisting of four conventional and four unconventional resources. Coal has formed the map dominated by Asia Pacific, North America and Europe. The development of new energy has initially formed three major maps of Europe, North America and Asia Pacific (Zou et al. 2016, 92, 2018, 95).

2.1 Map of World Fossil Energy

2.1.1 Map of Fossil Energy Resources

The global fossil energy mainly includes oil, natural gas and coal, and with the deepening of theoretical understanding and the significant improvement of exploration technology, the new map of fossil energy resources in the world has been reshaped. The development of unconventional oil and gas has reshaped the map of traditional oil and gas resource.

1. Petroleum

The world's proved oil reserves are abundant, mainly in the Middle East, Asia and Europe, the Americas and Asia Pacific (Zou et al. 2016, 2018). With the ratio of reserves over production of 54.5, it has been found that remaining oil can still sustain

C. Zou, *New Energy*, https://doi.org/10.1007/978-981-15-2728-9_2

human demand for more than 50 years. According to the forecast of the Research Institute of Petroleum Exploration and Development, the world's conventional recoverable oil reserves are 535 billion tons, mainly distributed in the Middle East, Russia, North America and South America; recoverable unconventional oil resources are 420.9 billion tons, slightly lower than Conventional oil.

According to data of BP (2018), global oil reserves in 2017 are 169.7 billion tons, of which the Middle East accounted for 47.6%, Central and South America accounted for 19.5%, North America accounted for 13.3%, Europe accounted for 9.3%, Africa accounted for 7.5%, Asia and the Pacific accounted for 2.8%. Global oil reserves increased by 30.4% compared to the year of 2000. The growing reserves were mainly from Central and South America, which had an oil reserves of 13 billion tons in 2000 and 33.01 billion tons in 2017, an increase of 154% (Fig. 2.1). The countries with the largest oil reserves in the world are mainly Venezuela, Saudi Arabia, Canada, Iran and Iraq, whose reserves are 30.3 billion tons, 26.6 billion tons, 16.9 billion tons, 15.7 billion tons and 14.9 billion tons, respectively. In general, the world's oil reserves are located in the three major areas with huge reserves including the Middle East, Central and South America and North America.

2. Natural Gas

The recoverable conventional natural gas resources of the world are 471 trillion cubic meters, mainly located in the four major regions of Middle East, Russia, North America and South America (Zou et al. 2016, 2018). Since 2000, with the improvement of understanding and development technology, the unconventional natural gas, represented by shale gas in North America, has achieved large-scale development. The latest estimate of global recoverable non-conventional natural gas resources are about 4000 trillion cubic meters, roughly 8 times the amount of conventional natural gas resources.

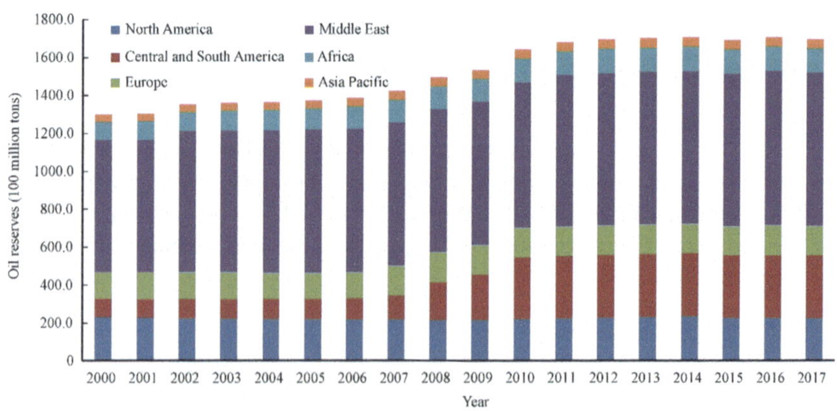

Fig. 2.1 Global oil reserves distribution changes from 2000 to 2017

According to data of BP (2018), Global natural gas reserves are 193.5 trillion cubic meters in 2017, of which the Middle East accounted for 40.9%, Europe accounted for 32.1%, Asia and the Pacific accounted for 10%, Africa accounted for 7.1%, North America accounted for 5.6% and Central and South America accounted for 4.3%. Global natural gas reserves increased by 38.9% compared to the year of 2000. The four regions with the largest growing rate are Asia-Pacific, North America, Europe and the Middle East, with natural gas reserves of 11.8 trillion cubic meters, 7.5 trillion cubic meters, 41.4 trillion cubic meters and 59 trillion cubic meters in 2000, respectively. In 2017, the reserves of them increased to 19.3 trillion cubic meters, 10.8 trillion cubic meters, 62.2 trillion cubic meters and 79.1 trillion cubic meters respectively (Fig. 2.2). The countries with the largest natural gas reserves in the world are Russia, Iran, Qatar, Turkmenistan and the United States, whose reserves are 33.2 trillion cubic meters, 35 trillion cubic meters, 24.9 trillion cubic meters, 19.5 trillion cubic meters and 8.7 trillion cubic meters, respectively. In general, the world's natural gas reserves have formed three major reserve centers in the Middle East, Europe and Russia.

3. Coal

Coal is the world's richest fossil energy, with a total of more than 100 trillion tons of resources, mainly distributed in three areas, Europe and Eurasia, Asia Pacific and North America (Zou et al. 2016). According to data of BP (2018), global coal reserves are expected to be 1.035 trillion tons in 2019, of which Asia Pacific accounted for 41%, Europe accounted for 31.3%, Central and South America accounted for 1.4% and the Middle East and Africa accounted for 1.4%. The countries with the largest coal reserves in the world are dominated by the United States, China, Russia, Australia and India, whose reserves are 250.9 billion tons, 138.8 billion tons, 160.3

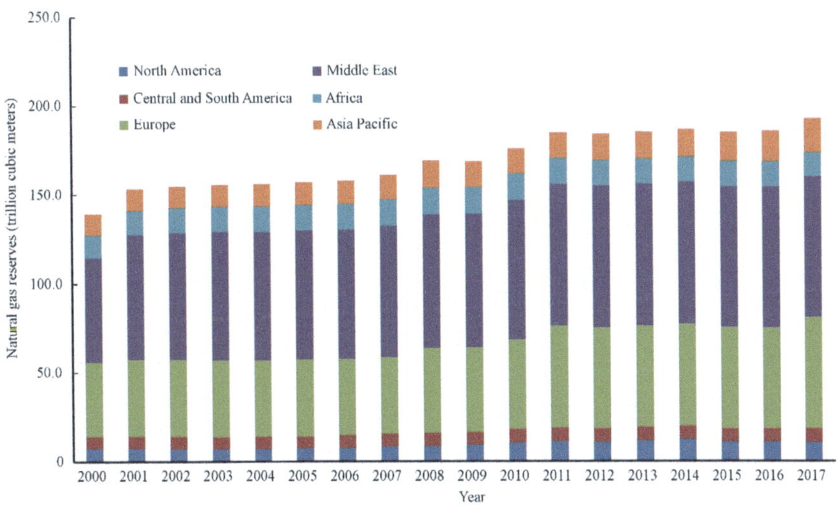

Fig. 2.2 Global natural gas reserve distribution changes from 2000 to 2017

billion tons, 144.8 billion tons and 97.7 billion tons, respectively. In general, the world's coal reserves have formed three major reserve centers, e.g. North America, Asia Pacific and Europe.

2.1.2 Map of Fossil Energy Production

The unconventional oil and gas revolution caused by technological progress is pushing the world oil and gas production structure to be profoundly changed. Over the past 10 years, world oil production has grown steadily and natural gas production has grown relatively rapidly (Zou et al. 2016, 2018). Under the influence of the expansion of coal production capacity in emerging economies such as China, the imbalance of world coal production has increased, and the situation of coal production dominated by Asia-Pacific has been strengthened.

1. Petroleum

Global oil production has entered a stable period, basically forming three major oil production centers including the Middle East, North America and Russia. Global oil production was 4.387 billion tons in 2017, increased by 21.3% compared to the year of 2000. Of these, oil production in the Middle East, North America, Europe, Central and South America, Asia Pacific and Africa in 2017 were 1.481 billion tons, 917 million tons, 717 million tons, 368 million tons, 521 million tons and 383 million tons, respectively, accounting for 33.8%, 20.9%, 16.3%, 8.4%, 11.9% and 8.7%. Compared with the year of 2000, oil production in the Middle East, North America, Europe, Central and South America, Asia Pacific and Africa in 2017 increased by 28.8%, 42.7%, 18.3%, 6.7%, −1.5% and 3.4%, respectively (Fig. 2.3). The 5 countries with

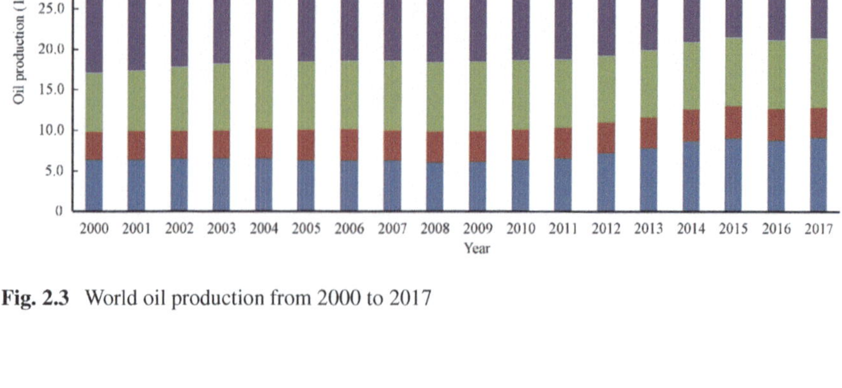

Fig. 2.3 World oil production from 2000 to 2017

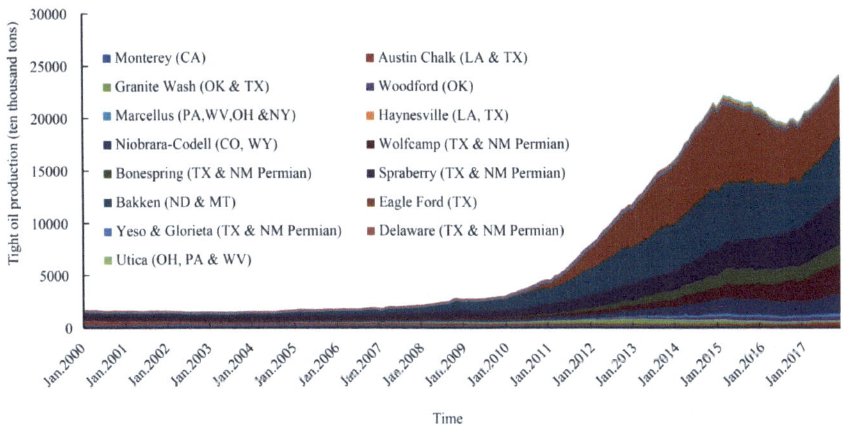

Fig. 2.4 Tight oil production of U.S. from 2000 to 2017

the highest oil production in 2017 were the United States, Saudi Arabia, Russia, Canada and Iran, with production of 571 million tons, 561 million tons, 554 million tons, 236 million tons and 234 million tons, respectively.

Since 2000, major breakthroughs have been made in exploration and development of unconventional oil in North America. With the rapid development of unconventional oils such as tight oil and oil sands, oil production increased by 37.4% in North American and the unconventional oil has become a major growing point of global oil production. The United States produced less than 20 million tons of tight oil in 2000 and 220 million tons of tight oil in the United States in 2017 (Fig. 2.4). Tight oil production not only takes the place of the rapid decline of conventional oil production, but also keeps U.S. oil production growing.

2. Natural Gas

Global natural gas production development has entered a new era, basically formed four major natural gas production centers including Europe, North America, the Middle East and Asia-Pacific. Global gas production was 3.6803 trillion cubic meters in 2017, and global gas production grew by 53% compared to the year of 2000. Of these, natural gas production in Europe, North America, the Middle East, Asia Pacific, Africa and Central and South America in 2017 was 1.0574 trillion cubic meters, 951.5 billion cubic meters, 659.9 billion cubic meters, 607.5 billion cubic meters, 225 billion cubic meters and 179 billion cubic meters, respectively, accounting for 28.7%, 25.8%, 17.9%, 16.5%, 6.1% and 5%, respectively. Natural gas production in Europe, North America, the Middle East, Asia Pacific, Africa and Central and South America increased by 13.1%, 27%, 213%, 119%, 69.6% and 75.8% respectively in 2017 compared to the year of 2000 (Fig. 2.5). The five countries with the highest natural gas production were the United States, Russia, Iran, Canada and Qatar, with production of 739.5 billion cubic meters, 635.6 billion cubic meters, 223.9 billion cubic meters, 176.3 billion cubic meters and 175.7 billion cubic meters, respectively.

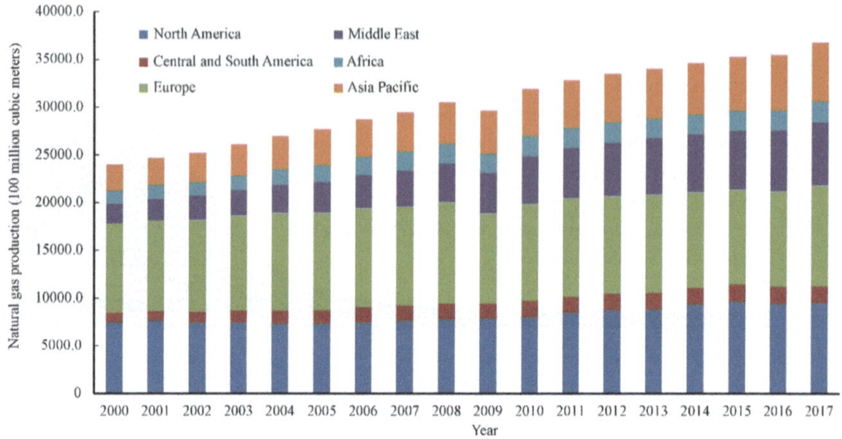

Fig. 2.5 World natural gas production from 2000 to 2017

Natural gas production in the Middle East, Asia Pacific and North America contributed 41.2%, 31.1% and 24.6% of global gas production growth, respectively, as the development of north area—South Pas gas fields in the Middle East accelerated natural gas development in the Asia-Pacific region and rapid growth continued in shale gas production in North America. Since 2000, North America has made major breakthroughs in unconventional natural gas exploration and development, represented by shale gas, gas production increased by 26.6% in North American. natural gas production of U.S. was 739.5 billion cubic meters in 2017, of which shale gas production was 477.2 billion cubic meters, become the main part of natural gas production (Fig. 2.6).

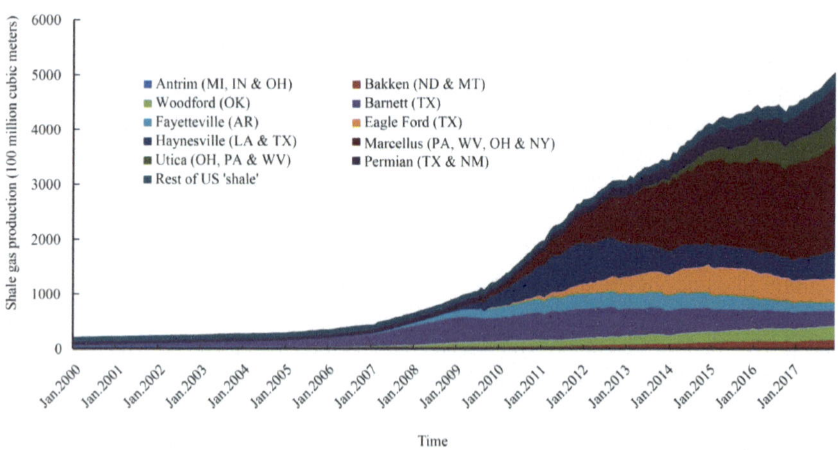

Fig. 2.6 Shale gas production in the United States from 2000 to 2017

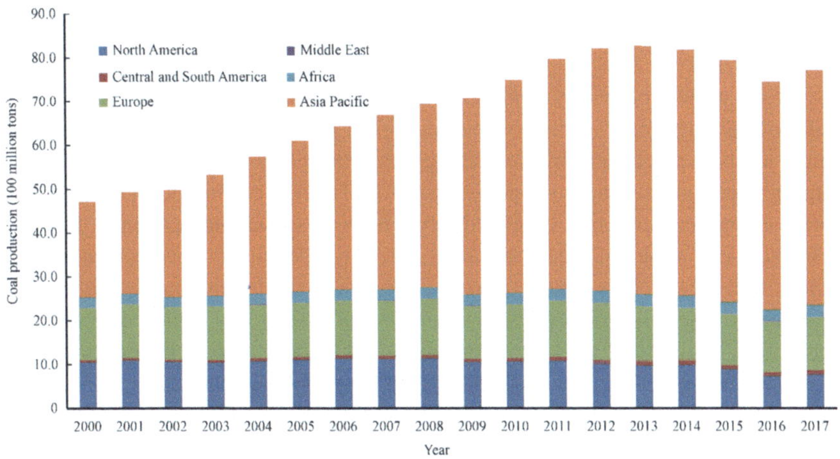

Fig. 2.7 World coal production from 2000 to 2017

3. Coal

Affected by the expansion of coal production capacity in emerging economies such as China, the imbalance of world coal production has increased, the dominant role of coal production in Asia-Pacific is strengthed, the world's coal production peaked in 2013, showing a downward trend (Zou et al. 2016). Global coal production was 7.73 billion tons in 2017, and global coal production grew by 63.5% compared to the year of 2000. Of these, coal production in Asia and the Pacific, Europe, North America, Africa, Central and South America and the Middle East in 2017 was 5.36 billion tons, 1.22 billion tons, 770 million tons, 270 million tons, 100 million tons and 1.6 million tons, respectively, accounting for 69.4%, 15.8%, 10%, 3.5%, 1.3% and 0, respectively. Coal production in Asia Pacific, Europe, North America, Africa, Central and South America and the Middle East increased by 145%, 2.4%, −28.8%, 17.4%, 85.7% and 6.9% respectively in 2017 compared to the year of 2000 (Fig. 2.7).

The 5 countries with the highest coal production in 2017 were China, India, the United States, Australia and Indonesia, with production of 3.52 billion tons, 716 million tons, 702 million tons, 481 million tons and 461 million tons, respectively.

Coal production in Asia Pacific, North America and Europe is in the dominant position, and China accounts for half of the world's coal production. It began to decline after global coal production peaked at 8.275 billion tons in 2013. Four of the world's top five coal producers were in the Asia-Pacific region in 2017, and China is the major producer of global coal production, accounting for 45.5% of global coal production.

2.1.3 *Map of Fossil Energy Consumption*

The strong growth of energy demand in emerging economies and the approaching limits of the carrying capacity of the ecological environment have forced human beings to choose between different energy varieties. This choice directly and profoundly affects and reshapes the new map of fossil energy consumption in the world. Global energy consumption is related to the level of social and economic development and the difficulty of access to resources. Energy demand in developed countries such as the United States and Europe has remained stable, and energy demand in Asia-Pacific emerging economies has grown rapidly, the fossil energy consumption map is developing from "three major locations" of North America, Europe and Asia-Pacific to "two poles" of eastern and western hemispheres (Zou et al. 2016, 2018). Total global energy consumption reached 13.51 billion tons of oil equivalent in 2017, an increase of 43.9% over the year of 2000. Among them, energy consumption in the Asia-Pacific region reached 5.74 billion tons of oil equivalent in 2017, an increase of 116.7% over the year of 2000, making it the main driver of global energy consumption growth. Energy consumption in Europe and North America in 2017 was 2.95 billion tons oil equivalent and 2.77 billion tons oil equivalent, respectively, an increase of 4.8% and 0.6% from the year of 2000, respectively, and generally tends to be stable (Fig. 2.8).

1. Petroleum

Global oil consumption has generally stabilized, basically forming the three major oil consumption centers, Asia-Pacific, North America and Europe's. Global oil consumption in 2017 was 4.622 billion tons, an increase of 24.5% in global oil consumption compared to the year of 2000. Of these, oil consumption in Asia and the Pacific, North America, Europe, the Middle East, Central and South America and Africa in

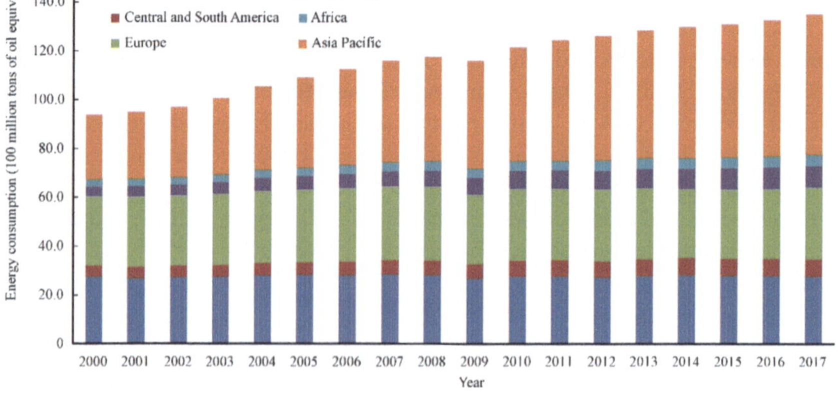

Fig. 2.8 World energy consumption of the from 2000 to 2017

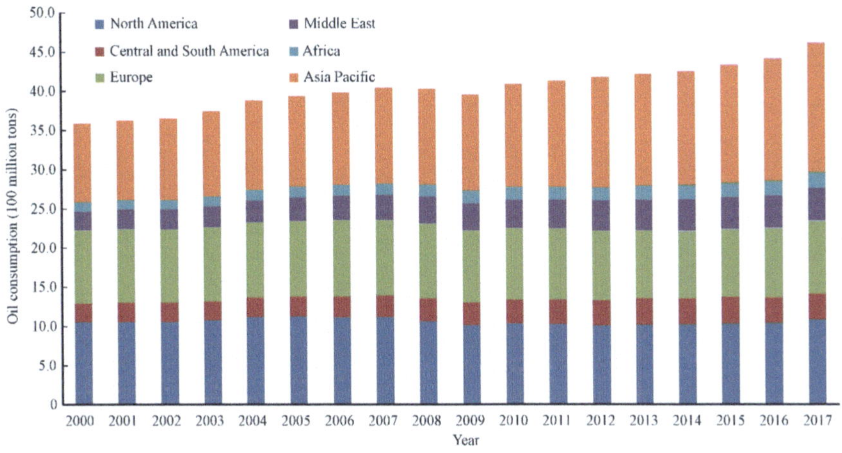

Fig. 2.9 World oil consumption from 2000 to 2017

2017 was 1.711 billion tons, 1.06 billion tons, 941 million tons, 400 million tons, 320 million tons and 190 million tons, respectively, accounting for 37%, 22.9%, 20.4, 8.7%, 6.9% and 4.1%, respectively. Compared to the year of 2000, oil consumption in Asia Pacific, North America, Europe, the Middle East, Central and South America and Africa increased by 60.2%, −4.3%, 3.1%, 66.8%, 33.8% and 59.8% respectively (Fig. 2.9) in 2017. The five countries with the highest oil consumption in 2017 were the United States, China, India, Japan and Saudi Arabia, with consumption of 910 million tons, 596 million tons, 221 million tons, 181 million tons and 166 million tons, respectively.

The imbalance between oil production and consumption has contributed to the global oil trade. In general, Asia Pacific, North America and Europe are both the center of global oil consumption and the centers of global oil import. According to data of BP (2018), the total global oil trade in 2017 was 2.184 billion tons. Among them, China, India and Japan in the Asia-Pacific region, imported 470 million tons, 211 million tons and 163 million tons in 2017, respectively, accounting for 38.6% of global oil imports; Oil imports for European amounted to 516 million tons, accounting for 23.6% of global oil imports; and oil imports of U.S. were 3.94 billion tons, accounting for 18% of global oil imports.

2. Natural Gas

The global natural gas consumption generally shows a rapid growth trend, basically formed three major natural gas consumption centers, Europe, North America and Asia-Pacific. Global gas consumption was 3.6703 trillion cubic meters in 2017, and global gas consumption increased by 51.8% per cent compared to the year of 2000. Of these, natural gas consumption in Europe, North America, Asia-Pacific, Middle East, Central and South America and Africa in 2017 was 1.1063 trillion cubic meters, 942.8 billion cubic meters, 769.6 billion cubic meters, 536.4 billion cubic meters, 173.4 billion cubic meters and 141.8 billion cubic meters, respectively,

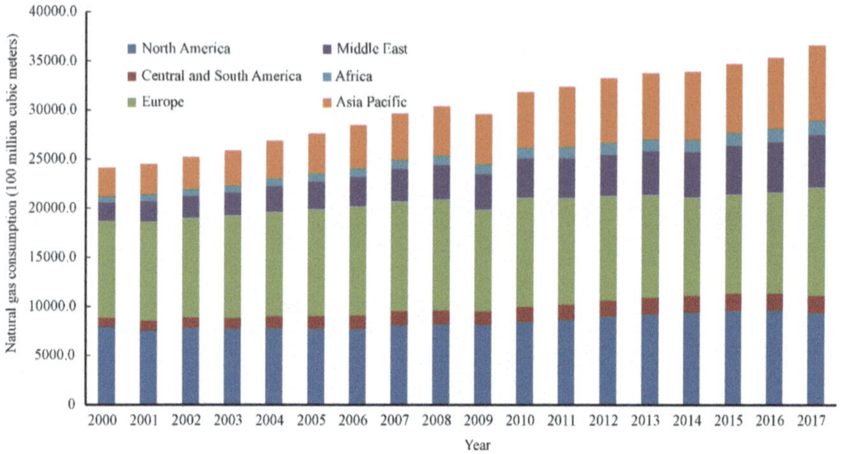

Fig. 2.10 World natural gas consumption from 2000 to 2017

accounting for 30.1%, 25.7%, 21.0%, 14.6%, 4.7% and 3.9%, respectively. Natural gas consumption in Europe, North America, Asia-Pacific, the Middle East, Central and South America and Africa in 2017 increased by 12.4%, 18.7%, 161.1%, 181.6%, 80.3% and 146.2%, respectively, compared to the year of 2000 (Fig. 2.10). The five countries with the highest natural gas consumption in 2017 were the United States, Russia, China, Iran and Japan, with consumption of 739.5 billion cubic meters, 424.8 billion cubic meters, 240.4 billion cubic meters, 214.4 billion cubic meters and 117.1 billion cubic meters, respectively.

The development of emerging economies such as China and India in the Asia-Pacific region has led to a significant increase in demand for natural gas. Gas production growth is well below consumption growth due to limited resources, gas trade is active in the Asia-Pacific region, and the global gas trade center is moving eastward. From 2007 to 2017, the growth of natural gas consumption in the Asia-Pacific region was 100.7 billion cubic meters higher than the production growth, while in the Middle East and North America, natural gas production growth was 81.7 billion cubic meters and 13.5 billion cubic meters, respectively, than consumption growth. According to data of BP (2018), North America, Asia Pacific and Europe are the three major centers of global gas trade. Among them, natural gas imports in North America and Europe largely are basically stable, and Asia Pacific is a major region with growing global gas trade. From 2007 to 2017, global natural gas trade volume increased from 776.1 billion cubic meters to 1.1341 trillion cubic meters, an increase of 46.1%. Among them, natural gas transporting through pipeline increased from 546.7 billion cubic meters to 740.7 billion cubic meters, an increase of 35.5%; LNG (liquefied natural gas) grew from 226.4 billion cubic meters to 393.4 billion cubic meters, an increase of 73.8%.

The breakthrough in LNG technology has freed the gas trade from distance restrictions, and the Asia-Pacific region has become the most important region for the growth of natural gas trade. Global LNG trade volume was 226.4 billion cubic meters in 2007, accounting for 29.2% of total trade; in 2017, the global LNG trade volume was 393.4 billion cubic metres, accounting for 34.7% of total trade volume. Natural gas trade in North America, Europe and Asia and the Pacific in 2007 was 154.9 billion cubic metres, 429.1 billion cubic metres and 165.1 billion cubic metres, respectively, and LNG trade accounted for 15.5%, 12.4% and 90% of natural gas trade, respectively. In 2017, natural gas trade in North America, Europe and Asia- Pacific was 155.9 billion cubic meters, 551.4 billion cubic meters and 346.5 billion cubic meters, respectively, LNG trade accounted for 5.9%, 11.9% and 81.8% of natural gas trade, respectively.

3. Coal

The global coal consumption is generally showing a slow decreasing trend. Similar to the structure of coal production, Asia and Pacific play a major role in coal consumption. Global coal consumption was 3.732 billion tons of oil equivalent in 2017, and global coal consumption increased by 58.41% compared to the year of 2000. Among them, coal consumption in Asia and Pacific, Europe, North America, Africa, Central and South America and the Middle East in 2017 was 2.75 billion tons of oil equivalent, 450 million tons of oil equivalent, 390 million tons of oil equivalent, 100 million tons of oil equivalent, 30 million tons of oil equivalent and 10 million tons of oil equivalent, respectively, accounting for 73.7%, 12.1%, 10.4%, 2.7%, 0.8% and 0.3%, respectively. Compared to the year of 2000, coal consumption in Asia Pacific, Europe, North America, Africa, Central and South America and the Middle East in 2017 increased by 140.7%, −13.7%, 36.2%, 15.8%, 66.1% and 20.5%, respectively (Fig. 2.11). The five countries with the highest coal consumption in 2017 were China, the United States, India, Japan and Russia, with consumption of 1.89 billion tons of oil equivalent, 332 million tons of oil equivalent, 412 million tons of oil equivalent, 120 million tons of oil equivalent and 87 million tons of oil equivalent, respectively.

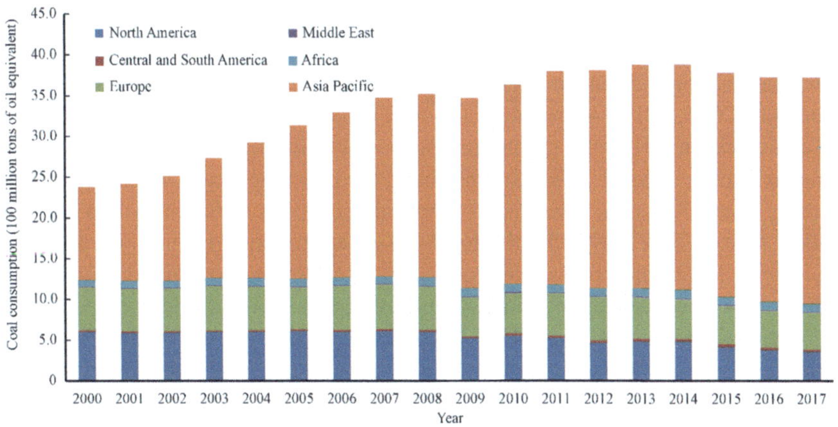

Fig. 2.11 World coal consumption from 2000 to 2017

Despite the abundance coal resources in Asia and Pacific, North America and Europe, consumption showed a downward trend after global coal consumption peaked at 3.89 billion tons of oil equivalent in 2014. China's coal consumption accounted for half of global coal consumption. Three of the world's top five coal producers are in the Asia and Pacific region and China is the mainstay of global coal consumption, accounting for 50.7% of global coal consumption.

2.2 New Energy Map of the World

2.2.1 Map of Nuclear Power

Affected by the nuclear power accidents, the global nuclear power development shows different trends (Zou et al. 2016, 2018). The United States, Japan and South Korea are gradually slowing down the development of nuclear power, China and other emerging countries are actively developing nuclear power due to strong demand of clean energy development. By the end of 2017, the number of nuclear power plants operating in the United States had been reduced from more than 100 to the current 60, with an installed capacity of 99 GW. Japan was affected by the accident of Fukushima nuclear power plant in 2011, nuclear power production was stopped for some time, and nuclear power was not restarted until 2015. Nuclear power consumption in 2016 is only 1/10 of that before the accident. In 2017, South Korea announced that it will completely cancel the new nuclear power plant construction plan under preparation, no longer extend the design life of existing nuclear power plants. Asia is the most active region for nuclear power development. As of 2017, of the 61 nuclear power units under construction in the world, there are 27 nuclear power units under construction located in Asia, accounting for 44.26%; and the installed capacity of nuclear power units in Asia is 30.21 GW, accounting for 49% of the installed capacity of nuclear power units under construction worldwide.

The world nuclear power development is stable, forming three major nuclear power production and consumption center in Europe, North America and Asia Pacific (Zou et al. 2016, 2018). World nuclear power consumption was 596 million tons of oil equivalent in 2017, and global nuclear power consumption increased by 2.1% compared to the year of 2000. Among them, the consumption of nuclear power in Europe, North America, Asia Pacific, Central and South America, Africa and the Middle East in 2017 was 258 million tons of oil equivalent, 216 million tons of oil equivalent, 112 million tons of oil equivalent, 5 million tons of oil equivalent, 3.6 million tons of oil equivalent and 1.46 million tons of oil equivalent, respectively, accounting for 43.3%, 36.2%, 18.7%, 0.8%, 0.6% and 0.3%, respectively. Compared with the year of 2000, nuclear power consumption in Europe, North America, Asia Pacific, Central and South America and Africa in 2017 increased by −3.3%, 9.2%, −1.4%, 80.3% and 15%, respectively (Fig. 2.12). The five countries with the highest nuclear power consumption in 2017 were the United States, France, China, Russia and

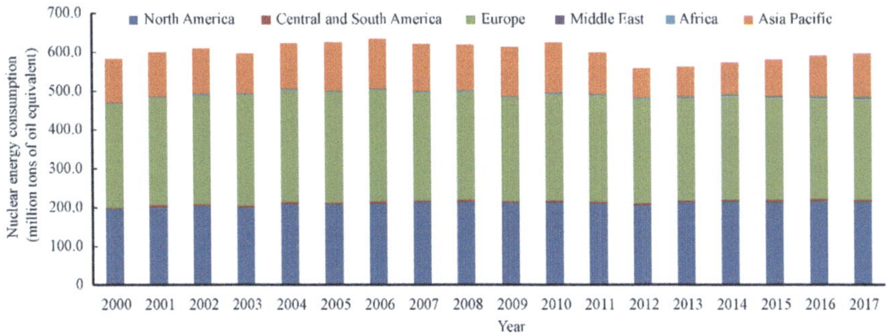

Fig. 2.12 Nuclear power consumption of the world from 2000 to 2017

South Korea, with consumption of 192 million tons of oil equivalent, 90 million tons of oil equivalent, 56 million tons of oil equivalent, 46 million tons of oil equivalent and 34 million tons of oil equivalent, respectively.

2.2.2 Map of Hydropower

The world hydropower technology has matured, and the industry development is mainly controlled by the distribution condition of hydropower resources. The hydropower has formed four major regions of Asia-Pacific, Europe, North America and Central and South America (Zou et al. 2016, 2018). The total installed capacity of hydropower in the world reached 1,246 GW in 2017, with a total power generation of about 4060 TWh (920 million tons of oil equivalent). Hydropower development in the United States and Canada is a world leader, with hydroelectric installed capacity of 102 GW and 79 GW (excluding pumped storage), respectively. The United States Government encouraged the development of hydropower by enacting two bills in 2014 to simplify the approval process for the construction of small-scale hydropower projects on existing hydropower infrastructure, and to increase the standard for licensing exemptions from 5 to 10 MW for hydropower projects built in existing hydropower facilities or hydropower potential sites.

The rapid development of hydropower in the world has resulted in four major hydropower production and consumption centers in Asia Pacific, Europe, Central and South America and North America, of which Asia and Pacific is the main center for the growth of hydropower consumption. World hydropower consumption was 920 million tons of oil equivalent in 2017, and global hydropower consumption increased by 52.8% compared to the year of 2000. Of these, hydropower consumption in Asia and Pacific, Europe, North America, Central and South America, Africa and the Middle East in 2017 was 372 million tons of oil equivalent, 187 million tons of oil equivalent, 164 million tons of oil equivalent, 162 million tons of oil equivalent, 29 million tons of oil equivalent and 4.5 million tons of oil equivalent, respectively,

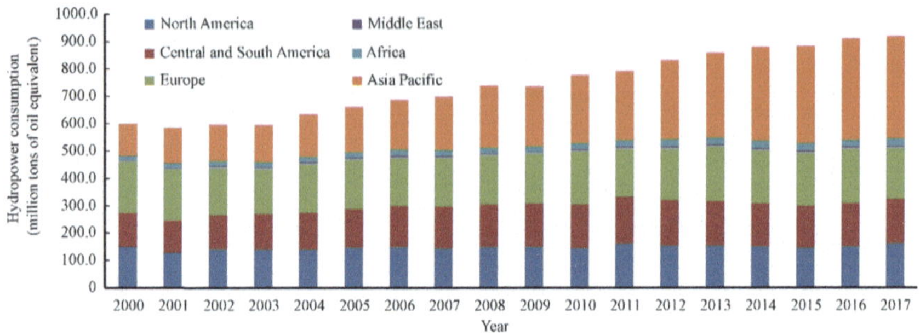

Fig. 2.13 World hydropower consumption from 2000 to 2017

accounting for 40.4%, 20.4%, 17.9%, 17.7%, 3.2% and 0.5%, respectively. Compared to the year of 2000, hydropower consumption in Asia and Pacific, Europe, North America, Central and South America, Africa and the Middle East in 2017 increased by 216.4%, −1.1%, 9.4%, 29%, 71.9% and 145.6%, respectively (Fig. 2.13). The five countries with the highest hydropower consumption in 2017 were China, Canada, Brazil, the United States and Russia, with consumption of 261 million tons of oil equivalent, 90 million tons of oil equivalent, 84 million tons of oil equivalent, 67 million tons of oil equivalent and 42 million tons of oil equivalent, respectively.

2.2.3 Map of Other Renewable Energy

Other renewable energy sources, such as solar energy and wind energy, will grow rapidly and their proportion will continue to rise. Renewable power generation has become a major means of energy utilization, leading the future of renewable energy development. With the continuous progress of the development and utilization of renewable energy resources such as wind and solar energy, the three renewable energy maps of Asia Pacific, Europe and North America have initially formed (Zou et al. 2016, 2018). The other renewable energy consumption of the world was 487 million tons of oil equivalent in 2017, and other global renewable energy consumption increased by 893.5% compared to the year of 2000. Among them, other renewable energy consumption of Asia Pacific, Europe, North America, Central and South America, Africa and the Middle East in 2017 was 175 million tons of oil equivalent, 163 million tons of oil equivalent, 110 million tons of oil equivalent, 33 million tons of oil equivalent, 5.5 million tons of oil equivalent and 1.4 million tons of oil equivalent, respectively, accounting for 36%, 33. 4%, 22.5%, 6.7%, 1.1% and 0.3%, respectively. Other renewable energy consumption in Asia Pacific, Europe, North America, Central and South America, Africa and the Middle East increased by 1590%, 985.8%, 450.8%, 874.7%, 1221.7% and 13781.1% in 2017, respectively, compared to the year of 2000. The top five countries with the highest consumption of other renewable energy in 2017 were China, the United States, Germany, Japan

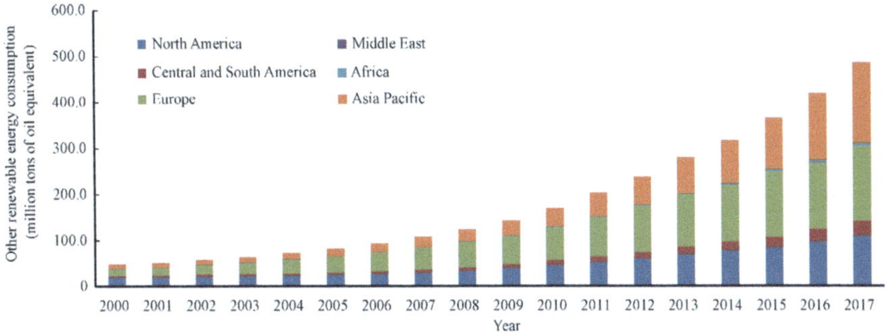

Fig. 2.14 Other renewable energy consumption in the world from 2000 to 2017

and Brazil, with consumption of 107 million tons of oil equivalent, 95 million tons of oil equivalent, 45 million tons of oil equivalent, 22 million tons of oil equivalent and 22 million tons of oil equivalent, respectively (Fig. 2.14).

1. Wind Power

As a technologically mature and environmentally friendly renewable energy, wind power has been widely developed and applied on a large scale worldwide (State Grid Energy Research Institute 2017). Wind power has surpassed traditional hydropower to become the largest renewable energy in the United States in the 2016, and the cost of wind power in the United States has fallen by almost 66% in the past 7 years. In Germany, onshore wind power has become the cheapest resource of energy in the entire energy system. Wind power accounted for 11.6% of electricity consumption across Europe in 2017, with Denmark's wind power as a share of electricity consumption continuing to increase by 4% to 44.4%, Germany to 20.8% and the UK to 13.5%. Global onshore wind power costs were already significantly lower than fossil energy resources in 2017, and are gradually approaching hydropower, about 6 cents/kwh. The average cost of new onshore wind power in 2017 was 4 cents/kwh. IRENA (International Renewable Energy Agency) expects the cost of the world's lowest-cost wind power project to reach 3 cents/kwh in 2019, making it one of the most economical green electricity.

Wind Power of the world has formed three major production and consumption centers in Europe, Asia Pacific and North America. World wind power consumption was 254 million tons of oil equivalent in 2017, and global wind power consumption increased by 3,465% compared to the year of 2000. Among them, wind power consumption in Europe, Asia Pacific, North America, Central and South America, Africa and the Middle East in 2017 was 87 million tons of oil equivalent, 83 million tons of oil equivalent, 68 million tons of oil equivalent, 13 million tons of oil equivalent, 3 million tons of oil equivalent and 0.2 million tons of oil equivalent, respectively, accounting for 34.3%, 32.7%, 26.8%, 5.1%, 1% and 0.1%, respectively. Compared to the year of 2000, wind power consumption in Europe, Asia Pacific, North America, Central and South America, Africa and the Middle East increased by 1612%, 14,209%, 4,975%, 22,603%, 4,958% and 2,343%, respectively (Fig. 2.15). The top

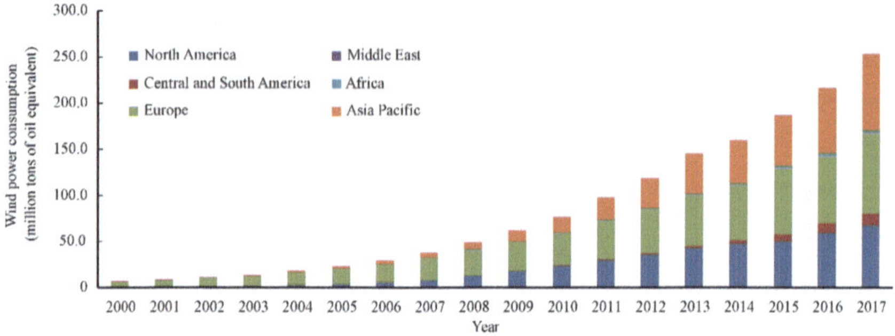

Fig. 2.15 World wind power consumption from 2000 to 2017

five countries with the highest wind power consumption in 2017 were China, the United States, Germany, India and Spain, with consumption of 65 million tons of oil equivalent, 58 million tons of oil equivalent, 24 million tons of oil equivalent, 12 million tons of oil equivalent and 11 million tons of oil equivalent, respectively.

2. Solar Energy

The development of global solar energy utilization market is diversified, and the major countries still dominate. The global solar market surged by 26% in 2017, surpassing 100 GW of photovoltaic installed capacity for the first time. According to data of GTM Research (Renewable Energy Advisory Agency), 106 GW of new photovoltaic capacity is expected to be added in 2018. China, the United States, India and Japan will continue to dominate the demand in 2018, but their share of the global overall market will fall from 82% in 2017 to 72% in 2018. The number of countries installing 1 GW or more per year will increase from the current 9 countries to 14 countries. In countries such as Brazil, Egypt, Mexico, the Netherlands and Spain, the installed capacity of solar energy exceeded 1 GW for the first time in 2018.

The world's solar energy has formed three major production and consumption centers in Asia Pacific, Europe and North America. World solar energy consumption in 2017 was 100 million tons of oil equivalent, increased by 383 times, compared to the year of 2000. Among them, solar consumption in Asia Pacific, Europe, North America, Central and South America, Africa and the Middle East was 49 million tons of oil equivalent, 28 million tons of oil equivalent, 19 million tons of oil equivalent, 2 million tons of oil equivalent, 1.3 million tons of oil equivalent and 1.1 million tons of oil equivalent, respectively, accounting for 48.8%, 28.3%, 18.5%, 2%, 1.3% and 1.1%, respectively (Fig. 2.16). The five countries with the highest solar consumption in 2017 were China, the United States, Japan, France and Italy, with consumption of 25 million tons of oil equivalent, 18 million tons of oil equivalent, 14 million tons of oil equivalent, 9 million tons of oil equivalent and 60 million tons of oil equivalent, respectively.

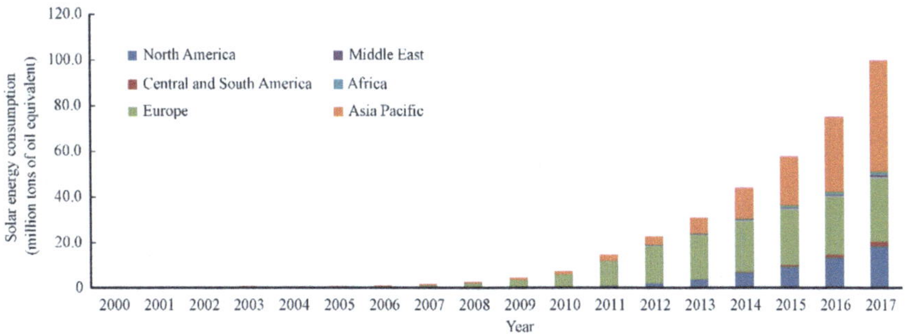

Fig. 2.16 World solar consumption from 2000 to 2017

3. Geothermal Energy

Geothermal utilization technology is mature, and the exploitation and utilization are restricted by the distribution of geothermal resources. The global geothermal resources are mainly distributed in the four tropical regions: one is the Circum-Pacific geothermal belt, which is the collision boundary between the world's largest Pacific plate and the Americas, Eurasian, Indian Ocean plate; the second is the Mediterranean-Himalaya geothermal belt, which is the collision boundary between the Eurasian plate and the African plate and the Indian Ocean plate; the third is the Atlantic mid-ridge geothermal belt, which is the rift of the Atlantic Ocean plate; and the fourth is Red Sea-Gulf of Aden-East African Rift Valley geothermal zone, which includes geothermal fields in Djibouti, Ethiopia, Kenya and other countries. According to data the relevant departments, the total calorific value of rocks and liquids at a depth of 5000 m from the surface of the earth and above 15 °C is estimated to be about 14.5×10^{25} J, equivalent to the heat of 4948 trillion tons of standard coal.

In 2017, the total installed capacity of geothermal energy in the world was 14,305 MW, an increase of 56.4% compared with the year of 2000. In 2016, the top five countries with the highest geothermal energy utilization were the United States, Philippines, Indonesia, New Zealand and Italy, with total geothermal installed capacity of 3,719 MW, 1,928 MW, 1,860 MW, 978 MW and 916 MW, respectively (Fig. 2.17).

2.3 China's Fossil Energy Map

As the largest developing country in the world, China's social economy has undergone tremendous changes after years of rapid development. The existing conditions of abundant coal resources and relatively insufficient oil and gas resources in China make the energy production and consumption have their own characteristics. China's fossil energy industry has developed steadily, coal has a dominant position in fossil

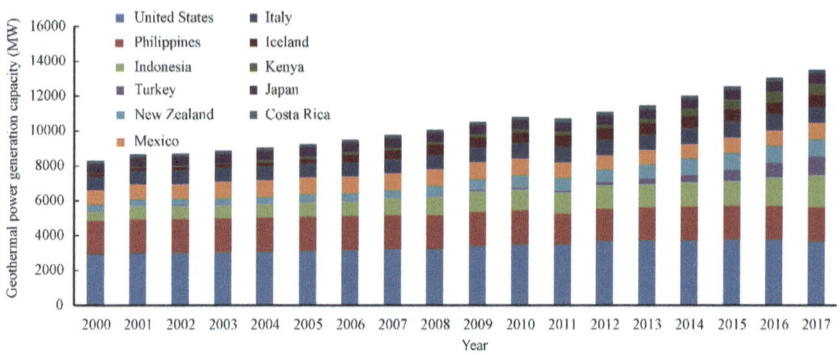

Fig. 2.17 Geothermal energy utilization in major countries of the world from 2000 to 2017

energy, oil production tends to be stable, and natural gas production grows rapidly. China's total fossil energy production reached 2.067 billion tons of oil equivalent in 2017, among them, coal, oil and natural gas accounting for 84.5%, 9.3% and 6.2%, respectively.

2.3.1 Map of Fossil Energy Resources

China's fossil energy resources are relatively rich in coal resources, relatively lack of oil and gas resources. Rich coal and insufficient oil and gas is the basic national conditions of China's energy resources (Zou et al. 2016). China's total coal resources is about 5 trillion tons, the overall pattern is more in west and less in east, richer in the north and poorer in the south. The total resources of Shanxi, Inner Mongolia, Shaanxi, Xinjiang, Guizhou, Ningxia and other six provinces/autonomous regions reached 4.19 trillion tons, accounting for 84% of the country's total coal resources. By the end of 2017, China's proved coal reserves were 138.8 billion tons, accounting for 13.4% of global proved coal reserves.

According to data of the Ministry of Land and Resources (2016), geological resources of petroleum in China are 125.7 billion tons, and recoverable resources are 30.1 billion tons. By the end of 2017, the proved oil reserves were 2.57 billion tons, accounting for 1.5% of the world's proved oil reserves (Fig. 2.18); geological resources of natural gas are 90.3 trillion cubic meters, recoverable resources are 50.1 trillion cubic meters, and proven reserves of natural gas are 5.5 trillion cubic meters, accounting for 2.9% of the world's proved natural gas reserves (Fig. 2.19). China's unconventional oil and gas resources have considerable potential. The geological resources of shale gas buried above the depth of 4500 m in China are 122 trillion cubic meters, the amount of recoverable resources are 22 trillion cubic meters. Until now, the cumulative proved geological reserves are 1.0456 trillion cubic meters, the ratio of proved resources over total recoverable resources is only 0.5%. The

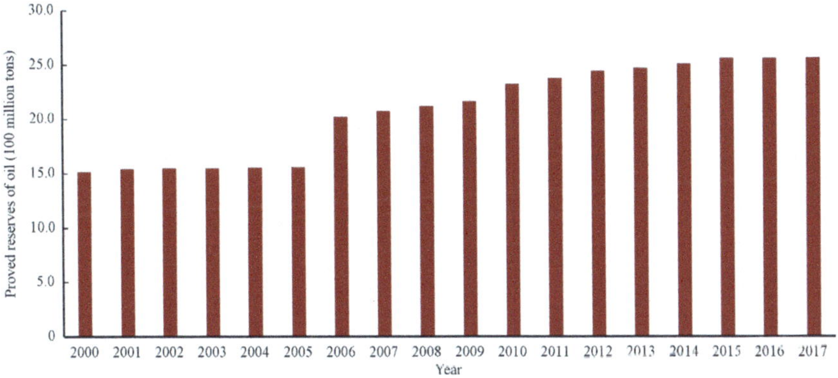

Fig. 2.18 Proved reserves of petroleum in China from 2000 to 2017

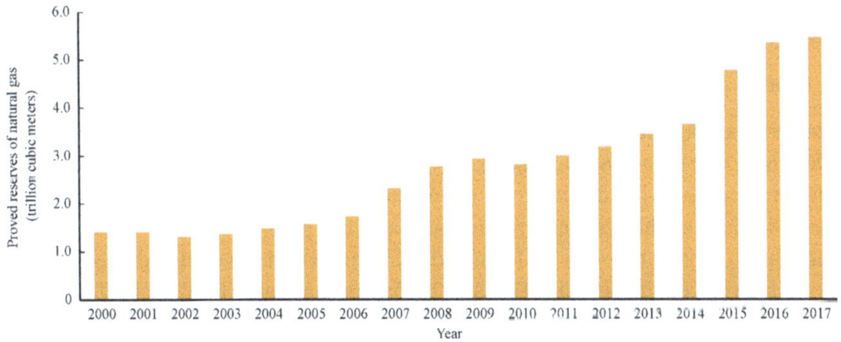

Fig. 2.19 Proved reserves of natural gas in China from 2000 to 2017

geological resources of shallow coalbed methane buried above 2000 m are 30 trillion cubic meters, the amount of recoverable resources are 12.5 trillion cubic meters, and the cumulative proved geological reserves are 629.3 billion cubic meters. The proved ratio is only 2.1%. With the improvement of theoretical understanding and engineering technology, there is potential room for further growth of oil and gas resources.

2.3.2 *Map of Fossil Energy Production*

China's fossil energy production has grown steadily, and coal production has been a dominator. In general, coal is overcapacity, oil production is stable and natural gas production grows rapidly (Zou et al. 2016). China's total fossil energy production

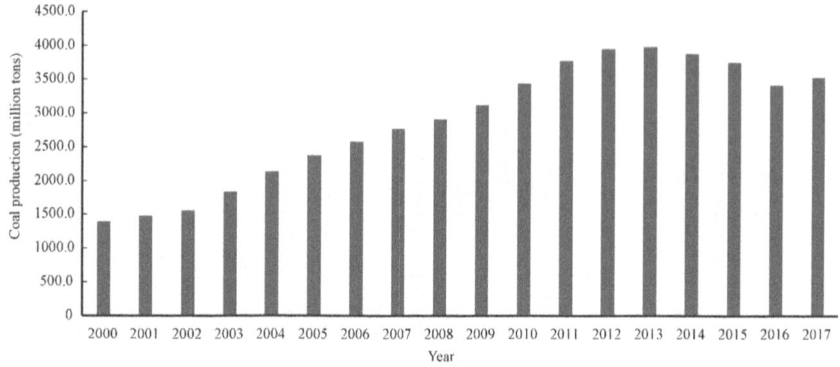

Fig. 2.20 China's coal production from 2000 to 2017

reached 2.067 billion oil equivalents in 2017, of which coal, oil and natural gas accounting for 84.5%, 9.3% and 6.2%, respectively.

1. Coal

After production of China coal peaked in 2013, the general trend of coal production is declining due to reduced demand. China's coal production in 2017 was 3.52 billion tons, an increase of 110 million tons, an increase of 3.2% over the same period of last year (Fig. 2.20). China's coal production reached a peak of 397,000 tons in 2013. Since then, demand has slowed down and coal production has declined year by year. The year-over-year declines in 2014 and 2015 were 2.5% and 3.3%, respectively, and coal production fell to 3.41 billion tons in 2016 after the start of supply-side reforms in the coal industry, down 9% from the same period of last year.

According to data of the National Energy Agency (2017), there are 3,907 registered coal production mines in China, with a capacity of 3.336 billion tons/year, distributed in 26 provinces and autonomous regions. The production capacity of registered coal mines in three provinces of Shanxi, Inner Mongolia and Shaanxi is 905 million tons/year, 822 million tons/year, 382 million tons/year, respectively, accounting for 63.22% of total coal production in China. In addition, there are 1,156 coal mines approved (under reviewing) and under constructions (including 83 producing coal mines under synchronous reconstruction, renovation), with production capacity of 1.062 billion tons/year, of which the newly built coal mine production capacity is 442 million tons/year, coal mine production capacity caused by resource integration is 414 million tons/year, coal mine production capacity caused by technical transformation is 0.85 Billion tons/year, the newly expanded coal mine production capacity is 121 million tons/year. There are 230 coal mines that have been built and put into joint trial operation, with a production capacity of 363 million tons/year, of which the newly built coal mine capacity is 224 million tons/year, the coal mine production capacity caused by resource integration is 98 million tons/year, the coal mine production capacity caused by the technical transformation is 20 million tons/year, and the newly expanded coal mine production capacity is 21 million tons/year.

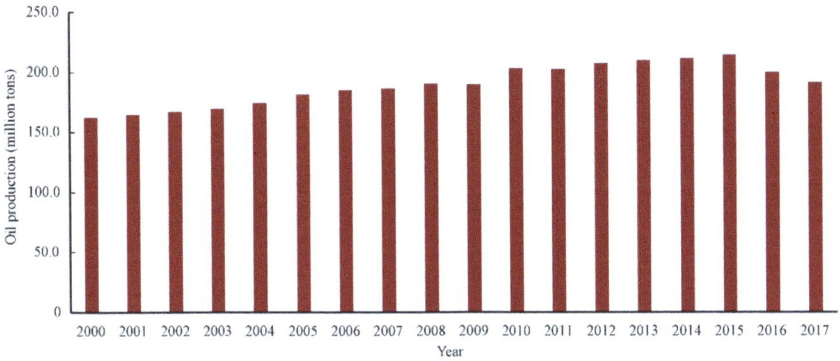

Fig. 2.21 Oil production in China from 2000 to 2017

2. Petroleum

In general, China's oil production tends to be stable, maintaining a level of around 200 million tons/year since 2010. After more than 60 years of development, China has basically formed six oil production bases with an annual output of more than 10 million tons in Bohai Bay Basin, Songliao Basin, Ordos Basin, Junggar Basin, Tarim Basin and Pearl River Estuary Basin (Hou et al. 2018). Oil production has exceeded 200 million tons since 2010 and has remained at around 200 million tons/year in recent years. However, the pressure to maintain 200 million tons/year in the late stage is relatively high. China's oil production in 2017 was 192 million tons (Fig. 2.21).

The production of large and medium-sized oilfields in the east China is gradually declining, and it is more and more difficult to keep the production of them stable. The production of Daqing, Shengli and Liaohe oilfields in the east has entered the decline stage. As the oilfield with the largest oil production, Daqing oilfield achieved a high and stable production of 50 million tons for 27 consecutive years from 1976 to 2002, and then began to decline. The oil production in 2017 was 34 million tons. The reservoir quality of Changqing, Karamay and other oilfields in the west China is poor, and the production growth is more difficult. The formations of Changqing oilfield are tight, and oil production exceeded 20 million tons since 2011. At present, the annual stable production is about 24 million tons.

3. Natural Gas

China's conventional natural gas entered a period of continuous growth, and unconventional natural gas entered a leap-forward development period. Compared to the year of 2016, national gas production increased by 10% in 2017 to 148 billion cubic meters (Fig. 2.22), mainly from the Sichuan Basin, Tarim Basin and the eastern area of the South China Sea. China's natural gas production will continue to rise as the reserves of conventional and unconventional natural gas continue to grow rapidly.

Conventional natural gas production is about to reach its peak and enter a stable development period. Since 2000, China's conventional natural gas production has continued to grow rapidly, from 28 billion cubic meters in 2000 to 95 billion cubic

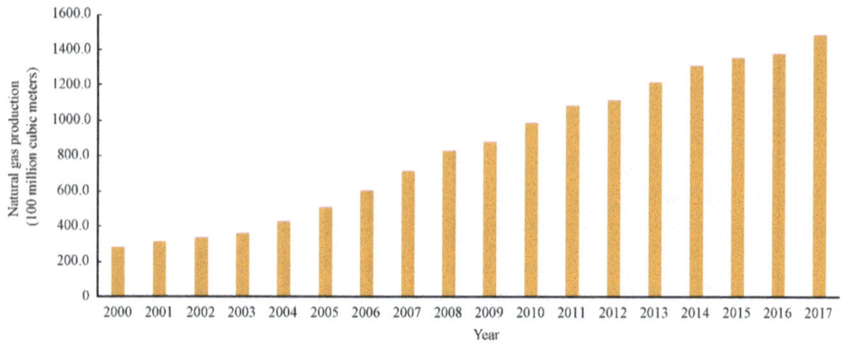

Fig. 2.22 Natural gas production in China in 2000–2017

meters in 2014, with an average annual increase of 4.8 billion cubic meters. From 2014 to 2017, conventional natural gas production remained at around 95 billion cubic meters. The production was 99.8 billion cubic meters in 2017, accounting for 67.43% of total natural gas production and being the main part of natural gas production.

The development of unconventional natural gas, such as tight gas, coalbed methane and shale gas, has made breakthroughs successively, and the production continued to grow rapidly, and it has become the main driving force of natural gas production growth. Compared with the year of 2016, China's unconventional natural gas production increased by 14% to 48.2 billion cubic meters in 2017, accounting for 32.57% of total natural gas production. In 2005, the production of tight gas made a breakthrough. The natural gas production was more than 30 billion cubic meters in 2012. In 2015, natural gas production reached 35 billion cubic meters and started to be stable. In 2017, the natural gas production was 34.3 billion cubic meters, accounting for 23.2% of total natural gas production. After the breakthrough of coalbed methane development in 2006, the production growth was relatively slow due to various factors. In 2017, the production of coalbed methane was 4.9 billion cubic meters in, accounting for 3.3% of total natural gas production. After a breakthrough of shale gas development in 2013, the production grew rapidly. In 2017, the production reached 9 billion cubic meters, accounting for 6.1% of total natural gas production.

2.3.3 Map of Fossil Energy Consumption

For a long time, the proportion of coal in China's energy consumption structure is too high, and the proportion of oil and natural gas consumption is low. For China's primary energy consumption structure in 2017, coal accounted for 60.4%, oil accounted for 19.4%, natural gas accounted for 6.6%. From the average level of the world's

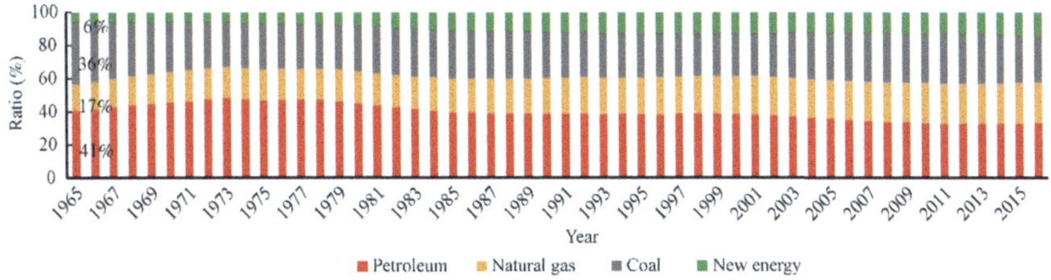

Fig. 2.23 Changes to primary energy consumption structures of the world from 1965 to 2016

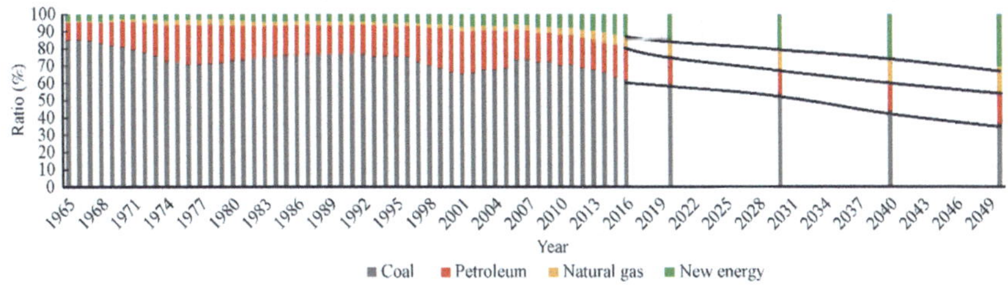

Fig. 2.24 Changes to primary energy consumption structure of China from 1965 to 2050

energy structure, China's coal-based energy consumption structure will reach the world's level in 1965 by 2050 (Figs. 2.23 and 2.24).

1. Coal

In recent years, affected by air pollution problems, the government has strictly controlled coal consumption, and China's coal consumption has generally indicated a decreasing trend. China accounts for half of global coal consumption. Global coal consumption totaled 3.732 billion tons of oil equivalent in 2017, while China's coal consumption amounted to 1.893 billion tons of oil equivalent, accounting for 50.72%. Coal will continue to dominate China's energy consumption over a long period of time and it will be difficult to change in the short term. In China's energy consumption structure in 2017, coal was the dominator and accounted for 60.4%. According to the "13th Five-Year Plan", China's coal consumption will decrease to 58% of total energy consumption by 2020.

Imported coal is an important supplement to China's coal supply, with imported coal price and the low-cost advantage of shipping, make China's coal import volume stay at high level. Since 2009, China has changed from a net exporter of coal to a net importer of coal. Coal imports continued to climb from 126 million tons in 2009 to 327 million tons in 2013. In 2014, China's coal imports fell sharply, influenced by factors such as the narrowing price difference of import coal, the increase in coal import tariffs and the total import volume restrictions. In 2016, affected by the comprehensive factors such as supply side reform, coal supply was tightly balanced, and China's coal imports increased significantly. In 2016, China's coal imports were 255.51 million tons, an increase of 25.2% compared with the same period in 2015.

Coal imports in the first half of 2017 were 133 million tons, an increase of 23.50% over the same period of last year.

2. Petroleum

China's oil consumption continues to grow, the degree of external dependence continues to increase (Zou et al. 2018). For the whole year of 2017, China's apparent consumption of oil was 610 million tons, a year-over-year increase of 6.1%, and the degree of external dependence of oil reached 67.79% (Liu and Jiang 2018). From the standpoint of the consumption of refined oil products, with the recovery of demand in industry, transportation and other fields, China's refined oil consumption in 2017 has returned to the growth range, and the growth rate has changed from negative growth in the previous year to positive growth, with the stability of kerosene demand and the continuous increase of diesel demand being the main factors for the recovery of the growth rate of apparent consumption of refined oil. According to data from the China Petroleum and Chemical Industry Federation, the apparent consumption of refined oil products in China for the whole year of 2017 was 330 million tons, a year-over-year increase of 3.5%. The growth rate of the whole year was 1.3% higher than the first three quarters of 2017, however, the growth rate last year decreased by 1.0% compared with the same term of last year.

In 2017, the import volume of crude oil exceeded 400 million tons, driven by factors such as the reduction of domestic crude oil production and the increase of crude oil processing capacity. The ratio of crude oil imports to domestic production continued to rise from 1.3:1 in 2012 to 2.2:1 in 2017. China's crude oil imports for the whole year of 2017 were 8.4 million barrels per day, surpassing 7.9 million barrels/day in the United States for the first time, making it the world's largest importer of crude oil. According to the data of S&P Global Platts, China's crude oil imports from OPEC members in 2017 were 4.7 million barrels/day, a year-over-year increase of 7.1% to 234.22 million tons, but the market share fell from 57.4% in 2016 to 55.8%.

3. Natural Gas

China's natural gas consumption is growing rapidly, and the degree of external dependence shows an accelerated trend (Zou et al. 2018). After China's natural gas consumption exceeded production in 2007, the degree of external dependence continued to increase. China's natural gas consumption was 235.2 billion cubic meters in 2017, a year-over-year increase of15.3%, well above the growth rate of natural gas production, and natural gas consumption has entered a period of rapid growth. Domestic natural gas production is far from meeting the growth of natural gas consumption. In 2017, gas imports reached 92.6 billion cubic meters, and the degree of external dependence reached 39.4%.

LNG imports exceed natural gas imports by pipeline, becoming the main force to meet the domestic natural gas demand. Natural gas imports by pipeline is restricted by resource countries and imports are growing steadily. Gas imports by pipeline comes mainly from Turkmenistan, Myanmar, Uzbekistan and Kazakhstan, with a gas supply of 42.7 billion cubic meters in 2017, an increase of 10.9% over the year of 2016.

Driven by consumer demand, LNG imports grew rapidly, with imports reaching 49.9 billion cubic meters in 2017, a year-over-year increase of 39%. LNG comes mainly from Australia, Qatar, Indonesia and other countries, of which Australia accounts for 46%.

2.4 Map of China's New Energy

In recent years, China's new energy industry has developed rapidly, and its proportion in the primary energy structure has been expanded to become an important part of energy (Zou et al. 2016). During the "13th Five-Year Plan", the goal of renewable energy development in 2020 is to exceed the bottom line, that is, installed capacity of solar photovoltaics will increase from 110 million kilowatts to 200 million kilowatts, installed capacity of wind power will increase from 210 million kilowatts to 350 million kilowatts, installed capacity of biomass energy generation will increase from 15 million kilowatts to 30 million kilowatts, a total increase of 500 million kilowatts (National Energy Administration 2017).

2.4.1 Map of Hydropower

China is rich in hydropower resources and the pace of hydropower development is accelerating (Zou et al. 2016). Since the 21st century, with the operation of the Three Gorges, South-to-North Water Transfer Project as a milestone, China has entered a new stage of independent innovation and leading development. The construction projects such as Xiaowan, Longtan, Shuibuya, Jinping-I and other projects have been completed, and construction technology constantly set a world record. At this stage, China pays more attention to the safety of giant engineering and ultra-high dam, to environmental protection, and is in an international leading position in many fields (National Development and Reform Commission 2017). At the same time, China also fully participates in the international water conservancy and hydropower construction market, with more than half of the international market share.

By the end of 2017, China's hydropower installed capacity was 341 million kilowatts, accounting for about 19.2% of China's full-scale power generation equipment installed capacity of 1.777 billion, and annual power generation of hydropower was about 1.16 trillion kilowatt-hours, accounting for 18% of annual electricity generation of 6.48 trillion kilowatt-hours, an increase of 4.2 times compared with the year of 2000.

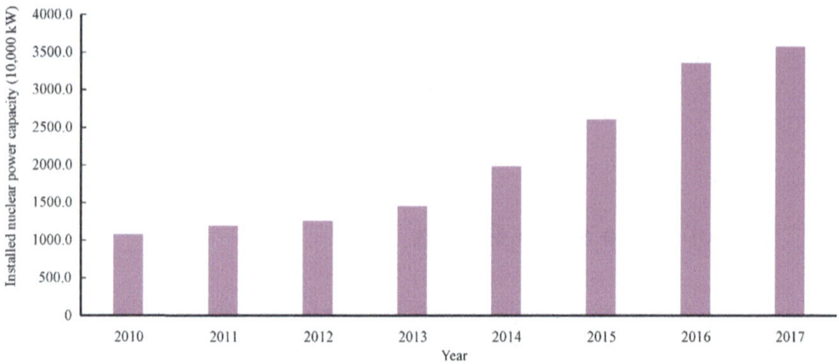

Fig. 2.25 Installed nuclear power capacity of China from 2010 to 2017

2.4.2 Map of Nuclear Power

The utilization rate of nuclear power equipment in China is declining, and the problem of nuclear power consumption and acceptance still exists. However, China's nuclear power generation continues to grow, and the proportion of nuclear power generation in the country is also increasing. The scale of China's nuclear power is relatively small (Zou et al. 2016). In recent years, the pace of nuclear power construction has accelerated, and the scale of nuclear power under construction ranks first in the world (Fig. 2.25).

In 2017, China's installed nuclear power capacity slowed down. The projects such as Hongyanhe nuclear power, Ningde nuclear power, Fuqing nuclear power, Fangchenggang nuclear power and Changjiang nuclear power have encountered the problem of consumption and acceptance, especially Hongyanhe nuclear power. Three of the four units of the average annual equipment utilization rate was less than 60%. From 2010 to 2017, the average utilization hours of nuclear power equipment in China showed a general declining trend. The average utilization hours of nuclear power equipment in China was 7,108 h in 2017.

2.4.3 Map of Other Renewable Energy

1. Wind Power

Under the strong motivation of national policy, China's wind power industry is booming. In 2017, the newly installed capacity in China was 19.66 million kilowatts, a year-over-year decline of 15.9%; the cumulative installed capacity reached 188 million kilowatts, a year-over-year increase of 11.7%, and the growth rate slowed down (Fig. 2.26). The main reason for the slowdown in growth is the low utilization rate of onshore wind power, resulting in a drop of 19% in China's onshore wind power

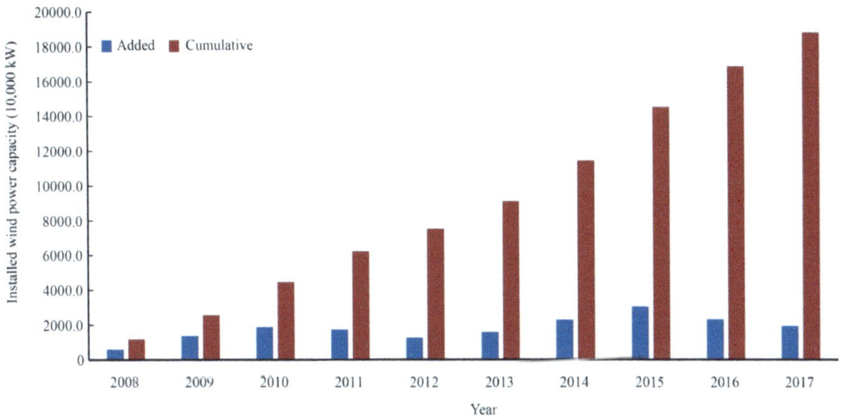

Fig. 2.26 Installed wind power capacity of China from 2008 to 2017

installed capacity to 18.5 GW in 2017. Despite the slowdown in growth, China's wind power new installed capacity and cumulative installed capacity are firmly ranked first in the world. Among them, the newly added installed capacity accounted for 37.40% of the global share in 2017, which was 12,643 MW higher than the second-ranked United States; the cumulative installed capacity accounted for 34.88% of the global share, which was 2.11 times that of the second-ranked United States.

2. Solar Energy

According to data of the market research firm (HIS Markit), the installed capacity of solar photovoltaic will fall, influenced by new government policy and regulations. With an annual installed capacity of more than 96 GW in 2017, the installed capacity of solar photovoltaic is expected to reach 105 GW in 2018, an increase of 9.4% over a year earlier. Public utilities and distributed power generation are the two largest market segments in China. The installed capacity of the project is expected to fall as the government sets the upper limit of development for these two sectors. As a result, global photovoltaic installed capacity is expected to fall from 113 to 105 GW in 2018.

3. Geothermal Energy

China is a country with relatively abundant geothermal resources (Lin et al. 2013), accounting for about 7.9% of the world's total geothermal resources, recoverable reserves equivalent to 462.65 billion tons of standard coal. It is the most abundant southwest China, with proved geothermal energy up to 2 204.45 MW, accounting for 51.05% of the total amount of discovered geothermal energy that can be used. Followed by North and South Central, the discovered geothermal energy that can be utilized reached 745.33 MW and 685.75 MW, respectively, accounting for 17.27% and 15.89% of the proved total amount of geothermal energy available in China; East

China is in the third position, accounting for 9.92% of the total geothermal energy in China; while in the northeast and northwest regions, the proved geothermal energy available accounted for only 2.53 and 3.34% of the total geothermal energy in China. China's geothermal utilization is mainly based on direct utilization, and there are five geothermal power stations with a total power of 27.78 MW.

References

BP. 2018. BP statistical review of the world energy.

Hou, Qijun, Haiqing He, Jianzhong Li, et al. 2018. Recent progress and prospect of oil and gas exploration by PetroChina Company Limited. *China Petroleum Exploration* 32 (1): 1–13.

Lin, Wenjing, Zhiming Liu, Yuli Wang, et al. 2013. The assessment of geothermal resources potential of China. *Geology in China* 40 (1): 312–321.

Liu, Chaoquan, and Xuefeng Jiang. 2018. *Domestic and foreign oil and gas industry development report of 2017*. Beijing: Petroleum Industry Press.

National Development and Reform Commission. 2017. The 13th five-year plan for power development.

National Energy Administration. 2017. Guidance on the implementation of the "13th Five-Year Plan" for renewable energy development.

State Grid Energy Research Institute. 2017. *China new energy power generation analysis report*, 1–35. Beijing: China Electric Power Press.

Zou, Caineng, Qun Zhao, Guosheng Zhang, et al. 2016. Energy revolution: From a fossil energy era to a new energy era. *Natural Gas Industry* 36 (1): 1–10.

Zou, Caineng, et al. 2018. Energy revolution and oil company transformation strategy of new era. *Journal of Beijing Petroleum Managers Training Institute* 25 (4): 3–15.

Chapter 3
Trend of Energy Development

According to the basic law of energy development, the world energy is entering a new stage of the third conversion of from coal, oil and gas to new energy, and a new structure of "four major components" incluidng oil, natural gas, coal and new energy is forming (Fig. 3.1) (Zou et al. 2016; Zou 2018). Faced with the objective condition of China's oil and gas resource endowment and the severe challenges, it is necessary to set the safe peak value of oil and gas consumption and maintain the external dependence of oil and gas within a scientific and safe upper limit. Take natural gas and new energy as the biggest strategic choice of China's energy revolution, accelerate the development of natural gas, coal gas, hydrate and hydrogen, accelerate

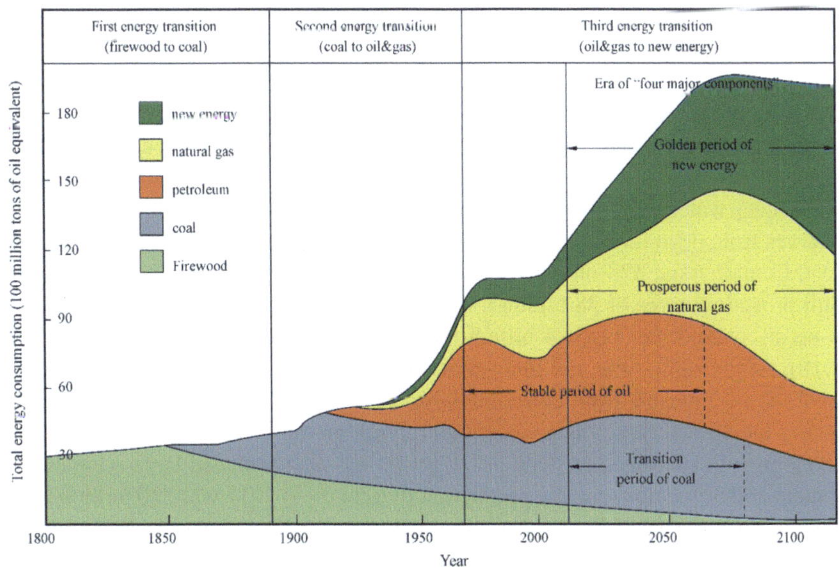

Fig. 3.1 Trends and forecasts of global energy consumption

C. Zou, *New Energy*, https://doi.org/10.1007/978-981-15-2728-9_3

the clean utilization of coal and low cost of new energy to achieve the "two large scale production" ahead of schedule, accelerate the transition of energy structure from energy dominated by coal to "three major components" consisting of coal, oil and gas, new energy. China's large oil companies are responsible for making good top-level design of the industrial structure transition, conducting energy foresight and guidance research, pushing "three greater leap" such as from conventional oil and gas to unconventional, from domestic oil and gas to foreign countries, from oil and gas industry to the new energy. The goal to achieve is to change the international oil company to the international energy company and to ensure the security of national energy supply.

3.1 Direction of World Energy Development

3.1.1 Oil Development Enters a Stable Stage

As a result of the continuous innovation of theory, technology and methods, the peak oil production theory proposed by Harbert in 1956 has been convinced wrong. The peak of world oil production is rising continuously, and the occurrence of peak production is delayed again and again. It is likely the real peak production will happen in the middle of the 21st century, and the oil industry life cycle may also be more than 300 years. Since 1986, the world's oil production has generally shown steady growth trend, oil production peak should appear around 2040 based on comprehensive judgement considering multiple factors, and the peak production will be about 5 billion tons (Fig. 3.2).

With the development of petroleum industry, the world's conventional petroleum exploration extends to deep water, deep formations and North Pole. From 2000 to 2012, the world's newly added proved oil reserves were 69.8 billion tons. Among them, the newly added reserves of deep water accounted for 28%, mainly distributed in four deep water areas of Brazil, Australia, West Africa, the Gulf of Mexico; the new reserves from deep formations accounted for 16%, mainly distributed in the Middle East, Central Asia; 423 oil and gas fields has been discovered in North Pole area, with proved reserves of 38 billion tons of oil equivalent, and the potential reserves to be discovered will be 56.4 billion tons of oil equivalent (Zou et al. 2016; Zou 2018). At the same time, the innovation of ideas and technological breakthroughs drive the oil industry from conventional to unconventional leapfrog development. Taking the United States as an example, relying on the theoretical technology and development experience of shale gas, tight oil has also achieved large-scale development and utilization, and the production of tight oil in 2017 was 220 million tons, accounting for 54% of the total oil production in the United States. According to the forecast of the International Energy Agency (IEA) in 2013, recoverable tight oil reserves of 42 countries around the world are 44.9 billion tons on the basis of present

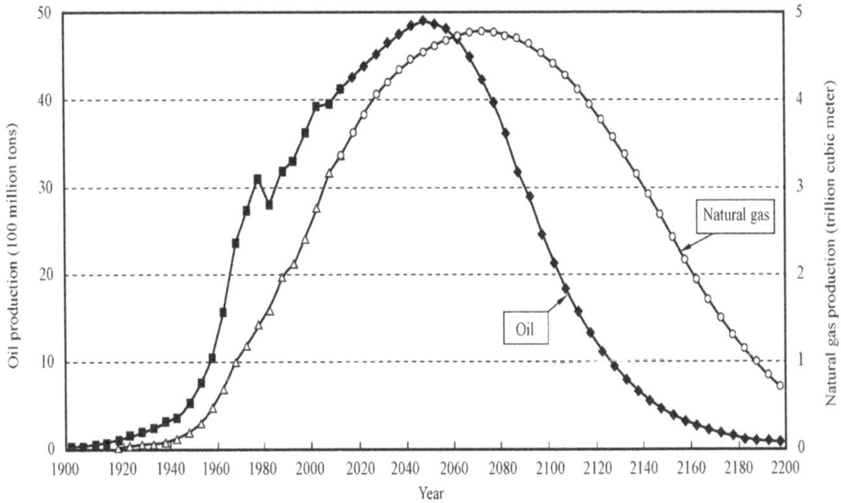

Fig. 3.2 Forecast of global oil and gas production growth trend

technology, unconventional oil and gas is expected to become a new area of future oil development.

Overall, the global oil reserves are generally abundant, and the ratio of storage over production has been maintained above 50, especially, the ratios of storage over production in Central/South America and the Middle East are as high as 120 and 70, respectively and the development potential is still very large. Global oil production continued to grow steadily, with an average growth rate of 8% in the last 10 years, which has entered a stable period of development.

3.1.2 Natural Gas Development Enters a Prosperous Stage

Natural gas has entered a stage of rapid development, is a bridge or associated transition from fossil energy to new energy, will play a major role in the future sustainable development of energy in the world (Zou et al. 2016). Natural gas is a very realistic low-cost, clean and environmentally friendly "three A energy" (available, affordable and acceptable). Over the past 50 years, its share of the global energy consumption structure has jumped from 16 to 24%, the fastest-growing fossil energy resource in the global energy structure.

According to a number of research institutions such as the United States Geological Survey (USGS), EIA (Energy Information Administration), the International Energy Agency (IEA) and CEDIGAZ (Independent Natural Gas Information Research Institute), the global remaining recoverable resources of conventional natural gas and three types of unconventional natural gas such as tight gas, shale gas, coalbed methane are more than 800 trillion cubic meters, which is estimated to produce for about 250 years according to the current production rate.

The breakthrough and development of unconventional natural gas is expected to significantly increase the scale of world natural gas production and extend the life cycle of natural gas industry (Zou et al. 2016). Since the 1970s, the rapid development of tight sandstone gas, coalbed methane and shale gas in the United States has effectively compensated for the decline in the production of conventional natural gas, especially the recent rapid development of shale gas (Chen et al. 2016), which has helped the US natural gas production hit a record high. Once again, the United States has become the world's largest gas producer. In 2017, natural gas net exports of the United States were 50 Billion cubic meters, which is changing the global energy supply pattern.

In general, natural gas has a huge amount of resources and reserves, will play a more important role in the energy structure, is the most realistic, accessible clean energy, which is enough to guarantee the global market demand for a long time. The global natural gas reserves are sufficient. At the end of 2017, the ratio of reserve over production was 52.3. Natural gas reserves and production will grow rapidly and enter its peak. The International pipeline network and LNG related facilities tend to be complete, which solves the problem of long-distance transportation of natural gas, and natural gas production has the basis of rapid growth. The peak of global gas production is expected to occur around 2060, with peak a production of about 5 trillion cubic meters.

3.1.3 Coal Development Enters a Transition Period

Coal, as the cheapest fossil energy source, will continue to play an important role in the world's energy structure (Zou et al. 2016). With the increasing demand for human ecological environment protection, coal utilization will be changed to an efficient and clean direction. More than half of the world's coal resources are used to generate electricity, efficient and clean coal power generation is the main direction of coal resources utilization.

The global discovered coal reserves are abundant, the ratio of reserves over production is up to 134. The discovered and proved coal reserves can also be produced and utilized for more than 130 years. The proportion of coal in primary energy consumption has generally declined. However, due to the difference of resource endowment and economic development in various countries, coal is still one of the important fossil energy resources in a short period of time. In 2017, the proportion of coal in the global primary energy consumption fell to 28%. The proportion of China's coal consumption fell to 60.4%, and energy consumption per unit of GDP (gross domestic product) fell 5% year-over-year. Of the three major coal-producing areas in the world, coal production and consumption in North America, Europe and Eurasia showed a downward trend.

The proportion of coal in the world's primary energy consumption structure will be further reduced (Zou 2018), but it is still growing in some areas. The share of coal in the global energy consumption structure is expected to fall from 28% in 2017 to 26% in 2022. By 2022, growth in coal consumption will mostly occur in India, Southeast Asia and several other countries in Asia. Demand for coal fell in Europe, Canada, the United States and China. Over the past three years, 440 gigawatts of coal-fired power stations around the world have been cancelled or delayed, of which more than 250 gigawatts are in China. The IEA predicts a structural and slow decline in coal demand will appear, accompanied by some fluctuations associated with short-term market demand. Global coal demand in 2022 will reach 5.53 billion tons of standard coal, which is basically equivalent to the current level. Coal power generation will grow by 1.2% each year from 2016 to 2022, but its share in the electricity structure will fall below 36%.

The coal demand center is changing. The coal consumption in most parts of the EU will continue to decline significantly. The coal consumption in Polish and German is the major sector of coal consumption in the EU, accounting for more than half in the EU. In Poland, coal demand is expected to remain stable in 2022; in Germany, coal demand is generally declining, and coal consumption remains highly sensitive to the relative prices of coal, natural gas and carbon dioxide. Pakistan is rich in lignite resources, with the continuous development of the economy, the coal produced by themselves can not meet domestic demand. The IEA predicts that Pakistan's coal demand will grow by more than three times from 2016 to 2022, with imported coal accounting for half of its consumption.

3.1.4 New Energy Development Gradually Enters the Golden Period

New energy refers to all kinds of energy, such as solar energy, geothermal energy, wind energy, hydrogen energy, ocean energy, biomass energy and nuclear fusion energy, which are just beginning to be exploited or are being actively studied in addition to traditional energy resources. The rapid development of new energy technologies, Internet plus, artificial intelligence and new materials and other technologies continue to progress, to promote the rapid development of new energy industries, new energy development has been in a breakthrough period and is gradually entering the golden period (Zou et al. 2016; Zou 2018). To develop new energy is the key to achieving low-carbon development, and the pace of development and utilization of new energy has accelerated, which has become a new driving force for global energy growth. With the progress of technology, the cost of new energy development and utilization has been declining, and it has already been more competitive than fossil energy.

It has become a consensus to strengthen science and technology of new energy, the arrival of the new energy revolution is likely to exceed expectations (Zou et al.

2016), especially the reduction of new energy generation costs and battery energy storage technology breakthrough will strongly speed up the arrival of the new energy era. On the 125th anniversary of its founding, the Science magazine has unveiled 125 of the most challenging scientific issues, including the substitution of oil, nuclear fusion and other energy issues. McKinsey proposed 12 major revolutionary technologies that determine the future economy in 2025, renewable energy and energy storage technologies are two important revolutionary technologies. On December 12, 2015, the 21st session of the United Nations Climate Change Conference adopted the Paris Agreement, proposed that the global average temperature is controlled under 2 °C higher than that before industrialization, and net zero greenhouse gas emissions tare achieved in the second half of the 21st century. The achievement of this goal puts higher requirements ahead for new energy development. The network big data system based on artificial intelligence will play an important role in the optimal allocation of energy structure. By developing distributed power grid structure, promoting the construction of intelligent energy network system of Internet plus energy network, and making rational use of the interchangeable coupling of hydrogen energy-electric energy, it will fundamentally solve the existing power grid's consumption and acceptance of renewable energy generation, further enhance the efficiency of the development and utilization of renewable energy, and finally free human beings from dependence on fossil energy.

New energy resources, represented by renewable energy, will continue to accelerate development. Forecasts of the share of renewable energy in the primary energy consumption structure by agencies vary from country to country in the world, but are developing more rapidly overall. BP and ExxonMobil believe renewable energy will account for less than 15% of primary energy consumption from 2030 to 2040. In the 450 ppm scenario predicted by the IEA, that is, when the possibility of controlling temperature increase under 2 Ca is 50%, the proportion of renewable energy in the primary energy consumption structure will reach a 27% in 2035. For 2DS scenario, in which the probability of temperature rise being controlled within 2 °C is 80%, the proportion of renewable energy in the primary energy consumption structure is expected to reach 40% in 2050. Green Peace Organization believes that renewable energy accounts for 80% of primary energy consumption in 2050, and WWF even predicts that the proportion of new energy will reach 95% in 2050.

3.2 "Four Production Revolutions" of China's Energy

China's energy demand is likely to peak around 2030, with about 4.4 billion tons of oil equivalent. China has a large population base and a relatively lack of oil and gas resources. Therefore, it is necessary to seek a low energy consumption and sustainable development path based on the national situation. Referring to the developed countries such as the United Kingdom, Germany, France and Japan, the per capita energy consumption in is 2.9–3.5 tons of oil equivalent. Taking China's economic and demographic development into consideration, it is estimated that per capita energy consumption should be controlled below the level of 3.0 tons of oil equivalent. According to statistics from the National Health and Family Planning Commission (2016), the population of China will reach a peak of 1.45 billion in 2030 and 1.38 billion in 2050. According to this forecast, China's energy consumption will reach a peak of 4.4 billion tons of oil equivalent in 2030, and will decline to 4 billion tons of oil equivalent in 2050.

In the new period of dealing with global climate change and vigorously developing low-carbon energy, China needs to speed up the "four production revolutions" of conventional-unconventional gas, coalbed methane and coal to gas, hydrate and hydrogen, accelerate the early arrival of "two scalable production" such as cleaner utilization of coal and new energy, and speed up transformation of the energy structure from "one dominator" of coal to "three major components" consisting of coal, oil and gas, new energy (Zou 2018). Taking the development opportunity of unconventional oil and gas revolution, try to increase production and extend the life cycle of petroleum industry, lead the revolution of low cost management by pushing low oil price forward, ensure the China's oil and gas supply available from multiple ways and much safer, reconstruct the energy map and new political pattern of powerful countries. At present, the focus of China's energy development is to realize the conventional-unconventional oil and gas production revolution, the clean revolution of coal development, the speed revolution of new energy development, in order to finally realize the structural revolution of China's energy development.

3.2.1 Revolution of Oil and Gas Production

China have generally entered the period of both conventional and unconventional development, and the revolution of oil and gas production revolution is presented to ensure national oil and gas security. Conventional oil production will continue to decline, unconventional oil development represented by tight oil needs to strengthen the realization of low-cost strategy; it is difficult to increase production a lot from conventional natural gas, so the conventional gas production will be stable generally, unconventional natural gas development is the direction of future development, LNG will be the main force of natural gas supply.

(1) The development of unconventional oil resources will play an important role in China's oil production, and tight oil is the main direction of future development. Compared with traditional resource countries, China's oil production costs are higher. The global oil and gas supply pattern is in the transition period, and the long lasting low oil price has caused a strong impact on domestic oil production. China's external dependence of oil reached 67.79% in 2017, and experts predict it will reach break 70% by the end of 2018. A significant reduction in production would significantly increase oil dependence on the outside world and increase national oil security risks. Therefore, the fluctuation of oil price, the reduction of production in old oilfields and the low recovery factor of non-conventional oil will affect the long-term stability of

China's oil production. However, from the view of the overall situation of national oil security, it is necessary to make effort to stabilize domestic oil production and conduct the technological revolution.

Tight oil exploration and development has made important breakthroughs in the major basins, establishing three tight oil areas with reserves over 1 billion tons and six tight oil areas with reserves over 100 million tons, launching eight tight oil development pilot areas in Ordos, Songliao, Junggar and other basins, building a production capacity of more than 1 million tons/year. Continental shale oil resources have great potential, the development of unconventional oil resources in Songliao, Ordos, Junggar and other basins demonstration project construction should be accelerated, and the continental shale oil revolution in Ordos and other basins should be pushed forward. According to more optimistic forecast, domestic oil production will maintain stable for a certain period, it is expected that the total production of conventional oil will be 170 million tons, non-conventional oil production will reach 30 million tons, external dependence ratio will be 66% in 2020; it is expected the total production of conventional oil will be 150 million tons, non-conventional oil production will reach 50 million tons, external dependence ratio will be of 67% (Fig. 3.3).

(2) China's natural gas in entering a new era. Under the guidance and strong support of national policies, the future natural gas industry will be promising, and it will play an irreplaceable role in the transition of China's energy structure. Natural gas will play an irreplaceable role in the transition of "making natural gas available around China and making China beautiful". Natural gas strategic plan should be based on the characteristics of China's natural gas resources, taking the advantage of upstream and downstream dual-wheel drive, persisting in the theoretical

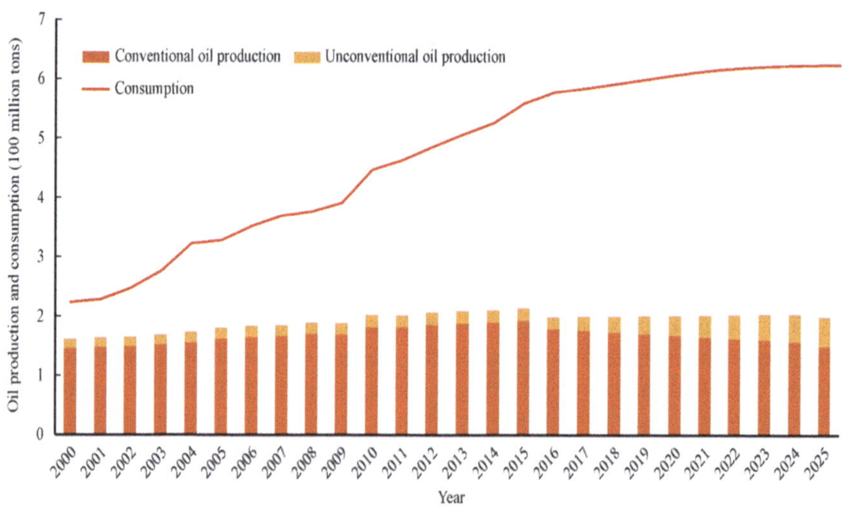

Fig. 3.3 China's oil production and consumption forecast

and technological innovation of exploration and development, continuously accelerating upstream development, making full use of domestic and foreign resources, strengthening the natural gas supply system, while strengthening policy guidance and infrastructure investment, and constantly replacing traditional energy with natural gas, expanding the space of natural gas consumption and utilization, and realizing the large-scale development of natural gas industry chain.

China's natural gas demand will show continuous and rapid growth trend, China's natural gas demand will be 650 billion to 700 billion cubic meters in 2050. Based on the comprehensive analysis of population, economy, resources, environment and politics, it is concluded that China's natural gas demand will reach 350 billion cubic meters in 2020, accounting for 10% of primary energy consumption structure, and natural gas demand will be between 550 billion and 600 billion cubic meters in 2030, accounting for 12% of primary energy consumption structure; Natural gas demand will be between 650 billion and 700 billion cubic metres in 2050, accounting for 15% of the energy consumption structure (Fig. 3.4). The peak of China's natural gas production under three scenarios in 2030 will reach 180 billion cubic meters, 200 billion cubic meters and 220 billion cubic meters, respectively.

① In the low case, natural gas production was 180 billion cubic meters in 2030. When there are no significant discoveries in conventional natural gas exploration, production will remain at around 100 billion cubic meters in 2020, then the production will decrease to 85 billion cubic meters in 2030 and 40 billion cubic meters in 2050. Unconventional gas production has risen steadily in existing areas. Tight gas production is generally stable and is expected to be 40 billion cubic meters in 2020 and stablishing at this level, which is expected to remain at 40 billion cubic meters in 2030 years and to decrease to 30 billion cubic meters in 2050; Marine shale gas is the major role of production growth and is expected to reach 180–200 billion cubic meters in 2020, reach 40 billion cubic meters in 2030, reach 50 billion cubic meters in 2050. Coalbed methane production continues to grow steadily, which is expected to reach 6 billion to 10 billion cubic meters in 2020, reach 20 billion cubic meters in 2030, reach 30 billion cubic meter meters in 2050 (Fig. 3.4).

② In the medium case, the natural gas production is 200 billion cubic meters in 2030. If more conventional natural gas exploration is found, production will remain at around 100 billion cubic meters in 2020, remain at 100 billion cubic meters in 2030 and remain at 60 billion cubic meters in 2050. Unconventional gas production in the existing field is growing relatively rapidly. Tight gas production is generally stable, which is expected to be 40 billion cubic meters in 2020, 45 billion cubic meters in 2030, 35 billion cubic meters in 2050. When shale gas from marine, continental and land-ocean transition deposition is fully and effective developed, production is expected to reach 18–20 billion cubic meters in 2020, reach 45 billion cubic meters in 2030, and reach 60 billion cubic meters in 2050. Coalbed methane production continues to grow steadily, which is expected to reach 6 billion to 10 billion cubic meters in 2020, reach 20 billion cubic meters in 2030, reach 30 billion cubic meters in 2050 (Fig. 3.4).

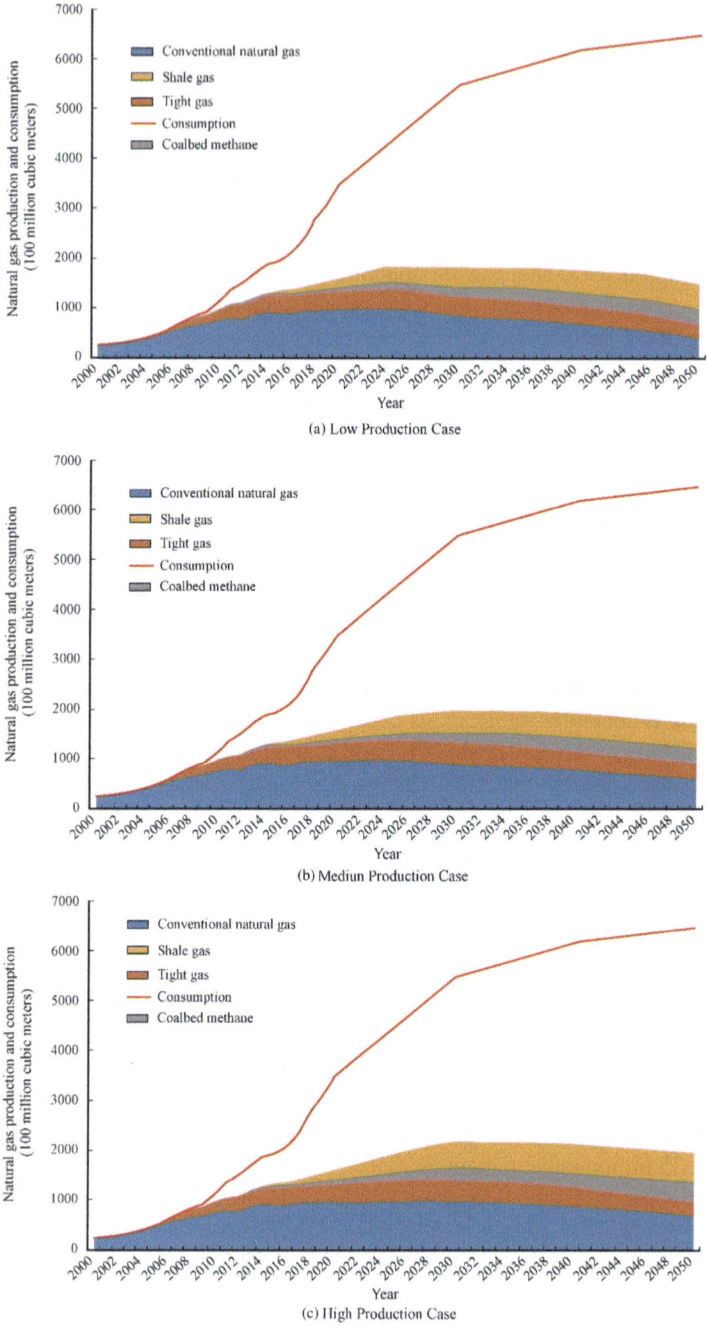

Fig. 3.4 Forecast of natural gas production and consumption in China from 2000 to 2050

③ In the high case, natural gas production will be more than 220 billion cubic meters in 2030. In the event of a major discovery of conventional natural gas exploration, production will remain at around 100 billion cubic metres in 2020, remain at 100 billion cubic metres in 2030 and 70 billion cubic metres in 2050. Unconventional gas production is growing relatively rapidly in existing areas: Tight gas production is generally stable, with an estimated 40 billion cubic metres in 2020, 45 billion cubic metres in 2030 and 35 billion cubic metres in 2050; and the full and effective development of shale gas from marine, continental and land-ocean transition phase will make the production reach 180 billion to 200 billion cubic meters in 2020, reach 50 billion cubic meters in 2030, reach 60 billion cubic meters in 2050. Coalbed methane production continued to grow steadily, which is expected to reach 6 billion to 100 billion cubic meters in 2020, reach 25 billion cubic meters in 2030, reach 40 billion cubic meters in 2050 (Fig. 3.4).

(3) After 2020, imports will become the major role of natural gas supply. Taking natural gas import methods, path, resources and politics into consideration, the limit of import supply capacity of onshore pipeline gas is 160 billion cubic meters, and the remaining gas demand gap is mainly imported by LNG. Natural gas pipeline transmission distance is restricted, onshore pipeline gas is mainly imported from Russia and two natural gas producing areas in Central Asia, in addition, the Middle East has become a potential target for natural gas imports. Taking the routes, resources, output, politics and other factors of land import into account, according to the current two major pipeline gas import routes in Russia and Central Asia, the supply capacity can reach 75 billion cubic meters in 2020, 120 billion cubic meters in 2030 and 140 billion cubic meters in 2050; If the Middle East pipeline gas import route can successfully be constructed, supply capacity is expected to reach 135 billion cubic meters in 2030, and reach 160 billion cubic meters in 2050.

LNG has become the main way to meet the natural gas demand gap. In the future, the global supply of natural gas is relatively supportive, diversified LNG supply market will be helpful to meeting China's natural gas demand gap. At present, the capacity of LNG receiving stations built and under construction in China is 69.4 million tons of oil equivalent. In 2020, the receiving capacity by 80% load is estimated to be 75 billion cubic meters. If the national natural gas consumption demand is met, LNG needing to import will be 200 billion to 250 billion cubic meters in 2030, which will account for 36–45% of China's natural gas consumption, accounting for 52–66% of natural gas imports; and LNG imports will grow to between 300 billion and 350 billion cubic meters in 2050, accounting for 46–53% of China's natural gas consumption and 66–77% of natural gas imports.

(4) Strive to speed up the development of hydrate resources and accelerate the expansion of natural gas supply space. In 2011, the U.S. Department of Energy predicted a global hydrate resource is 20,000 trillion cubic meters, equivalent to 20 trillion tons of oil equivalent (widely cited by the international scientific community), which is 20 times the amount of conventional oil and gas resources. The development of hydrates has entered the field test phase onshore and offshore, and only Japan and

China have conducted offshore hydrate production test. China has completed a high-resolution earthquake of 167,000 km, drilled 88 wells, delineated two mineral fields with reserves over 100 billion cubic meters, whose controlled resources are 123.1 billion cubic meters and 150 billion cubic meters, respectively. The initial evaluated amount of hydrate resources in China's offshore are about 80 billion tons of oil equivalent. On May 25, 2017, relying on the "Offshore Oil 708" deepwater project survey vessel in the 3rd Spot located in Shenhu area of the northern South China Sea, making use of the completely independent technology, operation and equipment, the world's first successful pilot production operation of shallow offshore non-diagenetic hydrate solid state fluidization was conducted. On November 3 of the same year, the State Council officially approved the natural gas hydrates as new mineral species, which will greatly promote the exploration and development of natural gas hydrates in China into a new stage of development. Hydrates have huge amounts of resources and reserves, and the hydrate production revolution may be more shocking than the shale gas revolution. However, it is still necessary to strengthen the research of "sweet spot" evaluation and industrialization technology, plan to implement a larger scope, greater precision of the sea hydrate exploration, providing the target area for the future industrialization projects. At the same time, strengthen international cooperation of producing hydrate located in the global offshore areas and polar areas, including the "sweet spot" evaluation and industrialization pilot, pushing the hydrate revolution forward and achieving overseas adequate gas supply capacity.

3.2.2 Revolution of Clean Utilization of Coal

Based on the national status of energy resources with rich coal but insufficient oil and gas in China, it is necessary to achieve clean utilization through the coal anti-pollution revolution and speed up breakthrough in advance of industrial technology coal clean utilization and industrial technology of new energy working as a main role. The breakthrough of industrial technology for the coal clean utilization which will prolong the industrial life cycle of coal resources under the premise of cleanliness is the key issue of China's future energy development (Zou et al. 2016; Zou 2018).

China's primary energy consumption structure with coal as the main role will not produce fundamental changes in the short term. Based on the national conditions, to reduce the direct combustion of bulk coal, strengthen the efficient and clean utilization of coal is the key to solve the energy environment problems. With the advance of industrialization and the improvement of people's living standard, the growth rate of electricity demand accelerates, which is restricted by the structure and distribution of energy resources, and coal has become the main energy resource of China's electric power industry. Electric power is the main part of coal resource consumption, and to realize efficient and clean coal power generation is the key to coal clean utilization. In order to achieve low-carbon clean utilization of coal, it is necessary to upgrade the existing units according to local and factory's conditions, and to solve the problems of high energy consumption and excessive pollutant emissions due to factors such as long service time, aging of the unit, backward design and manufacturing technology problem.

Coal is used directly as terminal energy, and the energy utilization efficiency is low and the problem of environmental pollution is serious. The emission of sulfur dioxide, nitrogen oxides and fine particulate matter produced by coal combustion accounts for 80%, 60% and 70% of the total amount of the whole country respectively, and the emission of sulfur dioxide and soot from the direct combustion of coal per 1 kg is 4 times and 8 times that of its power generation respectively, among which the direct combustion pollution of bulk coal is particularly serious. Taking 2013 as an example, the direct combustion consumption of coal in China is about 900 million tons, accounting for 24% of the total coal consumption, the sulfur dioxide produced is roughly equivalent to that of 2 billion tons of electric coal combustion, and the emissions of fine particulate matter are close to 3 times that of electric coal. Countermeasures to reduce direct and inefficient combustion of coal include: (1) speeding up the elimination of backward production capacity of small domestic steel, cement and other industries, and realizing industrial upgrading; (2) accelerating the pace of urbanization in China, achieving central urban and rural heating, gas supply, and actively developing small renewable energy networks to achieve green development in China's vast urban and rural areas. The efficient and clean utilization of coal is the key to solve China's environmental problems.

China's coal industry should adapt to the new requirements of environmental protection, energy conservation, safety, occupational health in the new era. It must further increase the strength of production de-overcapacity, unswervingly to eliminate unsafe production capacity, not environmentally friendly capacity, poor quality production capacity, inefficient production capacity, while vigorously to develop safe production capacity, environmental friendly capacity, high-quality production capacity, efficient production capacity. We should try to develop safe production and occupational health of enterprises, and promote enterprises to enter the healthy track of green development, safe development and sustainable development. The development direction of clean utilization of coal coal: first, it can generate electricity, the other is coal methanol, and the third is large-scale hydrogen production. In 2017, 11 of the 15 refining integration projects under construction approved by the National Development and Reform Commission and the National Energy Agency used coal for hydrogen production. With the promotion of domestic refined oil products, most of the new domestic refineries have chosen the full hydrogenation process to meet the requirements of key technical and economic indicators such as light oil production, product quality, comprehensive commodity rate and other, greatly increasing the demand for hydrogen, and promoting the development of hydrogen production market.

At present, the thermal efficiency of coal power generator sets can be increased to about 50% by means of large capacity and high parameter coal power generation, large circulating fluidized bed power generation, integrated gasification combined cycle power generation and other technologies. Judging from the implementation effect of Shenhua Sanhe thermal power plant and national electric Langfang thermal power plant, the optimal value of its chimney dust emission concentration has reached

0.23 mg per cubic metre, sulfur dioxide emission is stable within 20 mg per cubic metre, nitrogen oxides is stable at about 30 mg per cubic metre, and much lower than the gas generation emission standard, the cost of generating electricity per kilowatt-hour only increases by CNY 0.01–0.02. If its cost is allowed to increase to about CNY 0.1, coal power generation is already fully capable of achieving ultra-clean emissions. According to data released by the National Energy Administration, 580 million kilowatts of ultra-low emission transformation of coal power plants have been completed nationwide, exceeding the target of the power plant's ultra-low emission transition plan; By 2020, China will have completed the ultra-low emission upgrading of coal power plants to achieve clean coal power generation.

3.2.3 Revolution of Speeding up New Energy

The accelerated arrival of the new energy era is likely to exceed expectations, and we need to take the initiative to grasp the speed revolution of new energy sources and become strategic participants and important contributors to the global new energy revolution. Guided by national policies, energy departments should grasp the domestic and international energy development situation, comprehensively and strategically arrange the national energy security pattern, accelerate the arrival of a new era of "three major components" of coal, oil and gas and new energy, and change coal as "energy dominator" in the primary energy consumption structure as soon as possible.

(1) To increase the development of new energy resources, such as renewable energy and hydropower, is the key to realizing the development of low carbon energy. According to the United Nations Framework Convention on Climate Change, the Chinese government is committed to keeping carbon dioxide emissions below 10 billion tons/year from 2016 to 2020, and to increase the share of non-fossil energy in primary energy consumption to 20% by 2030, and carbon dioxide emissions will reach the peak.

There is great potential for renewable energy development in China (Beijing: National Energy Administration 2017). It is estimated that by 2020, China's wind power generation will reach 450 billion kilowatts, accounting for 5.3% of total electricity, is expected to exceed 10% in 2030, China's solar energy capacity will exceed 100 million kilowatts, and is expected to surpass the United States in 2030; the national hydroelectric installed capacity will reach 360 million kilowatts, reaching 450 million to 500 million kilowatts in 2030. The contradiction between the rapid growth of installed capacity of renewable energy generation and the shortage of power grid accepting capacity is the bottleneck that limits the development of new energy sources. By optimizing the structure of power grid and developing the combination of energy storage and multi-energy grid technologies, the proportion of renewable energy in power grid can be improved. The storage technology of electric power is the key to determine whether the new energy can revolutionize the traditional fossil energy (Zou et al. 2016).

(2) To develop the development and utilization of nuclear energy resources in a prudent and orderly manner on the premise of ensuring absolute security. As a result of several major events in the world, the utilization and construction of nuclear power has been in an awkward situation, in recent years, it has shown a slow recovery and development trend. China's nuclear power development needs to adopt advanced and mature technology on the basis of safety, focusing on the future and developing in moderate and orderly manner. In recent years, the construction cycle of nuclear power has been extended, and the growth rate of installed capacity has slowed down noticeably. In 2017, China's nuclear power installed capacity reached 35.82 million kilowatts, completing 28.5% of the planning target. The scale of nuclear power production was significantly reduced, with only 2 units put into operation in 2017, totaling 2.18 million kilowatts. In short, the development of nuclear power should always put safety first, take geographical location, energy demand, technical level and other factors into account, in the overall strategic framework of national energy to achieve healthy and orderly development.

According to the Medium and Long-Term Development Plan for Nuclear Power (2011–2020) and the nuclear power "13th Five-Year Plan", it is preliminarily predicted that the overall installed capacity of China's nuclear power will exceed 68 million kilowatts by 2022, with an undergoing capacity of more than 30 million kilowatts, and an average annual compound growth rate of 13.25% in from 2017 to 2022. There is still some room for the development of nuclear power industry.

(3) To pay attention to the development of hydrogen energy industry, which is expected to solve the problem of energy storage in the development of new energy resources. Hydrogen energy is considered to be the most promising cleanest secondary energy resources in the 21st century, and it is technically easy to achieve energy-hydrogen interchangeability, which can solve the energy storage problem of distributed power generation in renewable energy resources. China is the country with the world's largest hydrogen utilization, and has kept the world's number one for nine consecutive years since it first exceeded 10 million tons in 2009. In 2016, the National Development and Reform Commission, the National Energy Agency and other parties jointly released the Action Plan for Innovation in the Energy Technology Revolution (2016–2030) (Development and Reform of Energy [2016] No. 513), proposed a road map for key innovative actions for the energy technology revolution, and deployed 15 specific tasks. Among them, hydrogen energy and fuel cell technology innovation, marking the hydrogen energy industry has been incorporated into China's national strategy.

China's hydrogen refueling stations have grown rapidly in recent years, and a total of 31 hydrogen stations are currently under construction or have been built, of which 12 of them have been built and put into operation, located in 8 provinces and cities such as Beijing, Shanghai, Guangdong, Jiangsu, Henan, Hubei, Liaoning and Sichuan. There has been an explosive growth in the application of hydrogen energy in the field of transportation in China. The industrial park, represented by Yunfu, Foshan, Guangdong, has been rapidly established, with only one hydrogen fuel cell project in Yunfu, Guangdong Province, reaching an annual output capacity of 20,000 sets of electric reactors and 5,000 hydrogen fuel cell vehicles. Nanhai District of Foshan clearly stated in ten-year plan for the development of new energy vehicle industry that by 2025, Nanhai District will promote 5,000 fuel cell forklift trucks, 10,000 fuel cell passenger cars, 5,000 fuel cell passenger buses.

3.2.4 Revolution of Energy Structure

The energy structure of "one major component and three non-major components", of which coal is the major component and oil, natural gas, new energy accounts for a small proportion, cannot meet the development of ecological civilization, it is necessary to speed up the arrival of coal, oil and gas, new energy "three major components" era, and to complete the energy structure revolution (Zou et al. 2016; Zou 2018). From the point of view of energy consumption, the world is in an era dominated by oil and gas energy, and China is in an era dominated by coal energy. To understand the domestic and foreign energy development situation, and comprehensively and strategically plan national energy security, and with the continuous progress of oil and gas theory and technology, abundant reserves of oil and gas resources are discovered and the overall production supply is relatively loose, so, domestic energy structure adjustment enters a rare period of historical opportunity. At present, the domestic primary energy consumption structure of coal, oil and gas and new energy accounted for about 61%, 26% and 13%, respectively, the ratio of high carbon fossil energy consumption is large, leading to serious environmental pollution, the development of clean energy has become an urgent need of the people. Therefore, it is necessary to change the primary energy consumption structure dominated by coal as soon as possible. Under the premise of considering energy security, energy departments conduct all-round strategic plan at home and abroad, speed up the adjustment of energy structure, realize the "three major components" consisting of coal, oil and gas and new energy as soon as possible, and finally realize the structural revolution of China's energy development.

It is time to speed up "three capabilities" construction such as domestic oil and gas production capacity, pipeline transportation capability, LNG and oil storage capacity. Try our best to enhance domestic natural gas production capacity, achieving stable production increase of three conventional natural gas production base, e.g. Southwest, Changqing, Tarim; to promote strongly natural gas production rapid growth of marine shale gas, low-level coalbed methane and other unconventional, and to actively promote the gas hydrate "sweet areas" technology industrialization test in South China Sea. We will vigorously improve the onshore pipeline gas transmission capacity, and actively promote the new routes of onshore pipeline gas import. In order to ensure the stable supply of pipeline gas, it is necessary to actively participate in the development of supply-side natural gas projects under the background of "One Belt One Way", and at the same time explore the new routes of building pipeline gas import, and promote the diversification of pipeline gas import. We should make every effort to enhance the strategic position of LNG and gas storage, taking the domestic market, geography and environment into consideration, and construct the related infrastructure of LNG and gas storage in an orderly manner to ensure the demand of natural gas consumption.

Based on artificial intelligence and big data, the characteristics of oil and gas supply and consumption are analyzed, and the peak warning system of oil and gas safety consumption in China is established. According to natural gas consumption of 550 billion to 6,000 billion cubic meters in 2030, it is predicted that domestic gas production will be 180 billion to 220 billion cubic meters, onshore pipeline gas imports will reach 160 billion cubic meters, LNG imports will reach 200 billion to 250 billion cubic meters, external dependence ratio of natural gas will be over 64%. In view of the characteristics of natural gas resources, a high proportion of external dependence ratio will increase the risk of natural gas supply, the "gas shortage" of natural gas supply in the winter of 2017 has fully reflected the security risks of China's natural gas imports. Therefore, it is necessary to closely track the domestic and foreign natural gas production, consumption, climate, transmission path, inventory and politics on the basis of artificial intelligence and big data analysis, establish the peak warning system of safe consumption, avoid the security risk of natural gas supply, and adjust the import route in time.

Starting from China's conditions of rich coal but insufficient oil and gas energy resources, we will achieve early breakthrough of industrial technology for coal clean utilization and new energy. Based on the historical law of energy development and China's energy structure, it is necessary to reduce and shorten the proportion and time of oil and gas in China's energy structure, and speed up the historical process of converting fossil energy into new energy sources in order to reduce the long-term dependence on oil and gas resources and the safety pressure brought by them. Therefore, it is necessary to quickly break through the industrial technology of coal clean utilization, quickly break through the low-cost technology of major roles of new energy, accelerate the early arrival of scalable production era of new energy, to ensure China's energy production and consumption to be strategic, low-carbon and long-lasting security.

References

Chen, Jianjun, Nan Wang, Hongjun Tang, et al. 2016. Impact of sustained low oil prices on China's oil & gas industry system and coping strategies. *Natural Gas Industry* 36 (3): 1–6.

National Energy Administration. 2017. Guidance on the implementation of the "13thfive-year plan" for renewable energy development.

Zou, Caineng. 2018. Energy revolution and oil company transformation strategy of new era. *Journal of Beijing Petroleum Managers Training Institute* 25 (4): 3–15.

Zou, Caineng, Qun Zhao, Guosheng Zhang, et al. 2016. Energy revolution: From a fossil energy era to a new energy era. *Natural Gas Industry* 36 (1): 1–10.

Part II
Revolution of New Energy

Chapter 4
Revolutionary Energy Technology

The third scientific and technological revolution is another major leap in the field of science and technology after the steam technology revolution and the power technology revolution in the history of human civilization. The major symbol of it is the invention and application of atomic energy, electronic computers, space technology and bioengineering, involving information technology, new energy technologies, materials technology, biotechnology, space technology and marine technology in many fields of information control technology revolution, with the characteristics of technical grouping, complexity, mutual penetration and information sharing.

Historical development shows that every scientific and technological revolution is bound to be accompanied by the energy revolution. At present, to push the revolutionary transition of energy forward has become an important orientation of the energy strategy of the powerful countries, and the development of new energy is one of the core tasks. To promote the development of new energy sources, we can promote the transition of energy structure and even economic structure, which will have a profound impact on the national economy, and the future direction of energy industry will change from energy resource type to energy science and technology. The key to promote the energy revolution lies in technological innovation, mastering advanced and revolutionary energy technology and changing technological advantages into industrial advantages and economic advantages.

Under the trend of global energy reform, China must focus on building the competitiveness of advanced, revolutionary energy technologies and low-carbon development advantages, and develop cutting-edge technologies such as hydrogen energy, energy storage, new graphene materials, nuclear fusion, and build smart energy, promote the deep integration of energy Internet and distributed energy technologies, smart grid technologies and energy storage technologies, occupy the commanding heights of energy technology, and lead the process of global energy technology innovation and development.

C. Zou, *New Energy*, https://doi.org/10.1007/978-981-15-2728-9_4

4.1 Hydrogen Energy and Fuel Cells

The secondary energy is the intermediate link between energy and end users, and electric energy is currently the most widely used secondary energy sources. Hydrogen, as a carrier of chemical energy, is becoming as an important secondary energy resource as electricity. The hydrogen energy industry chain mainly includes the production, storage, transportation and application of hydrogen. Hydrogen can be widely used in traditional fields, but also in emerging hydrogen energy vehicles and hydrogen power generation (including thermoelectric cogeneration supply for distributed power generation, power storage, etc.). Hydrogen energy technology, with large-scale low-cost hydrogen production and safe hydrogen storage as the main body, will promote the transition of power industry and the development of emerging industries such as new energy vehicles and distributed supply, with the potential to change the energy structure and realize the transition of the whole industrial chain from the end of energy supply to the consumer end. Fuel cell is a key technology related to hydrogen energy utilization, just like the relationship between oil and internal combustion engine, electricity and electric motor.

4.1.1 Hydrogen Production Technology

Hydrogen production technology is one of the key technologies in the process of hydrogen energy utilization. Hydrogen resource is very wide, and there are many methods to produce it. At present, natural gas (CH_4), crude oil (hydrocarbon) or coal are used as main raw materials, and react with water vapor at high temperature by steam conversion method, partial oxidation method, gasification method and other processes. The main technical route of hydrogen production is shown in Fig. 4.1.

Industrial hydrogen production mainly includes the following four methods: First, the use of fossil fuels to produce hydrogen, the second is to extract hydrogen from chemical by-products, the third is to use methanol and methane from organisms to produce hydrogen, and the last is to use natural energy such as solar energy, wind energy to produce hydrogen from electrolytic water. The key technologies mainly include solar photothermal/coal synergistic hydrogen production technology, biomass reorganizing hydrogen production technology, photocatalytic hydrogen production and other technologies.

1. Solar photothermal/coal collaborative hydrogen production technology

Solar photothermal/coal collaborative hydrogen production technology refers to the use of solar focusing and clustering high-temperature thermal technology, to provide high-temperature thermal energy. It can replace some standard coal and heat the system to above 700 °C by solar energy to provide heat for the gas furnace raw materials, thereby producing carbon monoxide (CO), hydrogen (H_2) and so on, the resulting product further acts with water (H_2O) in the hydrogen generator to produce

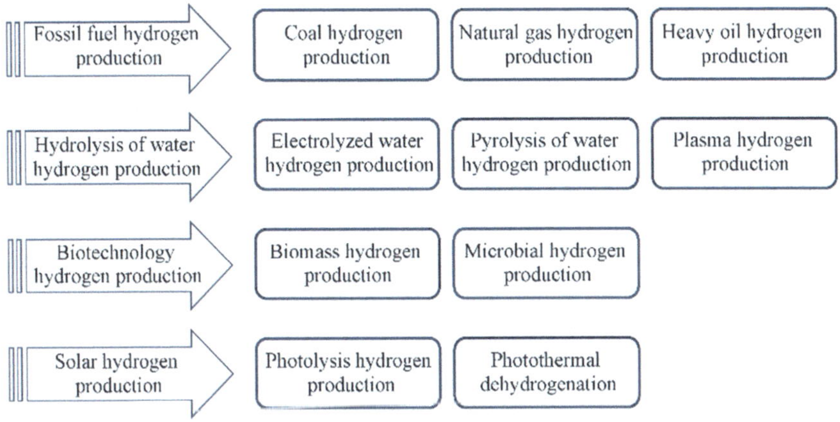

Fig. 4.1 Main technical roadmap of hydrogen production

hydrogen, and uses a heat exchange pre-heating device to collect the heat released by the reaction and recycle carbon dioxide (CO_2). There are two major feasible technical routes (Figs. 4.2 and 4.3).

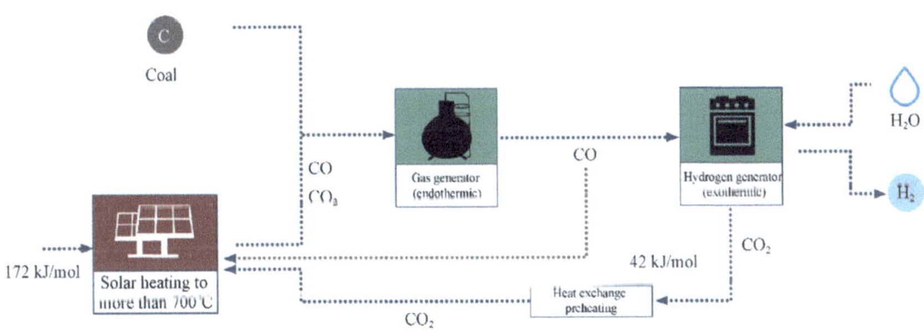

Fig. 4.2 Route I of solar photothermal/coal collaborative hydrogen production technology

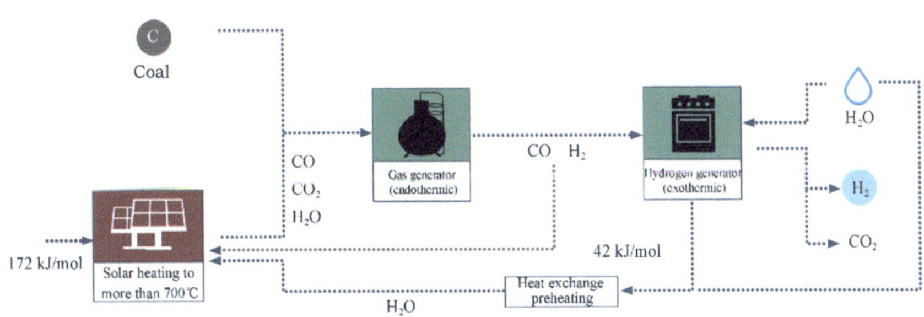

Fig. 4.3 Route II of solar photothermal/coal collaborative hydrogen production technology

1) Solar high temperature heat collection technology

In solar photothermal/coal collaborative hydrogen production technology, one of the key technologies is solar high temperature heat collecting technology.

Solar energy is a low-density, intermittent, constantly changing spatial distribution energy. The solar collectors can be used for collection and utilization of solar energy to more effectively absorb solar radiation and obtain high-temperature heat energy. The collector should adopt the technology of focusing, tracking and other technologies, and the solar focusing device can increase the solar radiation intensity per unit area, save the heat-absorbing material and improve the heat utilization efficiency. The solar tracking system can make sunlight always shine vertically on the receiving surface, and the received solar radiation will be greatly increased.

Solar high temperature heat collecting technology can be divided into two categories, e.g. centralized system and distributed system. Among them, solar photothermal/coal synergistic hydrogen production technology utilizes a centralized system (Fig. 4.4), which usually consists of a planar mirror, a tracking structure, a scaffold and so on, which has a heliostat that are always aimed at the sun, and the incident light is reflected to the receiver at the top of the tower near the center of the site. The heat transfer medium in the receiver reaches high temperature and can be transmitted to the steam generator on the ground through the pipeline to generate high temperature steam.

Solar high temperature heat collection technology has been relatively mature. In April 1982, the United States built a tower solar thermal power system called the "Sun No. 1" in the desert near Barstow in southern California. The system consists of a mirror array including a ring consisting of 1818 mirrors and a tower receiver up to 85.5 m. The unit was modified in 1992 to demonstrate molten salt receivers and heat storage devices. After that, the "Sun No. 2" system was built and connected to the grid in 1996.

Japan uses an improved molten salt energy storage method, that is, the solar radiant energy obtained on a sunny day is fed into a melt of salt (such as nitrate, etc.), so that the salts absorb heat and melt, and when the molten salt solidifies, heat energy is released to produce steam at night and the next day, thus avoiding interruptions in

Fig. 4.4 Centralized tower collection system

the power generation process and building a capacity of 10,000-kW large-scale solar radiant power station.

China's solar concentrating heat collecting project has also made progress. Nanjing built first 70 MW tower solar thermal power generation demonstration project in China, and successfully connected to the power grid in October 2005. China and Germany plan to establish tower solar heat collecting stations in various regions of China to make hydrogen using solar photothermal/coal collaborative technology.

2) Advantages of solar photothermal/coal collaborative hydrogen production technology

Traditional surface coal gasification hydrogen production is to gasify coal first to obtain hydrogen and carbon monoxide as the main components of gaseous products, and then to obtain a certain purity of the product of hydrogen through purification, carbon monoxide conversion, separation, purification and other treatment. Although this method produces a large amount of hydrogen with low cost, there are many drawbacks: (1) Environmental pollution. Hydrogen produced by coal hydrogen production process will be accompanied by carbon dioxide emissions, exacerbating the greenhouse effect and causing environmental pollution. (2) Energy waste. In the production of hydrogen accompanied by a large loss of heat, these energies are not fully utilized and are directly dissipated. (3) Hydrogen purity problem. If the production of hydrogen is used in fuel cells, because of the high sulfur content, the platinum catalyst of the fuel cell is easily poisoned, and the fuel cell stack is damaged.

China's vast coal resources are mainly distributed in Inner Mongolia, Shaanxi, Xinjiang, Qinghai, Ningxia, Shanxi, Hebei and other regions, which are rich in solar energy resources. Using solar high temperature heat collecting technology and coal to produce hydrogen and carbon monoxide, and based on this, methane (natural gas), methanol, ethanol and other basic raw materials can be produced and transported to all parts of the country conveniently, replacing coal, imported oil and natural gas. Compared with the traditional coal hydrogen production method, it not only reduces carbon dioxide emissions, but also rationally utilizes energy, reduces coal pollution, and enhances the safety of energy supply.

Compared with the traditional coal hydrogen production process, solar photothermal/coal collaborative hydrogen production technology can reduce the usage of coal by about 50%. In areas with good sun exposure, photothermal focusing on 1 m^2 of land can provide about 250 kWh/year of high temperature thermal energy, calculated on the basis of 8,134 kWh each ton of standard coal, the high-temperature heat of 1 m^2 of land heat provided by the focus can replace more than 30 kg of standard coal, and 1 km^2 of photothermal focus can replace about 30,000 tons of standard coal.

3) Direct hydrogen production using solar high temperature heat collecting technology

The use of solar high temperature heat collecting technology can directly obtain hydrogen (STCH), and there is no greenhouse gas and environmental pollution problems, but the current technology is not mature, hydrogen production efficiency is relatively low, the cost is also higher. With the development of technology, solar thermochemical hydrogen production technology will be widely used in industrial production.

The principle of STCH is to use the solar collector to achieve the temperature required by the solar energy aggregation reaction. In the reactor, the heat is decomposed by the metal oxides to convert the solar energy into chemical energy, and then the metal hydrolysis is performed to decompose water to produce hydrogen. The process of STCH can be divided into direct circulation and hybrid cycle, direct circulation only uses centralized solar energy, hybrid cycle needs to use an additional electrically driven electrolytic cell as part of the water decomposition circulatory system. Usually the direct circulatory system is relatively simple, but requires a higher operating temperature; The disadvantages of the hybrid system include the increased complexity and additional requirements of the electrical input (Hydrogen Production Tech Team Roadmap, USDRIVE 2017).

The high temperatures required to decompose water through the direct STCH process require the use of centralized solar energy, which is used to focus sunlight on thermal reactors to produce temperatures up to 2000 °C, including two forms of central tower and modular disc type. A central tower is a central receiver/reaction tower with a heliostats mirror that deploys a central STCH reactor in a solar receiver surrounded by an appropriate size mirror field (solar tracking mirror field); The modular disc is using smaller STCH reactor modules, each of which is connected to a tracking disc concentrator (Fig. 4.5).

Various studies related to solar photothermal technologies are currently under way in countries. German scientists used 149 xenon lights to form a huge artificial sun, known as Synlight, which looks like a giant hive, nearly 14 m high and 16 m wide, designed to mimic the sun and produce light energy to create environmentally friendly hydrogen. According to the German National Aerospace Research Center (DLR),

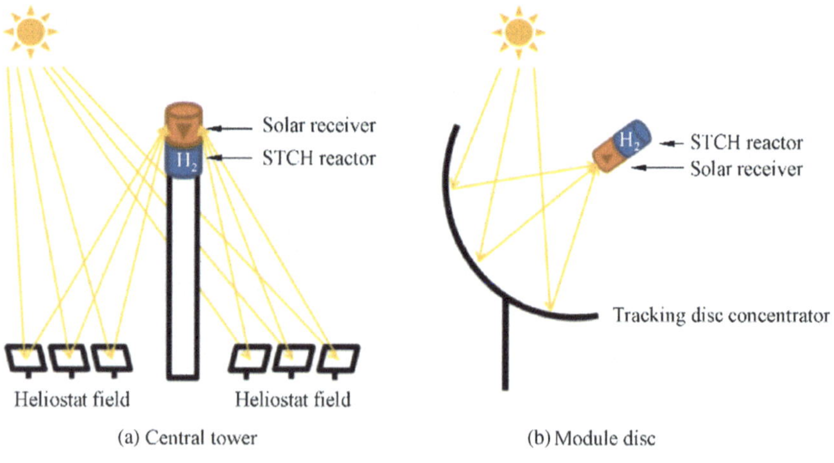

Fig. 4.5 Two mirror-based hydrogen production methods

the Synlight system was officially launched on March 23, 2017, with lights projected onto a focal plane of 20 cm × 20 cm, resulting in radiation intensity of 10,000 times that of the same area as sunlight exposure and temperature can reach up to 3,000 °C, which is approximately 2–3 times the temperature of the blast furnace and the power up is 350 kW. Bernard Hoffschmidt, director of the DLR Solar Research Institute, said the Synlight system was designed to take experiments in small labs to a new level, and once the technology for hydrogen production with 350-kW artificial light is mature, the process could be magnified 10 times to reach the technology level of the power plant. The target is expected to be achieved within 10 years, but the ultimate goal of the research center is to achieve the use of natural light to make hydrogen. At present, the current use of Synlight system is very expensive, and the electricity consumption of four hours is equivalent to the consumption a family of four for one year. The researchers hope the simulator will lead to faster progress in solar fuel manufacturing. In addition, the Synlight system can also promote studies on the use of solar energy to produce chemical raw materials, reduce carbon emissions and so on.

2. Biomass reforming hydrogen production technology

Biomass reforming hydrogen production technology can be divided into two categories: one is to use biomass as raw material using thermal physicochemical principles and technology to produce hydrogen, such as biomass gasification hydrogen production, supercritical conversion hydrogen production, pyrolysis hydrogen production, the other is the use of biological pathways to achieve hydrogen production, such as direct biological photolysis, indirect biological photolysis, light fermentation, photosynthetic heterotrophic bacteria water-gas transfer reaction to synthesize hydrogen, dark fermentation and microbial fuel cell technology. Simple compounds such as methane, methanol, and ethanol of the biomass fermentation product can also be converted to hydrogen by a chemical reforming process. At present, research on biomass hydrogen production focuses on how to efficiently and economically convert and utilize biomass (Yu and Xiao 2006).

1) Biomass high temperature decomposition pyrolysis hydrogen production

Biomass pyrolysis is a thermochemical process in which biomass reacts under high temperature and anaerobic conditions. Thermal cracking includes slow cracking and rapid cracking. The rapid cracking of biomass to make biological oil and re-reorganize hydrogen production is the focus of this technology. The increase of pyrolysis efficiency and production depends on the improvement of equipment and process, the selection of catalyst and the optimization of reaction parameters. The American Renewable Energy Laboratory (NREL) has pioneered a series of studies in this area and has achieved results. At present, the biomass pyrolysis reactor at home and abroad mainly includes mechanical contact reactor, indirect reactor and hybrid reactor.

2) Methanol conversion for hydrogen production

Methanol conversion hydrogen production refers to the conversion of biomass or waste into methanol through microbial fermentation, and then producing hydrogen through the reorganization, the main technology includes methanol cracking hydrogen production and methanol reforming hydrogen production. At present, the research hotspots mainly focus on the improvement of catalyst structure and the selection of new catalyst. Huber et al. published the results of catalytic hydrogen production of biomass based hydrocarbon in the journal Science, and found that the non-rare metals such as Raney Ni–Sn as catalysts were not only more economical than platinum, but also reduced methane production and increased hydrogen production under similar hydrogen production result. In recent years, researchers have also carried out investigations on methanol reforming in the water phase. For example, Shabaker et al. have carried out the hydrogen production technology of liquid phase methanol reforming with Pt/Al_2O_3 (platinum/alumina) as catalyst. Boukis et al. studied methanol hydrogen production from supercritical water reforming, the results show that the main products are hydrogen and a small amount of carbon dioxide. The conversion of methanol reached 99.9% without the addition of a catalyst. It was found that the inner wall of the nickel alloy had an impact on the reaction, and the oxidation of the inner wall could be improved in advance. The reaction rate and the concentration of carbon dioxide are lowered.

3) Methane conversion for hydrogen production

Methane conversion to hydrogen production refers to the use of waste and biomass as raw materials for anaerobic digestion to produce methane, and then convert to hydrogen, mainly producing by methane catalytic thermal cracking hydrogen production and methane reforming hydrogen production. Hydrogen production from methane is one of the most studied technologies in hydrogen production technology. However, most of the research has recently reported on the hydrogen production from methane conversion of natural gas, and the co-reformation of methane and natural gas produced by anaerobic digestion has also been reported. Hydrogen production of methane around new technologies to increase methane conversion rate is a hot research topic, such as using plasma to increase reaction temperature; determining optimal reaction parameters and improving equipment; using new catalysts, Ni (nickel) and CO (Carbon monoxide), Pd (palladium), Pt (platinum), Rh (ruthenium), Ru (ruthenium), Ir (Iridium) and other transition metals and precious metal supported catalysts; Ochoa et al. found the use of Li_2ZrO_3 (lithium zirconate) as an adsorbent in the methane steam reforming technology can increase the production of hydrogen through adsorption kinetics and reactor simulation.

4) Ethanol conversion for hydrogen production

Ethanol conversion hydrogen production refers to the conversion of biomass or waste into ethanol by microbial fermentation, and then producing hydrogen by reforming. The production of hydrogen by ethanol catalytic reforming is one of the most popular techniques in the field of hydrogen production. Hydrogen making from ethanol is not only good for environmental protection but also increases the use of renewable energy, but the technology is still in the laboratory development stage. At present, the research on hydrogen production by ethanol catalytic reforming mainly focuses on the selection and improvement of catalysts. The conversion efficiency and hydrogen production of ethanol vary greatly due to different catalysts, reaction conditions and preparation methods of catalysts. Benito et al. proposed a biomass based ethanol reforming hydrogen production mechanism with ICP0503 as catalyst, the catalyst ICP0503 catalytic effect is stable, and the gas produced by the catalyst reforming may not need purification treatment, can be directly used in fuel cells.

3. Photocatalytic hydrogen production technology

The principle of photocatalytic hydrogen production technology is that light quantum can break the hydrogen bond in the molecules of hydrogen-containing compounds, such as water, and produce hydrogen. At present, photocatalytic hydrogen production efficiency is low, how to choose catalyst is the key to photocatalytic hydrogen production at low cost.

Photocatalytic Hydrogen Production Principle:

$$2\,(2a^*) + H_2O \rightarrow \text{Hydrogen} + 1/2O_2 + 2\,(2a)$$

Among them, 2a is not only the electron donor, but also the electron acceptor. Under the excitation of light energy, electrons can be transferred to the water molecule, and H^+ (hydrogen ions) can be converted into hydrogen.

Three kinds of photocatalytic hydrogen production methods are usually used to produce hydrogen: Z-type hydrogen production system, photoelectric catalytic hydrogen production system, suspension system and so on.

1) Z-type hydrogen production system

The photosynthesis Z process consists of two different primary photoreactions. Sayama et al. uses RUO_2–WO_3 (cerium oxide–tungsten trioxide) as a catalyst, Fe^{3+}/Fe^{2+} (iron/ferrous ion) as an electron relay, visible radiation (<460 nm), Fe^{3+} was reduced to Fe^{2+}, UV Light (<280 nm) radiation, Fe^{2+} reacts with H^+ to form hydrogen, and the ratio of H_2 (hydrogen) to O_2 (oxygen) is 2:1. In the Z process of the simulated photosynthesis, the electronic relay body can be recycled, and in addition, the electronic relay body also has I^-/IO^{3-}. TiO_2 (titanium dioxide) is coated with a WO_3 film, WO_3 absorbs blue light to generate holes for oxidizing water, and DSSC-TiO_2 absorbs transmitted green and red light to produce highly active conductive electron-reduction hydrogen. The advantages of the Z-type hydrogen production system include: the use of the entire visible light; overcoming the limitations of a single semiconductor; the theoretical efficiency is about 47%, and the actual light conversion efficiency has reached 8%.

2) Electrochemical catalytic hydrogen production system

The electron-cavity pairs are excited by the photoelectrode to act as a redox agent and participate in the electrochemical reaction (Fig. 4.6).

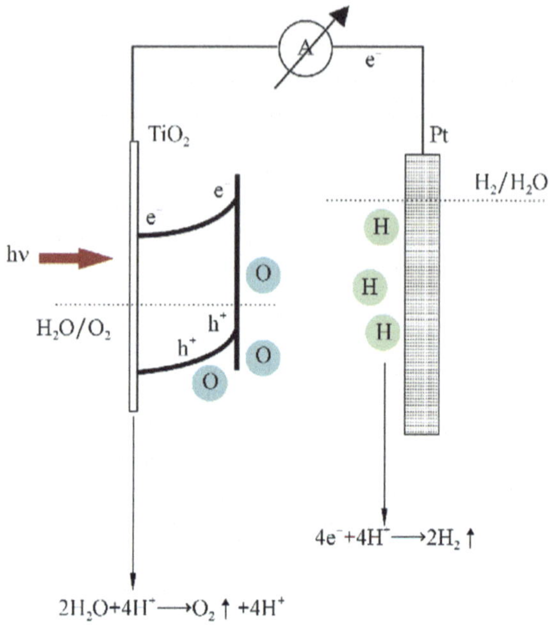

Fig. 4.6 Photochemical catalytic hydrogen production system

Photoexcitation process $TiO_2 + hv \rightarrow h^+ + e^-$
Oxidation reaction on photoelectrode $H_2O + 2h^- \rightarrow 1/2O_2 + 2H^+$
Cathodic reaction on the electrode $2H^+ + 2e^- \rightarrow H_2$
Total photolysis reaction $H_2O + hv \rightarrow 1/2O_2 + H_2$

3) Suspension system photocatalytic hydrogen production

In the photocatalytic hydrogen production system, exploring high-efficiency, stable and economical photocatalytic materials is the key to the practical application of photocatalytic hydrogen production. The catalyst can greatly improve its catalytic performance after certain modification. The commonly used modification methods mainly include size quantization, ion doping, semiconductor composite, and precious metal deposition (Xi 2012).

(1) Size quantization

Nanoparticles with small size have a large volume fraction on the surface and more active spots on the surface, which makes them the basic conditions for the catalyst. The positive shift of valence band potential means that nano-semiconductor particles have stronger oxidation or reduction ability. Nano TiO_2 particles are the most promising photocatalyst.

(2) Ion doping

Ion doping produces ion defects, which can be a capture trap for carriers and extend their life. Due to the different ion size, the crystal structure will be distorted, resulting in increased crystal asymmetry and improved cavitation separation. The results show that the catalytic activity can be improved significantly by doping the metal ions of the Group II and Group VI in the TiO_2.

(3) Semiconductor composite

The composite semiconductor has two different energy levels of the conduction band and the valence band, which can greatly shift the absorption wavelength toward red light, and the crystal structure of the composite semiconductor also improves the photocatalytic activity. In addition, the composite semiconductor can effectively absorb light of different wavelengths in the light source, and the catalytic activity is increased due to an increase in the utilization rate of the light source.

(4) Precious metal deposition

Pt (platinum), Ag (silver), Ir (iridium), Au (gold), Ru (ruthenium), Pd (palladium), Rh (ruthenium) and the like of Group VIII are commonly used for precious metal deposition. The appropriate amount of precious metal deposition on the surface of the catalyst facilitates the efficient separation of photogenerated electrons and holes, and also reduces the overvoltage of the reduction reaction, thereby increasing the catalytic activity. When the surface of the semiconductor is in contact with the metal, it helps the carrier to redistribute. Since the Schottky barrier becomes an effective trap for trapping the excited electrons, the photogenerated carriers are separated, suppressing the recombination of electrons and holes. The effective separation of electrons and holes increases the photon efficiency of the catalyst.

4.1.2 Hydrogen Storage Technology

Hydrogen storage technology is currently the main technical obstacle to hydrogen energy development. Hydrogen is characterized by its flammability, explosiveness, and difficulty in high-density storage. How to achieve high-density hydrogen storage and increase the hydrogen storage density per unit mass (i.e. the ratio of hydrogen storage mass to the total mass of the entire hydrogen storage unit) is the main indicator for measuring the advanced hydrogen storage technology, and is also a prerequisite for large-scale commercial application of hydrogen energy.

(1) Hydrogen storage

Hydrogen can be stored in the form of gases, liquids, solids, compounds, and the like. At present, hydrogen storage methods include high-pressure gaseous hydrogen storage, low-temperature liquid hydrogen storage, metal hydride hydrogen storage, organic compound hydrogen storage, microsphere hydrogen storage, carbon nanomaterial hydrogen storage and so on.

1) High-pressure gaseous hydrogen storage

High pressure gaseous hydrogen storage is the most common and direct method of hydrogen storage. The high pressure gaseous hydrogen storage device has fixed hydrogen storage tank, gas cylinder and cylinder group, long pipe cylinder and long tube bundle, and a vehicle hydrogen storage bottle and so on.

Fixed hydrogen storage tank: in recent years, China has achieved remarkable results in the research and development of fixed hydrogen storage tanks. On the basis of overcoming a number of key technologiesit has successfully developed a kind of anti-explosion, anti-explosion, defect dispersion and operation status by using its own patented technology. Multi-functional, multi-layer high-pressure hydrogen storage tanks that can be monitored online, the current pressure levels can reach 45, 77 and 98 MPa, and the relevant technical indicators have reached the international leading level.

High pressure gas cylinders: The operating temperature range is −50–60 °C. China's commonly used cylinders have a volume of 40 L (water volume) and a working pressure of 15 MPa. However, this cylinder can only store 6 m^3 of gas and about 0.485 kg of gas. The mass of hydrogen storage is less than 1% of the cylinder's own mass, with low efficiency and high transportation costs. Most commonly used cylinders abroad are 50 L (water volume), the typical working pressure range is 15–40 MPa, and the commonly working pressure is 20 MPa.

Long pipe cylinder group and long tube bundle: they are installed and operated in some hydrogen production plants, hydrogen enterprises and hydrogenation stations. At present, this kind of hydrogen storage equipment is generally used in hydrogen refueling stations that have been built and under construction in China.

2) Low-temperature liquid hydrogen storage

When the gaseous hydrogen cools to 253 °C, it becomes liquid. In various hydrogen storage methods, liquid hydrogen storage can achieve high storage volume density and mass density. Liquid hydrogen Storage has a mass fraction of 5–7.5% and a volume capacity of approximately 0.04 kg/L. Liquid hydrogen storage tanks and liquid hydrogen trailers have been successfully manufactured in China and are used in aerospace and other fields, and a liquid hydrogen tanker of 15 m^3 can transport about 1.05 tons (about 12,000 m^3) of hydrogen.

Although hydrogen liquefaction is a mature technology, due to some special physical properties of hydrogen, such as the existence of Joule-Thomson transition temperature, low boiling point (20.4 K), liquid positive-secondary spontaneous conversion and exothermic, hydrogen liquefaction cost is high, and storage and transportation have some difficulties. Hydrogen liquefaction is expensive, to liquefy hydrogen, first of all, it has to be pre-cooled to Joule-Thomson transition temperature, and then obtain liquid hydrogen through the refrigeration cycle. The energy consumption of liquefaction is 4–10 kwh/kg, which is almost equivalent to one-third liquid hydrogen costs; liquid hydrogen storage containers are large, requiring excellent insulation device to insulate and avoid boiling vaporization.

3) Metal hydride hydrogen storage

The principle of metal hydride hydrogen storage is that the hydrogen molecule is in contact with the hydrogen storage alloy, adsorbed on the alloy surface, and the H–H bond is removed, and becomes the atomic adsorbed hydrogen. There are three main processes for metal hydride hydrogen storage principle.

(1) The metal absorbs a small amount of hydrogen first to form a hydrogen-containing solid solution—Mhx (α phase), the structure of the alloy remains unchanged, and hydrogen exists in the form of atoms.

$$M + x/2H_2 \underset{\text{Abs.}}{\overset{\text{Des.}}{\rightleftarrows}} MH_x + \Delta H$$

(2) The solid solution further reacts with hydrogen to produce phase change (structural change), resulting in hydride phase–Mhy (β phase).

$$\frac{2}{y-x}MH_x + H_2 \longleftrightarrow \frac{2}{y-x}MH_y + Q$$

(3) To continue to increase hydrogen pressure, the hydrogen content inside the metal increased slightly.

Metallic hydride is divided into four categories, titanium system, magnesium system, zirconium system and rare earth, according to the distribution ratio of AB_5, AB_2, A_2B and AB.

Titanium Type: Typical representative TiFe, first discovered by the National Laboratory of Brookhaven in the United States.

Magnesium Type: Typical representative Mg_2Ni, first reported by the National Laboratory of Brookhaven in the United States.

Zirconium Type: Typical representative $TiMn_{1.5}H_{2.5}$, first discovered by Pana sonic, Japan.

Rare earth Type: Typical representative $LaNi_5$, first developed by the Philips laboratory in Netherlands.

4) Organic compound hydrogen storage

The hydrogen storage technology of organic compounds is to use the hydrogenation and dehydrogenation reaction between unsaturated liquid aromatic hydrocarbons and corresponding cycloalkane, which can be recycled multiple times. As an organic liquid hydrogen storage technology for hydrogen storage carriers, hydrogen is chemically added to the carrier to form a stable hydride liquid, and through ordinary storage and transportation similar to petroleum products, the hydrogen in the carrier is released for use by the user to obtain energy when it reaches the client. The carrier after dehydrogenation can be reused and recycled repeatedly.

For example, hydrogenation using benzene to obtain cyclohexane has a high hydrogen storage density, mass fraction can reach 7.2%, its conversion reaction is as follows:

$$C_6H_6 + 3H_2 \longleftrightarrow C_6H_{12}$$

China University of Geology (Wuhan) has successfully overcome the technical bottleneck that hydrogen energy is difficult to store and release under normal temperature and pressure by using unsaturated aromatic compounds to catalyze hydrogenation, and realized the transportation of hydrogen energy liquid at normal temperature and pressure. The technical performance index of hydrogen storage material exceeds the standard of automobile hydrogen storage materials promulgated by the United States Department of Energy. Experimental results show that the melting point of hydrogen storage molecules can be as low as −20 °C, achieving high efficiency catalytic hydrogenation at about 150 °C, and storage and transportation at room temperature and pressure; catalytic dehydrogenation temperature is lower than 200 °C, the purity of hydrogen produced by dehydrogenation process can be as high as 99.99%, and does not produce CO, NH_3 and other gases; Hydrogen storage material has high cycle life and strong reversibility (above 2,000 times); hydrogen storage mass fraction is greater than 5.5%, volume capacity is greater than 50 kg/m^3.

5) Microsphere hydrogen storage

Microsphere hydrogen storage refers to the use of glass microspheres to store hydrogen. The spheres have a diameter of 200–500 μm, wall thickness of 0.2–0.5 μm, and are heated to 300–400 °C. Under high hydrogen pressure (80 MPa), hydrogen enters the sphere through the wall, when cooling to room temperature, high-pressure gas is trapped due to the significantly reduced diffusion rate, and hydrogen can be recovered by heating. Microspheres have a high capacity of hydrogen absorption, 5–10% by mass fraction. The problem is that when the pressure is too high, the sphere will often make the sphere broken, and the absorption of hydrogen in the sphere will slowly leak at the ambient temperature. Studies have also shown that aluminosilicate zeolite crystals also have channels and pores suitable for the size of hydrogen molecules, but, like hydrogen storage in glass microspheres, they must be operated at very high pressures. At present, the hydrogen storage technology of microspheres is still in the exploring stage.

6) Carbon nanomaterial hydrogen storage

Hydrogen storage of carbon nanomaterials refers to the use of new carbon nanomaterials such as carbon nanotubes for hydrogen storage. Carbon nanotubes are a new type of nano-scale carbon nanomaterial with a complete molecular structure. It is a seamless, hollow tube made of a layer of carbon atoms of graphite. Carbon nanotubes have excellent hydrogen storage properties, but carbon nanotubes are expensive, and their scalable production methods have not yet been solved.

2. Transportation of hydrogen

At present, there are three kinds of hydrogen transportation methods, such as high pressure gaseous hydrogen, liquid hydrogen and pipeline transportation. These three methods have been applied at home and abroad. At present, the basic research work of compatibility of pipeline steel and high-pressure hydrogen, pipeline operation and management, etc. is under way, which will create conditions for the construction of long-distance, high-pressure large-scale hydrogen transportation pipelines.

Europe has the longest hydrogen pipeline: hydrogen has a long history of long-distance transportation by pipeline, with about 1,500-km hydrogen pipelines, and the world's longest hydrogen pipeline is built between France and Belgium, about 400-km long. The earliest long-distance hydrogen pipeline was built in Ruhr, Germany, in 1938, with a total of 208 km, the diameter of the hydrogen pipeline between 0.15 and 0.30 m, and the rated hydrogen transport pressure is about 2.5 MPa.

Hydrogen pipeline in the United States: the existing hydrogen pipeline in the United States is 720 km (2.08 million km of natural gas pipeline), and the total length of hydrogen pipeline between Houston and Arthur Port in Texas State, United States is 225 km.

Hydrogen pipeline in China: hydrogen transmission pipeline project from Jiyuan to Luoyang of Sinopec Luoyang Refining and Chemical Company, the horizontal length of the pipeline is 25 km, the annual gas transmission capacity is 104,400 tons; the Wuhai-Yinchuan coke oven gas transmission pipeline has a total length of 216.4 km. The gas volume is 1.61 billion m^3/year, passing through Inner Mongolia, Ningxia and Wuhai, Shizuishan and Yinchuan. The main line runs through multiple highways, railways and the Yellow River. The gas source is the coke oven gas from Wuhai Qianlishan Industrial Park and Xilaifeng Industrial Park and the hydrogen mixture from LNG plant and clean oil plant in the industrial park.

4.1.3 Fuel Cell

Fuel cell refers to a power generation device that converts chemical energy present in fuel and oxidizer directly into electrical energy. In the 1839, Grove in the United Kingdom invented a simple hydrogen-oxygen fuel cell with platinum black as an electrode catalyst, and in 1889 Mood and Langer introduced the name "fuel cell" for the first time. In the 21st century, fuel cells became an important high-tech technology that developed rapidly and was able to obtain electricity from hydrogen and oxygen. In the past twenty or thirty years, as a result of the lack of primary energy and environmental pollution problems, the development and utilization of new clean renewable energy has become increasingly high. Fuel cell has the characteristics of no pollution, high efficiency, wide application, no noise and continuous operation and modular power device, which is widely recognized by all countries in the world. Fuel cells are used in a wide range of applications, both in military, space and power plants, as well as in motor vehicles, mobile equipment, residential families and other fields (Wu 2014; Yi et al. 2018).

1. Fundamental principles of fuel cells

Fuel cells are composed of anode, cathode and ionic conductive electrolytes, the working principle is similar to that of ordinary chemical batteries. The fuel is oxidized, at the anode and the oxidant is reduced at the cathode, and electrons flow from the anode to the cathode through a load to form a circuit. The difference between a fuel cell and a conventional battery is that it requires continuous input of fuel and oxidant into the battery during operation, and the fuel cell continuously supplies electric energy as long as it is continuously supplied.

The main components of the fuel cell include electrodes, electrolyte diaphragms and a collector of current. It is also necessary to match the corresponding auxiliary systems, including the reaction agent supply system, heat discharge system, drainage system, electrical performance control system and safety device. The electrode of fuel cell is the electrochemical reaction site of fuel oxidation reaction and oxidant reduction reaction, and the key factor affecting its performance is the function of catalyst.

2. Categories of fuel cell

Fuel cells can be divided into several types according to different classification methods.

According to electrolyte type, fuel cells can be divided into alkali type, phosphoric acid type, polymer type, molten carbonate type, solid electrolyte type.

By means of fuel treatment, fuel cells can be divided into direct, indirect and regenerative types. Among them, direct fuel cells can be divided into three types of low temperature, medium and high temperature according to the difference of temperature: low temperature fuel cell including alkaline fuel cell (AFC, operating temperature is 100 °C), solid polymer proton membrane fuel cell (PEMFC, also known as proton exchange membrane fuel cell, operating temperature is within 100 °C) and phosphoric acid fuel cells (PAFC, operating temperature is 200 °C); high-temperature fuel cells include molten carbonate fuel cells (MCFC, operating temperature is 650 °C) and solid oxidized fuel cells (SOFC, operating temperature is 1000 °C). Indirect fuel cells include recombinant fuel cells and biofuel cells. Regenerative fuel cells have light, electricity, heat, radiation-chemical fuel cells and so on.

In the chronological development sequence, fuel cells can be divided into three types: phosphoric acid fuel cell (PAFC), known as the first generation of fuel cells; molten carbonate fuel cell (MCFC), known as a second-generation fuel cell; a solid oxidized fuel cell (SOFC), known as a third-generation fuel cell. These batteries are requiring combustible gases as fuel for their power generation. NASA has long used hydrogen fuel cells to provide power for spaceships, and hydrogen fuel cells will be a cleaner form of commercial power generation in the future.

1) Hydrogen fuel cell

Hydrogen fuel cells are electrochemical power generation device that convert chemical energy directly into electrical energy, using hydrogen as a chemical element to make batteries that store energy. The fundamental principle is the inverse reaction of electrolytic water, which supplies hydrogen and oxygen to the anode and cathode respectively. After hydrogen diffuses through the anode and reacts with the electrolyte, the electrons are released to the cathode through an external load.

The electrodes of hydrogen fuel cells are made of special porous materials and are a key technology of hydrogen fuel cells, which not only provide a large contact surface for gases and electrolytes, but also plays a catalytic role in the chemical reaction of batteries.

Proton exchange membrane fuel cell (PEMFC) technology is one of the most mature technologies in the world to synthesize hydrogen and oxygen from the air and release electricity.

2) Phosphoric acid fuel cell

The basic composition and reaction principle of phosphoric acid fuel cell (PAFC) is that fuel gas or urban gas is added to the modifier after adding water vapor, which converts fuel into a mixture of hydrogen, carbon monoxide and water vapor, and carbon monoxide and water are further converted into hydrogen and carbon dioxide by catalyst in the mobile reactor. After such treatment, the fuel gas enters the cathode (fuel electrode) of the fuel reactor, while the oxygen is transported to the cathode (air electrode) of the fuel reactor for chemical reaction, and the action of the catalyst is used to generate electricity and heat quickly. The electrolyte of PAFC is a concentrated phosphoric acid aqueous solution, while the electrolyte of PEMFC is the membrane of the proton conductive polymer system. The electrodes are made of porous frame of carbon, in order to promote the reaction, Pt is used as catalyst, and the carbon monoxide in the fuel gas will cause poisoning and reduce the electrode performance. For this reason, the amount of carbon monoxide contained in fuel gases must be limited in the application of PAFC and PEMFC, especially for low temperature working PEMFC should be strictly limited.

3) Molten carbonate fuel cells and solid oxidation fuel cells

Compared to PAFC and PEMFC, high-temperature fuel cells-molten carbonate fuel cells (MCFC) and solid oxidized fuel cells (SOFC) do not require catalysts, and gas from coal gasification with carbon monoxide as the main component can be used directly as fuel, and it also has the characteristics of being easy to use its high-quality exhaust gas to form combined cycle power generation.

During the operation of the MCFC, the oxygen and carbon dioxide of the air electrode combine with the electrons to form CO_3^{2-} (carbonate ion). The CO_3^{2-} moves through the electrolyte and the diaphragm to the fuel electrode side, and combines with the hydrogen supplied as fuel to emit electrons. At the same time, water and carbon dioxide are generated. The chemical reaction formula is as follows.

Fuel electrode: $H_2+CO_3^{2-}=\!=\!=H_2+O+2e^-+CO_2$

Air electrode: $CO_2+1/2O_2+2e^-=\!=\!=CO_3^{2-}$

Overall: $H_2+1/2O_2=\!=\!=H_2O$

In the above reaction, the electrons are the same as in the phosphoric acid fuel cell (PAFC), which is discharged from the fuel electrode, returned to the air electrode through an external circuit, and the fuel cell power generation is realized by the uninterrupted flow of electrons in the external circuit. In addition, the biggest feature of MCFC is that it must have CO_3^{2-} which is helpful for the reaction. Therefore, the supplied oxidant gas must contain carbon dioxide gas. The inside of the battery is filled with a catalyst, which will modify the methane, a main component of natural gas, inside the battery. When the fuel is gas, the main component of carbon monoxide reacts with water to generate hydrogen, so carbon monoxide can be used equivalently as a fuel. In order to achieve greater output, the diaphragm is usually made of Ni (nickel) and stainless steel.

SOFC is mainly composed of ceramic materials. The electrolyte is usually made of zirconium oxide (ZrO_2), which is a conductor of O^{2-}, and Y_2O_3 (yttrium oxide) acts as a stabilizer to form stabilized zirconia (YSZ). The fuel electrode is made of Ni and YSZ composite porous bodies to form a cermet electrode, the air electrode is made of lanthanum manganese oxide ($LaMnO_3$), and the membrane is made of lanthanum oxide ($LaCrO_3$). During the operation of the SOFC, the fuel hydrogen passes through the electrolyte and reacts with O^{2-} to form H_2O and e^-; the air electrode generates O^{2-} from oxygen and e^-. The overall effect is the same as other fuel cells, which produce water from hydrogen and oxygen. In SOFC, because it is a high-temperature working type, the main component CH_4 of natural gas can be directly converted into hydrogen for utilization without any other catalyst inside the cell, and the main component of the gas, carbon monoxide, can be directly used as a fuel.

3. Application of fuel cells

At present, the electrodes of high temperature fuel cell are mainly made of catalyst material, such as stable zirconia electrode (YSZ) of solid oxidation fuel cell (SOFC) and nickel oxide electrode of molten carbonate fuel cell (MCFC), while the low temperature fuel cell is mainly composed of a thin layer catalyst material supported by the gas diffusion layer, such as platinum fuel electrodes of phosphoric acid fuel cells (PAFC) and proton exchange membrane fuel cells (PEMFC).

Fuel cells have many advantages, the most important of which are low pollution and high energy efficiency. Fuel cells use energy in a way that is significantly better than the internal combustion engine which emits a lot of hazardous waste gas, most of which is water. Some fuel cells emit carbon dioxide, but their content is much lower than gasoline emissions (about 1/6 of gasoline). Therefore, fuel cells can not only improve the situation of air pollution, but may even give mankind a clean sky in the future. The fuel cell converts the chemical energy of the fuel into electrical energy by electrochemical reaction, avoiding the combustion process, does not need to do work to outside through the Carnot cycle, unlike the normal thermal power generator, does not change the energy form of the boiler, steam turbine and generator, and avoid the loss of intermediate conversion, it also avoids the efficiency limitation of the Carnot cycle work mode. The fuel-electric energy conversion efficiency of the fuel cell system is 40–60%. For example, the conversion efficiency of the power of the hydrocarbon fuel power generation system can reach 40–50%; the system efficiency of direct use of hydrogen can exceed 50%; When the gas turbine is used together with the power facilities, the overall efficiency can exceed 60%; if the waste heat discharged from the battery is recycled, the fuel energy utilization rate can exceed 85%. The efficiency of thermal power generation and nuclear power is 30–40%, and the efficiency of automobile internal combustion engines is only 10–20%. It can be seen that the lowest efficiency of fuel cells is higher than that of the best heat engines.

The fuel cell is small in size and easy to use flexibly. Large thermal power stations must have a large enough unit size to achieve satisfactory efficiency. In general, power plants equipped with giant units are limited by various conditions and cannot be adjacent to users, so the centralized power generation is distributed to the user by the power grid. The huge unit has insufficient flexibility to generate electricity and cannot adapt to the immediate needs of users. The user's power load changes sometimes appear as peaks, and sometimes appear as trough. The corresponding grid load also shows peaks and troughs. In order to cope with changes in electrical load, it is possible to use a spare part of the unit or to construct a pumped storage power station, which is generally at the expense of the efficiency of the grid. Nearly 70% of the combustion energy of traditional thermal power stations is consumed in huge equipment such as boilers and steam turbine generators, and emits a large amount of harmful substances when burned. The use of fuel cells to generate electricity directly converts the chemical energy of the fuel into electrical energy. The fuel cell is equivalent to a chemical power station, does not require combustion, has no rotating parts, and is structurally simple, so its footprint is small and the construction period is short, and the generated power can directly enter businesses, restaurants, hotels and families.

Fuel cells also have the characteristics of fast response and high quality of operation. Fuel cells provide power based on electrochemical reactions, which are far more responsive than the heat engine system and can be transformed from the lowest power to the rated power in a few seconds. Because electrochemical reaction determines the power generation capacity, high power generation efficiency can be maintained at full load or partial load, and high power generation efficiency can be maintained regardless of the size of the device, with a strong overload capacity.

Fuel cells are independent, reliable and safe. Fuel cell Whether it is made into a centralized large-scale power generation equipment or a decentralized small equipment, its power reactor independence is unmatched by the heat engine system, it is because of its independence, fuel cell has a high safety. The power supply system using fuel cell is superior to the thermoelectric supply system, both from the point of view of fault recovery and from the point of view of grid stability. And in terms of national defense and energy independence, the safety of fuel cells is self-evident. On the one hand, since the oil crisis of the 1970s, the dependence of major industrial countries on oil has continued to increase and is largely dependent on the supply of oil-exporting countries, and if fuel cell systems are used as a major source of energy, the dependence on oil for daily energy consumption will be significantly reduced and more oil reserves can be used for war preparation. On the other hand, the independent fuel cell systems require less power grids, and even if the power grid is damaged, it is easy to re-establish electricity supply on an isolated basis, and the thermoelectric power grid system is difficult to achieve.

Hydrogen is still the ultimate direction of fuel in a wide variety of fuel cells in modern times, but with different catalysts and reaction conditions, many organic compounds can be used as fuel for fuel cells, such as methanol, natural gas or gasoline, and even biomass. In 2014, American scientists reported on Nature Communications, a journal of Nature, the development of a low-temperature fuel cell directly based on biomass, which uses only solar or waste heat to make straw, sawdust, algae and even organic fertilizers are converted into electricity, and the energy density is nearly 100 times higher than that of cellulose-based microbial fuel cells. Using electricity from renewable sources such as solar energy and wind in nature, water can be electrolyzed to produce hydrogen, which is then supplied to fuel cells. So, water can also be seen as an unmodified fuel, through the water cycle, the entire energy supply chain can achieve complete zero emissions, the real realization of energy cleanliness.

Fuel cell technology is still a developing technology in the rising period, and some shortcomings are obstacles that cannot be avoided by current technology promotion. First of all, the life of fuel cells needs to be strengthened. The latest development in China is a breakthrough of 5,000 h. The life expectancy of foreign advanced technology is only about 8,000 h. This life level together with the expensive cost has formed a roadblock for fuel cell technology promotion. Secondly, the fuel cell startup performance is not good enough, and the starting speed is not as good as the internal combustion engine. Reactivity can be achieved by increasing electrode activity, increasing operating temperature, and reaction control parameters, but improving stability must avoid the occurrence of side effects. Reactivity and stability are often not achieved like not having fish and bear's paw at the same time. Once again, it is subject to hydrogen storage technology. The hydrogen fuel of FCV (fuel cell vehicle) is mainly compressed hydrogen, and the carrying capacity of the vehicle body is limited. The filling amount is only 2.5–3.5 kg each time, which is not enough to meet the endurance of 480–650 km in one-way running of automobiles nowadays. The liquid hydrogen system that maintains hydrogen at $-253\ ^\circ$C has been tested successfully, but has major drawbacks: about 1/3 of the electrical energy must be used to maintain the low temperature of the tank, keep the hydrogen in a liquid state, and the hydrogen lost from the gap due to evaporation is 5% of the total stock. Finally, the lack of hydrogen fuel infrastructure is also an important factor for the rapid promotion of fuel cells. Although hydrogen has been used for many years in the industry and has an economic scale, there are only more than 300 hydrogen charging stations in the world, and it is completely incomparable with the number of gas stations.

In recent years, developed countries have proposed various solutions for reducing greenhouse gas emissions. Among them, the development of fuel cells has become a consensus. Many countries have focused on the development of large-scale fuel cells as a key research project, and the corporate sector has also invested heavily in the research and development of fuel cell technology, and has achieved many important results, making fuel cells soon replace traditional generators and internal combustion engines and widely used in power generation and automobiles. Nowadays, in North America, Japan and Europe, fuel cell power generation is rapidly entering the stage of industrial scale application with the momentum of catching up, and will become the fourth-generation power generation mode after the thermal power, hydropower and nuclear power in the 21st century. It is not difficult to imagine that with the gradual improvement of the hydrogen energy industry chain, the realization of the widespread application of fuel cell technology will be just within reach.

4.2 Energy Storage Technology

The development of innovative energy storage technology plays an important role in accelerating the development of new energy industries. With the extensive access of renewable energy and distributed energy in the large power grid, as well as the popularization and application of micro-grid and electric vehicles, energy storage technology will become the key node of the whole energy Internet. At present, energy storage methods are mainly divided into physical energy storage (such as pumped energy storage, compressed air energy storage, flywheel energy storage, etc.), chemical energy storage (such as lead-acid batteries, redox liquid flow batteries, sodium sulfur batteries, lithium-ion batteries) and electromagnetic energy storage (such as superconducting electromagnetic energy storage, supercapacitor energy storage, etc.). According to the characteristics of various energy storage technologies, flywheel energy storage, superconducting electromagnetic energy storage and supercapacitor storage can be suitable for applications that require short-time and large pulse power, while electrochemical energy storage and compressed air storage are suitable for large-scale, high-capacity applications such as regulating system peak and merging renewable energy.

4.2.1 Energy Storage Battery

The rise of the energy Internet will significantly boost demand for energy storage. For all kinds of energy storage modes, chemical energy storage has obvious advantages in energy density, response speed, energy loss and safety, so it is widely recognized by the industry to convert energy into chemical energy for storage. Among chemical energy storage technology, it is also represented by electrochemical energy storage technology. Electrochemical energy storage has the advantages of flexible configuration of power and energy according to different application requirements, fast response speed, not limited by external conditions such as geographical resources, suitable for mass production and large-scale application, and has broad development prospects in power storage. Electrochemical energy storage technology has been used in both peak shaving and valley filling, improving the stability of power grids and micro-network renewable power generation. Electrochemical energy storage mainly includes liquid flow batteries, lithium-ion batteries, sodium sulfur batteries (Zebra batteries), lead acid (carbon) batteries, nickel metal hydride batteries, liquid metal batteries and fuel cells.

At present, lithium-ion batteries are the most widely used, but it is necessary to further improve the safety performance of batteries and reduce battery costs; lead-carbon batteries are expected to become an important technology in the development of large-scale energy storage systems, but their manufacturing technology and cathode hydrogen evolution problems still need to be further optimized and improved; the liquid flow battery has a wide range of applications in improving the grid connection rate of renewable energy and balancing the stability of the power grid. However, it is necessary to further optimize key materials and reduce costs. The manufacturing cost and safety performance of sodium-sulfur batteries still need to be studied intensively. More efforts of the research and development of Zebra battery are needed to realize the localization of battery as soon as possible; liquid metal battery is a new cheap and efficient battery system, which has low energy storage cost and long service life. With the breakthrough of technology, it is expected to be better in the field of energy storage in the future application; Since hydrogen is the cleanest energy source, fuel cell is the cleanest energy conversion device. The combination of fuel cell and clean hydrogen production technology will become the most competitive energy supply technology in the future. The future energy sector will form a situation dominated by hydrogen fuel cells.

1. Lithium-ion battery

A lithium-ion battery is a secondary battery (rechargeable battery) that relies mainly on lithium ions (Li^+) to move between the positive and negative electrodes. During the charge and discharge process, Li^+ is intercalated and deintercalated between the two electrodes: Li^+ is deintercalated from the positive electrode during charging, and the negative electrode is in a lithium-rich state through the electrolyte embedded in the negative electrode; the opposite is true when discharging.

According to the different electrolyte materials, lithium-ion batteries can be divided into two categories, liquid lithium-ion batteries (LIB) and polymer lithium-ion batteries (LIP). The difference between the two is mainly due to the difference of electrolytes. For liquid lithium ion batteries, if electrolyte is liquid substances, it is called liquid lithium-ion battery; if electrolyte is an organic polymer, it is called a polymer lithium ion battery, and polymer lithium-ion battery is currently the most common battery. The polymer lithium ion battery can be further divided into a solid polymer electrolyte lithium ion battery, a gel polymer electrolyte lithium ion battery, and the like. Since the solid electrolyte is used instead of the liquid electrolyte, the polymer lithium ion battery has advantages of being thinner, arbitrary area, and arbitrary shape as compared with the liquid lithium ion battery.

Based on the different cathode materials, lithium-ion batteries can be divided into lithium cobalt acid, lithium manganese acid, lithium nickel acid, ternary materials, lithium iron phosphate and other battery types. According to the different anode materials, lithium-ion batteries can be divided into graphite, soft carbon, hard carbon, lithium titanate, silicon base and other battery types.

1) Lithium iron phosphate battery

Lithium iron phosphate battery is a lithium-ion battery that has developed rapidly in recent years, the main applications include large electric vehicles, light electric vehicles, power tools, remote control toys, solar and wind power storage equipment, UPS (uninterruptible power supply) and special lamps, small medical equipment and portable instruments. Lithium iron phosphate battery has the advantages of good output performance, good safety performance, good cycle life, good charging performance, high energy density, high environmental protection evaluation and good economic benefits. However, there are still some disadvantages to be overcome. During the sintering process of preparing lithium iron phosphate, iron oxide may be reduced to elemental iron under high temperature reducing, while elemental iron may cause micro short circuit of the battery; lithium iron phosphate vibration density and compaction density is very low, resulting in the battery energy density is difficult to improve; the process cost of preparing lithium iron phosphate product is relatively high; the product consistency is poor, and the synthesis reaction of lithium iron phosphate is a complex multiphase reaction, including solid phase phosphate, the oxide of iron as well as the lithium salt, the precursor of carbon, and the reducing gas phase, it is difficult to ensure uniformity of the reaction.

2) Lithium cobalt acid battery

Lithium cobalt acid battery is a lithium battery with superior electrochemical properties, which has the characteristics of stable structure, high specific capacity and outstanding comprehensive performance. However, the main limitation is that the cost is very high, because cobalt is a scarce strategic metal. The application of lithium cobalt oxide battery is relatively small, mainly used for small and medium-sized batteries, as well as small electronic devices such as notebook computers and mobile phones, with a nominal voltage of 3.7 V.

3) Ternary lithium battery

The ternary lithium battery refers to a lithium ion battery using a nickel-cobalt-manganese ternary positive electrode material as a positive electrode material. Compared with lithium iron phosphate batteries, ternary lithium batteries have better overall performance, higher energy density, and higher volumetric energy. The biggest advantage of the ternary lithium battery is the high energy density of the battery. The storage density is usually above 200 W-h/kg, which is greater than 90–120 W-h/kg of lithium iron phosphate. It is more conducive to lightweight design and more suitable for the current new energy passenger car market demand for cruising mileage. However, the ternary lithium battery also has shortcomings. The decomposition temperature of the ternary lithium battery is between 250 and 350 ° C. At this time, the internal chemical components begin to decompose, releasing oxygen molecules, and the electrolyte will burn rapidly under the action of high temperature, increasing the risk of spontaneous combustion and explosion of the battery, therefore, the ternary lithium battery needs to do more work of adequate protection, overcharged, over-discharged, over-temperature protected and over-current protected during the design process. With the advancement of technology, especially the use of ceramic diaphragms, the safety of ternary lithium batteries has been improved, and due to its excellent comprehensive performance, it has been widely used in the market. The ternary lithium battery surpassed the original one of the original lithium iron phosphate batteries in the field of power batteries. In the first eight months of 2017, the ternary/manganese lithium battery installed capacity of 6.4 GW was the largest, accounting for 55%, and the installed capacity of lithium iron phosphate battery was second, with 5.1 GW, accounting for 44.1%.

Lithium-ion battery has many outstanding advantages, such as high operating voltage, large capacity, small self-discharge and good circulation, and is the representative of modern high-performance battery. In recent years, the production and sales of new energy vehicles have shown a blowout growth, with power batteries as part or all of the power of electric vehicles, because of the significant advantages of efficient energy saving and off-site emissions, is currently the main direction of the development of new energy vehicles. China Automotive Engineering Society published the "Energy Saving and New Energy Vehicle Technology Roadmap" for China's power battery technology has drawn a blueprint for development. The roadmap proposes that by 2020, the power cell monomer of a pure electric vehicle would reach 350, 400 W-h/kg in 2025 and 500 W-h/kg in 2030. The existing system of lithium-ion power battery technology will be optimized in the near middle term, to meet the needs of large-scale development of new energy vehicle, focusing on the development of new lithium-ion power batteries, at the same time, developing a new system of power battery forward-looking research and development; in the medium and long term, while continuously optimizing and upgrading the new lithium-ion power battery, reduce costs significantly and realize the new system power battery practical and large-scale application. It can be seen that lithium-ion batteries will remain the mainstream products of power batteries for a considerable period of time in the future.

Lithium-ion batteries also have extraordinary performance in the field of energy storage. Since 2010, China has established lithium-ion battery energy storage systems in Anxi, Ningde, Zhengzhou in Henan, Dongguan in Guangdong and Changzhou in Jiangsu Province. It has been successfully used to cut peaks and fill valleys and improve the ability of the grid to accept wind power. Lithium-ion battery energy storage systems of different sizes have also been established in Washington, California, New York, and Michigan to cut peaks and fill valleys, improve grid reliability, and realize microgrid renewable power generation. Countries such as Chile and South Korea also use lithium-ion battery energy storage technology for grid frequency modulation and improved power quality.

In the future development of lithium-ion batteries, it is necessary to further improve the specific capacity, cycle life and safety performance, while reducing product costs. (1) The development of anode materials that can output higher voltages is one of the important ways to improve the energy density of materials. The significant advantage of high voltages is that when the battery is assembled into a group, the rated output voltage can be reached using smaller monomer batteries in series, which simplifies the control unit of the battery pack. The most typical high-voltage characteristic material is $LiNi_{0.5}Mn_{1.5}O_4$, whose derivative materials include $LiTiMnO_4$, $LiCoMnO_4$, and olivine structure phosphate/fluorine phosphate, such as $LiCoPO_4$, $LiNiPO_4$, $LiVPO_4F$, etc. (2) Improving the specific capacity of anode materials is also an important means to improve battery quality. Because the specific capacity of the cathode material of the lithium-ion battery is much higher than that of the anode material, the anode material has a greater effect on the energy density of the whole battery. At the current level, if the specific capacity of the anode material is doubled, the energy density of the whole battery can be increased by 57%. Among nickel-cobalt-manganese ternary material, nickel is the main active element, and the higher the content, the larger the material capacity. A development direction high-capacity anode material is the development of high nickel ternary or multi-system. In high nickel multi-component systems, multiple materials (NCA or NCM811) with nickel content above 80% have the obvious energy density advantage. A battery made of these materials matching to a suitable high-capacity negative electrode and electrolyte, its energy density can reach 300 W-h/kg or more. At present, the high-nickel multi-material mass production technology is mainly controlled by a few Japanese and Korean companies. In addition, the lithium-rich material $[xLi_2MnO_3{\cdot}(1-x)LiMO_2(M = Fe, Co, Mn, Ni)]$ has a specific discharge capacity of more than 230 mAh/g, an operating voltage is more than 3.5 v, and has a high specific capacity and thermal stability. It has good development potential, good cycle performance, wide charging and discharging voltage range, low price and environmental friendliness. (3) The cathode material of lithium-ion battery will continue to develop in the direction of low cost, high specific energy and high safety, and graphite materials (including artificial graphite, natural graphite and intermediate phase carbon microspheres) are still the mainstream choice of lithium-ion power battery at present. New high-capacity cathode materials such as silicon base will gradually mature, and high-power density and high-safety cathode materials represented by lithium titanate will be more widely used in hybrid electric vehicles and other fields. The medium and long-term silicon-based cathode materials will completely replace other cathode materials and have become the industry consensus. The silicon-based cathode material is considered to be one of the best choices for greatly increasing the energy density of lithium batteries, and its theoretical specific capacity can reach 4,000 mAh/g or more.

Lithium batteries still have certain problems in terms of battery life and safety, and all solid-state batteries are expected to solve these problems better. The all-solid-state battery is designed with no electrolyte, and its energy density, charging rate, safety and production cost will be significantly ahead of existing lithium-ion batteries. The energy density of solid-state batteries can reach twice that of existing lithium batteries, meaning solid-state batteries of the same size and quality will achieve twice the cruising mileage of current lithium-electric models, and the fast-charging technology of all-solid-state batteries is also worth looking forward to.

2. Lead carbon battery

Lead carbon battery is a capacitive lead-acid battery that has been improved from the traditional lead-acid battery. Lead-acid batteries are rich in raw materials, low cost and good safe, but the cycle life is short. It is found that the addition of activated carbon to the cathode of lead-acid batteries can significantly improve the performance of lead-acid batteries, such as 8 times higher charging speed, 3 times higher discharge power and 6 times longer cycle life. New energy and energy conservation areas.

The basic principle of lead carbon batteries is to incorporate high specific surface carbon materials (such as activated carbon, activated carbon fibers, carbon aerogels or carbon nanotubes) into the lead anode to achieve high conductivity of high surface carbon materials and the dispersion of the active material, to increase the utilization rate of the lead active material and to inhibit the growth of lead sulfate crystals. As a new type of super battery, lead carbon battery combines the technology of lead-acid battery and supercapacitor, which is a dual-function energy storage battery with both capacitive and battery characteristics, which not only gives full play to the advantages of super capacitor instantaneous power high capacity charging, but also plays the energy advantage of lead-acid battery, and has good charge and discharge performance.

The rapid development of new energy has also put forward new requirements for battery technology. Lead carbon battery is the most advanced technology in the field of lead-acid batteries, its high power density, long cycle life and low price and other characteristics make it become the focus of international new energy storage industry development, which has a very broad application prospects in photovoltaic power plant energy storage, wind power storage and power grid peak regulation and other energy storage areas. At present, lead carbon battery energy storage technology has a wide range of demonstration applications at home and abroad. Around 2011, the United States established a power grid-level lead carbon battery energy storage project with a capacity of 3 MW/1–4 MW in North America, and a lead carbon battery energy storage system with a capacity of 15 MW/10 MW in Oahu and Maui, Hawaii, respectively, and a 10 MW/20 MWh lead-carbon battery wind energy storage system for grid-assisted energy storage, frequency regulation and energy demand management; Australia is operating lead carbon battery energy storage system at 3 MW/1.6 MWh in the King Island project to ensure that new energy is connected to the power grid; energy storage systems with lead carbon battery technology as the core in some regions of Colombia and Antarctica; 14 micro-scales net energy storage project also uses lead-carbon battery energy storage technology in Hebei, Qinghai, Tibet, Zhejiang, etc. of China.

3. Liquid flow battery

Liquid flow battery, commonly referred to as redox liquid flow battery, is a new high-performance large-scale electrochemical energy storage device which is stored separately by positive and negative electrode electrolyte. The active substance is present in the electrolyte, which realizes the spatial isolation of the electrode from the active material. Battery power is determined by the size of the electrode and the number of batteries in the reactor, while the battery capacity is determined by the concentration and volume of the electrolyte, so the battery power and capacity can be designed separately, flexible and convenient. The structure principle is shown in Fig. 4.7.

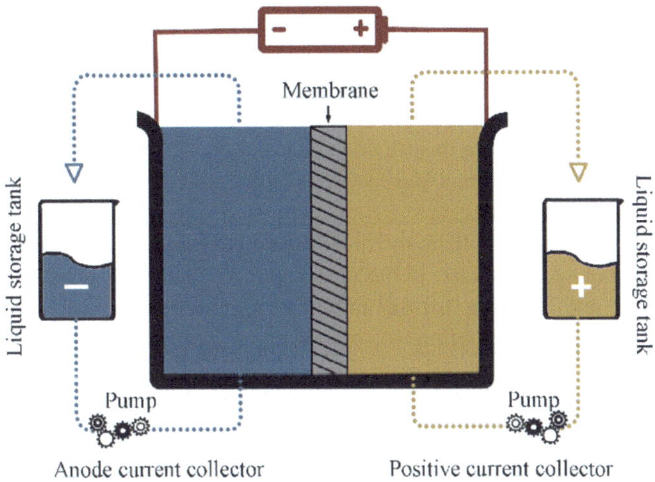

Fig. 4.7 Diagram of redox liquid flow battery

The liquid flow battery was first designed by NASA and was patented and published by Thaller L. H. in 1974. Professor Maria of Australia studied the use of vanadium sulfate as a positive and negative electrode electrolyte and applied for a patent in 1986. At present, the more mature liquid flow battery system includes iron chromium system, iron and titanium system, vanadium bromine system and total vanadium system. Among them, the positive and negative electrode active materials of all vanadium redox liquid flow batteries are vanadium, which can avoid the cross-contamination of elements caused by the diffusion of active substances through the ion exchange membrane, and is currently the main commercial development technology direction.

The all-vanadium liquid flow battery is used as an energy storage system and has the following characteristics.

(1) The output power of the battery depends on the size of the battery stack. The energy storage capacity depends on the volume and concentration of the electrolyte, so its design is very flexible, when the output power is certain, in order to increase the energy storage capacity, as long as the electrolyte storage tank volume or the concentration of electrolyte are increased.
(2) The active material of the vanadium battery is present in liquid, and electrolyte ion is only one type of vanadium ion, so there is no phase change in other batteries during charging and discharging, and the battery has a long service life.
(3) It is with good charge and discharge performance, can be deeply discharged without damaging the battery.
(4) The self-discharge is low, and the electrolyte in the storage tank has no self-discharge phenomenon when the system is in the off mode.

(5) The vanadium battery has a large degree of freedom in site selection, and the system can be fully closed and operated, with features of pollution free, simple maintenance and low operating cost.
(6) The battery system has no potential explosion or fire hazard and is highly safe.
(7) Battery components are mostly cheap carbon materials, engineering plastics. The materials are abundant and easy to recycle, and no precious metal is needed as an electrode catalyst.
(8) Energy efficiency is high, up to 75–80%, and very high cost performance.
(9) The starting speed is fast. If the electrolyte is filled with electrolyte, it can be started within 2 min. It takes only 0.02 s to switch between charging and discharging during operation.

As a newer type of battery, the total vanadium liquid flow battery is still imperfect, especially the energy density is very low. The current advanced full vanadium liquid flow battery energy density is only about 40 W-h/kg, only close to the lead-acid battery of 35 W-h/kg; and secondly, the working conditions are stricter, must be limited to the temperature range between the 5 and 45 °C.

Liquid flow batteries will play an important role in improving the access of renewable energy into the network and balancing the stability of the power grid, which has received extensive attention at home and abroad. At present, the main problems facing the development of liquid flow battery are to develop high performance electrolyte, optimize diaphragm and plate material, further reduce cost and improve performance, so as to better promote its industrialization development. The United States has adopted liquid flow battery as an important development direction of energy storage technology in the Energy Storage Development Plan in 2011, and funded the establishment of 12 liquid-flow battery energy storage system in the following year. In 2015, Canada launched 4 liquid-flow battery energy storage projects in Ontario Province; China established the world's largest 5 MWh/10 MWh Woniushi wind farm full vanadium liquid energy storage system in Liaoning, in 2012, which is the first to achieve the demonstration application of this technology at home and abroad, and the full vanadium liquid flow battery national energy storage demonstration project of 200 MWh/800 MWh is under construction.

4. Sodium battery

1) Sodium sulfur battery

Sodium sulfur batteries were invented by Ford in the United States in 1967. The positive and negative electrode materials of sodium sulfur battery are molten sulfur and sodium respectively, the electrolyte is alumina ceramic tube, and the working temperature is 300–350 °C. The principle of sodium sulfur batteries is that sodium ions react reversible with sulfur permeating through the electrolyte diaphragm, thus releasing and storing energy.

Sodium sulfur batteries have the following remarkable characteristics: the specific energy is higher (that is, the effective electrical energy of the unit mass or unit volume of the battery), the theoretical specific energy is 760 W-h/kg, the actual value is more than 150 W-h/kg, which is 3–4 times of the lead-acid battery; It can discharge large current and high power, its discharge current density can generally reach 200–300 mA/cm^2, and can instantly release 3 times of its inherent energy; The charging and discharging efficiency is high, due to the use of solid electrolytes, there is no self-discharge and side reaction of the liquid electrolyte secondary battery, and the charge and discharge current efficiency is almost 100%. At the same time, sodium-sulfur batteries also have shortcomings, the working temperature is 300–350 °C, so the battery needs certain heating and heat preservation when working. At present, high-performance vacuum insulation technology can be used to solve this problem.

Sodium sulfur batteries are emerging chemical energy batteries, and because of their high energy capacity, many countries use them on electric bicycles or electric vehicles. At the same time, due to some limitations of the battery itself, sodium sulfur batteries are also commonly used in the energy storage aspects of power stations to give full play to its own advantages. Japan's NGK Corporation, the world's only supplier of sodium and sulfur batteries, worked with Tokyo Electric Power Co., Ltd. to develop the application of sodium sulfur batteries for energy storage in the 1980s, and in the late 1990s, it successfully developed a megawatt-hour level energy storage system, mainly for peak shaving and valley filling, auxiliary standby and stable power grids. In 2002, the United States built a 100-kW/500 kVA demonstration power plant in Ohio using the sodium sulfur battery provided by NGK company. In 2006, a sodium sulfur energy storage power station was established in West Virginia State, which successfully ensured the supply of electricity to the residents in the surrounding area. In China, sodium sulfur batteries have also received more and more attention. At present, Shanghai Electric Power Company and Chinese Academy of Sciences have carried out a series of studies on the scalable preparation and consistency control of β-alumina ceramic tube electrolytes, and have successfully developed 650 A-h cell monomers and established 2 MW battery monomer pilot assembly line. The sodium sulfur battery energy storage system at 100 kW-h/800 kW-h was successfully applied to the smart Grid project of Shanghai World Expo Park in 2010. In 2013, Shanghai Electric Power Company successively passed the acceptance of three sodium-sulfur battery energy storage projects. However, sodium-sulfur batteries have higher manufacturing costs, poor rate performance, limited practical lifespan, and large safety hazards, which severely limit their application in energy storage systems.

2) Sodium-nickel chloride (Zebra) battery

Sodium-nickel chloride (Zebra) batteries are a new type of high-energy battery composed of alumina solid electrolytes with sodium ion conduction, and Zebra batteries are developed on the basis of the development of sodium sulfur batteries (NaS batteries). Zebra is an abbreviation for zero emission battery research activity, which represents a zero-emission, pollution-free green power supply.

Zebra batteries are very similar to the structure of sodium sulfur batteries, the battery cathode active substances are metal sodium, solid electrolytes are β-Al_2O_3 ceramic materials, but the difference is Zebra battery with solid porous nickel dichloride ($NiCl_2$) or the like as a positive electrode and a liquid $NaAlCl_4$ as a secondary electrolyte. Zebra Battery has an energy density of 100 W-h/kg, with a long service life, low energy storage costs, good resistance to overcharge and overdischarge characteristics, high safety performance.

Zebra battery reaction is: $2Na + NiCl_2 \rightarrow 2NaCl + Ni$

According to the reaction, the electrode reaction process of Zebra battery is: in the discharge, the electron loads from the sodium cathode to nickel dichloride anode through the external circuit, while the sodium ion reacts with nickel dichloride through the β-Al_2O_3 solid electrolyte ceramic tube to produce sodium chloride and nickel; when charging, the electrode process under the action of external power supply is exactly the opposite of the discharge.

Zebra batteries have similar advantages of NaS batteries, such as high open circuit voltage (2.58 V at 300 °C), high specific energy [theoretical specific energy is 790 W-h/kg, actually up to 100 W-h/kg), high specific power [it reaches 150 W-h/kg at 80% (DOD)], high energy conversion efficiency (with 100% Coulomb efficiency, no self-discharge), independent capacity and discharge rate (the internal resistance of the battery is basically ohmic resistance)], quick charging (battery can reach 50% discharge capacity after 30 min), long cycle life (storage life is more than 5 years, charge and discharge cycle life is more than 1000 cycles), maintenance-free (full seal structure, no outside environmental temperature effects) and so on.

Zebra batteries also have their own unique advantages, as follows.

(1) There is no liquid sodium in the battery preparation process, and the operation is relatively simple and safe. Since the preparation is usually in a discharge state, that is, nickel and sodium chloride (NaCl) are used as a positive electrode material, sodium metal is generated at the negative electrode by the first charge.
(2) The battery connection can be arranged in any way such as in series and in parallel arrangement combination. Even when a small amount of battery damage occurs inside the battery pack (generally less than 5% of the total number of batteries), it can continue to work without replacement.
(3) The battery can withstand repeated cycles of heat and cold. According to reports, the battery showed no signs of capacity and life decline after 100 cycles of thermal cycling. Since the nickel-based mixture of the positive electrode has a higher coefficient of thermal expansion than the β-Al_2O_3 ceramic tube, it shrinks away from the β-Al_2O_3 tube upon cooling and solidification, and there is no problem of stress generation.
(4) The battery's corrosion resistance is correspondingly enhanced. Because the positive electrode active material with relatively strong corrosivity in the battery structure is placed in the β-Al_2O_3 ceramic tube, the anti-corrosion requirement for the battery metal shell material is reduced, the range of material selection is expanded, and the battery manufacturing cost is also reduced.
(5) The battery has a relatively wide operating temperature range. The melting point of $NaAlCl_4$ is 157 °C, and the temperature at which the solid eutectic point of solid NaCl and $NiCl_2$ is 570 °C. In theory, Zebra battery operating temperature can be in the range of 157–570 °C, taking into account the actual effective power

output of the battery, it is generally considered that the 270–350 °C temperature range is more suitable for Zebra batteries.

Zebra batteries have broad prospects for development, in addition to being used in electric vehicles, they are also very suitable for energy storage batteries and so on. Zebra batteries have been successfully used in cars such as Mercedes-Benz, BMW and Clio. MAS-DEA in Switzerland and GE in the US have already implemented the Zebra battery industrial application of tubular design. Zebra batteries are also good in communication and military application prospects. In the future, it is necessary to further enhance the energy density and power density of Zebra batteries, and promote the localization and commercial application of Zebra batteries.

4.2.2 Compressed Air Energy Storage

Renewable energy limits its own development to a large extent due to its intermittent and unstable nature. An effective way to solve this problem is to develop large-scale energy storage systems. Energy storage systems that have been operating in large-scale (over 100 MW) commercial systems to date include both pumping stations and compressed air. Compressed air energy storage refers to the use of electric energy for compressed air energy storing mode during low load period of power grid, sealing high air pressure in abandoned mines, subsided submarine gas storage tanks, caves, abandoned oil and gas wells or new gas storage wells, releasing compressed air to drive steam turbines for electricity during peak load period. Compressed air energy storage has the characteristics of large scale, high efficiency and high reliability, and is considered to be the most energy storage mode with market potential.

1. Principle of Compressed air energy storage

The compressed air energy storage system mainly includes two parts: a surface energy storage power station and an underground air storage gas storage. The underground air storage is an important part of the compressed air energy storage system and the key to ensuring its operational performance and reliability. Underground air storage reservoirs generally utilize geological structures which have been developed by gas and oil wells, underground aquifers, salt karst caves that have been mined or specially drilled, and underground spaces such as underground caverns that are manually excavated in hard rock. Among them, the common forms are salt karst caverns and hard rock caverns. The basic working principle of the compressed air energy storage system is derived from the gas turbine system (Figs. 4.8 and 4.9). The difference is that the compressor of the gas turbine works simultaneously with the turbine, and the compressor consumes part of the turbine work to compress the air. The compressed air energy storage system is divided into two processes, energy storage and energy release. When in the low electricity consumption, excess power (from thermal power plants, nuclear power plants or renewable energy power stations) is used to drive the compressor to generate high-pressure air and store energy; when

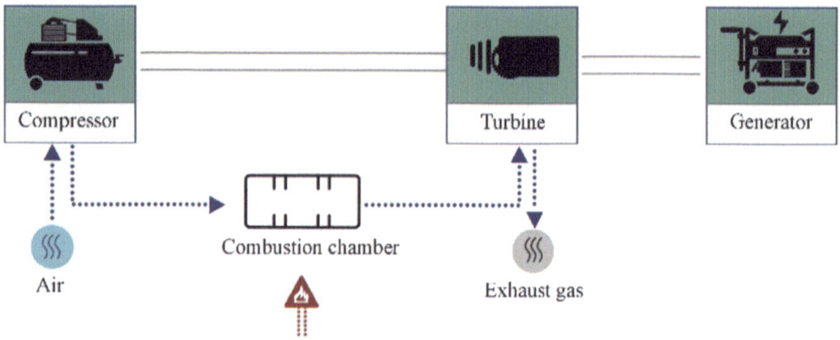

Fig 4.8 Diagram of gas turbine system

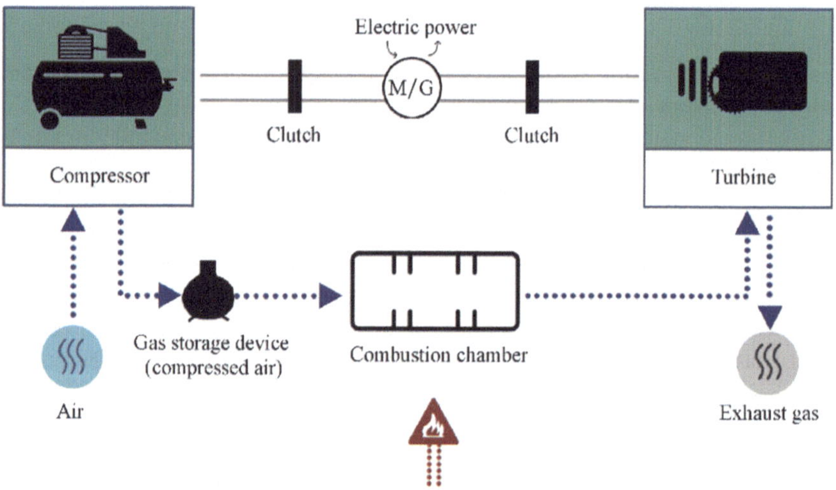

Fig. 4.9 Diagram of compressed air energy storage system

at the peak of power consumption, compressed air obtains thermal energy through the combustion chamber, and then enters the turbine to perform work to generate electricity. Due to the time-sharing work during energy storage and energy release, no compressor consumes turbine work during the energy output process. Therefore, at the peak of power consumption, the compressed air energy storage station can provide more work than the gas turbine system that consumes the same fuel.

The compressed air energy storage system has six main components: (1) motor/generator (connecting compressor and turbine through clutch); (2) multistage compressor (isothermal compression or multistage compression intermediate cooling); (3) Multi-stage turboexpander re-heating equipment; (4) control system (controlling power plant conversion energy storage and energy release work mode, etc.);

(5) auxiliary equipment (such as fuel tanks, cooling systems, mechanical transmission systems and heat exchangers, etc.); (6) underground or surface gas storage units (including some piping and accessories, etc.).

Compared with other energy storage technologies, the compressed air energy storage system has the characteristics of fast off/start speed (±27% of maximum load per minute), large capacity, long working time, good economic performance and many charging and discharging cycles. They include: (1) Compressed air energy storage system can build large power station (>100 MW), second only to pumping station; compressed air energy storage system can continue to work for hours or even days, long working hours. (2) The construction cost and operating cost of compressed air energy storage system are relatively low, much lower than that of sodium sulfur battery or liquid flow battery, and also lower than that of pumped energy storage station, which is very economical. (3) Compressed air energy storage system has a long service life, and can store and release tens of thousands of times, service life can be up to 40–50 years, and its efficiency can reach about 70%, which is close to pumped energy storage power station.

According to the heat source of the compressed air energy storage system, it can be divided into a traditional compressed air energy storage system and a compressed air energy storage system with heat storage.

Traditional compressed air energy storage power plants need to burn fossil fuels. The basic working principle is shown in Fig. 4.10. The system mainly includes underground salt cavern gas storage, heat exchange system, air compression system, motor/generator and turbine, etc. The above structural principles are used in the Huntorf power station in Germany and the McIntosh power station in the United

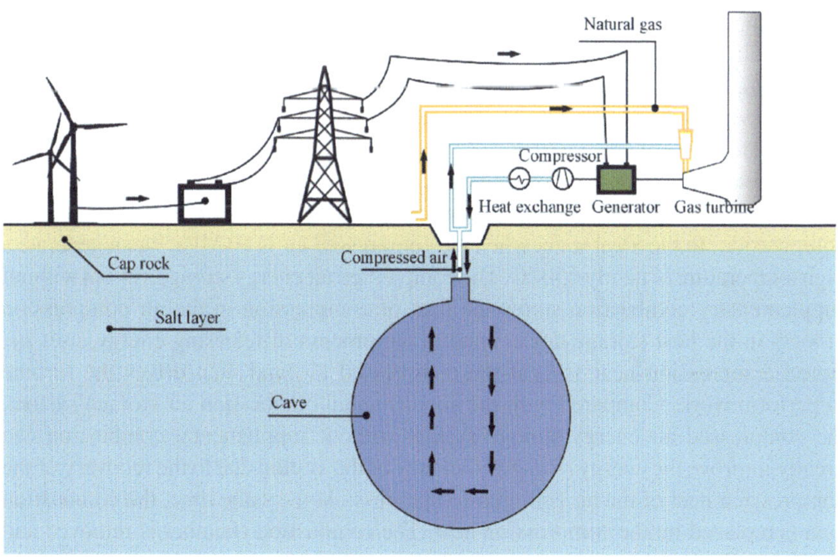

Fig. 4.10 Compressed air energy storage system with supplementary combustion

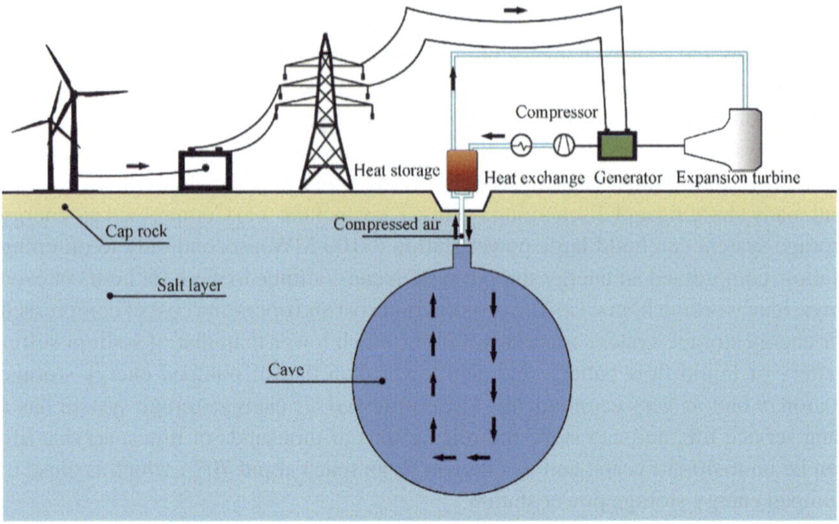

Fig. 4.11 Compressed air energy storage system without supplementary combustion

States. In order to reduce the pollution discharge of high-pressure combustion, for the multi-stage expansion turbine, the combustion chamber can also be installed after the first-stage expansion, and the waste heat of the exhaust gas can be used to heat the initial compressed air, and the temperature can also reach 400–600 °C. This can reduce the generation of pollutants, while making full use of waste heat, reducing system fuel consumption and improving system efficiency.

A compressed air storage system with heat storage, or a non-combustion compressed air energy storage system (Fig. 4.11), also known as the Advance Adiabatic Compressed Air Energy Storage System (AACAES), the heat storage material should have a large specific heat capacity, a wide temperature range, environmentally friendly and other characteristics.

The compression process of air in a compressed air energy storage system with heat storage is close to the adiabatic process, producing a large amount of heat of compression. In the ideal state, when the compressed air is 10 MPa, it can generate a high temperature of nearly 650 °C. The compressed air energy storage system without supplementary combustion stores the heat of compression in the air compression process in the heat storage device, and in the process of releasing energy, uses the stored compression heat to heat the compressed air, and then drives the turbine to perform work. Compared with the supplemental combustion air storage system, the compressed air energy storage system without supplemental combustion can greatly improve the energy storage efficiency of the system due to the recovery of the compression heat of the air compression process. At the same time, the combustion heat is replaced by the compression heat. The combustion chamber is removed and zero emission requirements are met. The main disadvantage of this system is that the initial investment cost of the system will increase by 20–30% compared to the traditional compressed air energy storage station due to the addition of heat storage devices.

2. Development status

In the 1949, Stal Laval proposed the use of compressed air energy storage. At present, there are ten compressed air energy storage power stations in operation or under construction worldwide (five in operation and five in construction) with an installed capacity of 1.26 GW (435 MW of operation and 825 MW under construction) (Table 4.1).

Two large-scale compressed air energy storage stations have been put into commercial operation, located in Germany and the United States.

The first compressed air energy storage device was the Huntorf power station in German, which was put into commercial operation in 1978, located 40-km northwest of Bremen. It is still in operation and is the world's largest capacity of compressed air energy storage power station. The compressor power of the unit is 60 MW and the output power of the release energy is 290 MW. The system stores compressed air in an abandoned mine of 600–800 m underground. The total volume of the mine is 310,000 m^3, and the pressure of compressed air is up to 10 MPa. The unit can be continuously inflated for 8 h and continuously generate electricity for 2 h. The power station was launched more than 5,000 times during the period from 1979 to 1991, with an average starting reliability of 97.6% and the actual operating efficiency is about 42%.

The second is the McIntosh compressed air energy storage power station in Alabama, USA, which was put into commercial operation in 1991. The energy storage unit's compressor unit has a power of 50 MW and a generating capacity of 110 MW. The gas storage cave is 450 m underground with a total volume of 560,000 m^3 and a compressed air storage pressure of 7.5 MPa. It can achieve continuous 41 h of air compression and 26 h of power generation, and the unit takes about 9 min from start to full load. The power station is controlled remotely by energy control center of the Alabama Power Company, and the actual operating efficiency is about 54%. Ohio has been building a 2,700 MW large compressed air energy storage commercial power station since 2001. It consists of nine 300 MW units. The compressed air is stored in a cavern of 670 m underground, and the volume of the gas storage cave is 9.57 million m^3.

In 2001, Japan launched the Kamisunagawa compressed air energy storage demonstration project in Sorachi, Hokkaido, with an output of 2 MW. It is an industrial test intermediate unit for the development of 400 MW units in Japan. It uses abandoned coal mine pits (about 450 m underground) as gas storage caves with a maximum pressure of 8 MPa.

ABB of Switzerland (now integrated into Alstom) is developing a combined cycle compressed air energy storage system. The energy storage system has a power generation capacity of 422 MW, an air pressure of 3.3 MPa, and a system inflation time of 8 h. The gas storage cave is a water sealed, hard rock geology unit.

Table 4.1 Development status of compressed air energy storage in countries around the world

Name	Country	Capacity (MWh)	Power scale (MW)	Gas storage device	Efficiency (%)	Application
Huntorf	Germany	580	290	Mine cave of 310,000 m^3	42	Peak shaving, frequency modulation, spinning reserve, black start
McIntosh	US	2860	110	Salt layer of 283,000 m^3	54	Peak shaving, frequency modulation, spinning reserve
PGECAU	US	3000	300	Underground storage tank	Confidential	Peak shaving, frequency modulation, spinning reserve, smoothing renewable energy
ATK	US	0.06	0.08	Gas storage tank	Confidential	Electrical bill management, power supply on the construction site
Texas	US	500	1	Cave	Confidential	Smoothing renewable energy
Apex	US	33,285	317	Cave	Under construction	Smoothing renewable energy, Frequency modulation, black start, climbing support

(continued)

Table 4.1 (continued)

Name	Country	Capacity (MWh)	Power scale (MW)	Gas storage device	Efficiency (%)	Application
SustainX	US	1.5	1.5	Gas storage tank	Under construction	Smoothing renewable energy, climbing support, transmission block relief, spare
NextGen	US	40.5	9	Gas storage tank	Under construction	Peak shaving, frequency modulation, spinning reserve, smoothing renewable energy, black start
Highview	UK	2.45	0.35	Gas storage tank	Under construction	Peak shaving, frequency modulation, spinning reserve, smoothing renewable energy,
Adele	Germany	1000	200	Gas storage tank	Under construction	Peak shaving, power supply

China's research and development of compressed air energy storage systems started late, mostly concentrated on theoretical and small experimental levels. The Institute of Engineering Thermophysics of the Chinese Academy of Sciences is constructing an advanced compressed air energy storage demonstration system. The Jintan salt cave compressed air energy storage power generation national demonstration project is the first salt cave compressed air energy storage project in China. It was included in the National Experimental Demonstration Energy Project by the National Energy Administration in May 2017. The project uses non-combustion compressed air from Tsinghua University. Energy storage technology relies on Jintan underground salt cave resources to build and operate a 60 MW energy storage power generation system with a storage capacity of 300 MWh, an investment scale of CNY 500 million, and a system electricity-electricity conversion efficiency of 61.8%.

3. Prospect of application

Compressed air energy storage system is a mature and feasible energy storage method. It has broad application prospects in the fields of power production, transportation and consumption, including: (1) Peak shaving and valley filling. Power generation companies can use compressed air energy storage systems to store power at low power consumption and release it at peak power consumption to achieve peak-filling. (2) Balance the power load. The compressed air energy storage system can reach the full load working state from start-up within a few minutes, which is much lower than the starting time of the ordinary coal power station, so it is more suitable as an electric load balancing device. (3) Demand side power management. In areas where peak and valley differential electricity prices are implemented, demand-side users can use compressed air energy storage systems to store low-valley and low-cost electricity, and then use them at peak high-priced periods, thereby saving electricity costs and achieving greater economic benefits. (4) Applied to renewable energy. Intermittent renewable energy sources can be "spliced" together with a compressed air energy storage system to create a stable power supply. (5) Backup power supply. The compressed air energy storage system can be built near the power station or the user as a backup power source for line maintenance, failures, or emergencies.

Compressors and expanders are the core components of compressed air energy storage systems, and their performance has a decisive impact on the performance of the entire system. Although the compressed air storage cycle is similar to that of a gas turbine, the gas turbine compressor pressure ratio is typically less than 20, while the compressed air energy storage system requires a compressor pressure ratio of 40–80 or even higher. Therefore, the compressors of large-scale compressed air energy storage power plants generally adopt the structure of multi-stage compression, cascade and post-stage cooling of axial flow and centrifugal compressor. For example, the Huntorf power station uses a compressor of this type, with a compressor cost of about US$170/kW. For compressed air storage systems with heat storage, the pressure ratio is usually higher than that of the traditional large compressed air storage power station, and the heat storage unit needs to be added, so the cost is higher than US$170/kW. Small compressed air energy storage systems require high space flexibility. In order to reduce the volume of the gas storage device, the storage pressure of the air is generally higher. At the same time, due to the small flow rate of the system, single or multi-stage reciprocating compressors are used. (about 20 kW) is more appropriate. Reciprocating compressors can provide pressures up to 300 bars and have an investment cost of US$500 to US$1,500/kW.

Compressed air energy storage system expanders also have a much higher expansion ratio than conventional gas turbines, and thus generally employ a multi-stage expansion plus intermediate reheat configuration. For example, the expander at the Huntorf power station consists of two stages, the first stage expanding from 4 to 11 bars and then fully expanding through the second stage. Due to the high pressure, the first stage turbine cannot directly apply a conventional gas turbine, and the power station uses a modified steam turbine as the first stage turbine. For large power plants, the investment cost of a turbo expander is approximately US$185/kW. Small compressed air energy storage systems can use turbine components of micro gas turbine, reciprocating expanders or screw-type air engines. For example, the Mercuy50 gas turbine, but its turbine needs to work at about 10 bars and 1,150 °C. Therefore, a front turbine was installed in front of the Mercuy50 gas turbine to reduce the air pressure from 80 to 10 bars before entering the gas turbine combustion chamber. The total cost of the part of system for work (combustion chamber, front turbine, turbine, and residual heat exchanger) is approximately US$430/kW. The screw-type air engine technology is relatively mature, and its working pressure is generally lower than 13 bars, but the efficiency is low, and the cost is US$500/kW to US$2,400/kW. Small high-pressure reciprocating expanders are still in the research stage, and there are currently no market-oriented products.

Large compressed air storage systems require large volumes of compressed air, usually in underground salt mines, hard rock caves or porous caves. The two power stations that have been operated by using underground salt caves with a volume of 310,000 m^3 (Huntorf) and 560,000 m^3 (McIntosh) respectively. The daily air leakage is only 1/106 to 1/105. The investment cost is US$1 to US$2/kWh (the energy that can be generated by gas storage). If a newly excavated hard rock cave is used, the investment cost is higher (about US$30/kWh); if the existing cave is rebuilt, the cost can be greatly reduced (about US$10/kWh), but in the process of gas storage, there is a leak problem in caves and cement pipelines. Porous caves such as saline-alkali aquifers have a low investment cost of only US$0.11/kWh, and power stations in Iowa, USA, will use this cave to store gas. For micro-miniature compressed air energy storage systems, the use of surface high pressure gas storage containers can get rid of the dependence on gas storage caves.

Analysis of the compressed air energy storage system shows that the underground air storage is an important part of the system. The advantage of developing this system in oil companies lies in the construction and site selection of underground air storage, mainly reflected in the following three aspects: First, the formation of a certain scale of underground gas storage professional team, with the experience of underground gas storage construction. As of 2017, China has built 25 underground gas storages. Second, there are considerable abandoned oil and gas wells and abandoned coal mines in the oil and gas in mineral rights areas. These abandoned wells can be used to build underground air storage. Third, background of geological theory, drilling and completion, oil and gas field development, oil and gas storage and transportation, etc., can carry out research on site selection, injection and integrity of underground compressed gas storage.

4.3 Graphene Materials

Graphene is a two-dimensional carbon nanomaterial composed of six angular honeycomb lattices by carbon atoms in a certain structural form. Graphene has excellent optical, electrical, mechanical and other characteristics, and has important application prospects in materials science, new energy and other aspects. It is considered to be a revolutionary material in the future.

4.3.1 Graphene Material Overview

Graphite is a rich mineral in nature, and its single layer structure is graphene. In 2004, the physicists of the University of Manchester, André Heim and Konstantin Novoselov, used mechanical micro-stress techniques to successfully separate graphene from graphite. The integer quantum Hall effect and the quantum Hall effect at normal temperature were found in single-layer and double-layer graphene systems. The two scientists also won the 2010 Nobel Prize in Physics for their outstanding contributions to the groundbreaking experiments of two-dimensional graphene materials.

The basic structural unit of graphene is the most stable benzene six-membered ring in organic materials, and is the most ideal two-dimensional nanomaterial. Graphene is currently the thinnest and hardest nanomaterial known, and a single layer of graphene has only one carbon atom thickness. Graphene has the following remarkable features.

1. Electrical and thermal conductivity

The graphene structure is very stable, and the connection between the carbon atoms in the graphene is very flexible. When an external mechanical force is applied, the carbon atom plane is bent and deformed, so that the carbon atoms do not have to be rearranged to adapt to the external force, thereby maintaining structural stability. The ideal graphene structure is a planar hexagonal dot matrix, which can be regarded as a layer of stripped graphite molecules, each of which is sp^2 hybridized and contributes the remaining electrons in p orbital to form a large π bond. The π electrons can move freely, giving the graphene good conductivity. When electrons in graphene move in orbit, they do not scatter due to lattice defects or introduction of foreign atoms. Since the interaction between atoms is very strong, at normal temperature, even if the surrounding carbon atoms collide, the interference of electrons in graphene is very small. The velocity of electrons in graphene reaches 1/300 of the speed of light, far exceeding the speed of electron propagation in a general conductor.

Graphene has a thermal conductivity up to 5,300 W/(m K), which is higher than the thermal conductivity of a single-walled carbon nanotubes (Table 4.2). The thermal conductivity of single layer graphene is closely related to the layer bandwidth, defect density and edge roughness, and the graphene layer is characterized by thermal anisotropy along the planar direction, and the thermal conductivity decreases gradually with the increase of temperature at room temperature.

Table 4.2 Comparison of thermal conductivity of graphene with other materials

Material type	Thermal conductivity K [W/(m K)]
Single layer graphene	4840–5300
Single-wall carbon nanotube	Around 3500
Diamond	900–2320
Gold	317
Silver	429
Copper	401
Aluminum	237
Reduced graphene oxide	114–2187

2. Mechanical properties

Graphene is the most known comprehensive mechanical properties of the material. Its hardness is similar to that of diamond, its toughness is good, and it can be bent and folded (Table 4.3). The study found that the maximum pressure that can be tolerated per 100 nm distance before the graphene sample particles began to fragment, reached approximately 2.9 μN. It is estimated that this result is equivalent to applying a pressure of 55 N to break the 1-μm long graphene. If it is possible to produce graphene with a thickness equivalent to that of an ordinary food plastic bag (thickness of about 100 nm), it is necessary to apply about 20,000 N of pressure to tear it off.

3. Optical properties

Graphene is a new material of a single-layer sheet-like structure composed of carbon atoms, which is almost completely transparent and has excellent light transmittance in the near-infrared and visible-light bands. Single-layer graphene (thickness of 0.33 nm) can absorb about 2.3% of visible light, but graphene is generally used in single/multilayer form, light transmittance is better than indium oxide glass; It is completely transparent in the infrared spectrum, and can produce new high-efficiency solar energy battery; It can emit fluorescence, its photoelectric interaction is more prominent, with light adjustment and light control effect.

4. Large specific surface area

Graphene has only one carbon atom thickness, so graphene has an extra large specific surface area. The ideal single-layer graphene has a specific surface area of 2,630 m^2/g,

Table 4.3 Mechanical parameters of graphene

Graphene type	Young's modulus E (TPa)	Breakdown strength σ (GPa)
Single-layer	1.02	130
Double-layer	1.04	126
Three-layer	0.98	101

while the ordinary activated carbon has a specific surface area of only 1,500 m^2/g. The surface area makes graphene a potential energy storage material.

4.3.2 Preparation Methods of Graphene

At present, there are no methods and ways suitable for large-scale production of graphene, which also leads to the high cost of graphene. Generally, the preparation method of graphene is divided into two types of physical methods and chemical methods. The physical method is to obtain graphene from graphite or similar materials with high lattice completeness, and the obtained graphene has a size of more than 80 nm, mainly including mechanical stripping method, epitaxial growth method, heating SiC (silicon carbide) method, explosion method, etc. The chemical method is to obtain graphene by small molecule synthesis or solution separation, and the obtained graphene has a size of 10 nm or less, mainly including graphite intercalation method, thermal expansion stripping method, electrochemical method, vapor deposition method, graphite oxide reduction method, ball grinding method and so on.

At present, the main manufacturing methods are mechanical stripping method, epitaxial growth method, gas phase deposition method, graphite oxide reduction method and so on (Table 4.4). These preparation methods have their own advantages and disadvantages, such as simple mechanical stripping method, high-quality graphene can be obtained, but the repeatability is poor, production and success ratio are very low; epitaxial growth method can obtain a large area of single-layer graphene, but the preparation conditions of this method is harsh, high temperature and high vacuum are required, and graphene is difficult to transfer out from the substrate. The graphene prepared by vapor deposition has a relatively complete crystal structure, and the graphene has a large area. It exhibits obvious application advantages in the fields of transparent electrodes and electronic equipment, but disadvantages of this method are low yield, high cost, difficulty in transferring graphene, etc.; Graphene prepared by graphite oxide reduction method can not completely eliminate oxygen-containing functional groups, and the prepared graphene has disadvantages such as defects and poor conductivity.

1. Mechanical stripping method

The mechanical stripping method uses the friction and relative motion between the object and graphene to obtain the graphene thin layer material. The method is simple to operate, and the obtained graphene usually maintains a complete crystal structure. In the 2004, the University of Manchester first peeled off a single layer of graphene from highly directional pyrolysis graphite (HUPG) by mechanical method. The microgrooves are etched on the surface of the HUPG by oxygen plasma, transferred to a glass substrate with a photoresist, and then repeatedly peeled off with a transparent tape. The thickness of the HUPG is gradually reduced, and a thin layer is left on the substrate, including a single layer of graphene; the glass substrate

Table 4.4 Comparison of preparation methods of main graphene

Method	Size	Graphene quality	Cost	Suitable for industrialization or not?	Advantages and disadvantages
Mechanical stripping	Small and medium size	Molecular structure is relatively complete	Lower	Pretty difficult	High quality graphene for low cost and uncontrollable size for mass production
Epitaxial growth	Large size	Sheets are not easily separated from SiC	Higher	Small batch production	Large preparation area; high temperature process; not easy to control; SiC substrate expensive, difficult to transfer from SiC substrate
Vapor deposition	Large size	With complete structure and good quality	Higher	Mass production	Complete structure; large area; high temperature process; transfer may be defective
Graphite oxide reduction	Large size	Molecular structure is more easily destroyed	Lower	Mass production	Low cost preparation; high graphene yield; difficulty in preparing high quality graphene with less defects

with the microchip attached is placed in an acetone solution for ultrasonication, and then a single crystal silicon wafer is placed in the solution, and the single layer of graphene is adsorbed on the surface of the silicon wafer under van der Waals force. The mechanical method has been simplified in later developments, such as directly removing a layer of graphite from the HUPG with tape and then repeatedly pasting between the tapes. The graphite sheet will become thinner and thinner, which also contains a single layer of graphene, and then the tape is applied to the substrate and the single layer of graphene is transferred to the substrate. In addition, there are many other new mechanical methods, such as mechanical pressure methods, rolling friction methods. The biggest advantage of mechanically preparing single-layer graphene is that the process is simple, the production cost is low, and the quality of the sample is high, but the yield is low and uncontrollable, and it is difficult to find a single layer of graphene from a large area of thick layer, and at the same time, there will be a little glue located on the area of sample and the surface cleanliness will not be high.

2. Epitaxial growth method

The silicon carbide epitaxial growth method refers to heating a SiC single crystal at a high temperature, so that Si (silicon) atoms on the SiC surface are evaporated to leave the surface, and the remaining C (carbon) atoms are reconstructed by an independent form, thereby obtaining a SiC-based substrate graphene. The metal catalytic epitaxial growth method is to pass a hydrocarbon into a catalytically active transition metal substrate (such as Pt, Ir, Ru, Cu, etc.) under ultra-high vacuum conditions, and catalytically dehydrogenate the adsorbed gas by heating to make graphene. The gas can grow over the entire metal substrate during the adsorption process, and the growth process is a self-limiting process, that is, the substrate does not repeatedly absorb after adsorbing the gas. Therefore, the prepared graphene is mostly a single layer and can be a large area of the prepared uniform graphene.

3. Chemical vapor deposition method

Chemical vapor deposition (CVD) is a process technology in which a reaction substance reacts under gaseous conditions to produce solid matter deposited on a heated solid substrate surface, resulting in the production of solid materials, it is a method for making graphene films from gas phase deposition using carbon-containing organic gases as raw materials.

Chemical vapor deposition has made a new breakthrough in the large-scale preparation of graphene. Metal nickel with a thickness of 300 nm is deposited on a SiO_2/Si substrate, and then the sample is placed in a quartz tube and heated in an argon atmosphere to 1000 °C, a flowing mixed gas (containing methane, hydrogen and argon) is introduced, and finally a sample is rapidly cooled (cooling rate of 10 °C/s) to room temperature under an argon atmosphere to obtain a graphene film. The nickel is etched away with a solvent to float the graphene film on the surface of the solution, and then the graphene can be transferred to any desired substrate, and a patterned graphene film can be prepared by forming a nickel layer pattern.

The CVD method can produce large-area high-quality graphene, but the ideal substrate material, monocrystalline nickel is too expensive and the process is complicated, which may become an important factor affecting the industrial production of graphene. At present, graphene obtained by this method can be compared with graphene prepared by mechanical stripping in some properties (such as transport properties), but the quantum Hall effect and other properties of the latter are not observed in graphene prepared by CVD method. At the same time, the electrical properties of graphene prepared by CVD method are greatly affected by the substrate material.

4. Graphite oxide reduction method

Preparation of graphene by graphite reduction method, graphite oxide is prepared firstly. The graphite powder is first dispersed in a strong oxidizing mixed acid, such as concentrated nitric acid and concentrated sulfuric acid, and then an oxidant such as potassium permanganate or potassium chlorate is added to obtain graphite oxide, which is then subjected to ultrasonic treatment. Graphene oxide is obtained and finally reduced to obtain graphene.

Graphite itself is a hydrophobic material and a large number of structural defects are formed during the oxidation process. These defects cannot be completely eliminated even after annealing at 1,100 °C. A large amount of hydroxyl, carboxyl, epoxy and other groups remain on the surface and edge of the graphite oxide. The electronic structure change caused by defects causes the graphene to be converted from a conductor to a semiconductor, which seriously affects the electrical properties of graphene and restricts its application. At the same time, due to the presence of these oxygen-containing functional groups, graphene is easily dispersed in a solvent, and graphene is functionalized and easily reacts with many substances, making graphene oxide the basis for preparing graphene functional composite materials. The methods for preparing graphite oxide generally include the Standenmaier method, the Brodie method and the Hummers method. The basic principle is that the graphite is first treated with a strong protonic acid to form a graphite intercalation compound, which is then oxidized by adding a strong oxidizing agent. The graphite oxide reduction method includes chemical liquid phase reduction, thermal reduction, plasma reduction, hydrogen arc discharge stripping, supercritical water reduction, photoreduction, solvent thermal reduction, microwave reduction, and the like.

5. Other methods

In addition to the above-mentioned preparation methods, there are a carbon nanotube cutting method, a high temperature and high-pressure growth method, an arc method, an organic synthesis method, and the like. How to comprehensively apply the advantages of various graphene preparation methods, to synthetically use the advantages of various graphene preparation methods, to complement each other, to solve the problem of difficult solubility and instability of graphene, and to improve the structure and electrical properties, etc. is a hotspot and a difficult point in future research, and opens up new prospects for the synthesis of graphene in the future.

4.3.3 Application of Graphene in New Energy

Graphene has the following advantages in the development of lithium-ion batteries: (1) the graphene sheet is flexible and can effectively buffer the volume expansion of the metal electrode material; (2) the excellent electrical conductivity of graphene can enhance the electron transmission capacity of the metal electrode material; (3) the activated nuclear point on the surface of graphene can control the growth of metal oxide particles on the surface to maintain nanometer size and improve the rate performance of the material; (4) the specific capacity of the composite material is larger than that of pure graphene. (5) nanoparticles of metal or metal oxide can protect the surface layer of graphene, prevent the electrolyte from intercalating into the graphene sheet layer and cause electrode material to peel off; (6) adding graphene to the battery electrode material can greatly improve charging efficiency and improve battery capacity. In addition, the new graphene materials do not rely on platinum or other precious metals, which can effectively reduce costs and environmental impact.

1. Applied to positive electrode material

For the positive electrode material, the main role of graphene is to improve the electronic conductivity of the material and improve the ion diffusion properties of the material. Many anode materials are less conductive, such as $LiFePO_4$, $LiMn_2O_4$, and ternary materials. The combination of graphene and anode material can provide a conductive network. Due to the excellent conductivity of graphene, it can be used less than traditional conductive additives and has excellent performance. In addition, graphene has a large specific surface area. If the active material is well dispersed on the graphene surface, the diffusion distance of lithium ions can be effectively shortened, which is advantageous for the improvement of the magnification performance, and also improves the utilization efficiency of the active material, then increase the specific capacity. Graphene and anode materials are used in a variety of graphenes, including graphene, graphene oxide, heteroatom-doped graphene, and even graphene and other carbon materials such as carbon nanotubes CNT composite. The morphology of the composites is also different, such as sandwich morphology and three-dimensional graphene foam. It is generally believed that the final topographical characteristics of the composite determine its electrochemical properties.

2. Applied to cathode materials

For the lithium ion battery cathode material, the role of graphene is to simultaneously increase the specific capacity and magnification performance. Currently, graphite is commonly used as a negative electrode for commercial lithium ion batteries. Although the graphite cathode is safe, its capacity is low, only 372 mAh/g. At present, high specific capacity silicon-based, germanium-based, tin-based, metal oxides and nitrides have received extensive attention and research. These negative electrode materials can provide much higher specific capacity than graphite, but there is a common problem that these negative electrodes have a huge volume change during the charging and discharging process. During the cycle process, the huge volume expansion causes the electrode to pulverize and fall off, resulting in capacity attenuation and cycle stability deterioration. Among them, silicon has the largest theoretical specific capacity, 4,200 mAh/g, but also has the largest volume expansion coefficient of 420%. Silicon cathodes are the most concerned in the research of lithium ion battery cathode materials.

Graphene coated silicon cathode is a new idea. The principle includes: the conductivity of silicon is relatively poor, and the excellent conductivity of graphene is used to provide an excellent conductive network. The improvement of conductivity can also improve the utilization efficiency of active materials and increase the specific capacity. The strength and toughness of graphene should be utilized to deal with the problem of volume expansion for the improvement of the stability of the cycle; using graphene as the carrier of the active material, the conductive additive and the binder can be omitted, thereby reducing the redundancy quality in the electrode; in addition, the graphene can also prevent the aggregation of the nanoparticles of active material.

At present, most of the negative electrode materials for lithium ion batteries are Sn (tin) and its oxide SnO_2 (tin oxide). The non-macroscopic and electrochemical properties of graphene composites were investigated by preparing synthetic tin-based/graphene composites. During the charging and discharging process, the unique graphene structure can make the battery structure more stable, increase the conductivity, and provide additional lithium ion storage space, so that the battery capacity does not change, but the battery life is increased. Therefore, the tin oxide/graphene nanocomposite has a large reversible capacity and good charge and discharge performance. However, due to the difference in preparation methods, the optimal composite ratio is also different, so the exploration of the synergistic mechanism needs further research. The results show that the tin oxide/graphene nanocomposites have better battery capacity and performance than other tin-based compounds. Tin-based compounds using spherical nanoparticle structures are significantly superior to electrode properties of nanolayer structures. Graphene composites of tin oxide spherical particles have broad prospects in the application of negative electrode materials for high performance lithium ion batteries.

Silicon and lithium ions can form Li_4Si with a theoretical charging specific capacity of 4,200 mAh/g and a low discharge voltage. Silicon has a very rich natural reserve and has good prospects as a negative electrode material for lithium ion batteries. However, the unmodified silicon-based material has poor cycle stability, and the volume effect is relatively serious during charging and discharging. In order to buffer a large volume change to some extent, it is necessary to coat and nano-crystallize the silicon material to improve the volume effect. Compared with other carbon material modification methods, the introduction of graphene has the advantages of preventing aggregation of silicon nanoparticles, reducing volume change of materials, and effectively increasing lithium storage capacity and electron velocity.

Because oxides and compounds of transition metal have high lithium storage capacity, they have also become one of the research directions of new lithium ion battery cathode materials. In the process of charging and discharging, such materials also have the problems of significant volume change and low electron migration rate, so graphene can be used to modify such materials, thereby improving the electrochemical performance of the materials.

3. Applied to lithium-sulfur batteries

Lithium-sulfur batteries produce intermediate polysulfides during operation and dissolve in the electrolyte. On the one hand, these dissolved polysulfides cannot be completely converted to the final product at the end of the cycle, resulting in a certain amount of capacity loss; on the other hand, the long-chain lithium polysulfide dissolved in the electrolyte will spread from the positive electrode to the negative electrode under the concentration gradient, and reduction reaction with lithium metal of the negative electrode will occur to form Li_2S_2, Li_2S and short-chain lithium polysulfide:

$$2Li + Li_2S_n \rightarrow Li_2S + Li_2S_{n-1}$$
$$2Li + Li_2S_n \rightarrow Li_2S_2 + Li_2S_{n-2}$$

Li_2S_2 and Li_2S will further react with the long-chain lithium polysulfide which subsequently diffuses to the surface of the negative electrode to form short-chain polysulfide lithium:

$$Li_2S + Li_2S_n \rightarrow Li_2S_k + Li_2S_{n-k+1}$$
$$Li_2S_2 + Li_2S_n \rightarrow Li_2S_k + Li_2S_{n-k+2}$$

These short-chain lithium polysulfides will again diffuse back to the sulfur positive electrode under the concentration gradient and be oxidized to long-chain lithium polysulfide. This phenomenon in which polysulfide diffuses back and forth between the positive and negative electrodes of a battery is called a shuttle effect. It not only consumes the active material of the positive electrode, but also causes the negative electrode to be corroded and generates a Li_2S_2, Li_2S passivation layer, which reduces the coulombic efficiency of the battery.

The idea to solve this problem is to add a conductive material to disperse the active material to make it a good conductor; the second is to use a porous material as a carrier of sulfur to bind the polysulfide to reduce the shuttle effect. Graphene has such properties, it has good electrical conductivity and large specific surface area. It has abundant functional groups on the surface to fix sulfur nanoparticles. It is an excellent carrier material and can play an important role in solving the problem of lithium-sulfur batteries.

4. Applied to Lithium-air batteries

Lithium-air batteries, also known as lithium-oxygen batteries, have the highest theoretical energy density in battery systems, reaching 11,680 W-h/kg (excluding oxygen mass), much higher than lithium-ion batteries. The positive electrode material (air electrode) is a key component of the lithium air battery, and functions to provide a reaction interface, distribute and collect current, provide an oxygen transmission channel, accommodate a discharge product, load a catalyst and so on. The reaction mechanism of the lithium air battery requires that the positive electrode be a conductive porous material, and the size, shape, structure and surface chemical properties of the pore directly determine the discharge performance of the battery. Graphene has good conductivity, large specific surface area and modifiable surface. These factors determine the frame of graphene as a anode material for lithium air battery.

Although a large specific surface can provide more reaction spots, it is not as small as possible. The discharge product Li_2O_2 covers the surface of the electrode and fills the pores between the electrode particles, hindering the transport of oxygen, lithium and electrons. In order to solve this problem, a scheme of grading a porous material has appeared, and the positive electrode material contains pores of two sizes, and the pore size distribution is bimodal. Large pores are used to retain Li_2O_2, so that the channels through which oxygen enters are not blocked, while small pores are used to provide a reaction interface. Graphene oxide can be easily formed into a porous material having such a pore characteristic by a microemulsion method, and at the same time, a functional group on the surface of the graphene oxide can also function as fixing Li_2O_2.

5. Other applications

Graphene has important application prospects in proton exchange membrane fuel cells. In recent years, significant progress has been made in the study of carbon-doped non-platinum non-precious metal fuel cell catalysts. Non-platinum non-precious metal catalysts are considered to be the hope for large-scale commercialization of proton exchange membrane fuel cells. Graphene has ultra-thin layer structure and the large specific surface area, if the doping and modulation of graphene can be realized, it is hoped to produce a kind of doped carbon-based non-platinum non-precious metal catalyst which is far superior to the current carbon material. In addition, graphene has a much higher specific surface area than conventional carbon black-based materials, making it a potential carrier material for the preparation of highly dispersed low platinum catalysts.

Graphene has an extremely large specific surface area and is an ideal target for making supercapacitor electrode materials. However, graphene is prone to aggregation due to the effect of interlayer van der Waals forces. The biggest technical problem of graphene as an electrode material is to overcome the aggregation of graphene and ensure that there are interconnected pores inside the material, so that the electrolyte is in full contact with the electrode surface. Therefore, graphene is used in double electric layer supercapacitors. The specific surface, pore structure and electrical conductivity need to be considered in the design of the preparation method. These factors directly determine the electrochemical performance of the supercapacitor.

The chips currently used in computers are generally silicon-based, and there is a phenomenon of heat generation during the calculation. Therefore, the silicon substrate can perform only a certain number of operations per second at room temperature, and graphene has good thermal conductivity and electron migration rate. The movement of electrons in it is almost without any resistance, and it runs much faster than a computer using silicon devices. Combining the characteristics and advantages of both silicon and graphene, such as good thermal conductivity, electron migration rate, electrical conductivity, and large specific surface area, the study of silicon atom doped graphene nanoribbons can broaden further applications in the field of nano-electronic devices.

4.4 Energy Decarbonization Technology

The development direction of clean, low carbon and carbon free is the general trend of energy development. The traditional energy decarbonization mainly involves underground coal gasification technology, carbon dioxide capture and storage technology, traditional fossil energy cleaning technology and so on.

4.4.1 Underground Coal Gasification Technology

Underground coal gasification refers to the process of controlled combustion of coal underground and producing combustible gases (such as hydrogen, carbon monoxide, etc.) through the thermal action of coal and chemical action. The underground gasification of coal redefines the concept of "clean coal", which fundamentally changes the way coal is exploited, not only improves the efficiency of coal mining and utilization, but also overcomes the negative environmental impacts caused by coal mining and application.

1. Fundamental principle of underground coal gasification

The underground coal gasification process is mainly realized in the gasification channel of the underground gasification furnace. The whole gasification process can be divided into three stages of oxidation, reduction and dry distillation, which respectively correspond to three reaction zones of oxidation, reduction and dry distillation. From the perspective of chemical reaction, there are no strict limits in the three reaction zones. The oxidation zone and the reduction zone also have coal pyrolysis reactions. The division of the reaction zone only indicates the relative strength of oxidation, reduction and pyrolysis reactions in the gasification channel. After these three stages, the generated combustible components are mainly gas containing hydrogen, carbon monoxide and methane, and the gasification reaction zone gradually moves to the gas outlet, thereby maintaining the continuous progress of the gasification reaction, and the coal wall of the gasification channel (gasification working face) is continuously burned and pushed forward, leaving the remaining ash and residue left in the goaf.

Underground gasifier plays an important role in the process of underground coal gasification process, and the types of underground gasifier are mainly divided into three kinds, such as well type, no well type and mixed type.

1) Well

The construction of a well type gasifier firstly drills a wellbore from the ground, then develops a flat roadway underground, and connects the underground gas generator to the ground with a wellbore and a flat roadway. The coal bed is ignited in the alley, blasted from one wellbore, and the gas is discharged from the other wellbore (Fig. 4.12).

2) No-well

The underground gasifier of no-well type refers to an underground gas generator that uses a hole to expose the coal bed and uses special technology to establish a gasification passage in the coal bed. The well-free gasifier can be divided into several basic furnace types such as a V-shaped furnace, a blind hole furnace, and a U-shaped furnace according to the relative positions of the inlet and outlet points and the gasification passage.

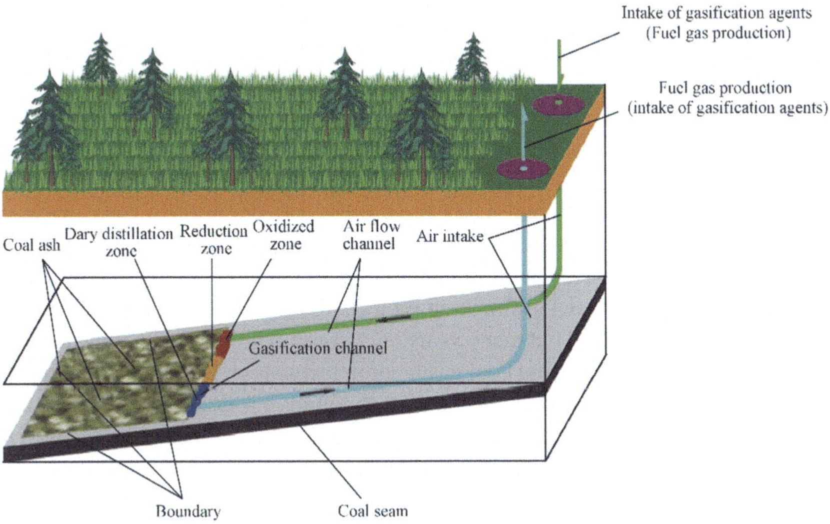

Fig 4.12 Diagram of underground coal gasification

3) Mixed

Hybrid gasification furnace refers to exposing coal bed by drilling holes at the surface or exposing coal bed by using wellbore to install pipelines. The artificially excavated coal roadway is used as a gasification passage, and the gas passages (manually excavated coal lanes) are used to connect the gasification passages and the boreholes or the pipe, which constitutes the gasifier.

2. Foreign underground coal gasification technology

In 1888, Mendeleev proposed the idea of underground coal gasification. Up to now, there are about 33 underground coal gasification projects implemented abroad, mainly in the former Soviet Union, the United States, Australia, South Africa, Canada and Europe.

In the 1970s and 1980s, the United States conducted 38 trials at several project sites by Lawrence National Laboratory and other research departments, and formed a backward underground coal gasification technology, and positioned the technology as a national energy security emergency. The reserve technology of the period was launched when the national energy encountered a crisis, but there was no series of optimization studies on the economic benefits of the technology.

The Skochinsky National Mining Institute of the former Soviet Union developed underground gasification technology and applied it to multiple mines in the former Soviet Union, confirming the feasibility of the technology. Up to now, there have been six underground gasification projects in the former Soviet Union. Among them, the Anglin project in Uzbekistan is the only underground coal gasification project that is still in operation in the world.

Canada cooperates with the United States, based on the United States underground coal gasification technology, forming UCG™ technology through R&D and field trials. This technology basically represents the mainstream of underground gasification technology in Western countries, and has been directly or indirectly applied to underground gasification tests in countries such as Australia, South Africa and India.

Australian underground coal gasification work was developed on the basis of the introduction of Canadian technology. Up to now, Australia's underground coal gasification industrial test industry chain is relatively complete, represented by Linc's Chinchilla underground coal gasification project.

South Africa has carried out many years of coal-to-oil projects with a certain coal chemical base, but underground coal gasification technology is basically directly or indirectly introduced to Canada's UCG™ technology. From 2006 to 2012, South Africa conducted relevant industrial tests in Eskom and other places, and planned to implement larger-scale industrial tests, but no substantive work has been carried out so far.

The countries that conduct research on underground coal gasification technology include the United Kingdom, Pakistan, Bulgaria, Vietnam and Poland; the EU, India and other countries and regions are concerned about the development of this technology. Since the first underground gasification test was carried out in the world in 1912, the UK has once again approved 612.74 million m^2 of coal mining rights in the Beihai Basin for underground gasification research. The Polish Central Mining Institute and the Australian company Linc are responsible for the EU's HUGE2 underground gasification project, which is a follow-up study of the EU's deep coalbed gasification HUGE project. Pakistan has conducted preliminary experiments in the desert in cooperation with the China University of Mining and Technology. The Indian government is drafting a underground coal gasification policy.

In terms of implementation effect, Uzbekistan's Anglin and Australia's Chinchilla project can best reflect the comprehensive level of the industry. The Anglin underground coal gasification station was built in 1961 and a well-free type was used. The gasification station converts underground brown coal into power gas, blending with heavy oil for use in Anglin thermal power station. The gasification station's production capacity is 1 million m^3 per day. The Chinchilla project began its demonstration in 1999 and was completely stopped in 2013 due to environmental impacts. One of the technical features of the project is the realization of negative pressure gasification and direct gasification of gas into synthetic oil. It is the first representative project of forming coal gasification, gas purification, synthetic petroleum.

3. Underground coal gasification technology in China

The development of underground coal gasification technology in China is mainly divided into the following stages: to absorb and digest Soviet technology before the 1980s; the introduction of American technology in the 1990s; and the absorption of digestion of Western and former Soviet technologies with some innovations at the same time in the early 21st century. Gradually, the theory and technology represented by "long channel, large section, two stages, forward and reverse" and reverse combustion gasification were formed. The underground coal gasification technology has gradually shifted from laboratory experimental research and field test research to industrial demonstration production and application, and developed underground coal gasification technology with independent intellectual property rights. At present, the projects with better industrial demonstrations mainly include the underground coal gasification project with well of Xinkuang Group and the underground coal gasification project without well of Xin'ao Group.

The underground coal gasification of the Xinkuang Group began its experimental research work in 1999. It was successfully ignited in March 2000. In July of the same year, it officially supplied more than 10,000 families with gas. From 2001 to 2002, Xiezhuang Gasification Station and Ezhuang Gasification Station (Phase I) were successively built, and ignited successfully. At present, the daily gas production reaches 100,000 m^3, and the gas heat value reaches 11.26 MJ/m^3. In 2002, the underground coalbed methane applied for the national "863" plan—"study on the stability control technology of underground coal gasification", which was approved by the Ministry of Science and Technology and was included in the national "863" plan and test base.

In January 2007, Xin'ao Group invested more than CNY 200 million to build Wulanchabu Xin'ao Gasification Coal Mining Technology Co., Ltd., together with China University of Mining and Technology and Angren Gasification Station in Uzbekistan to carry out study of "no-well underground coal gasification Pilot pro ject". In October 2007, China's first 150,000 m^3 per day of no-well coal underground gasification test system and production system was successfully ignited. The test site has the ability to supply heat, generate electricity, and produce chemical raw materials. It has obtained a number of innovative research results and declared 9 patents. The research innovatively constructed a new structure underground gasification furnace with "L-shaped backward surface expansion", and creatively developed gasification channel penetration technology, gasification channel dredging technology and well-free gasification. The gas production cost is only about 40% of ground gasificatio. By the end of 2011, the Xin'ao Group Wulanchabu Gasification Station has been in continuous operation for four years. The third test furnace has been operating stably for 900 days. The calorific value and composition are stable. The generator is continuously operated for 780 days, and the air continuous gasification production gas 300,000 m^3/day, oxygen-rich continuous gasification production gas 150,000 m^3/day, which meets the requirements of industrial production. After the completion of the test function at the end of 2014, the gas production was stopped, and considered optimization before making a large-scale study.

4.4.2 Carbon Dioxide Capture and Storage Technology

Carbon dioxide capture, oil displacement and storage technology refers to the injection of carbon dioxide into the oil zone, which makes the viscosity of crude oil reduce after mixing with carbon dioxide, and a part of carbon dioxide be sealed underground in the process of continuously extracting crude oil. After the carbon dioxide is injected into the oil layer, 50–60% of the oil is permanently sealed underground. The remaining 40–50% return to the surface with the associated gas in the oil field. Through the capture and purification of crude oil associated with gas, the carbon dioxide in the associated gas can be recovered, and the recovered carbon dioxide can be reinjected on the spot to further reduce the cost of carbon dioxide flooding. In this process, the oil recovery rate is increased and the production is increased, the carbon dioxide emissions are reduced, and the atmospheric environment is protected.

The main mechanism of carbon dioxide flooding is as follows: (1) carbon dioxide mixed into crude oil can reduce the oil-water interfacial tension and significantly reduce the viscosity of the crude oil, so as to reduce the resistance during the displacement process, thereby improving oil recovery; (2) carbon dioxide is mixed into crude oil. In the middle, the volume of crude oil is expanded, the internal kinetic energy of crude oil is increased, and the fluidity of crude oil is increased. (3) carbon dioxide is mixed into crude oil, which occupies a certain pore space in the oil zone, increases the pressure of the oil zone, and enhances the oil displacement effect.

PetroChina has formed its own carbon capture, utilization and storage (CCUS) technology. In recent years, the progress of CCUS engineering has been accelerated. Through technology research and development and oilfield practice, three carbon dioxide gas injection stations have been built in Jilin Oilfield, and CCS-EOR (carbon capture and storage-enhanced oil recovery) technology has been successfully applied to store carbon dioxide for 1.1 million tons, the storage rate at this stage is maintained above 96%, which guarantees the cleaner production of Changling gas field with high carbon dioxide content. CO_2 flooding has increased oil by 100,000 tons, achieving a win-win situation in controlling carbon dioxide greenhouse gas emissions and providing energy for increasing oil recovery.

Sinopec has carried out CCUS technology development and application demonstration using flue gas from coal power plant, natural gas, refinery gas for carbon dioxide capture source and carbon dioxide flooding as the main resource utilization mode. Through technical research and on-site application. Sinopec has formed a carbon dioxide capture, purification and liquefaction technology for coal power plants with completely independent intellectual property rights. In 2012, the "12th Five-Year" National Science and Technology Support Program "development and application demonstration project for flue gas carbon dioxide capture, displacement and storage technology for large coal power plants" was launched in Shengli Oilfield (Zhang 2017).

CCUS technology involves the capture, transportation and storage of carbon dioxide. As far as the entire industrial chain is concerned, high cost is an important factor restricting its development. This is because the main component of flue gas is nitrogen, and the content of carbon dioxide is relatively low, resulting in high separation energy consumption and capture cost. For the petroleum and petrochemical industry, in order to meet the needs of oil development and improve oil recovery, most of the gas sources come from the carbon dioxide in the flue gas of coal power plants. The captured carbon dioxide is transported to the corresponding oil wells through pipelines, and the cost is high. Therefore, as far as the CCUS industrial chain of the petrochemical industry, it is necessary to make full use of the carbon dioxide source inside the oil field, local materials and reduce transportation costs; it is also necessary to strengthen cooperation and integration with the power industry and expand the source of carbon capture supply.

Gas injection technology is also an effective method to improve coalbed methane recovery. The theoretical basis is that the theoretical maximum recovery factor for the adsorption of binary gases in the coal bed is higher than the theoretical maximum recovery of pure methane in the case of the same gas content. The basic idea of deep coal bed carbon dioxide—ECBM (enhanced coalbed methane recovery) technology is that carbon dioxide or its mixed gas obtained from concentrated emission sources is injected into the coal bed with appropriate sealing conditions underground from the injection well. Carbon dioxide is easier to adsorb to the surface of the coal bed and replace more methane, so production and recovery factor of the single CBM well is increased, while a large amount of greenhouse gases is buried in the coal bed. The study found that in the CO_2-ECBM implementation, the desorption rate of methane at unit pressure drop with injected CO_2 is significantly higher than the desorption rate at unit pressure drop for pure methane desorption. Compared with the international status, China has a big gap in the theory of carbon dioxide replacement CBM and storage, carbon dioxide capture, utilization and storage experience, carbon dioxide storage monitoring and early warning.

4.4.3 Clean Technology of Traditional Fossil Energy

The cleansing of traditional fossil energy is the most important development direction of clean energy. As fossil energy is more mature in development, transportation, energy storage and application, compared to large-scale application of new energy, cleaner utilization of fossil energy is relatively more cost-effective. The main directions include integrated gasification combined cycle power generation, circulating fluidized bed combustion technology, supercritical and ultra-supercritical thermal power units, and upgrading of refined oil products.

1. Integrated Gasification Combined Cycle Power Generation

Integrated Gasification Combined Cycle Power Generation (IGCC) is an advanced power system that combines coal gasification technology with an efficient combined cycle. IGCC technology combines clean coal gasification technology with an efficient gas-steam combined cycle power generation system. Power generation efficiency, and excellent environmental performance, is a promising clean coal power generation technology (Fig. 4.13).

IGCC power plants can also be combined with carbon capture, integrating pre-combustion carbon capture technology with IGCC to separate carbon dioxide before combustion and collect it for easy processing and reuse for efficient, low-carbon green energy conversion (Zhai et al. 2014). At present, the technology is still in the

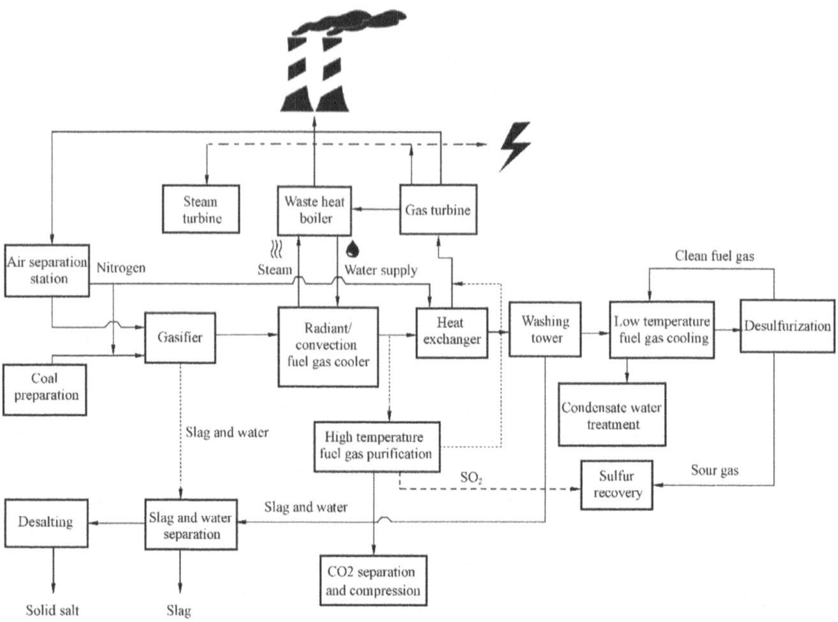

Fig 4.13 Gasification combined cycle carbon capture power generation process

optimization stage, its construction and operation costs are still too high, and the operational stability is not satisfactory.

The IGCC consists of two parts, namely the gasification and purification part of coal and the gas-steam combined cycle power generation part. The main equipment of the first part includes gasifier, air separation unit and gas purification equipment (including sulfur recovery unit); the main equipment of the second part includes gas turbine power generation system, waste heat boiler and steam turbine power generation system. The process of IGCC is as follows: the coal is gasified into a medium-low calorific value gas, which is purified to remove pollutants such as sulfides, nitrides and dust from the gas, and becomes a clean gaseous fuel, which is then sent to the combustion chamber of the gas turbine for combustion. The gas working medium is heated to drive the gas turbine to work, and the gas turbine exhaust gas enters the waste heat boiler to heat the water supply, and the superheated steam generates the steam turbine to perform work.

At the current state of the art, the net efficiency of IGCC power generation can reach 43–45% and is expected to reach higher in the future. The discharge of pollutants is only 1/10 of that of conventional coal power stations, the desulfurization efficiency is as high as 99%, and the sulfur dioxide emissions are about 25 mg/m^3, which is much lower than the emission standard is 1200 mg/m^3 and the emission of nitrogen oxides is only 15–20% of that of conventional power plants, while the water consumption is only 1/3 to 1/2 of conventional power stations. Therefore, the development of IGCC technology is of great significance to environmental protection.

There are currently eight IGCC power stations in the world, and Huaneng Group's IGCC power station is the sixth of them. The two new IGCC stations are the Edwardsport power station in Indiana, and the Camper power station in Mississippi, USA. The Edwardsport station has an installed capacity of 618 MW and a unit cost of approximately US$5,750/kW. The unit cost of the Camper power station has exceeded US$11,600/kW. The investment in IGCC power station is not as expected, with the gradual progress of technology and lower costs, as well as the size of the power plant, the unit cost is also reduced. Due to complicated design and unstable technology, the construction cost and operating cost of IGCC power station are both high.

Huaneng Group built China's first and currently the only integrated gasification combined cycle power station in the Lingang Industrial Zone of Tianjin Binhai New Area. The project started construction in July 2009 and was completed and put into operation in 2012, with an installed capacity of 265,000 kW. The investment in the entire IGCC power station reached CNY 13,000–15,000 per kW. The on-grid price of the IGCC power station is about CNY 0.5 per kWh, but the electricity price is close to CNY 0.9 per kWh. Huaneng Group's IGCC power station can achieve dust less than 0.6 mg/m^3, sulfur dioxide less than 0.9 mg/m^3, nitrogen oxides less than 50 mg/m^3, better or closer to national emission standards of gas power plant, and nitrogen oxides still have room for further reduction.

2. Circulating fluidized bed (CFB) combustion technology

Circulating fluidized bed (CFB) combustion technology is a clean coal combustion technology developed in the past 20 years. It has outstanding advantages such as wide fuel adaptability, high combustion efficiency, low nitrogen oxide emission, low-cost desulfurization in limestone furnace, large load regulation ratio and fast load regulation. After 2000, circulating fluidized bed boilers developed from the thermoelectric field to the power generation field, and successively developed 135–300 MW subcritical circulating fluidized bed boilers. By 2005, domestic circulating fluidized bed technology basically occupied the domestic market and exported to foreign countries. So far, China's circulating fluidized bed combustion boilers have a power generation capacity of nearly 100 million kW, and the total number of circulating fluidized bed boilers is more than 3,000, making it the world's number one.

3. Supercritical and ultra-supercritical thermal power units

Supercritical and ultra-supercritical thermal power units have significant energy saving and environmental improvement effects. Compared with supercritical units, ultra-supercritical units can increase thermal efficiency by 1.2–4% and save 6,000 tons of high-quality coal in one year. Ultra-supercritical units and supercritical units in thermal power plants refer to the pressure of the working fluid in the boiler. The steam temperature in the furnace is not lower than 593 °C or the steam pressure is not lower than 31 MPa, which is called ultra-supercritical. Future thermal power construction will mainly develop high-efficiency, high-parameter supercritical (SC) and ultra-supercritical (USC) thermal power units, which have been widely studied and applied in developed countries.

The United States is the earliest country to develop supercritical power generation technology. The world's first ultra-supercritical parameter unit (125 MW, 31.03 MPa, 621 °C/565 °C/538 °C) was put into operation in the United States in 1957. The supercritical units put into operation in the United States accounted for more than 30% of large thermal power units, with a capacity of 500,000–800,000 kW. It has the world's largest single-machine capacity and highest steam parameters of supercritical units, with a maximum single-machine capacity of 1,300 MW. The highest steam parameters are 34.5 MPa and 649 °C/566 °C/566 °C. In recent years, GE has also designed and manufactured ultra-supercritical units with steam parameters of 26.6 MPa/577 °C/600 °C and 25 MPa/600 °C/610 °C for Japan.

In 1963, the first 300,000 kW supercritical unit of the Soviet Union was put into operation, with a unit parameter of 23.5 MPa/580 °C/565 °C. After long-term experimental research, it has a complete set of supercritical technology and product series. Now there are more than 200 supercritical units, accounting for more than 50% of the total installed capacity. All of the units with a capacity of more than 300,000 kW use supercritical parameters. At present, the new generation of large ultra-supercritical units developed by Russia adopts parameters of 28–30 MPa/580–600 °C.

Japan's development of ultra-supercritical units started late, but soon transitioned from learning and imitation to independent research and development, and established a corresponding test platform at the same time, with rapid development and remarkable results. Japan's supercritical units account for more than 60% of the capacity of conventional thermal motor assembly machines. All of its 450,000-kW units use supercritical parameters. The two ultra-supercritical units that were initially put into operation were designed by Mitsubishi Corporation with an installed capacity of 700,000 kW. The steam parameter is 34.5 MPa/620 °C/650 °C.

China's supercritical and ultra-supercritical units have developed late. China began to introduce 300,000 and 600,000 kW of subcritical units from abroad in the late 1980s. The first supercritical unit was put into operation in Shanghai Shidongkou No. 2 Plant in June 1992 (2 × 600 MW, 25.4 MPa, 541 °C/569 °C). Since the introduction, digestion and absorption in the past 20 years, domestic supercritical power generation technology and domestic million kilowatts ultra-supercritical technology have been put into commercial operation since the beginning of the 21st century. This leaping development is the inevitable result of common progress and common development of the manufacturing industry of the power generation industry and power station equipment. Under the development and application of ultra-supercritical coal power generation technology, the overall level of China's power generation industry and power station equipment manufacturing industry has developed to a new level. At present, China has put into operation 101 sets of 1,000 MW ultra-supercritical units (Table 4.5), and ultra-supercritical coal technologies are constantly being explored and updated, such as mixed buring with biomass, district heating, carbon dioxide capture and combination with other technologies, which has become an important method to solve the further reduction of carbon dioxide in new coal generating units.

Table 4.5 Statistics of 1,000 MW ultra-supercritical unit production

Province/Municipality/Autonomous region	Number of units put into production	Province/Municipality/Autonomous region	Number of units put into production
Jiangsu	21	Hubei	4
Zhejiang	16	Fujian	3
Guangdong	14	Tianjin	2
Shandong	8	Liaoning	2
Anhui	8	Ningxia	2
Henan	7	Xinjiang	2
Shanghai	4	Chongqing	2
Guangxi	4	Jiangxi	2
Total	101		

4. Refined oil upgrade

In view of the composition of gasoline in China, it is mainly necessary to solve the problems of desulfurization, olefin reduction and so on, and to supplement the octane loss caused by this (Wang 2015).

1) Localization of catalytic gasoline hydrodesulfurization technology

PetroChina introduced Prime-G gasoline desulfurization technology in the early stage, which is applicable to the national IV standard, and the problem of lowering the olefin saturation and octane number when upgrading to the national V standard. Therefore, PetroChina developed DSO and GARDES technologies for catalytic gasoline hydrogenation with independent intellectual property rights, replacing foreign imported technologies, and independently completed processing package design and engineering design, and realized the localization of hydrodesulfurization technology.

PetroChina has introduced 12 sets of Prime-G technology, and independently developed 17 sets of technology (9 sets of DSO and 8 sets of GARDES), of which DSO technology has developed into M-DSO technology. DSO technology mainly uses desulfurization. M in M-DSO technology plays the role of isomerization and aromatization. It desulfurizes gasoline and guarantees octane number. The GARDES technology first desulfurizes and then supplements the octane number by isomerization and aromatization.

In 2010, Sinopec acquired the patented technology of Phillips catalytic gasoline desulfurization (S_Zorb), and innovated the device in this technology, forming a new generation of S_Zorb patented technology with its own characteristics, mainly solving the problem of long cycle operation. Sinopec has a total of 25 sets of S_Zorb technology equipment with a total processing capacity of more than 31 million tons/year.

2) Development of catalytic light gasoline etherification process technology

When the national IV standard was upgraded to the national V standard, although the purpose of desulfurization was achieved, the octane number was lost. PetroChina has developed a catalytic light gasoline etherification (LNE) process technology. According to the needs of different refineries, LNE-1, LNE-2 and LNE-3 series light gasoline etherification technologies have been developed. At present, 6 sets of devices have been built, with significant economic benefits (methanol cost 2,000 CNY/ton, gasoline price 5,000–6,000 CNY/ton), can increase the octane number in the case of reduced olefin content.

But etherification also has many problems. Ether-added substances are similar to ethanol gasoline. When the oxygen content is 2.7%, the emission is the lowest. Beyond this value, not only the calorific value but also side effects will increase the pollutant emissions.

3) Development of ionic liquid alkylation technology

Ionic liquid alkylation is a new generation of liquid alkylation technology. Compared with traditional sulfuric acid and hydrofluoric acid processes, carbon tetraalkylated gasoline has the characteristics of high octane number, no sulfur, no olefin, no aromatic hydrocarbons. It has the advantages of light corrosion, low environmental pollution, small investment and high profit. It is an ideal blending component and can fully meet the quality standards of National IV oil products. In September 2017, PetroChina's first ionic liquid alkylation unit was laid at the Golmud Refinery. It is expected to be completed and put into operation in 2018 with an annual output of 50,000 tons.

From an environmental point of view, future oil product upgrades may focus on vehicle diesel and marine fuel. Compared with the national IV gasoline standard, the diesel standard is rougher and there is still more room for improvement. Marine fuel is more harmful to the environment than diesel, and there is huge room for upgrading. The standard should be further improved. The national VI gasoline is one of the strict emission standards in the world. It has reached or exceeded the European standard. The future standard will develop towards olefin-free, which can be achieved by means of isomerization and alkylation.

4.5 Nuclear Fusion Power Generation

China's traditional energy performance is characterized by rich coal and insufficient oil and gas. In the face of international pressure to cope with climate change, it is necessary to make strategic adjustments to the energy structure. Therefore, comprehensively promoting the transition of energy production and utilization methods, accelerating the transformation of energy development mode, and taking the road of green and low-carbon development have become an important direction for China's energy development strategy in the future. Nuclear energy, especially with the characteristics of economic, clean, efficient and safe, has shown great development prospects to human beings. Judging from the general trend of nuclear power development, the current development of pressurized water reactors, the development of fast neutron reactors in the medium term, and the development of fusion reactors in the long run are the goals to achieve (Guan et al. 2012).

4.5.1 Methods of Using Nuclear Energy

The application of nuclear energy mainly includes nuclear fission and nuclear fusion. Nuclear fission still has problems such as raw material depletion and environmental pollution. Therefore, the final solution to the energy problem is attributed to the development of nuclear fusion technology. Nuclear fusion energy has abundant fuel reserves, and there is no serious radioactive pollution problem. It is an ideal new energy source for mankind in the future.

The study of hydrogen fusion by human beings stems from the understanding of solar energy sources. The inexhaustible energy in the sun comes from the nuclear fusion of hydrogen. The main component of the sun is hydrogen. Under the gravitational force, the density of hydrogen inside the sun is very large. Because of its large volume, the chance of heat and fuel particles passing through the surface is small, so the solar restraint performance is very good. At the same time, nuclear fusion does not occur under normal conditions, but the temperature inside the sun is always above 10 million °C, so the hydrogen fusion reaction can occur at such a high temperature and emit huge energy. It can be seen that all the energy that the sun is constantly flowing comes from the nuclear fusion reaction of hydrogen on the sun (Qiu and Yu 2017).

Nuclear fusion is the process of combining two lighter nuclei and releasing energy. The most easily realized fusion reaction in nature is the hydrogen isotope—the fusion of yttrium and lanthanum. The combination of deuterium (D) and tritium (T) forms a heavier 3He atom or 4He atom and releases a large amount of energy. However, there is no possibility of a high temperature of 15 million °C and a high pressure of 200 billion atmospheres on the earth. Therefore, when fusion occurs, the temperature can only be higher, and it is necessary to reach a high temperature of hundreds of millions of degrees Celsius in an instant.

Therefore, nuclear fusion reactions can only be achieved at extremely high temperatures. In order for the two nuclei to undergo a fusion reaction, they must be close enough to each other to reach the nuclear force between the nuclear nuclei and the nucleus, and then the nuclear force can "bond" them into a whole to form a new nucleus.

Since the nucleus is positively charged, as the two nucleuses get closer together, the Coulomb repulsion between them becomes larger and larger, and the Coulomb repulsion separates the two light nucleus. To overcome the Coulomb repulsion and make the two nuclei meet, in the case of deuterium (D) and tritium (T), their relative speed must be above 1,000 km/s, which is equivalent to a high temperature of 100 million °C. This fusion nuclear reaction, which occurs at very high temperatures, is also called a thermonuclear reaction. At such high temperatures, the material is completely ionized to form a high temperature plasma. The plasma is a fully ionized, electrically neutral gas. Due to the high temperature, the electrons have gained enough energy to get rid of the nuclear nucleus, and the nucleus is completely exposed, which can overcome the Coulomb repulsion.

4.5.2 Conditions of Nuclear Fusion Reaction

The realization of the fusion reactor needs to meet the following three necessary conditions (Piao and Zhao 2017).

(1) There must be a sufficiently high temperature. Nuclear fusion is carried out at extremely high temperatures, and the plasma of deuterium and tritium must be heated to more than 100 million °C. Therefore, the fusion reaction is also thermonuclear reaction, fusion power generation is controllable thermonuclear reaction.
(2) The density of the reaction particle (plasma) must be high. In order to obtain a certain amount of power, that is, to obtain a certain amount of power in each second, or to achieve a certain number of reactions per second, it is necessary to have a high reaction density.
(3) There must be a "container" that can form a longer constraint time. If the plasma is heated at a certain power and maintained at a high temperature, a "container" with excellent thermal insulation properties is required. To achieve this, a magnetic field body is utilized as a "container" for the plasma. The quality of the constrained performance is expressed by the average residence time of the plasma or energy, which is called the constraint time, which is the third necessary condition and has a long Confinement Time.

Among the three conditions of temperature, density and constraint time, the latter two play the role of the product. As long as the product of the density and the constraint time is equal, the same effect will be obtained. This feature was first discovered by Lawson of the United Kingdom, and the above product is called Lawson's Parameter. Using the Lawson parameter as the longitudinal coordinate, the temperature of the plasma as the horizontal coordinates, the drawing is called Lawson (Fig. 4.14), which shows the necessary plasma characteristics and the necessary conditions for the fusion reactor (Buchao et al. 2012).

As mentioned above, the fusion Power P_{fusion} is proportional to the triple integral of temperature T, density n and constraint time τ_E, and the theoretical prediction is that sufficient fusion power can be obtained if the value of the triple integral can reach the 21th power of 10.

If the charged particles (^4He) generated in the fusion reaction are confined inside the plasma by a strong magnetic field, they are used to heat the plasma to maintain the extremely high temperature required for the fusion reaction, so that it is no longer

Fig 4.14 Progress of fusion plasma (Lawson chart)

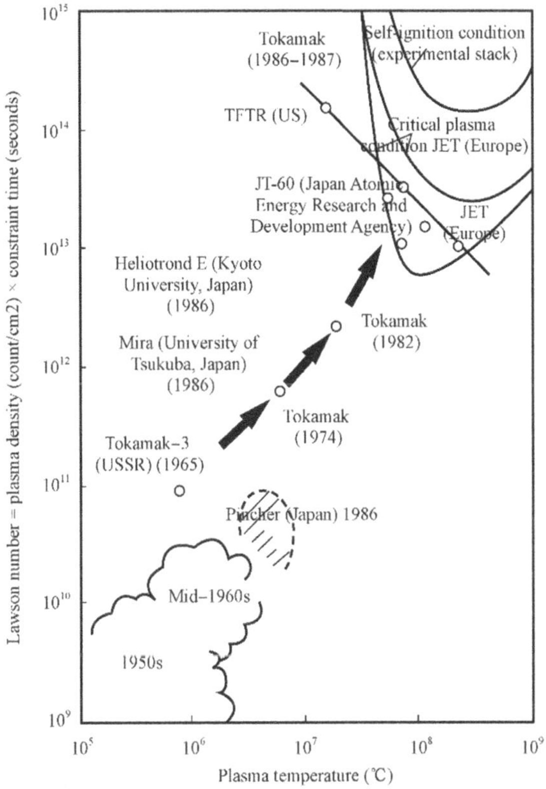

necessary to input energy from the outside, and the fusion reaction can also be carried on in a self-sustaining manner. At this point, the special "furnace" that "burns" the fusion fuel has already been lit. The scientific term for characterizing this concept is called fusion ignition, also known as thermonuclear ignition.

It should be pointed out that the realization of ignition is only the first goal of controlled nuclear fusion research. After obtaining sufficient fusion power, it is also necessary to pursue the benefits of energy and the power of fusion power. Therefore, the second goal of controlled nuclear fusion research is to make the output energy exceed the input energy, obtain the net fusion energy E fusion, and establish a nuclear fusion power station. Scientists refer to the first goal of achieving ignition as the verification of scientific feasibility, and the second goal is called verification engineering feasibility.

On the basis of pursuing the first step of the goal, to obtain the fusion energy E_{fusion}, is to extend the profitable fusion power and the combustion time t to obtain the fusion energy. There are three ways to achieve this goal, namely gravity constraint, magnetic constraint and inertia constraint.

1. Gravity constraint

The sun and other stars use gravitation (i.e., gravity) to constrain the high-temperature fusion fuel plasma to achieve a large number of fusion reactions. These planets are of great mass, so huge gravitation can bind the plasma together and prevent them from scattering. For example, the diameter of the sun is 109 times that of the earth, the mass of the sun is 330,000 times that of the earth, and its gravitation is also 330,000 times that of the earth. Although its composition is hydrogen, the volume and amount of hydrogen are so large. The gravitation between them is sufficient to constrain the hot plasma on the sun. Therefore, it is the universal gravitation of the sun that gathers these hydrogens together. The probability of the particles running out is very small, that is, the constraint time is extremely long. Therefore, the particles and energy are continuously making a steady stream of fusion reaction without any other means, this is the gravity constraint.

The principle of gravity constraint is that under the condition of large mass and large volume, high density and high constraint time can be realized under the constraint of gravity. At this time, even if the temperature is far lower than the level reached by the fusion experimental device on the earth, it can still be stable. A large number of fusion reactions occur, resulting in huge fusion solar energy.

There is a premise like gravity on the sun, that is, the volume must be very large. On Earth, it is impossible to build a reactor as large as the sun, and then a fusion reaction occurs continuously; it can only be achieved in experimental devices. Therefore, scientists have proposed other solutions, namely magnetic constraints and inertial constraints.

2. Inertial constraint

Inertial constraint is the use of energy provided by a driving form such as a laser to cause a nuclear fusion fuel in a target to form a plasma, and to undergo a fusion reaction by centripetal detonation to a high-temperature, high-density state, thereby causing a fusion reaction. The basic principle is to use a multi-channel high-power laser beam or other ion beam to bombard the target (the target is a glass ball or a plastic ball, which is mixed with the easiest to achieve fusion reaction of deuterium and tritium on the earth). The uniform thrust is generated around the ball, and the deuterium and tritium gas of the core is compressed, so that the deuterium and tritium mixed fuel in the ball is heated to a very high temperature in a very short time, and a fusion reaction occurs.

The basic idea of inertial constraint is to supplement the lack of constraint time by increasing the density of the plasma. The inertia constraint relies on the inertia of the plasma particles themselves to constrain the high-temperature fusion fuel plasma, thereby achieving a large number of fusion reactions. Its forefront includes the National Ignition Unit NIF, the French MJ Laser LMJ and the China Shenguang Series.

3. Magnetic constraint

Magnetic constraint is the use of magnetic field to constrain the characteristics of charged particles, constructing a special "magnetic container" to constrain the high-temperature fusion fuel plasma, so that it moves along the magnetic field lines, ensuring that the plasma does not escape, so that the plasma in the high temperature state continues to achieve a large number of fusion reactions within the magnetic field spontaneously and continuously.

The magnetic field can constrain the plasma, but if the magnetic field is linear and there is a loss of high temperature plasma at its ends, many early linear magnetic constraint fusion devices are stopped. Later, scientists were concerned that if a magnetic field was formed into a ring to form a toroidal magnetic field, then there was a gradient in the magnetic field perpendicular to the direction of the magnetic field, and the particles would have a gradient drift perpendicular to the direction of the magnetic field; at the same time, the particles move along the circular magnetic field. It is subjected to a centrifugal force in the direction of the curvature of the magnetic field, producing a curvature drift perpendicular to the direction of the centrifugal force and the applied magnetic field. Therefore, it is necessary to add a magnetic field along the polar direction so that the drift of the particles at different positions cancel each other out, thereby ensuring the stability of the constraint. The more mainstream magnetic constraint fusion devices in the world are Tokamaks and Stellarator. The main difference between the two designs is the way they produce polar magnetic fields. The Tokamak passes current through the plasma itself, while the Stellarator uses a magnet outside the device to create a helical magnetic field that surrounds the plasma. Tokamak has become the first choice for magnetic constrained fusion research in countries due to its relatively simple magnetic field design and good constrained performance. The Stellarator has its unique advantages in steady-state operation, but its complex magnetic field design and construction are also very challenging.

References

Buchao, Xu, Yan Wenbo, Hua Ming. 2012. A study on modes and routes of nuclear energy development under dual restraints from public security and low-carbon economic growth. *Sino-Global Energy* 47: 38–42.

Guan, Genzhi, Xiaoqiong Zuo, and Jianping Jia. 2012. Nuclear power technology. *Hydropower and New Energy No.* 1: 7–9.

Hydrogen Production Tech Team Roadmap, USDRIVE Nov 2017.

Jie, Yu., and Hong Xiao. 2006. Research progress in biomass hydrogen production technology. *China Biotechnology* 26 (5): 107–112.

Piao, Xuan, and Zhongxiu Zhao. 2017. The historical experience of world energy substitution and economic development is an important inspiration for China to restart nuclear power projects. *Modern Management Science* 2: 3–5.

Qiu, Weilin, Wen Yu. 2017. Research on the status quo and countermeasures of China's nuclear energy industry in the new era. *Market Analysis* 184–185.

Wang, Xinming. 2015. Influence of fuel quality on vehicle emission and economic analysis of upgrading fuel quality in China. *Bulletin of the Chinese Academy of Sciences* 30 (4): 536–539.

Wu, Sufang. 2014. Hydrogen Energy and Hydrogen Production Technology. Hangzhou: Zhejiang University Press, 86–93.

Xi, Haibin. 2012. Research progress in hydrogen production by photocatalysis. *Energy Research and Management*, 1005-7676, 04-0058-03.

Yi, Wenjing, Qi Liang, and Qingbing Zhao. 2018. Enhance the hydrogen application in China's energy system to accelerate the energy transition: Status and progress. *Environmental Protection* 2: 30–34.

Zhai, Mingyang. 2014. Ureent status and development of carbon capture in power generation industry. *Environmental Science and Technology* 27 (2): 66–67.

Zhang, Yindi. 2017. Research progress of CO_2 capture, utilization and storage (CCUS) technologies in petroleum and petrochemical industry. *Oil and Gas Storage and Transportation* 36 (6): 637–643.

Chapter 5
Energy Internet Technology

Energy Internet refers to a combination of advanced power and electronics technology, information technology and intelligent management technology, and a large number of new power networks, petroleum networks, natural gas networks, etc., which are composed of distributed energy gathering devices, distributed energy storage devices and various types of loads. The energy nodes are intelligently interconnected to achieve an energy reciprocal exchange and sharing network with two-way flow of energy.

The energy Internet is the product of the combination of Internet information technology and renewable energy. The development of the energy Internet will fundamentally change the dependence on traditional energy utilization models and promote the transition of traditional industries into emerging industries based on renewable energy and information networks, it is a fundamental change in the way of life in human society. The outline of the tenth five-year plan for national economic and social development states: "We will promote the deep integration of new technologies in the fields of energy and information, coordinate the construction of infrastructure networks such as energy, communications and transportation, and build a 'source-network-load-storage' coordinated development, integrated and complementary energy Internet". China is currently in a critical period of the energy revolution, and the National Energy Administration's "Guiding Opinions on Promoting the Development of "Internet Smart Energy" will trigger "chain reform" in various domains such as energy technology, production, supply, etc. and China's energy revolution is an inevitable choice driven by history. The new energy technologies represented by renewable energy, distributed power generation, energy storage, electric vehicles, etc., and the Internet technology sectors represented by the Internet of Things, big data, cloud computing, mobile Internet, etc. are developing rapidly and deeply integrated, and the energy Internet has become another important development direction in the energy field following the smart grid.

C. Zou, *New Energy*, https://doi.org/10.1007/978-981-15-2728-9_5

5.1 Distributed Energy

Distributed energy is a geographically dispersed, relatively close-to-user, relatively independent energy supply and demand system. The World Alliance for Decentralized Energy (WADE) defines distributed energy as: an efficient cooling/heating power supply system installed at the customer's end. The system can generate electricity at or near the consumer's location, efficiently use the waste energy generated by power generation to produce heat and electricity. Renewable energy systems of the field end include energy recycling systems that use on-site exhaust, waste heat, and excess differential pressure to generate electricity. Taking a power network with Internet topology as an example, traditional large-scale centralized energy supply stations (large-scale central power plants) are generally far away from users, and power is sent to end users through high-voltage transformation and transmission of large power grids, while distributed energy is that power generation and energy supply equipment are built on the near-user side, using clean energy such as natural gas and hydrogen, renewable energy such as local photovoltaics and wind energy, or combining industrial waste heat and residual pressure to meet the user's cold, heat, electricity, steam, domestic hot water and other energy load needs through energy cascade utilization.

Distributed energy has the characteristics of reasonable energy efficiency, low loss, less pollution, flexible operation, and good system benefits. However, it has certain difficulties in grid connection, power quality, capacity reserve and fuel supply. In terms of environmental protection, some of the pollution will be decentralized and resourced, and advanced energy conversion technologies will be adopted to minimize the emission of pollutants; in the energy transmission and utilization, the regional placement will be realized, the loss of long-distance transportation energy will be effectively reduced, and safety and flexibility in energy use will be increased. Distributed energy disperses emissions and facilitates the absorption of surrounding vegetation. Distributed energy relies on the most advanced information technology, using intelligent monitoring, networked group control and remote-control technology to achieve unattended on-site, while distributed energy relies on energy social service system with energy service company as the main role to achieve operation. The specialization of management to ensure the safe and reliable operation of various energy systems.

5.1.1 *Blockchain Technology*

The blockchain is a distributed database system with different nodes. It is an open bookkeeping system (Ledger), which can effectively verify the behavior and authenticity of all parties and parties involved. Its database is shared by all network nodes, updated and supervised by all users, but no user can own, control, or tamper with the database. A block of data blocks or data packets generated by cryptography is used to form a block, and each block data information is automatically time-stamped to calculate a data hash (Hash). A block contains the hash value of the previous block, starting from the Genesis Block and linking to the current area, thus forming a blockchain (Fig. 5.1).

Blockchain is a consensus agreement that follows a commonly recognized mechanism for direct, point-to-point information interaction without intermediate authority arbitration. Future blockchains will be widely used in financial transactions, public records, documents, private records, certification, physical assets and intangible assets. The blockchain builds trust through this consensus mechanism. With the blockchain, all nodes in the entire system can automatically and securely exchange data in a trusted environment. The blockchain can be automatically combined and enforced in real time, and the cost is very low. It can be point-to-point transactions. Without the need for a third-party platform, the transaction data will be permanently stored in the blockchain system. Once the transaction is reached, it is irrevocable.

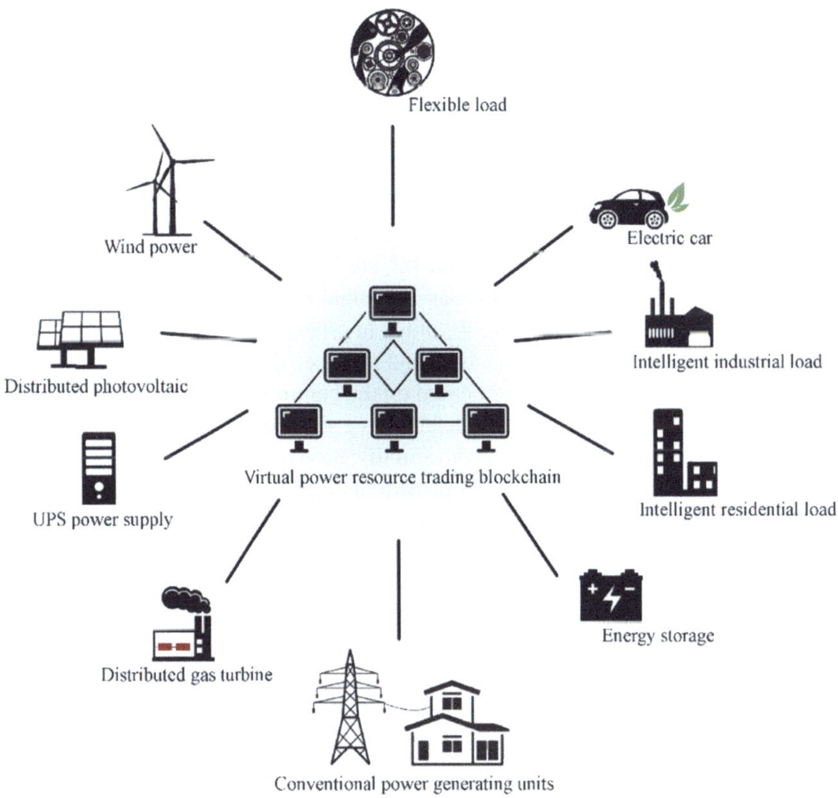

Fig. 5.1 Block chain technology diagram

Blockchain technology has been used in energy trading, and US energy company LO3 Energy has partnered with Bitcoin development company Consensus Systems to create an interactive grid platform based on blockchain systems for a small number of residents in the Brooklyn neighborhood of New York. Every green energy producer and consumer on the platform can freely trade green energy directly on the platform without relying on third parties, including carbon emissions trading on the world's networks. Community energy producers and consumers are able to conduct local energy transactions based on blockchain, balancing local production and consumption. Due to the encryption process and distributed storage, the confidentiality of users and transaction information can be guaranteed. In addition, the combination of microgrid control systems and blockchain technology will enable building roof PV system suppliers to return excess power to existing local grids in Brooklyn and receive payments directly from buyers. In essence, blockchain reduces transaction costs, especially the institutional costs of energy trading. If it is applied in the energy industry, it will greatly change the production and trading mode of the energy system. Transaction entities can cross the huge grid system and achieve point-to-point energy product production, trading, and energy infrastructure sharing.

The blockchain technology is used to build a carbon emission certification and trading platform, which will give each unit the carbon emission rights exclusive ID (identity), time stamp, and record in the blockchain, record the carbon emissions of the generator set in real time, carbon trading behavior and fines for over-standard companies.

Future digital information can be added to the blockchain. As long as it can be chained, information property rights can be clarified, protection conditions can be set, and transaction contracts can be automatically initiated and enforced without worrying about trust verification and trust execution, and trust is built on the blockchain rather than by a single organizational control, credibility can be cross-validated and supervised by multiple parties. This is an important manifestation of the decentralization and de-intermediation of blockchains. In other words, the trading of energy products can be packaged as information, and the electricity and oil and gas transactions within the block are completed based on the common market mechanism. If A and B reach a deal and A unilaterally defaults as a supplier, then it will be labeled as a non-integrity transaction. Based on the amplification effect of the huge blockchain system, A may find it difficult to find downstream customers in future transactions.

Blockchain technology has its own unique advantages, but it is still in the early stages of development and there are many problems. For example, an excessively large and daily increase in the storage space of a complete ledger is a huge challenge for a node that wants to run a complete blockchain; information distribution uses a full-network broadcast, which requires a good network performance. The transaction efficiency is low; the power is wasted; the single point node is upgraded by itself, and the upgrade cost is high. In addition, although the blockchain technology uses cryptography-related techniques and is highly secure, the entire blockchain network still has weakness in terms of privacy and security.

5.1.2 IoT and Energy Sharing

1. IoT

The Internet of Things (IoT) is another technological innovation after the computer and the Internet, representing the future development direction of computers and communications. The Internet of Things was first proposed in 1999, through information sensing devices such as radio frequency identification (RFID), infrared sensors, global positioning systems, laser scanners, gas sensors, etc., according to the agreed agreement, A network that connects items to the Internet for information exchange and communication for intelligent identification, location, tracking, monitoring, and management. In short, the Internet of Things is the Internet connected to the objects (Fig. 5.2).

The Internet of Things (IoT) is an information carrier based on the Internet and traditional telecommunication networks. It enables all the common physical objects that can be independently addressed to communicate with each other. It has three important characteristics: intelligence, advanced and interconnected. According to the definition of the International Telecommunication Union (ITU), the Internet of Things mainly addresses the interconnection between Things and Thing (T2T), Human to Thing (H2T), and Human to Human (H2H). However, unlike the traditional Internet, H2T refers to the connection between people using universal devices and items, which makes the connection of items more simplified, and H2H refers to the interconnection between people without relying on personal computers. Because the Internet does not take into account the problem of connecting anything, the Internet of Things is used to solve this traditional problem. As the name suggests, the Internet of Things

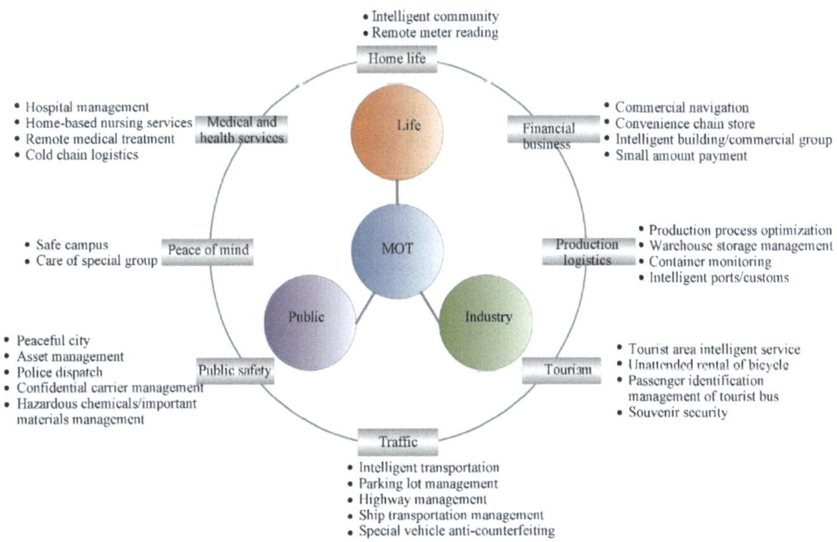

Fig. 5.2 IoT diagram

is a network of connected items. Many scholars often introduce the concept of M2M when discussing the Internet of Things, which can be interpreted as Man to Man, Man to Machine, Machine to Machine (Machine to Machine). In essence, the interaction between people and machines, machines and machines is mostly to achieve information exchange between people. Sensors and embedded systems have a significant position and role in the Internet of Things.

(1) Sensor technology: It is the key technology in computer application. So far, most computers have been dealing with digital signals. Since the computer, it has been necessary for the sensor to convert the analog signal into a digital signal before the computer can process it.
(2) RFID tags: It is a sensor technology. RFID Technology is a comprehensive technology integrating radio frequency technology and embedded technology, which has a broad application prospect in automatic identification and item logistics management.
(3) Embedded system technology: It is a complex technology integrating computer hardware and software, sensor technology, integrated circuit technology and electronic application technology. After decades of evolution, smart terminal products characterized by embedded systems can be seen everywhere, from the MP3 around the people to the satellite systems of aerospace. Embedded systems are changing people's lives, driving industrial production and the development of the defense industry. If the Internet of Things is likened to the human body, the sensor is equivalent to the human eye, nose, skin and other senses. The network is the nervous system, which is used to transmit information. The embedded system is the human brain, which sorting out the received information.

The Internet of Things has the characteristics of massive information, diversified access equipment types, and synergy and intelligence of object and object interoperability. Its success depends on many technologies, such as IoT architecture technology, unified identification and identification technology, network and communication technology, security and privacy protection technology, etc. The intelligent processing of the Internet of Things relies on advanced information processing technologies such as cloud computing and pattern recognition. Cloud computing can promote the realization of the Internet of Things and the smart earth from two aspects: First, cloud computing is the core of the Internet of Things; secondly, it promotes the intelligent integration of the Internet of Things and the Internet.

2. Energy Sharing

Jeremy Rivkin, an American energy economist, put forward the concept of "energy sharing", arguing in his book "The Third Industrial Revolution" that Internet technology would be combined with renewable energy, moving from a traditional centralized to an intelligent distribution, transforming the global grid into an energy-sharing network. This will be a new industrial revolution facilitated by the combination of energy and communication technologies, which will revolutionize human business models and social development.

In the energy sector, the spread of distributed energy provides people with the opportunity to participate in the production of clean energy, and people are willing to share these energy sources with each other in order to practice their own concept of sustainable development, which will gradually evolve into a new culture of energy sharing. The development of electric vehicles can rely on advances in information and energy technologies and inevitably integrate into a shared culture. Electric vehicles are the most suitable means of transportation to achieve travel sharing, sharing can not only solve the product limitations of this technology in the early stages of development, but also make its advantages to maximize. Information, energy, transportation sharing should be based on the network. It is necessary to form an information internet, energy Internet and transportation Internet, these networks are not only information, energy, transportation itself connectivity, but also mutual interaction and collaborative optimization through the data.

China is proposing to explore the construction of a global energy Internet to promote a global demand for electricity in a clean and green way. Developing the global energy Internet concept has become an important strategy for sustainable development.

Sunshine Power Internet (SPI) put forward the concept of global energy Internet trading platform, integrating the concepts of crowdfunding, group buying, flattening and de-intermediation commonly used in the Internet with successful business models such as Internet finance and people's ecological civilization values. Let the general public who wants to own a renewable energy power station come true through the platform. The platform not only serves investors around the world, but also serves operators of solar power plants around the world. It can also serve the needs of the world through these green power sources.

The State Grid Corporation has proposed the Global Energy Internet. The global energy Internet will be a strong smart grid with a UHV grid as the backbone grid channel, a clean energy source and a global interconnect.

Alibaba believes that the energy Internet is not only a tangible network, but also a physical connection between different energy sources. It should be an intangible network based on information and communication technology, integrated into every aspect of energy industry production, including the collection of energy data, sharing and processing of energy data, resulting in complete change of the energy industry business model. Through information and energy integration, an information-led, precisely controlled energy system is formed. The power system network is reconstructed on the principle of openness, interconnection and peer-to-peer sharing to improve the security of the power grid and the efficiency of power production, so that energy can be freely accessed and shared by any legal entity like the information in the network. Information is the key to it, not only the information transformation of the power grid, but also the product of the integration of multiple energy systems. Energy Internet is a multi-network system with power system as the core, based on the

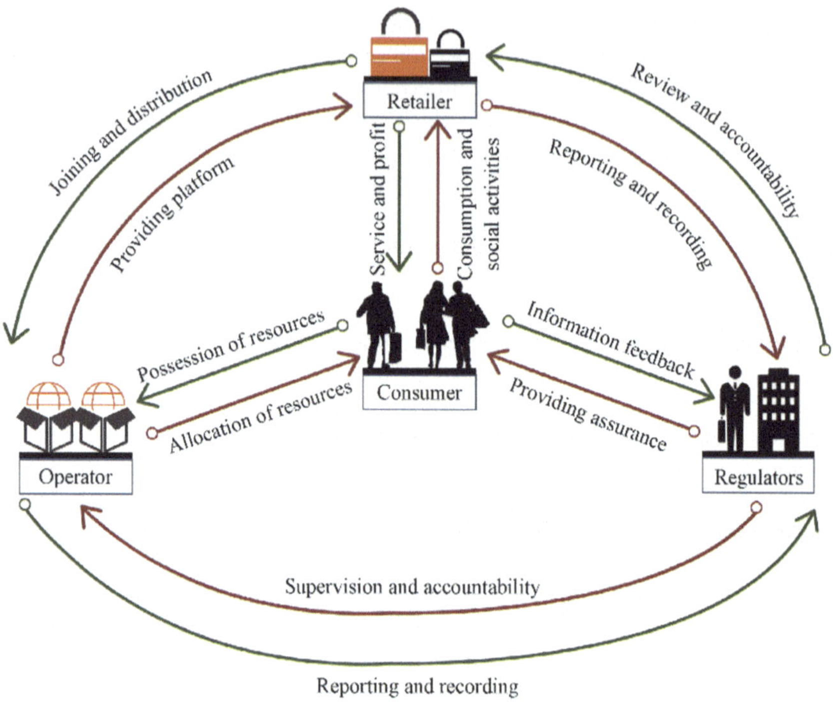

Fig. 5.3 Shared energy systems

internet and other cutting-edge information technology, with distributed renewable energy as the main, closely coupled with other systems such as natural gas networks and transportation networks to form complex multiple-network systems (Fig. 5.3).

The Internet revolution is the opening of the telephone line network by governments, allowing Internet companies to use the data packet exchange technology for information exchange on the telephone line network, which has made the Internet revolution. In the future, the power network and transportation network will be reopened, making it a new innovation platform, from information sharing to energy sharing and traffic sharing.

5.2 Power Energy Internet

Most clean energy sources can only be used efficiently if they are converted into electricity. As a high-quality, clean and efficient secondary energy source, electric energy can replace most energy needs and is an important terminal energy source. Therefore, building a global power energy Internet is the key to building a global energy Internet. The global power energy Internet is a strong smart grid with a UHV grid as the backbone grid (channel), to transport clean energy as the leading, and achieve global interconnection. The construction of the global power Energy Internet will make energy globally mutually beneficial, achieve staggered peak, mutual standby, and promote transformation and innovation in the energy sector as a traditional industry.

5.2.1 Associated Technologies

The core of power energy Internet includes two parts, Smart grid and UHV technology.

1. Smart Grid

Different institutions have different definitions for smart grids. According to the US Department of Energy's "Grid 2030", the smart grid refers to a fully automated power transmission network that monitors and controls each user and grid node to ensure two-way flow of information and electrical energy of all nodes from the power plant to the end user during the entire transmission and distribution process. The China Internet of Things School and Enterprise Alliance believes that the smart grid consists of many parts, which can be divided into intelligent substation, smart distribution network, smart energy meter, intelligent interactive terminal, intelligent dispatch, smart home appliances, intelligent power buildings, smart city power grid, intelligent power generation system, new energy storage system, etc. According to the European Technology Forum, a smart grid is a power transmission network that integrates all the actions of all users connected to the grid to effectively deliver sustained, economical and safe power. China Electric Power Research Institute of State Grid believes that smart grid is a new type of power grid based on physical grid, which integrates modern advanced sensor measurement technology, communication technology, information technology, computer technology and control technology with physical grid. The smart grid is designed to fully meet the user's demand for power and optimize resource allocation, ensure the safety, reliability and economy of the power supply, meet environmental protection constraints, ensure power quality, and adapt to the development of power market, so as to achieve reliable users, economical, clean and interactive power supply and value-added services.

1) Main features of Smart grid

Although the definition of smart grid has yet to be standardized, the characteristics of smart grid, which is different from conventional power grid, have gradually formed a consensus, the smart grid mainly has the following six aspects of the characteristics.

(1) Self-healing

Self-healing refers to isolating the problematic components of the grid from the system and, with little or no human intervention, can quickly restore the system to a normal state of operation, with little disruption to the user's power supply service. In essence, self-healing is the "immune system" of the smart grid. This is the most important feature of smart grids. Self-healing power grids conduct continuous online self-assessments to predict possible problems in the grid, identify existing or developing problems, and take immediate steps to control or correct them. Self-healing power grids ensure the reliability, safety, power quality and efficiency of the grid. Self-healing power grids will minimize the disruption of power supply services, fully apply data acquisition technology, implement decision support algorithms, avoid or limit the interruption of power supply, and quickly restore power supply services. A probabilistic risk assessment based on real-time measurements will identify the equipment, power plants and lines most likely to fail; real-time emergency analysis will determine the overall health level of the grid, trigger early warning that could lead to the development of the grid, and determine whether immediate inspection or appropriate measures are required; communications from local and remote devices will help analyze failures, voltage reduction, poor power quality, overload and other undesirable system states, take appropriate control action based on these analyses. Self-healing power grids often apply network design methods that connect multiple power supplies. In the event of a failure or other problem, advanced sensors in the power grid equipment identify the failure and communicate with nearby equipment to remove the faulty element or quickly switch the user to another reliable power supply, while the sensor also has the ability to detect the precursor of the fault, inform the system of the condition of the equipment before the failure actually occurs, the system will provide early warning information in a timely manner.

(2) Encourage and promote user participation

In smart grids, users will be an integral part of the power system. Encouraging and promoting user participation in the operation and management of power systems is another feature of smart grids. From a smart grid perspective, the user's needs are entirely another manageable resource that will help balance supply and demand and ensure the reliability of the system; from a user's point of view, electricity consumption is an economic option that obtains tangible benefits by participating in the operation and management of the grid and correcting the way it uses and buys electricity. In smart grids, users will adjust their consumption according to their power needs and the balance of power systems' ability to meet their needs. At the same time, the demand response (DR) program will meet the basic needs of users with more choices in energy purchases, reduce or transfer peak power requirements, enable power companies to minimize capital expenditure and operating expenses, reduce line losses and reduce the operation of inefficient peak-regulating power plants, while also providing significant environmental benefits. In the smart grid, the establishment of two-way real-time communication system with users is the basis to encourage and promote the active participation of users in the operation and management of power systems. Real-time notification to users of the cost of electricity consumption, real-time electricity prices, the status of the power grid, planned power outage information and some other services can be achieved, and users can also develop their own power use program based on this information.

(3) Defend against attack

Grid security requires a system-wide solution that reduces the vulnerability to physical and cyberattacks on the grid and quickly recovers from power outages. After being attacked, the smart grid will demonstrate the ability to recover quickly. The design and operation of smart grids have a strategy to prevent attacks, which can minimize the consequences of attacks and quickly restore power services. Smart grids can also withstand attacks on several parts of the power system and multiple coordinated attacks over time. Smart grid security strategies will include deterrence, prevention, detection, and response to minimize and mitigate the impact on the grid and economic development. Regardless of whether it is a physical attack or a cyber attack, the smart grid must strengthen the ability to resist risks by strengthening the close communication of major threat information between the power company and the government, emphasizing security risks in the power grid planning, and strengthening network security.

(4) Plug and play

The smart grid will safely and seamlessly allow different types of power generation and energy storage systems to access the system, simplifying the networking process, similar to plug-and-play, which poses a serious challenge to the power grid. Improved interconnect standards will make it easy to access a wide range of power generation and energy storage systems, ensuring that power generation and energy storage systems of varying capacities can be interconnected at all voltage levels, including distributed power sources such as photovoltaics and wind power, advanced battery systems, plug-in hybrid vehicles and fuel cells. It is easier and more economical for commercial users to install power generation equipment (including efficient cogeneration units) and power storage facilities. In smart grids, large centralized power plants, including environmentally friendly power sources, such as wind and large solar power plants and advanced nuclear power plants, will continue to play an important role. Strengthening the construction of transmission systems enables these large power plants to still deliver electricity over long distances. At the same time, access to a wide variety of distributed power sources reduces dependence on external energy sources, improving power supply reliability and power quality, especially for war and terrorist attacks.

(5) Real-time supply and demand interaction

In the smart grid, advanced equipment and a wide range of communication systems support the operation of the market in each time period and provide sufficient data for market participants, so the power market infrastructure and its technical support system are the key factors in booming the electricity market. Through the interaction of supply and demand in the market, the smart grid can most effectively manage parameters such as energy, capacity, capacity change rate, tidal blockage and other parameters, reduce tidal congestion, expand the market, and bring together more buyers and sellers. Users can feel the price increase through real-time quotation to reduce power demand, promote lower cost solutions, and promote the development of new technologies. New clean energy products will also provide more choices for the market.

(6) Asset optimization management

Smart grid optimizes and adjusts the management and operation of its grid assets to deliver the desired functionality at the lowest cost. This does not mean that assets will be used continuously to their limits, but to effectively manage what assets are needed and when they are needed. Each asset will be well integrated with all other assets to maximize its function and reduce costs. Smart grids will apply the latest technology to optimize the application of their assets. For example, through dynamic assessment techniques to enable assets to perform at their best capabilities, and through continuous monitoring and evaluation of their capabilities, assets can be used under greater load. The smart grid realizes the online status monitoring of the running equipment through the high-speed communication network to obtain the running state of the equipment, and gives the signal of the equipment to be repaired at the most appropriate time, realizes the state maintenance of the equipment, and makes the equipment run in an optimal state at the same time. The control device of the system can be adjusted to reduce losses and eliminate obstructions. Through these adjustments to the system control unit, the lowest cost energy delivery system is selected to increase operational efficiency. Optimal capacity, optimal conditions and optimal operation will greatly reduce the cost of grid operation. In addition, advanced information technology will provide a large amount of data and information and be integrated into existing enterprise-wide systems, greatly enhancing its capabilities to optimize operations and maintenance processes. This information will provide designers with better tools, create the best design, and provide planners with the data they need to improve capabilities and levels of their grid planning. Operation and maintenance costs, as well as grid construction investments, will be managed more effectively.

2) Development trend

Driven by the awareness of green energy conservation, smart grid has become a focus area for the development of countries around the world.

Japan plans to popularize the smart grid in 2030, and the government and the people are fully committed to building a smart grid overseas. In the battery field, the global market share of Japanese companies is aiming to reach 50% and get a market of about 10 trillion yen. The Ministry of Economy, Trade and Industry has established the "International Standardization Research Association for the Next Generation of Energy Systems", and Japan and the United States have established joint projects for smart grid experiments in Okinawa and Hawaii.

The United States has carried out corresponding work on power distribution and electricity utilization. Xcel Energy built the nation's first smart grid city in Burdell, and New York launched a superconducting network in 2010.

European countries are also vigorously developing smart grids. Italy installed and renovated 30 million smart meters and built an intelligent metering network; more than 1,000 small distributed power stations were built in the UK, with an average of 1 unit per 6.01 million people; the EU plans to select 25–30 cities to build so-called smart grids and space-insulated cities, the city's energy comes from the city's own garbage reuse, solar energy and wind energy, without relying on external energy. Germany has developed an E-Energy plan with a total investment of 140 million euros. During the period from 2009 to 2012, it will conduct empirical experiments on smart grids, as well as demonstration experiments on wind power and electric vehicles, and conduct test of Internet management on the investigation of electricity consumption. Major companies such as Siemens, SAP and ABB in Switzerland participated are all involved in the project.

During the "12th Five-Year Plan" period, China invested in the construction of "three-longitudinal and three-cross" UHV backbone grids and 13 long-distance tributary power transmission projects connecting large energy bases and main load centers, and initially built a core of world-class strong smart grid. The "Strong Smart Grid Technology Standard System Planning" formulated by the State Grid defines the roadmap for strong smart grid technology standards and is the world's first programmatic standard for guiding the development of smart grid technologies. By 2015, a strong smart grid with information, automation, and interactive features was basically established, forming three synchronous grids with North China, Central China, and East China as the receiving ends, and the northwest and northeast power grids as the sending ends, so that the resources allocation capacity of the grid, the economic operation efficiency, safety level, technology level and intelligence level have been comprehensively improved.

Smart grid is an inevitable trend in the development of grid technology. Technologies such as communication, computer and automation have been widely applied in the power grid and organically integrated with traditional power technologies, greatly improving the intelligence level of the power grid. The application of sensor technology and information technology in the power grid provides technical support for system state analysis and decision-making, making the grid self-healing possible. The mature development of dispatching technology, automation technology and flexible transmission technology provides basic guarantee for the development and utilization of renewable energy and distributed power. The improvement of communication network and the promotion and application of user information collection technology promote the two-way interaction between power grid and users. With the further development, application and integration of various new technologies with the physical grid, smart grids came into being.

The development of smart grid is an inevitable choice for social and economic development. To achieve clean energy development, delivery and consumption, the grid must increase its flexibility and compatibility. In order to withstand the increasingly frequent natural disasters and external disturbances, the power grid must rely on intelligent means to continuously improve its security defense capabilities and self-healing capabilities. In order to reduce operating costs and promote energy conservation and emission reduction, the grid operation must be more cost-effective, and intelligent control of the electrical equipment must be carried out to minimize the consumption of electricity. The rapid development of distributed generation, energy storage technology and electric vehicles has changed the traditional power supply and consumption mode, which has led to the continuous integration of power flow, information flow and business flow to meet the increasingly diverse needs of users.

1. Ultra high voltage

UHV (Ultra high voltage) power grid refers to the transmission network that the voltage level is AC 1,000 kV and DC plus or minus 800 kV and above. UHV technology has the characteristics of long distance, large capacity, low loss and cost efficient.

1) Development history

The study of UHV AC transmission technology began in the late 1960s. At that time, the power industry in Western industrial countries was in a period of rapid growth. The United States, the Soviet Union, Italy, Canada, Germany, Japan, Sweden and other countries have formulated their own plans to develop UHV according to their economic growth and electricity demand forecasts. U.S., Soviet Union, Japan, and Italy have built UHV test stations and test lines, specializing in UHV transmission and transformation technology and related power transmission and transformation equipment.

The Soviet Union began construction of a 1,150 kV transmission project from the late 1970s. In 1985, the Ekibastuz-Kokchetav-Kustantite high-voltage line was built, with a total length of 900 km and was put into operation at a voltage of 1,150 kV. By 1994, the UHV line was completed with a total length of 2634 km. The operation shows that the structure of the line and transformer substation used is basically reasonable. Major equipment such as ultra-high voltage transformers, reactors, and circuit breakers have withstood various operating conditions and have been operating normally since they were put into operation. In 1991, due to the collapse of the Soviet Union and the economic recession, the demand for electricity was clearly insufficient, resulting in the voltage decrease of UHV lines to 500 kV.

Japan is the second country in the world to use AC of millions of volts for electricity transmission. In order to meet the needs of large-scale atomic power stations in coastal areas to transmit power to the load center and to maximize the savings of line corridors, Japan began research on UHV transmission in 1973, which was not limited to the fact that the transmission capacity of UHV systems is 4–5 times that of 500 kV systems, as it can solve the problem that the short-circuit current of the 500 kV system is too large to be broken. For the selection of transmission voltage, Japan conducted a technical comparison study between 800 and 1,500 kV. Through a comprehensive comparison of various aspects, 1,000 kV was selected as the nominal voltage of the UHV system. A total of 426 km of Tokyo outer ring UHV transmission lines have been built. In order to ensure the reliable operation of the UHV system, Japan has built two UHV test and research bases in Shinohara and Akagi, and the operation is in good condition, which proves that UHV transmission and transformation equipment can meet the reliable operation of the system.

China began research of AC UHV transmission technology in 1986. The preliminary research includes data collection and analysis of UHV transmission at home and abroad, covering the demonstration of UHV voltage level, UHV transmission system, external insulation characteristics, electromagnetic environment, UHV transmission and transformation equipment and UHV transmission engineering. During the "8th Five-Year Plan" period, a preliminary study on the characteristics of UHV external insulation was carried out. The saturation performance of the long gap discharge was analyzed and discussed. The gap discharge between the conductor and the tower was tested experimentally. In 1994, China's first high-voltage transmission research line of millions of volts was built at the Wuhan High Voltage Research Institute. In order to meet the needs of the UHV test, in 1997, the research on the use of the power frequency test device to generate the long wave head operation wave was carried out. By transforming the power frequency test device, the voltage of long wave head operation wave was 2250 kV and the wave head time was 2800–5000 microseconds. At the same time, the research on the environmental impact of UHV lines was carried out. The results show that when 1 m diameter, 8 split wire are used, the ground static induction level of UHV lines is basically the same as that of 500 kV transmission lines. Radio interference levels are less than 500 kV transmission lines and audible noise is within public acceptable limits.

2) Significance of constructing UHV power grid

The construction of UHV power grids is related to national energy strategy and energy security, as well as the sustainable development of the national economy. It involves power planning, scientific research, design, construction, equipment manufacturing, and international cooperation. It is extremely challenging and leading and has great political significance, economic significance and the significance of technological innovation.

The construction of UHV grids is the fundamental guarantee for meeting the growing demand for electricity in the future. At present, China's economy continues to develop rapidly, the people's living standards are constantly improving, and power demand is growing rapidly. In China's power industry, the State Grid has an important task of optimizing energy resource allocation, ensuring national energy security and promoting national economic development. Accelerating the construction of a strong national grid with the UHV grid as the core is of great significance for achieving a wider range of resource optimization and allocation, promoting the efficient development and utilization of China's energy, and promoting sustainable economic and social development.

The construction of a UHV power grid is the basic way to improve the efficiency of China's energy development and utilization. China is the world's largest coal consumer. The regional differences in China's productivity development are very different, and the primary energy distribution is seriously uneven. Energy demand is mainly concentrated in the eastern and central economically developed regions, accounting for about three-quarters of the total demand; coal and hydropower resources for power generation are mainly distributed in the western and northern regions. This imbalance between energy distribution and consumption determines that energy resources must be optimally allocated nationwide, relying on large coal power bases and large hydropower bases to achieve on-site conversion of coal-fired power and large-scale development of hydropower, and to realize cross-regional, cross-watershed hydropower and coal power mutual protection, and large-scale transport of clean electric energy from the west and north to the central and eastern regions through construction of UHV power grids.

The construction of UHV power grid not only saves installed capacity, reduces network loss, reduces the disposal of water, improves utilization of thermal power equipment, saves land resources, but also improves the safe operation level of the power grid and avoids repeated construction of 500 kV power grid, with significant economic and social benefits. The large power grid not only plays an important role in resource optimization configuration, but also has many advantages in terms of safety, reliability, flexibility and cost efficiency. China has a vast territory, and the time difference and seasonal difference are very obvious. Coupled with the imbalance of regional economic development, the power load in different regions is highly complementary. UHV power grids have great potential in rationally utilizing energy, saving construction investment, reducing operating costs, reducing accidents and maintenance costs, and obtaining staggered peaks, peak regulation and compensation benefits between hydropower and coal power and cross watersheds.

Accelerating the construction of UHV power grids is of great significance for promoting the technological upgrading of China's power grid, driving the technological innovation of relevant domestic research and design departments, as well as manufacturing and construction enterprises, and improving the overall technical level and comprehensive competitiveness of electric power and related industries.

3) Challenges faced

China's UHV transmission technology has taken the lead in the world, and has achieved a series of innovations in the fields of large power grid control and protection, smart grid, clean energy access to the power grid, etc., but, at the same time, further researches such as reactive power balance, elimination of submerged power supply arc, overvoltage limit and insulation coordination and the manufacture of UHV equipment still need further research.

(1) Reactive power balance

The charging power of UHV lines is very large (about 5 times that of 500 kV lines of the same length), and the problem of reactive power balance is particularly prominent. The reactive power compensation of the reactor with fixed inductance value can limit the power frequency overvoltage during load shedding and the capacity rise effect during normal operation, but it may reduce the transmission capacity of the UHV line. In order to effectively solve this problem, it is necessary to focus on the technical requirements, parameters and the effects on the potential and power frequency and operating overvoltage of the controllable reactor.

(2) Elimination of submersible arc

The UHV line has large potential supply current, high recovery voltage, and difficult to extinguish the submerged power supply arc, which affects the no-current intermittent time and success rate of the single-phase reclosing. This requires studying the measures to quickly eliminate the submerged arc, to ensure that the fault phase extinguishes the potential arc after the circuit breakers at both ends are tripped. Japan uses a fast grounding switch to limit the single-phase reclosing time to 1 s, which solves the problem better.

(3) Overvoltage limit and insulation coordination

During the operation of the UHV transmission system, it will withstand the overvoltage caused by operational shock, fault impact, lightning impulse, etc. Studying the overvoltage and insulation coordination and choosing the correct and economical way to reduce the overvoltage level and insulation level of the equipment are very important for the safe operation of the system. The operating wave characteristics have a great influence on the size and cost of UHV equipment. If the saturation effect occurs, the size will increase non-linearly, which will make the cost too high. It is necessary to focus on new measures to reduce the overvoltage of UHV systems to a lower level, so that the requirements for internal overvoltage of UHV equipment in insulation coordination are reduced.

(4) Series capacitance compensation

In order to improve the transmission capacity of UHV lines and enhance the stability of the system, it is necessary to study the series compensation device and related parameters and technical requirements of UHV system.

(5) External insulation

The discharge characteristics of air gaps of various electrode structures in UHV lines and transformer substations, the discharge characteristics of external insulation of various types of transmission and transformation equipment, and the altitude correction coefficient at different altitudes need to be combined with Chinese characteristics. The lightning protection, anti-fouling and live working of UHV lines also need to combine the lightning activity along the line, the soil resistivity and the distribution of pollution sources. The special study of the principle of reasonable insulation cooperation, combined with China's live working methods, tool characteristics to study the minimum safety distance and combination gap, provide technical basis for design, operation and maintenance.

(6) UHV equipment

Domestic equipment manufacturers have developed and supplied ultra-high voltage power transmission and transformation equipment such as wires, fittings, arresters, capacitive voltage transformers and capacitors, and suspension insulators; and development of UHV transformers, reactors, high-speed grounding switches and switches equipment needs to be accelerated; UHV casing, GIS (gas insulated closed switchgear), controllable reactors and other equipment are difficult to manufacture, and there is no manufacturing capacity in China at present. In the early stage of UHV transmission project construction, it is still necessary to introduce foreign products. It can be expected that the development of UHV transmission technology will not only promote the development of the power grid, but also effectively promote and improve the manufacturing level and production capacity of China's high-voltage electrical appliances.

5.2.2 The Power Internet Model

The concept of the energy Internet was first seen in the book of "The Third Industrial Revolution" by American scholar Jeremy Rifkin. Rifkin's vision is to combine the various types of distributed generation equipment, energy storage equipment and controllable loads with ICT technology (information and communication technology) and smart grids to provide users with a clean and convenient energy supply and users can participate in the entire process of energy production, consumption and optimization. Since then, the energy Internet concept has been actively responded by many scholars and IT (information technology) companies, and has become a hot spot in the energy field. But like the integrated energy system, the energy Internet has not yet formed a unified definition. The broader concept now is that the energy Internet is a complex multi-network flow system with power systems at its core, based on the internet and other ICT technologies, with distributed renewable energy as the main primary source, and closely coupled with natural gas networks, transportation networks and other systems.

The Energy Internet is a centralized and distributed multi-energy system that is dominated by the grid, supported by renewable energy, based on electricity, and supplemented by other energy sources. The system realizes the interactive and orderly configuration of supply and demand of multiple energy systems through the operation platform with Internet technology as the core, thus constructing a new ecological energy system with economical, low carbon, intelligent, efficient and balanced development (Fig. 5.4). The energy Internet can be divided into three levels: physical infrastructure network, information data platform and value realization platform. The functional requirements and specific connotations of each level are as follows.

A physical infrastructure network can implement a multi-energy converged energy network. The power network is the main framework, which integrates gas and heat networks to cover the entire industrial chain of energy production, energy transmission, energy consumption, energy storage and energy conversion. Energy interconnection relies on a highly reliable and secure mainframe (grid, pipe network, road network); flexible and scalable; and plug-and-play support for distributed energy (production, storage, and consumption).

The information data platform is a system for realizing the physical integration of energy information. Information sharing of multiple energy systems, information

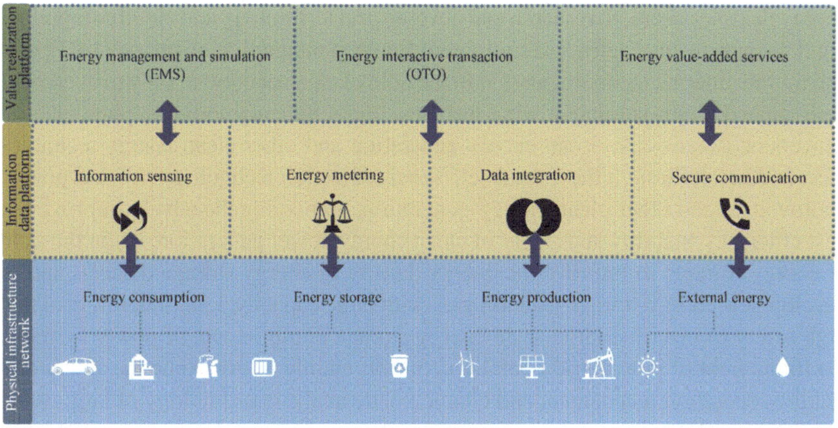

Fig. 5.4 Power grid-led energy internet architecture

flow and energy flow are closely coupled through the information physical integration system, and the information flow will run through the entire life cycle of the energy Internet, including its planning, design, construction, operation, utilization, monitoring, maintenance, and asset management, asset valuation and trading. The smart grid has done a lot of basic work in the integration of information physics systems, and has realized the effective combination of information flow and power flow of the main network.

The value realization platform can realize innovative mode energy operation. Innovative mode energy operation should make full use of Internet thinking, and use Internet technologies such as big data, cloud computing, and mobile Internet to maximize the utility of Internet energy producers, energy consumers, energy operators, and energy regulators. The important role of Internet is for stabilizing growth, promoting reform, adjusting structure, benefiting people's livelihood, and preventing risks.

To realize the energy Internet, we need to strengthen research in the technical fields of UHV, smart grid, clean energy power generation, energy storage, and grid operation control. (1) In the field of UHV: mainly to overcome the UHV DC transmission and transformation technology of plus or minus 1,100 kV, accelerate the development of equipment and engineering applications; break through the UHV DC submarine cable technology to solve the problem of trans-sea transmission; accelerate the development of extremely cold, UHV core equipment manufacturing and application technology under special circumstances such as extreme heat; break through technologies such as positive and negative 800 kV UHV multi-terminal flexible DC, ultra high voltage/UHV flexible DC, high voltage DC circuit breaker, and controllable series compensation and controllable parallel UHV flexible AC transmission technology such as reactors. (2) In the field of smart grid: mainly to overcome the intelligent technologies covering power generation, transmission, substation, power distribution, power consumption, and dispatching, to adapt to various types of clean energy flexible access; to research and develop grid technology and cloud computing, big data, new technologies and equipment for deep integration of Internet of Things and mobile internet technologies. (3) In the field of clean energy: mainly to overcome high-capacity high-parameter wind turbines, high-efficiency low-cost photoelectric conversion, large-scale ocean energy generation and other clean energy technologies, and break through the virtual self-synchronization technology of wind power, photovoltaic and other clean energy generation, and comprehensively improve the cost efficiency, stability and safety of clean energy power generation. (4) In the field of energy storage: to improve the energy density of energy storage devices, and to develop new energy storage technologies with large capacity, long life, low cost and high safety to meet the large-scale energy storage requirements of power systems. (5) In the field of power grid operation control: mainly to overcome the safety and stability operation mechanism and characteristic analysis technology of large-scale AC/DC hybrid power grid, build a new generation power grid simulation platform, improve calculation efficiency and calculation accuracy; break through grid fault diagnosis, recovery and automatic reconstruction. Such technologies can comprehensively improve the defense capabilities of large power grids under cascading failures, extreme disaster weather or external damage.

5.3 Pipeline Energy Internet

With the improvement of the electrification level of the whole society, the energy Internet that is widely discussed nowadays is based on the grid as the core platform, but the energy system still coexists in various forms outside the power grid, and almost all of the "off-grid" energy on both sides of supply and demand is fossil energy. Let fossil energy combine clean energy or renewable energy, and improve efficiency through the network of gas pipeline network. It is the necessary way to promote energy transformation at this stage, which can promote the clean and low-carbon transformation of fossil energy, and alleviate the current shortage of natural gas in China. Therefore, a new model of the energy Internet led by the gas pipeline network is proposed. Focusing on the core business of natural gas pipelines, we will use cloud computing, big data analytics, Internet of Things, mobile applications and other technologies to establish an intelligent pipeline energy internet system based on "Internet" smart energy.

5.3.1 Pipeline Energy Internet Overview

An important feature of the pipeline energy Internet dominated by the gas pipeline network is the combination of natural gas pipeline network and hydrogen energy. There are two main methods: one is a hydrogen-based centralized hydrogen energy

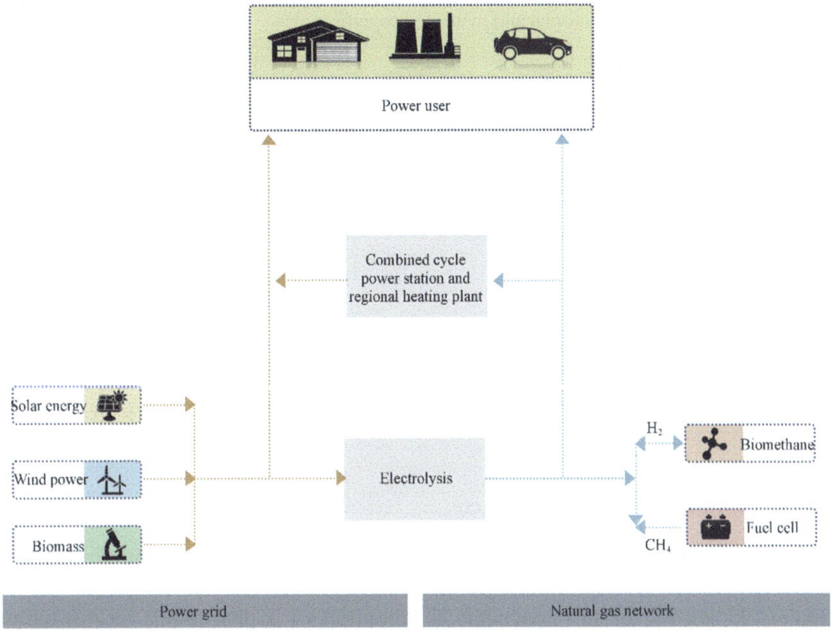

Fig. 5.5 Coupling relationship between power energy and hydrogen energy

supply system; the other is the use of natural gas pipeline network to transport hydrogen energy or organic compound hydrogen storage technology.

1. Centralized hydrogen energy supply system

The centralized hydrogen energy supply system (Fig. 5.5) includes clean wind power, electrolyzed water hydrogen production equipment, pressure hydrogen storage equipment, fuel cell FC or hydrogen fuel internal combustion engine HICE (Hydrogen Internal Combustion Engines), hydrogen transport, applications and so on. Through the control system to adjust the ratio of wind power on grid to hydrogen production, maximize the acceptance of abandonment wind power of wind power, alleviate the bottleneck problem of large-scale wind power grid, use abandoned wind power to electrolyze water to produce hydrogen and oxygen, and increase the storage density of hydrogen by pressure hydrogen storage. As a multi-purpose, high-density clean energy source, hydrogen can be fed back to the grid through FC or HICE to improve the quality of wind power on-grid. It can also be used as an energy carrier to enter industrial and commercial fields by vehicle or pipeline. This requires natural gas pipeline or hydrogen pipeline. The transportation of pipelines helps long-distance transportation of hydrogen to downstream applications such as hydrogen refueling stations.

Table 5.1 Comparison of hydrogen and methane properties

Properties	Hydrogen	Methane
Relative molecular mass	2.02	16.04
Critical temperature (Kelvin)	33.2	190.65
Critical pressure (MPa)	1.315	4.540
Density (kg/m^3)	0.0852	0.6801
Low calorific value (MJ/m^3)	10.23	34.04
High calorific value (MJ/m^3)	12.09	37.77
Flameout constant	0.60	0.67
Explosion limit (%)	18.2–58.9	5.7–14.0
Theoretical air requirement (m^3/m^3)	2.38	9.52
Coefficient of diffusion into the air (m^3/sec)	6.1×10^{-5}	1.6×10^{-5}

1) Pipeline transportation of hydrogen

For large amount, long-distance hydrogen transport, the use of piping is the most effective method. Hydrogen has been transported over long distances for more than 60 years. However, the physicochemical properties of hydrogen are very different from methane, the main component of natural gas (Table 5.1, where the density, low calorific value, and high calorific value are both at the status of 15 °C and the standard atmospheric pressure). The addition of hydrogen to the natural gas pipeline will change the nature of the gas in the pipeline, which will lead to risks in the pipeline, and affect the pipeline gas transmission efficiency and end users, mainly reflected in the following aspects.

(1) The effect of hydrogenation on pipelines.

Hydrogen can cause hydrogen damage to the pipeline, and the degree of hydrogen damage is related to hydrogen concentration, delivery pressure, and pipe material properties (Huang 2013; Zhang and Jiang 2015).

① Hydrogen embrittlement risk. Hydrogen embrittlement can change the mechanical properties of pipes, such as the hydrogen embrittlement of steel, which accelerates the cracking of tiny cracks. The risk of hydrogen embrittlement is difficult to calculate, not only related to the material of the pipeline, but also to the operating life of the pipeline. Generally, the higher the pressure of the pipeline, the longer the service life, the greater the risk of hydrogen embrittlement, and the conclusion can be drawn through complicated experiments.

② Risk of hydrogen permeation. Since the permeation rate of hydrogen is much larger than that of natural gas, the hydrogen transporting pipe produces hydrogen loss. In addition, the permeation ability is also related to the pipeline material. In cast iron pipes and fiber cement pipes, the risk of hydrogen permeation is high; in PE (polyethylene) pipes, the hydrogen permeation rate is five times higher than that of natural gas. The calculation shows that the permeation loss of hydrogen volume in the PE pipe for one year is close to 0.0005–0.001% of the volume of its transport gas.

③ When the concentration of hydrogen added to the natural gas pipeline is not less than 10%, the pipe adaptability analysis can be carried out from the aspects of pipeline steel grade, operating pressure, impurity elements, and hydrogen environmental toughness index according to the relevant requirements of ASMEB 31.12-2014. The corresponding measures that the pipeline needs to take under the conditions of adding different hydrogen concentrations. When the natural gas pipeline is added with a hydrogen concentration of less than 10%, it can be analyzed according to the results of European research. If the pipe grade is lower than X52 (including X52), it can be directly used for transportation; if the steel grade is higher than X52, it needs to be adaptively analyzed from the aspects of pipe strength, hardness, chemical element and toughness, and the corresponding measures that the pipeline needs to be taken under this condition is obtained.

(2) The effect of hydrogen on pipeline transport power after addition.

After hydrogen is added to the natural gas pipeline, if the pressure is constant, no matter how large the volume fraction of hydrogen is, the gas transmission power of the pipeline is reduced compared with the natural gas, that is, the gas transmission capacity of the pipeline is reduced. When the volume fraction of hydrogen in the mixed gas is more than 83%, the gas transmission power of the pipeline slightly rebounds, but the gas transmission capacity of the pipeline is still lower than that of the natural gas.

Increasing the delivery pressure is an effective way to increase the gas transmission power of the pipeline. Therefore, when designing the gas pipeline network, the design pressure should be appropriately increased to leave a certain margin for the operation pressure of the pipeline network to cope with future changes. Once the hydrogen-mixed natural gas is used on a large scale, it is only necessary to increase the operating pressure on the existing pipeline to meet the pipeline's gas transmission power requirements.

(3) The effect of hydrogen on the use of gas by end users

Any burning appliance is designed with a certain gas composition. When hydrogen is added to the natural gas and the gas composition changes, the combustion condition of the burner will change, which will affect the performance of the burning appliance and even cause the original burning appliance to malfunction. Therefore, interchangeability must be considered after hydrogen is added to natural gas.

According to the relationship between the combustion potential of the hydrogen-mixed natural gas and the Wobbe index, after the addition of hydrogen, the Wobbe number of the mixed gas tends to decrease, and the combustion potential sharply increases. It shows that with the increase of the volume fraction of hydrogen in the mixed gas, the heat load of the burning appliance decreases, and the flame propagation speed of the gas increases sharply, and the risk of tempering of the burning appliance increases sharply. In order to effectively solve the gas interchangeability and the adaptability of the burning appliance, according to the characteristic index, the volume fraction of hydrogen in the hydrogen-mixed natural gas must be less than 23%, that is, controlled in the replaceable region. In this way, it is ensured that the end user's burning appliances do not need to be changed after the hydrogen is added.

2) Current status of hydrogen pipeline at home and abroad

At present, the total length of pipelines used to transport industrial hydrogen has exceeded 4,000 km globally, the diameter is 0.25–0.30 m, the operating pressure is generally 1–3 MPa, and the gas transmission is 310–8900 kg/h, of which Germany has 208 km. The air liquefaction company of France has 880 km in Belgium, France and New Zealand, and the United States has reached 720 km.

The 24 km hydrogen pipeline from Jiyuan Industrial Park, Henan Province to Jili District of Luoyang is the hydrogen pipeline with the largest diameter, highest pressure and largest output in China. It passed the safety inspection of Henan Provincial Safety Supervision Bureau and was put into production in October 2016. The pipeline was built by the China Petroleum Pipeline Bureau and has been operating safely for nearly three years. The hydrogen pipeline is made of L245NS seamless steel pipe with a diameter of 508 mm, design pressure of 4 MPa, and an annual transportation capacity of 104,000 tons.

3) Hydrogen refueling technology

Hydrogen refueling stations are an important infrastructure for the supply of fuel hydrogen to fuel cell vehicles, and are highly valued by major automakers, energy suppliers and governments. In recent years, the construction of hydrogen refueling stations around the world has grown at a significant rate. Developed countries such as the United States, Japan and Germany have begun to gradually shift their hydrogen refueling stations from the demonstration and promotion phase to commercial operations.

As of January 2018, there are 327 hydrogen refueling stations in operation worldwide, of which 139 are located in Europe, 118 in Asia, one in Australia, one in South America and 68 in North America. Japan, the United States and Germany, as the three countries with the largest number of hydrogen refueling stations, have 91, 40 and 45 hydrogen refueling stations in operation. There are also 25 hydrogen refueling stations in Germany that are under construction or will be open soon. There are currently 12 hydrogen refueling stations in operation in China, which are located in 8 provinces and cities including Shanghai, Beijing, Guangdong, Jiangsu, Henan, Hubei, Liaoning and Sichuan.

Looking at the development and construction of international hydrogen refueling stations over the past 10 years, the following trends are mainly presented.

(1) The refilling pressure is increased from 35 to 70 MPa

In order to improve the driving mileage of fuel cell vehicles, the current fuel cell passenger cars basically use a 70 MPa on-board hydrogen supply system, and the filling pressure of the hydrogen refueling station is also increased from 35 MPa to 70 MPa. In recent years, the newly-built hydrogen refueling stations generally have a filling capacity of 70 MPa, and some hydrogen refueling stations with a capacity of only 35 MPa have been upgraded to 70 MPa hydrogen refueling stations through technological transformation. The 70 MPa hydrogen refueling station places high demands on the storage, compression and filling of ultra-high-pressure hydrogen. Relevant technical equipment, control strategies and standard regulations are in the stage of development, demonstration, standardization and demonstration.

(2) Hydrogen production method is changing to producing hydrogen from the renewable energy

As the number of global hydrogen refueling stations increases, the hydrogen source of the hydrogen refueling station is diversified. Countries generally choose various hydrogen production methods according to their own resource characteristics. For example, in areas with good chemical industry base, industrial by-product hydrogen is usually used as hydrogen source, while in areas with rich renewable energy, the renewable energy is used to produce hydrogen, and there are more and more demonstration projects using renewable energy to produce hydrogen. Although there is no advantage in using renewable energy to make hydrogen compared with the economics of traditional fossil energy hydrogen production methods, from the perspective of long-term energy development, it is strategic and forward-looking to actively develop hydrogen production methods based on renewable energy, which has been widely recognized.

(3) The station construction method is developed from a single hydrogen refueling station to a joint station such as hydrogen refueling/gas fueling, hydrogen refueling/charging

A practical problem in the development of hydrogen refueling stations is the difficulty in site selection and high construction costs. Research on safety shows that the safety level of the hydrogen refueling station is at the same level as that of the gas station and the natural gas station. Therefore, it is a good solution to combine the hydrogen refueling station with the gas station, such as the hydrogen refueling station in Frankfurt, Germany, the Shell hydrogen refueling station in Washington, DC, which are models for joint construction with oil and gas stations. With the rapid development of pure electric vehicles, further integration with electric vehicle charging stations is the latest development trend. For example, the German Brandenburg international airport hydrogen station in Berlin combines charging, hydrogen refueling, natural gas, biofuels and other filling facilities which provides fuel resupply for a variety of new energy vehicles, and its hydrogen comes from the nearby wind farm electrolytic produce hydrogen. It is believed that with the gradual replacement of traditional vehicles by new energy vehicles, the traditional gas station can also gradually complete the transition from gas station to gas station/hydrogen refueling and charging station to hydrogen refueling and charging station.

(4) Gradual transition to the network development

The networking of hydrogen refueling stations is an obvious development trend. With the rapid development of fuel cell vehicles, the demand for large-scale hydrogen infrastructure facilities is becoming more and more urgent. Developed countries such as the United States, Germany, and Japan have developed roadmaps and implementation plans for fuel cell vehicles and hydrogen infrastructure, and are gradually moving from the initial hydrogen energy highway or small hydrogen refueling network to a larger network and a wider geographical area.

2. Hydrogen energy supply system based on organic compound hydrogen storage technology

The hydrogen energy supply system based on organic compound hydrogen storage technology mainly refers to the direct use of natural gas pipelines to transport organic liquids containing hydrogen. The most important feature of organic liquid hydrogen storage materials is that they are generally liquid at room temperature, similar to gasoline, and can be transported and stored very conveniently. This compound acts as a carrier for hydrogen, which is chemically added to the carrier during the hydrogenation process to form a stable hydride liquid, passing through a common storage and transportation process similar to petroleum products. At the end users, the hydrogen-carrying organic liquid hydrogen storage material releases hydrogen through the catalytic reactor for use in the hydrogen fuel cell, and after dehydrogenation, the hydrogen storage carrier is returned to the storage tank, and the hydrogen storage station is replaced with a new hydrogen-carrying organic liquid for hydrogen storage.

The working principle of organic liquid hydrogen storage technology can be divided into three processes.

(1) Hydrogenation: hydrogen is added to the liquid hydrogen storage carrier by catalytic reaction to form a hydrogen storage compound of organic liquid which can be stored stably under normal temperature and pressure conditions.
(2) Transportation: the hydrogen-storing organic liquid after hydrogenation is transported to the supply terminal through an ordinary tank truck. It is pumped in a form similar to gasoline and diesel oil, and is simply and quickly filled into the organic liquid storage tank on the vehicle.
(3) Dehydrogenation: the dehydrogenation process of the hydrogen storage organic liquid is carried out in a hydrogen supply (dehydrogenation) unit. The fluid is sent to the dehydrogenation reaction device through metering pump, and a catalytic dehydrogenation reaction occurs under a certain temperature condition. After the reaction product is separated to gas and liquid, the hydrogen gas is transported to the fuel cell stack, and the liquid carrier after dehydrogenation is recovered by heat exchange and recycled.

At present, the bottleneck of this technology is how to develop a dehydrogenation catalyst with high conversion, high selectivity and stability. Since the reaction is a heterogeneous reaction with strong endotherm and is limited by equilibrium, it is necessary to select a suitable reaction mode and optimize the reaction conditions to solve the problems of heat transfer and mass transfer. In addition, it is necessary to solve the economic problems of the overall process of naphthenic hydrogen storage technology, such as reducing the amount of precious metals in the catalyst and improving the energy conversion efficiency of dehydrogenation by vehicle.

5.3.2 Power-to-Gas Technology

The so-called Power-to-gas technology is a technology that uses renewable energy or redundant power to produce hydrogen or synthesize methane. Hydrogen is produced by electrolysis or other methods of hydrogen production. The hydrogen produced can be directly injected into pipelines for transportation or other industrial fields; or converted to methane by methanation when combined with carbon dioxide produced in the atmosphere, biomass waste gas and industrial waste gas for subsequent transport and use. If the electricity used to electrolyze water comes from solar or wind energy, the power-to-gas technology can form a comprehensive utilization system for renewable energy in all applications (Shell Hydrogen study-energy of the future-sustainable mobility through fuel cells and H_2, 2017).

With the expansion and popularization of renewable energy, the surplus power of renewable energy has increased significantly, and the method deal with the natural output change of renewable energy are usually including output suppression, transmission network expansion, energy storage, and demand dynamics, which are usually implemented by the closed system in the power system. Power-to-Gas (PtG) is not that the power system handles natural change of the power output of renewable power alone, but an overall response from the energy system extended to the gas pipeline network and the transport sector for low carbonization purposes, and its industrial chain diagram is shown in Fig. 5.6.

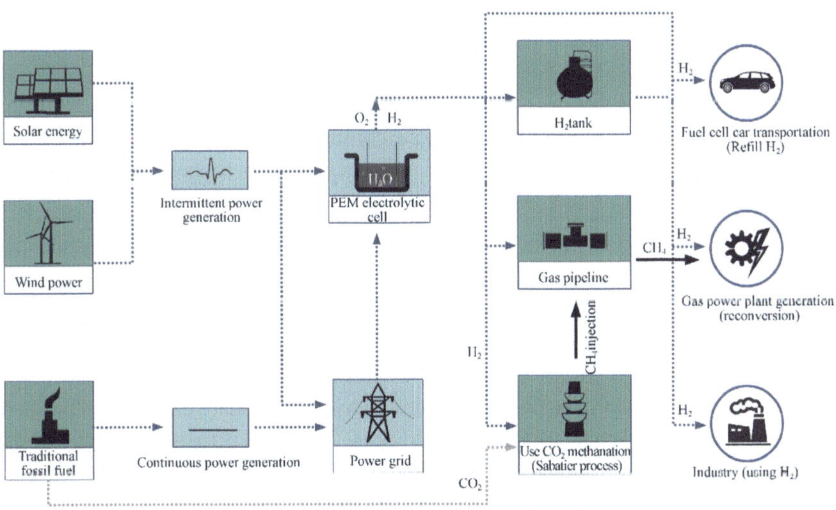

Fig. 5.6 Industrial chain of power-gas technology (power generation, conversion/storage and application)

1. Principles of power-to-gas

The power to hydrogen is only subjected to electrolyzed water reaction, and the reaction efficiency is 56–73%, which avoids the energy loss of the methanation reaction and reduces the infrastructure cost associated with the methanation reaction. However, hydrogen injection into existing natural gas pipelines poses risks in terms of pipes (such as hydrogen embrittlement and infiltration), so there are certain restrictions. The power-methane conversion includes two steps of electrolysis water reaction and methanation reaction, and the efficiency is 42% to 58%, which is lower than that of power to hydrogen. Methane can be injected directly into existing natural gas pipelines and storage facilities for long-distance energy transfer and large-scale storage (Li and Yin 2016).

The catalytic hydrogenation of carbon dioxide is an important way of recycling carbon dioxide, the main reaction products are CH_4 and H_2O, and the side reaction products are CO, C, CO_2, C_2H_6, $C_xH_yO_z$ and so on.

The methanation reaction is an exothermic reaction. From the thermodynamic point of view, the methanation of carbon dioxide at high temperature is restricted by the thermodynamic equilibrium, and the low temperature is more favorable for the forward reaction; from the kinetic point of view, the low temperature carbon dioxide methanation reaction has a low reaction rate. It can be seen that the key to achieving low temperature carbon dioxide methanation is a highly active catalyst. The carbon dioxide methanation catalyst is mainly a supported catalyst in which a Group VIII B metal such as Ni, CO, Rh, Ru, and Pd is used as an active component. At present, the efficiency of the methanation reaction is 75% to 80%. The synthesized methane can be mixed into the natural gas pipeline network as a main component of natural gas.

The sources of carbon dioxide, the reactants for the methanation reaction are varied, so the power-to-gas projects can be combined with biogas plants, sewage treatment plants, coal-fired power plants and other projects to save investment and achieve comprehensive benefits optimization.

(1) Separation of carbon dioxide from biogas as a source of carbon dioxide for power-to-gas. The main components of biogas are methane and carbon dioxide, the proportion of which are 55%-65%, 30%-40%, respectively.
The methane and carbon dioxide in the biogas are separated, the purified methane is injected into the natural gas network, and the carbon dioxide is used as the reactant of the power-to-gas. In addition, the heat generated by the methane reaction can be used by biogas plants and biogas purification devices. The device for separating carbon dioxide from biogas for power-to-gas is shown in Fig. 5.7.

(2) Carbon dioxide is captured from sewage treatment plants as a source of carbon dioxide for power-to-gas. Sewage treatment is one of the important sources of greenhouse gas emissions. The greenhouse gases emitted are mainly carbon dioxide, methane and nitrous oxide, of which carbon dioxide can be used as a source of carbon dioxide for power-to-gas. In addition, the oxygen and heat generated during the power-to-gas conversion process can be used by the sewage treatment plant. The

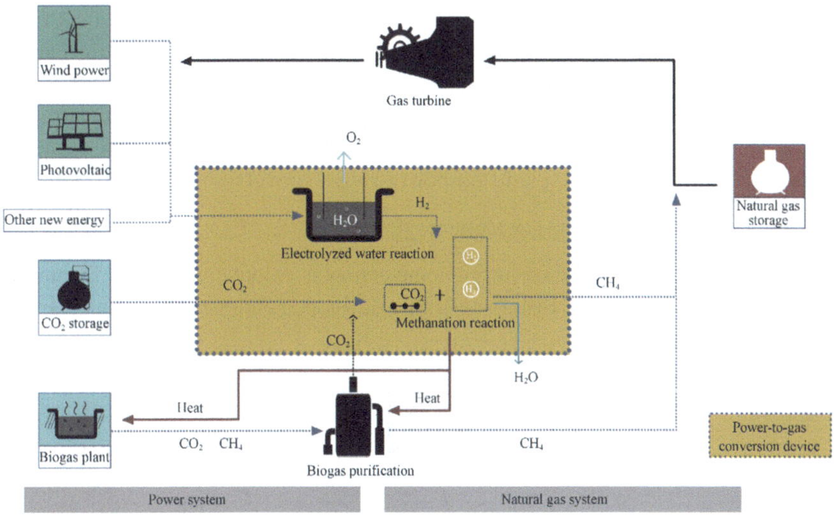

Fig. 5.7 Separation of carbon dioxide from biogas for power-to-gas devices

device for capturing carbon dioxide from a sewage treatment plant for power-to-gas conversion is shown in Fig. 5.8.

(3) Carbon dioxide is captured from the exhaust of coal-fired power plants as a source of carbon dioxide for power-to-gas. Carbon dioxide capture technology in coal-fired power plants is an important technical means to deal with climate change and reduce

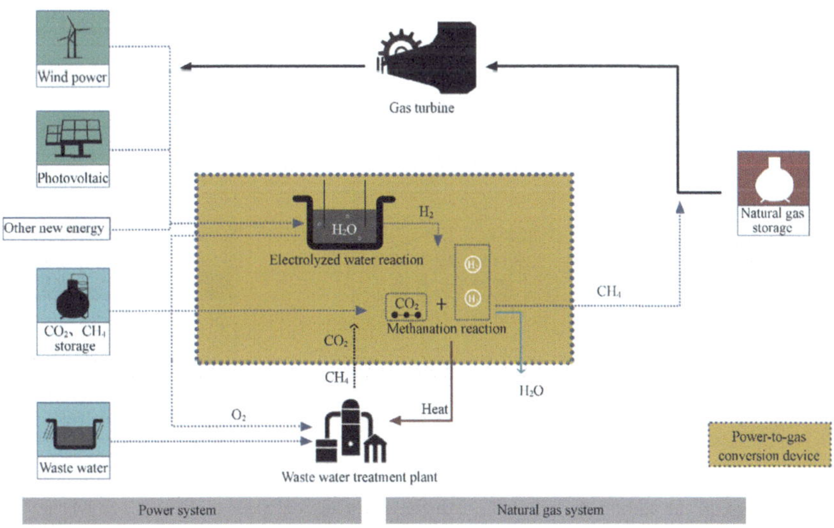

Fig. 5.8 Capture of carbon dioxide from sewage treatment plants for power-to-gas devices

greenhouse gas emissions. The increase in carbon capture rate leads to a decrease in the cycle thermal efficiency of the unit. Studies have shown that when the carbon capture rate is around 90%, the cycle thermal efficiency of the unit is reduced by about 9%. In addition, the heat generated during the power-to-gas conversion process can be used by the power plant. The device for capturing carbon dioxide from the waste gas of coal-fired power plants for power-to-gas conversion is shown in Fig. 5.9.

2. Applications of power-to-gas technology

The main technology of power-to-gas conversion is relatively mature, Denmark, Germany, the United States, Britain and other developed countries have successively

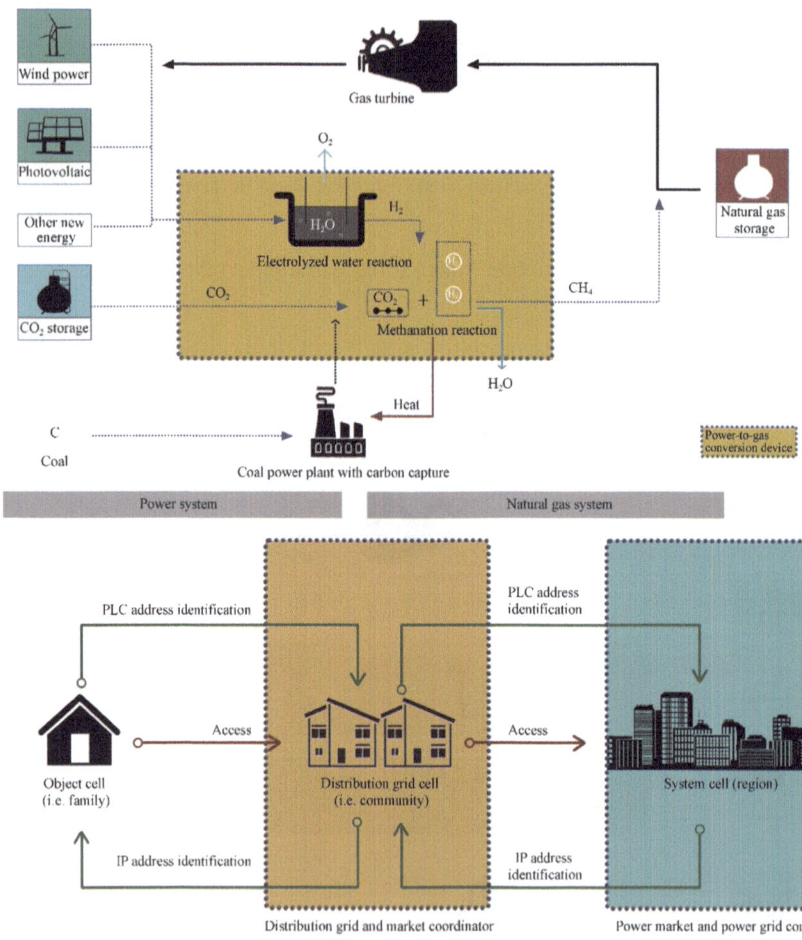

Fig. 5.9 Capture of carbon dioxide from exhaust gases in coal power plants for power-to-gas devices

carried out and completed the relevant projects. The key issue at present is how to improve the conversion efficiency and reduce costs.

1) Denmark

(1) Biocatalytic PtG-BioCat project. The project is located in Avedøre, Denmark, and started at the Avedøre wastewater treatment plant in July 2015 and became operational in April 2016. The source of carbon dioxide is pure carbon dioxide provided by an anaerobic digester (60% methane and 40% carbon dioxide) or from an upgraded system of conventional biogas. Hydrogen comes from the 1 MW alkaline electrolytic cell from Hydrogenics, which uses local excessive wind power to generate electricity, and by-product oxygen is recycled to the wastewater treatment process. In addition, the heat generated by methanation will also be integrated, and the resulting synthetic natural gas will be output to the user at a pressure of 4 bars. The project's overall budget is €6.7 million, partly supported by the ForskEL program (€3.7 million), which is supported by Danish power and natural gas transmission network operator Energinet.dk. The alliance is headed by Electrochaea, a methanation technology developer, and includes partners such as Hydrogenics, Audi, NEAS Energy, HMNGashandelA/S, BiofosA/S and Insero, etc.

(2) El-Opgraderet Biogas project in Haldor Topsoe. The goal is to build and operate a pilot plant in Foulum, Denmark, which purifies biogas by obtaining hydrogen from a 40 kW SOEC electrolytic cell. The project was launched in June 2013 for three years, but no information has been published so far. Haldor Topsoe has conducted extensive research on SOEC electrolysis of power-to-gas.

(3) Green Syn Fuels (2010–2011) project. The project analyzed the development of methanol and DME (dimethyl ether) as a green synthetic fuel through technical and economic calculations. The most popular concept is methanol synthesis based on wood gasification, which is supplied by a SOEC electrolytic cell. The total budget is €190,000, partly funded by the European Union Development Plan Danish Plan.

(4) Biogas SOEC (2011–2012) project. The project studied two upgrade routes based on SOEC electrolytic cells: direct methanation of biogas by the addition of hydrogen, and co-electrolysis of biogas and steam to produce carbon monoxide and hydrogen for subsequent methane formation. In addition, the presence of sulfur in the biogas supplied to the SOEC electrolytic cell was experimentally studied. The project has a budget of €180,000 (67% funded by the Forsk NG program).

(5) Methane Association Towards the Methane Society (2011–2012) project. Haldor Topsoe built a three-reactor pilot plant to purify biogas, which operated for 1,000 h at Aarhus University. The first two reactors reduced the sulfur content below 0.04 mg/L and the third reactor removed the remaining traces. In addition, before the last cleaning reactor enters, hydrogen is supplied to consume the oxygen contained in the biogas to prepare for the Sabatier reaction.

However, no methanation experiments were conducted. In addition, the ongoing El-opgraderet biogas project is the second phase of development for this research. The budget is 214,000 euros.

2) Germany

There are more than 30 European power-to-gas projects. For the utilization of power-to-gas products, the amount injected into natural gas pipeline network greatly exceed that used in the transportation. The amount of hydrogen and methane is almost the same, and the renewable energy power used is mostly variable load. There are more than 20 power-to-gas projects planned and under construction in Germany.

(1) Audi verification project in Germany. The project produces methane from hydrogen and carbon dioxide made from renewable energy, and carbon dioxide is supplied by neighboring bioenergy devices. The produced synthetic methane is transported to the compressed natural gas (CNG) station via the original natural gas pipeline network. There are already 1,000 CNG stations in Germany. Compared with hydrogen directly supplied to fuel cell vehicles, direct hydrogen supply must be built with new infrastructure, so transportation to CNG stations can reduce investment costs. Germany's MCPHY Company supplied a complete set of power-to-gas conversion equipment to Audi. This is the first power-to-gas project in Europe. It uses solar power to generate hydrogen through water electrolysis. The Audi plant takes the tail gas (carbon dioxide) emitted to the atmosphere by cars to react with hydrogen to obtain methane, and then injects the synthesized methane into the local natural gas. In the pipeline network, the whole process not only utilizes renewable energy, but also generates electricity under zero carbon emissions. At the same time, it also absorbs the tail gas emitted by the automobile to the atmosphere to synthesize methane, which protects the environment and reduces carbon emissions.
(2) Mainz Energy Zone Project. On July 2, 2015, the Mainz Energy Zone project jointly developed by Linde Group, Siemens AG of Germany, Municipality of Mainz, Germany, and Rheinmann University of Applied Sciences was officially launched. The project aims to use clean electrical energy sources (such as electricity generated by wind farms) to produce hydrogen and store it. It is reported that the total investment of the project is 17 million euros. After several years of construction, it has become the world's largest green hydrogen station.
(3) Thüga Group project. The project was supported by companies from 13 countries and several research institutes such as DVGW (German Gas and Water Industry Association), European Energy Research Institute, Fraunhofer Institute, etc., as well as the Ministry of economy, transportation and regional development in Hessen, Germany. The "Thüga Group" power-to-gas conversion plant was established in Frankfurt, Germany, to absorb the excess electricity generated by renewable energy and convert it into hydrogen storage in the natural gas pipeline network. The first phase of the project is from the end of 2013–2016,

with an investment of more than 15 million Euros. The project has completed its first sale of product. The second phase of the project is the use of hydrogen and carbon dioxide to synthesize natural gas and inject into the pipe network. The Fraunhofer Institute developed software that monitors the entire system in real time. The core equipment for power-to-gas conversion (such as proton exchange membrane electrolyzers, etc.) is supplied by ITM Power, UK.

3) The United States

(1) SoCalGas, the largest natural gas supplier in the United States, announced the implementation of the first power-to-gas conversion project in the United States. The project will introduce Proton On Site, the world's leading proton exchange membrane fuel cell technology. This will be the first energy storage demonstration project in the United States, and the actual data of the renewable energy system can be collected for future reference. The project will be powered by two proton exchange membrane electrolyzers rated at 7 kW and 60 kW, respectively, and the two electrolytic cells will be powered by local photovoltaic power. The hydrogen produced by the cell will be transported to the National Research Center's natural gas pipeline system. In this way, renewable electricity will be stored in hydrogen or natural gas for later use. To meet the requirements of large-scale energy storage, Proton has launched a megawatt electrolytic cell platform project. The new 2 MW M-series electrolytic cell produces 13 times more hydrogen than conventional electrolytic cells, which also reduced its production cost of hydrogen per kilogram by 50%.
(2) The University of Chicago has established an innovative company called Electrochaea to promote and apply power-to-gas conversion technology. The company has been selected as one of the world's 100 clean technology innovation companies. In June 2016, Electrochaea built a demonstration project called BioCat in Copenhagen. On this basis, Electrochaea is preparing to build a 10 MW project in Hungary, which will be the world's largest power-to-gas conversion project.
(3) Bifunctional material for carbon dioxide capture and conversion at Columbia University. Columbia University in New York, with the financial support of BASF, studied bifunctional materials that use renewable hydrogen to capture carbon dioxide from sources and convert them to synthetic natural gas at the same reactor and at the same temperature.

4) The United Kingdom

Keele University's Green Pioneer Energy Experiment, the HyDeploy project is the first test in the UK for a hybrid network of domestic hydrogen and natural gas. The project is led by the gas network Cadent and works with Northern Gas Networks and a team of technical experts. The HyDeploy project is supported by the OFGEM Network Innovation Competition with the goal of establishing a hybrid supply network of hydrogen and conventional natural gas. Beginning in 2019, Keele University will conduct a one-year real-time trial of the school's private gas network. With 1,2,000 students and staff, Keele University has 350 mixed-use buildings on campus, including on-campus housing, university facilities, and science parks. The school has reached a smaller town size and is a suitable testing ground. The project will determine the amount of hydrogen that users can safely use, minimize the impact of gas changes on users, and not change the user's existing household appliances. Although the natural gas used by "city" in the British natural gas network was composed of up to 60% hydrogen in the 1970s, it was not widely available. In the HyDeploy project, users will use hydrogen as their energy source for the first time at home. A part of purposes in the first phase of the HyDeploy project is a gas safety check in the houses and building of the test area. At this stage, a series of ordinary household gas appliances were also tested in the laboratory, and the effects of hydrogen on gas pipes of different materials were further studied. The results of the HyDeploy project can be extended to the entire UK natural gas network.

In addition, Japan, France, Switzerland, Sweden, Poland, Finland, Spain, Australia, Thailand and other countries have carried out experiments and projects related to power-to-gas conversion.

5.3.3 China's Pipeline Energy Internet

In 2016, the National Development and Reform Commission, the Energy Bureau and the Ministry of Industry and Information Technology jointly issued the "Guiding Opinions on Promoting the Development of Smart Energy with the Internet" (Development and Reform Energy [2016] No. 392), proposing to promote the deep integration of energy and information, and to build an organic, efficient, low-cost, sustainable, and controllable energy Internet system that promotes supply-side structural reforms and the energy revolution in the energy domain. Oil and gas pipeline transportation and intelligent pipe network are the main development directions in the future and an important part of smart energy. Through the construction of intelligent pipeline system, from the concept, strategy, management, technology and other aspects of the integration and innovation of the "Internet", give full play to the optimization and integration of modern information and data technology in the allocation of production factors, to achieve real-time sharing and optimization integration of information, knowledge, technology, truly realize the intelligent management of pipeline production operation, operation and maintenance guarantee and emergency response, form a new core competitive advantage and endogenous growth momentum, and achieve comprehensive transformation and upgrading.

PetroChina has initially completed the digital design system and the construction of a full life cycle database. Starting from the east line between China and Russia, the implementation of full digital handover, fully intelligent operation, and full life cycle management, and comprehensive promotion of smart pipeline construction, marks the stage of China's oil and gas pipelines entering the intelligent pipeline construction from digital pipelines. In the process of scale construction of oil and gas pipeline network, the standardization, modularization and informatization level of pipeline business have been continuously improved, and the transformation from traditional pipeline to digital pipeline has been completed, and digitalization of design, mechanization of construction, electronic procurement of materials, and management informationization have been realized. With the maturity of next-generation information technologies such as big data, cloud computing, and the Internet of Things, it has built a solid foundation for the construction of smart pipelines. Based on the same standard and pipeline digitalization, the smart pipeline integrates the pipeline life cycle data through the "end + cloud + big data" architecture, provides intelligent analysis and decision support, and realizes pipeline visualization, network and intelligent management. The intelligent pipeline has the characteristics of comprehensive data integration, perceptual interaction visualization, system integration and interconnection, accurate supply matching, intelligent and efficient operation, and predictable and early warning control. The realization of "five major changes" are as follows: the pipeline data is changing from dispersion, paper to centralization, digitalization; risk control mod is changing from passive to active, operational management is changing from human-oriented to system intelligence, resource allocation is changing from local optimization to overall optimization, and pipeline information system is changing from isolated to centralized integration. As the starting point of China's smart pipeline, Sino-Russian East Line Pipeline Project will implement the concept of full digital handover, fully intelligent operation and full life cycle management, formulate construction implementation plan, and finally realize full process management, full information management, and full digital handover.

Sinopec Sales Co., Ltd. carried out the construction and exploration of intelligent pipeline system according to the idea of "adapting to local conditions, overall planning, long-term integration and step-by-step implementation", and proposed the intelligent pipeline system architecture design and intelligent logistics platform construction plan. The intelligent pipeline system covering digital management, integrity management, operation management, emergency management, and integrated management has core functions such as data mining and sharing, integrity management and emergency management. It has completed the enterprise pipeline data center and steadily pushed optimizing and deepening of the business module forward, which has built a solid foundation for the establishment of the "Internet" smart pipe network.

Intelligent pipeline system is conducive to transforming pipeline management methods, and has a significant role in improving decision-making management efficiency and optimizing resource allocation. Around the core business of pipelines, it can realize intelligent analysis and early warning, rapidly collaborative response and safe and efficient operation, build an intelligent pipeline expert system based on improving the system data quality, and provide support for the transformation and upgrading of oil and gas pipelines in the smart energy era by using technologies such as cloud computing, big data analysis, Internet of Things, and mobile applications.

5.4 Foreign Energy Internet Development Model

On the basis of the deep integration of the Internet concept, advanced information technology and energy industry, through the multi-energy collaborative energy network, the energy system of information physics integration, and the energy operation of the innovative model, green, coordinated and efficient development will be realized, which will drive economic growth and support the energy revolution.

5.4.1 German Model—Supply Side Driven

The German energy Internet began with 6 E-Energy demonstration projects. Following the successful completion of the E-Energy project, the IRENE, Peer Energy Cloud, ZESMIT and Future Energy Grid projects began (Yin et al. 2015).

Background of the 6 demonstration projects of E-Energy are as follows: In 2000, Germany formulated a policy on energy transition with the goal of making energy utilization throughout Germany environmentally friendly, economical and safe. This goal laid the foundation for the German framework of the energy Internet. On the power market platform where consumers, service providers and grid operators participate, six major technologies are supported: hierarchical system operation, cellular system control, terminal energy conservation, improved primary energy utilization efficiency, production management, and load management.

The E-Energy project was developed under the three strategies implemented in Germany. Green T Pioneer: The National Action Plan, which began in 2008, aims to increase the penetration of ICT (Information and Communication Technology) in the energy sector and on the other hand reduce the energy consumption of the national IT (information technology) industry. Hightech Strategy 2020: Originated in 2006, it was formally formed in 2010 to help transform innovation into market-oriented products and accelerate the application of scientific research results. Digital Germany 2015: Started in 2010 with the aim of advancing the development of the German ICT industry and infiltrating into other industries.

With the support of government departments, BMWi and BMUB, E-Energy carried out demonstration projects in the whole German bidding demonstration area, and developed and demonstrated key technologies and business models in the energy Internet field from 2008 to 2013. There were 6 regions bidding successfully, and the government issued a total of 140 million euros in five years to support the demonstration project.

1. Main features

(1) The bidding area is not only a matter of the local government, but also a collective bidding jointly by the power grid company, equipment manufacturing enterprise and technology development enterprise. Each region has developed a detailed plan after conducting an analysis of the local energy economic infrastructure and corporate strength before it is eligible to participate in the bidding.

(2) The six demonstration areas are distributed in both north and south, and the proportion of renewable energy development in each region has local characteristics. For example, the eTelligence project in the north belongs to a large number of wind power feeding areas, while the Meregio project in the south has more photovoltaics. Demonstration projects in the area where each project is located are synchronously conducted with large cities and small villages.
(3) E-Energy's goal is not only to promote research and product development and demonstration. The deeper purpose is to establish communication links between power grids, enterprises, governments, and associations, and to vigorously promote the development of standards and the promotion of international standards.
(4) E-Energy had 6 demonstration areas, but the content of each regional project is different. In order to implement 6 E-Energy projects, the government has established a cross-project joint organization that integrates universities, enterprises, consulting, associations, the research institution and designated the leader unit is BAUM consulting company.

2. Specific cases

1) eTelligence Project

The eTelligence project was selected for the Cuxhaven Port, which has a small population, abundant wind energy resources, and a relatively large load. The project consists of a wind power plant, a photovoltaic power station, two cold storages, a cogeneration plant and 650 households. Typical adjustment measures include: (1) The automatic power regulation of the cold storage load with the fluctuation of the electricity price and the output of the wind power, realizing the deep integration of both the power generation facing power consumption and the power consumption facing power generation; (2) Introduction of the combination of sectional electricity price and dynamic electricity price, the benchmark electricity price between 3:00 and 20:00 is 39.8 Euro cents, and the rest time is 11.7 Euro cents. Between 8:00 and 20:00, make out preferential electricity prices and peak electricity prices depending on the load and new energy power generation; (3) The introduction of virtual power plants concept to centrally manage multiple types of distributed power sources and load conditions. From the above analysis, it can be concluded that the eTelligence project uses Internet technology to construct a complex energy regulation system, using load regulation to stabilize the intermittent and fluctuation of new energy output, improve the ability to absorb new energy, and build a regional Integrated energy market.

After several years of operation, the eTelligence project achieved good economic and social benefits, mainly reflected in: (1) The use of virtual power plants reduced the power imbalance caused by wind power output uncertainty by 16%. (2) The sectional price has saved the family 13% of the electricity. The dynamic electricity price has increased the load by 30% during the off-peak electricity price and the load has decreased by 20% during the peak electricity price. (3) As a production consumer of electric energy, the virtual power plant can reduce the cost of 8–10% according to the supply and demand of internal electricity and the regional electricity sellers. The heat-based cogeneration as the producer of electric energy realized the full sales of electricity, under the regulation of virtual power plants, its profits have also increased. (4) The Open IEC 61850 communication protocol standard based on the eTelligence project has been recognized by the German industry.

In the eTelligence project, virtual power plants that combine renewable energy with variable heat demand are more controllable. In contrast, cogeneration systems are more difficult to make market-based judgments.

2) E-DeMa project

The E-DeMa project is located in the cities of Mülheim and Krefeld in the Rhine-Ruhr region, focusing on the construction of distributed energy communities under differentiated power load densities.

The basic means is to integrate multiple roles such as users, power producers, electricity sellers, and equipment operators into one system, and perform virtual power transactions, including power and spare capacity. A total of 700 users participated in the E-DeMa project, and 13 of them installed a miniature cogeneration unit. At the heart of the E-DeMa project is power management through intelligent energy routers that enable both intelligent monitoring and demand response, as well as distributed power to the grid or other power users in the community. Smart energy routers are composed of photovoltaic inverters, household energy storage units or smart meters. According to the power generation of the power plant and user load conditions, the power transmission route is selected and distributed in the best path to transmit power. For the received power, the energy router recalculates the network load and user load changes, assigns a new physical address, and transmits it. For a complex network, the use of energy routers can improve the overall efficiency of the network and ensure the security and stability of the power grid. The first type of energy router is a typical direct load manual control, and the second type of energy router is to automatically control home appliances through the Internet to receive price signals. In the communication mode, the user's power consumption data in the first type of energy router is transmitted to the user terminal only through the communication channel within the house; the second type is completely automatic control, and the power data is transmitted to the terminal, and is also sent to the household power management system, adaptive adjustment and control are thus performed. The E-DeMa project is one of the tests in the EU's "Super Smart Grid" program. Currently, it is only a virtual power market test. Although its energy-saving effect has not directly affected the electricity price, residents can use the smart meter to understand the electricity consumption of each period, so as to adjust the electricity habit accordingly, to achieve the effect of saving electricity bills.

3) Meregio project

The Meregio project was built in two small rural towns in Göppingen and Fleamt in southern Germany. The local industry and commerce are more developed, and a large number of distributed renewable energy sources have been connected to the distribution network. Due to the weak grid structure of the distribution network, the access of distributed power sources has caused a series of problems in the power grid. Therefore, the project aims to better predict and allocate resources by sensing the load of each user, to better predict and allocate resources, thereby reducing grid congestion and improving the operational efficiency of the distribution network. There are 1,000 industrial and commercial enterprises participating in the project with home users.

The main measures of the Meregio project are: (1) In terms of electricity prices, a traffic light electricity price system is introduced, in which red indicates a high electricity price, yellow represents a medium electricity price, green indicates a low electricity price. In the first 3 months, users see that the price of electricity changes from red to green, which will increase electricity consumption by 25–35%. From yellow to green, it will increase electricity consumption by 10–22%. When the user adjusts the power usage behavior after 3 months, it gradually decreases and reaches a steady state. When the color changes from red to green, it will increase by 7–12%, and when the yellow color turns green, it will increase the power consumption by 4–7%. (2) In the customization of the user load curve, the smart meter uploads the user's real-time load data to the EnBW data center. After a period, the load characteristic curve of each household is established according to the statistical method, so that the grid operator can more accurately predict the load situation of each user at each point, and then calculate the load situation of the distribution network, locate the weak link of the distribution network, and display it with visualization technology. When the user's electricity behavior is abnormal, the user will receive a smart reminder in the personal account. (3) In terms of equipment renovation, the Meregio project installed a variable transformer in the substation to adjust the transformer according to the power situation of the renewable energy fed in the low-voltage network, thereby stabilizing the voltage level of the medium-voltage network, and the introduction of adjustable transformers resulted in a nearly double increase in the capacity of distributed energy acceptance.

4) Moma project

The Moma project is located in Mannheim, an industrial city in southern Germany. The city has many satellite cities and the energy supply is largely from distributed energy. The Moma project is a typical energy Internet demonstration project whose main goal is to build a virtual energy market for energy producers, consumers and network operators to study the impact of consumers on energy efficiency.

In this energy market, consumers are placed at the core and have the right to know how to reduce energy consumption and spending. The Moma project is mainly divided into three phases. The first one is the preliminary preparation stage, which is the achievability of the test technology for the subsequent test. The second test is the actual test stage. The new system called "Energy Manager" is tested. The system can realize automatic opening and closing of electrical appliances. In the third test phase, smart meters, energy housekeepers and web portals coordinate and cooperate to provide more free and diverse electricity price options to participating customers, encouraging them to use electricity during off-peak hours.

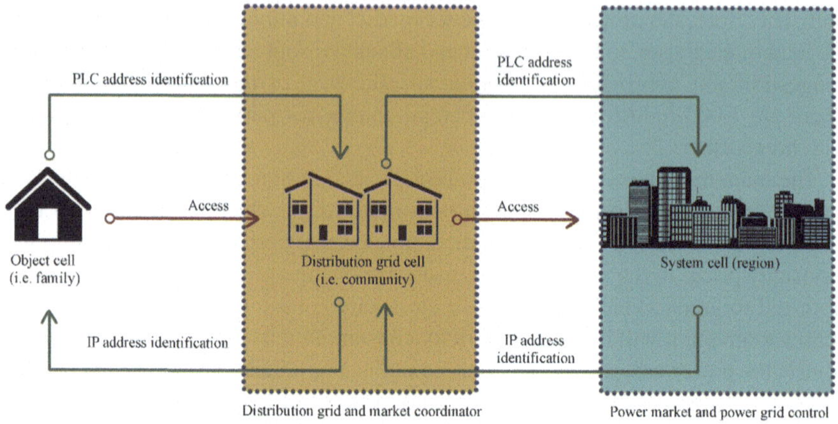

Fig. 5.10 Introduction to cell power grid

The main contribution of the Moma project is to propose the concept of a cell grid and divide the cell grid into three levels: physical cells, distribution network cells and system cells. A schematic diagram of the cell grid is shown in Fig. 5.10.

Each layer of cells in the cell grid is self-optimizing, trying to balance and adjust. Through the PLC communication (power line communication) between the cells and the superior cells, the superior cells recognize the lower cells through the Internet Protocol (IP), and the system cells have functions of dispatching and controlling the power grid, thereby completing the power market transactions. For system cells, only the amount of interface provided by the cells of the distribution network is concerned, and the photovoltaic panels and appliances in each household are not concerned. The residual power imbalance left after the equilibration of the physical cells and the distribution network cells is scheduled and complemented by the system cells. The advantages of cell division of the grid are as follows: (1) Try to make energy consumption close to the point and reduce transmission loss. (2) Ensure the safety of the power grid, when a cell power grid collapses, it will not cause the large power grid to collapse, and the cell power grid can be immediately shut down and restarted quickly. (3) Reduce the complexity of grid management caused by a large number of distributed devices, and partition and stratification are suitable for future energy development. (4) Improve the real-time performance of data processing through distributed data processing and storage. (5) Some cells can sometimes be self-sufficient and formally separated from the superior network. The power quality, voltage, frequency, and phase angle of the cell grid at each level are guaranteed to be within a certain range. Therefore, the structure of the cell grid has higher requirements for power generation than the traditional grid.

5) Smart Watts project

A total of 250 families participated in the Smart Watts project in Aachen, with the goal of using high-end, proven ICT technology to track every step of the value chain from production to consumption, and to communicate with users about the source of electricity and the level of power consumption of the appliances used by users. Consumers use smart meters to learn the real-time changing electricity price, and adjust the household electricity plan and electric vehicle charging plan according to the price of electricity. Users are free to choose their own electricity bills, and the electricity price in the package is also the time-of-use price. Through the data obtained by the smart socket, the user can not only view how much electricity is used by each power device on the electronic device, but also how much electricity is used; or the appliance can be controlled by the application program, and the program can be set by setting parameters and automatically determine the operation of the appliance. The results of the actual test data show that the load has increased by 10% during the lowest price period; the load has decreased by 5% during the highest price period. Another highlight of the Smart Watts project is the design of the EEBus. In response to the problem of multiple communication standards between various electrical devices in smart appliances, EEBus emerged as a communication translator, capable of translating current communication standards to sellers, power grids, power generators, users, home appliances dealers and other relevant departments.

6) Reg Mod Harz project

The Reg Mod Harz project is located in the Haci Mountain area of Germany. Its basic structure consists of two photovoltaic power plants, two wind farms and one biomass power generation with a total capacity of 86 MW. The production plan is determined by the predicted price of the market in the day-to-day market and the intraday market and the standby market. The goal of the Reg Mod Harz project is to coordinate decentralized renewable energy generation equipment such as wind, solar, biomass and pumped storage power plants to optimize the combined recycling utilization of renewable energy. Its core demonstration content is a virtual power station that integrates energy storage facilities, electric vehicles, renewable energy and smart household appliances on the electricity consumption side. It contains many energy demand elements that are closer to real life and can be called the prototype of the energy Internet. The main measures of the Reg Mod Harz project are: (1) Establish a household energy management system, the household appliance can be plugged and used to the system, the system determines the operating state of the household appliance according to the electricity price, and can also track the changes of renewable energy generation according to the user's load. The two-way interaction between load and new energy power generation can be achieved. (2) Ten power management units are installed in the distribution network to monitor operating parameters such as voltage and frequency of key nodes and locate weak links in the power grid. (3) Photovoltaics, wind turbines, biomass power generation, electric vehicles and energy storage devices together constitute a virtual power plant to participate in electricity market transactions. The typical results of the Reg Mod Harz project include three aspects: (a) Conducted the development and design of the Java-based open source software platform OGEMA, the implementation of standardized data structures and equipment services for external electrical equipment, independent of the manufacturer to support building automation and energy efficiency management, which enables plug-and-play of load devices in information transmission. (b) The virtual power plant directly participates in power trading, enriches the regulation and control means of the distribution network system, and provides a reference for the distributed energy system to participate in market regulation. (c) Based on the regulation of hydropower and energy storage equipment in the Haci area, the fluctuation and instability of power output such as wind turbines and photovoltaics are well smoothed, and it is proved that it is entirely possible to realize 100% of clean energy supply in regional power market for areas with abundant renewable energy.

The construction of the German Energy Internet Demonstration Project is a gradual process. Demonstration projects such as Peer Energy Cloud, ZESMIT, Future Energy Grid and Web 2 Energy have also achieved satisfactory results, but due to their basic concepts and related measures similar to the demonstration projects in E-Energy, which will not be discussed in detail here.

5.4.2 *Japanese Model—Demand Side Drive Type*

Hydrogen energy is a focus for Japan to promote energy transition and ensure energy security. In the "4th Energy Basic Plan", which was formulated in April 2014, Japan has clearly defined the strategic direction of accelerating the construction and development of a hydrogen energy society. According to this strategic goal, the Ministry of Economy, Trade and Industry of Japan formulated the "Strategic Roadmap for Hydrogen Energy and Fuel Cells" in June of the same year, and proposed a development roadmap for achieving the goal of the hydrogen energy society in three steps. In October of the same year, Tokyo Metropolitan announced the construction of the hydrogen energy society in Tokyo will be comprehensively promoted, and the Tokyo Olympic Games 2020 will be a grand event to showcase the development of Japan's hydrogen energy society to the world. In December of the same year, Toyota Motor Corporation took the lead in launching commercial mass-produced fuel cell vehicles around the world. At the end of 2015, sales of Japanese domestic fuel cell cogeneration units exceeded 150,000 units. Japan has designated 2015 as "the starting year of hydrogen energy". With the advancement of hydrogen energy production, storage, transportation and utilization technologies, Japan has significantly accelerated the pace of creating a hydrogen energy society in recent years. Following the staring year of hydrogen energy, according to the established strategic goals and roadmap, the year 2020 is defined as "Hydrogen Energy Olympic Year", the year 2025 is defined as "the first year of hydrogen going abroad", and the year 2030 is defined as "the first year of hydrogen fuel power generation." (International Hydrogen Industry Development Report 2017).

1. Overall Strategy

In June 2014, the Ministry of Economy, Trade and Industry of Japan issued the "Strategic Roadmap for Hydrogen Energy and Fuel Cells" in accordance with the Strategic Plan for Hydrogen Energy Social Development of the Fourth Energy Basic

Plan. In March 2015, the research and development of "Hydrogen Fuel Power Generation Research Report" was published; in March 2016, the "Strategic Roadmap for Hydrogen and Fuel Cells" was revised; in March 2017, further research and development of the "Carbon Free Hydrogen Fuel Research Report" was conducted. A series of frequently published documents constitutes the main contents of the overall strategy for the development of hydrogen energy society in Japan and its implementation roadmap.

1) Strategy and Policy

Hydrogen energy is recognized as the most promising clean energy in the 21st century. Japanese Prime Minister Shinzo Abe announced in his policy address in early 2017 that hydrogen energy is a trump card for Japan's energy security and global warming measures. Japan is about to kick off the future of hydrogen energy society. It will take the lead in realizing urban hydrogen energy power supply in the world, take the lead in using liquefied hydrogen energy ship to transport large-scale hydrogen energy, and take the lead in building an international hydrogen supply chain from production, transportation to consumption in the world. At the first meeting of the Cabinet ministers on renewable energy and hydrogen energy, held on April 11, 2018, Abe repeatedly stressed "to lead the world and achieve a hydrogen society", and will develop a basic strategy for universal access to green energy such as hydrogen, during the year.

The research and development of hydrogen energy technology in Japan has been more than 40 years old. After the oil crisis hit in the 1970s, Japan focused on developing oil alternative energy technologies. The "Sunshine Project" and the subsequent "Moonlight Project", which began in 1974, included hydrogen energy and fuel cells in the program's key research and development projects. The first wave of hydrogen energy development in Japan was the World Energy Network (WE-NET), which was implemented from 1993 to 2002. The project is expected to use hydrogen as a secondary energy carrier to promote the internationalization of renewable energy utilization. The second wave of hydrogen energy development boom is marked by the "Fuel Cell Practical Strategy Research Report" issued in January 2001. The goal of Japan's development of the hydrogen energy industry is shifting from large-scale utilization of overseas renewable energy to produce hydrogen, and then shipping the them to Japan as a fuel for power generation to the development and practical application of hydrogen fuel cells. From 2002 to 2011, the Japan Hydrogen & Fuel Cell Demonstration Project (JHFC) was implemented, with a focus on the reliability of fuel cells and refueling technologies. Marked by the introduction of the 2014 "Hydrogen Energy and Fuel Cell Strategic Roadmap", Japan has once again set off a third wave of hydrogen energy development and utilization.

Japan has long proposed the concept of "hydrogen society". Hydrogen society refers to hydrogen energy as a fuel widely used in social daily life and economic industry activities, and in the future, hydrogen energy will become the pillar of secondary energy together with electricity and heat. The Fourth Energy Basic Plan in 2014 proposed five measures to accelerate the construction and development strategy of a hydrogen energy society: (1) expanding the utilization and popularization of fixed fuel cells; and (2) accelerating the promotion of fuel cell vehicles; (3) strengthening the research and development of new technologies such as hydrogen fuel power generation to promote the utilization and popularization of hydrogen energy; (4) promoting the research and development of hydrogen production, hydrogen storage and hydrogen transport technology to establish a stable hydrogen fuel supply system; (5) conducting the research and working out the strategic development roadmap of realizing hydrogen energy societ. The plan also proposes to showcase the latest achievements in the development of the Japanese hydrogen energy industry through the 2020 Tokyo Olympic Games.

In recent years, the Japanese government has attached great importance to the construction and development of the hydrogen energy society, and the development of the hydrogen energy industry has become a national policy. It has occupied an important position in the government's economic growth strategy over the years. In 2013, the "Japan Rejuvenation Strategy" proposed to accelerate the development of the fuel cell market. the domestic fuel cell market will reach 5.3 million units by 2030, and at the same time, it will relax the control of fuel cell vehicles and hydrogen refueling stations, and promote the popularization of fuel cell vehicles at the fastest speed in the world. In 2014, the "Japan Revitalization Strategy" proposed to take measures to promote the development of hydrogen energy society based on the hydrogen energy society development roadmap after reviewing the progress of hydrogen energy production, storage, transportation, supply and utilization. In 2015, the "Japan Revitalization Strategy" proposed that in order to take advantage of the opportunity of the 2020 Olympic Games in Tokyo to promote economic growth, the government will accelerate the comprehensive reform of Japan's economic and social structure and implement the "2020 Reform Project", in which the production of carbon free hydrogen fuels using renewable energy has become one of the six key projects. The project will strengthen the technology research and development of hydrogen preparation and supply systems, use carbon dioxide capture and storage technology and renewable energy to produce hydrogen, establish a large-scale carbon free hydrocarbon fuel production system, and propose to use local rich renewable energy to produce hydrogen, and then transport to the big cities to build a hydrogen energy industry chain that balances supply and demand, and show the international community a "Japan Plan" to solve energy and environmental problems. In 2016, the "Japan Revitalization Strategy" proposed that in order to achieve the goal of total GDP of 600 trillion yen, it is necessary to implement the "Top Ten Engineering Strategy for Officials and People", and specifically set up market development goals of fuel cell vehicles and refueling stations. At the same time, the government will continue to promote the "2020 Reform Project", accelerate reform and technological innovation, and use renewable energy to produce carbon free hydrogen fuel. Hydrogen fuel transportation will focus on hydrogen liquefaction and hydrogen compound technology.

The Japanese government has adopted various policy measures to cultivate and develop the hydrogen energy industry. First, deregulating, lowering the threshold, smoothing various institutional obstacles, and activating private capital investment in the hydrogen energy industry; second, increasing investment in technology development, breaking through key technologies, setting up hydrogen energy demonstration cities and demonstration bases according to local conditions; third, promoting market transactions of zero-hydrocarbon fuels to fully reflect the environmental value attributes of zero-carbon hydrogen fuels through green power certificates, carbon trading and other systems; fourth, taking various preferential measures to encourage the promotion, utilization and popularization of hydrogen energy products and other zero hydrocarbon fuels, focusing on subsidies for household fuel cells, fuel cell vehicles and hydrogen refueling stations.

2) Road map for hydrogen energy development

The "Strategic Roadmap for Hydrogen and Fuel Cells", developed by the Japanese government, details the three-step strategy for the development of the Japanese hydrogen industry.

The first phase of the goal is to accelerate the expansion of the field of hydrogen energy utilization. The main task is to promote and popularize fixed fuel cells and fuel cell electric vehicles, and expand the market for hydrogen energy and fuel cells. The second phase of the goal is to use hydrogen fuel to generate electricity and build a large-scale hydrogen energy supply system. By 2030 or so, further expand the domestic hydrogen energy market demand, and develop and establish a supply system for overseas hydrogen production, hydrogen storage, and hydrogen transportation. The goal of the third phase to establish a zero-hydrocarbon fuel supply system. By 2040, using fossil energy carbon capture and storage technology to produce hydrogen, or using renewable energy to produce hydrogen, a carbon free hydrogen energy supply system throughout the life cycle will be established. The objectives for each phase are shown in Fig. 5.11.

The first phase focuses on expanding the fuel cell market. Firstly, in the field of household fuel cells: the market sales target will reach 1.4 million units by 2020 and 5.3 million units by 2030. To reduce the economic burden of household fuel cell end users, the price of PEFC standard machines will drop to 800,000 yen by 2019. The price of SOFC standard machine will drop to 1 million yen by 2021; the investment recovery period will be 7–8 years until 2020, and will be reduced to 5 years by 2030; SOFC industrial grade fuel cell with higher power generation efficiency is planned to put on market in 2017. Secondly, in the transportation sector: in 2016, fuel cell buses and forklifts will be put into operation and expanded to other transportation sectors such as ships and aviation; the market sales target for fuel cell vehicles will reach 40,000 units by 2020, about 200,000 units by 2025, and about 800,000 units by 2030; fuel cell vehicles will be popularized for the mass consumer market, and the price will be the same as that of the current hybrids by 2025. In 2016, 100 hydrogen refueling stations centered on the four metropolitan areas will be built, the hydrogen refueling station will double to 160 in 2015 by 2020, it will double again to 320 in 2025. In addition, 100 small-scale hydrogen refueling stations for renewable energy hydrogen production will also be established in 2020. Hydrogen energy prices will remain the same as existing hybrid vehicle fuel costs, while hydrogen station construction costs will be reduced continuously.

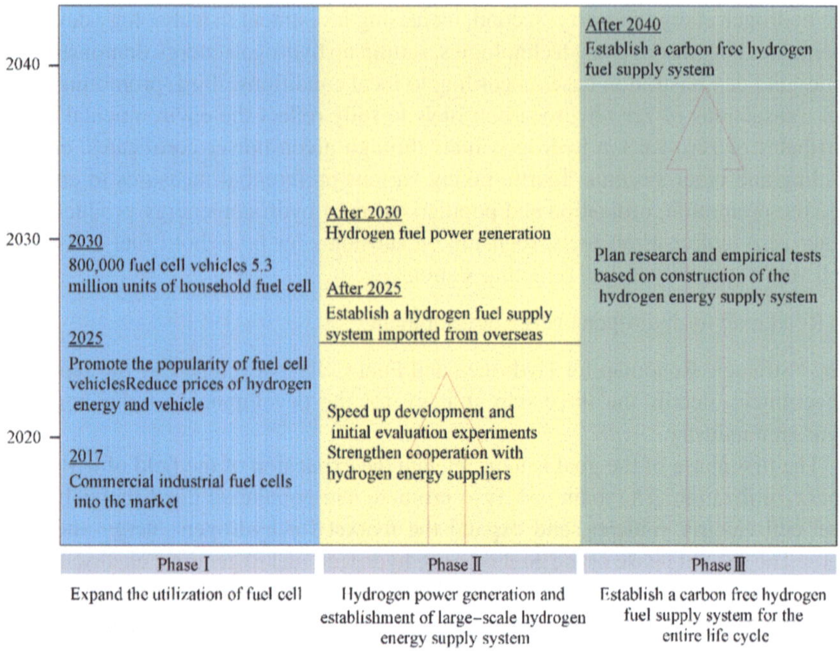

Fig. 5.11 Strategic road map for hydrogen energy and fuel cells

The second phase focuses on the realization of hydrogen fuel power generation and the establishment of a large-scale hydrogen energy supply system. By 2020, hydrogen fuel power generation will be mainly used for self-use or backup power supply market, and will be expanded to large-scale power generation enterprises by 2030. By 2030, comprehensive use of overseas untapped energy to produce hydrogen will fully develop hydrogen fuel power generation industry. The price of hydrogen purchased from overseas will be reduced to 30 yen/m^3 by 2030, and the cost of power generation will be reduced to less than 17 yen/kWh. The energy that has not been developed and utilized overseas refers to industrial by-product hydrogen, crude oil associated gas, lignite, etc. The principle is low price, stable supply and low environmental load. The storage and transportation method is based on the realistic and feasible principle, and the long-distance transportation is mainly based on methylcyclohexane and liquefaction.

The third stage is to comprehensively master low-cost, stable, reliable and clean hydrogen production technology by 2040, and establish a zero-emission hydrogen supply system for hydrogen production, hydrogen storage and hydrogen transportation. On March 7, 2017, the Ministry of Economy, Trade and Industry released the "Zero Hydrogen Fuel Research Report", which specifically described the roadmap for technological development to achieve this goal.

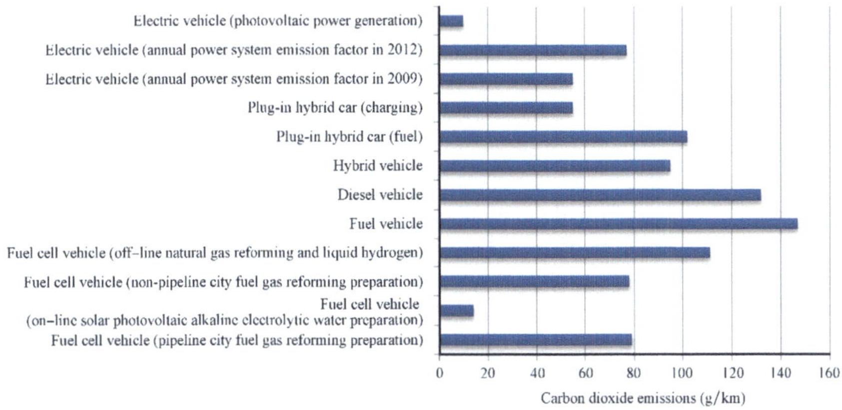

Fig. 5.12 Carbon dioxide emissions per kilometer for fuel cell vehicles

Zero-hydrocarbon fuels are evaluated for life cycle based on natural gas methane steam reforming. The calculation of emissions includes not only the hydrogen production phase, but also transportation, storage and refueling stages. From the life cycle emissions, less than 50% of the natural gas reforming emissions are defined as clean energy. The use of sludge biogas to produce hydrogen, electrolyzed brine to produce hydrogen, wind power and photovoltaic power generation to produce hydrogen is a category of clean energy. The strategic goal of the roadmap is to withdraw from the renewable energy fixed-price acquisition system by 2032. Japan will make large-scale use of renewable energy for power to produce hydrogen. On the one hand, it will actively develop low-cost zero-carbon hydrogen fuel, and on the other hand, improve hydrogen fuel-related legal system with zero carbon, and comprehensively promote the use and popularization of zero-hydrocarbon fuels. The carbon dioxide emissions of fuel cell vehicles are also essentially dependent on the hydrogen production process. As shown in Fig. 5.12, fuel cell vehicle emissions from hydrogen production from fuel gas and natural gas reformation are comparable to those of hybrid vehicles, even larger than those of pure electric vehicles. Only by using photovoltaic electrolyzed water to produce hydrogen, fuel cell vehicle emissions can achieve the best results, and emissions are only one-fifth of that of electric vehicles. Electric vehicles that use photovoltaic power directly have the lowest emissions because they reduce the transportation, storage, and refueling of hydrogen energy. Therefore, from the perspective of reducing carbon dioxide emissions, the best way is to maximize the use of renewable energy electric vehicles, followed by hydrogen fuel cell vehicles that convert renewable energy surplus power into hydrogen energy.

2. Supply side system

The hydrogen energy system mainly includes three aspects of hydrogen production, storage and transportation, and applications. The supply side mainly refers to hydrogen production, storage and transportation. Its strategic focus is to establish an overseas hydrogen energy supply system, a renewable energy hydrogen production system, and a regional hydrogen energy supply system to achieve the strategic goal of fully establishing a zero-hydrocarbon fuel supply system.

1) Establishment of overseas hydrogen energy supply system

Judging from the Japanese government's road map, Japan is more inclined to import hydrogen from overseas because of its natural resource endowment. Therefore, the strategic goal of Japan's hydrogen energy society development is to first establish an overseas hydrogen energy supply system. On the one hand, in order to control energy costs, the large-scale import of low-cost hydrogen energy can greatly expand the domestic supply of hydrogen fuel. On the other hand, it is to ensure the use of zero-emission clean hydrogen fuel. The overseas zero-carbon hydrogen production methods mainly choose the following two types: one is to use hydrogen from overseas cheap lignite to produce hydrogen, and the fossil fuel hydrogen production method using coal and natural gas reforming is still the most economical and realistic hydrogen production method, but carbon dioxide emissions will occur during producing hydrogen by this method, carbon capture and storage technology must be used to achieve emission reduction. The second is to use electrolyzed water to produce hydrogen in countries and regions with better renewable energy endowment conditions and lower power generation costs.

At present, Japan plans to use Australian lignite and Middle East natural gas as raw materials to produce hydrogen, while capturing and sequestering carbon dioxide for the main reason to reduce costs. As is known, lignite has a low grade and a large amount of water. It is easy to catch fire after drying, and transportation is very difficult. Generally, it has not been developed or utilized, or it is used for power generation with limited production. However, its advantages are large reserves, wide distribution, and low price. Hydrogen production from lignite is a low-cost preparation method. Kawasaki Heavy Industries is currently developing jointly with the Australian government. Because Australia's lignite reserves reach 38 billion tons, the cost of hydrogen production is 29.8 yen/m^3 (CIF price, cost plus insurance and freight), which can be used for power generation in Japan for 240 years.

The biggest difficulty in establishing an overseas hydrogen energy supply system is storage and transportation. Hydrogen at normal temperature and pressure is a gas, and the calorific value per unit volume of fuel is small. In order to increase the energy density, it is necessary to adopt a liquefaction method for transportation and storage. However, the boiling point of liquefied hydrogen is -253 °C, and it is difficult to avoid natural heat exchange at a low temperature in a closed storage tank, resulting in energy loss. Hydrogen is easy to explode, and it is easy to make the metal embrittlement. It must be stored and transported in a container that prevents metal embrittlement, high pressure resistance and leakage-proof. Since the hydrogen molecules are small, once the hydrogen is leaked and diffused quickly, it is necessary to add an odorant to make it easy to perceive. Large-scale long-distance transportation from overseas is not an easy task. Therefore, hydrogen fuel must be converted into an energy carrier to facilitate storage and transportation. The energy carrier refers to the main chemical substances that carry the energy for storage and transportation, including traditional energy materials such as petroleum, gasoline and natural gas, as well as chemicals such as hydrogen energy, methylcyclohexane, ammonia, methanol and magnesium metal.

The development and utilization of energy carriers is one of the top ten strategic innovation projects of the Japanese government. The cross-ministerial Strategic Innovation Promotion Program (SIP), initiated by Japan in 2014, is technology development project led by the Cabinet. Hydrogen energy carriers are one of the important research topics. The five-year plan from 2014 to 2018 sets five technical goals: renewable energy hydrogen production technology, ammonia decomposition hydrogen production technology (300 m^3/h), methyl cyclohexane hydrogen storage technology, ammonia fuel cell technology (1 kW level), ammonia turbine power generation technology (2 MW level), etc., focusing on the development of production and utilization technologies of liquefied hydrogen, methylcyclohexane and ammonia, with the aim of developing low-cost production and utilization of zero-hydrocarbon fuel technology. All of the above methods face key issues of improving efficiency and reducing costs. To this end, SIP's technical goal is to focus on the development of energy carriers such as ammonia, methylcyclohexane, and liquid hydrogen to achieve low-cost, high-efficiency hydrogen preparation and storage technologies. These technologies will be used in the 2020 Tokyo Olympic Games. In addition to using hydrogen fuel buses to transport athletes and spectators to the playing field, they also plan to use hydrogen fuel to generate electricity and heat to various facilities such as athletes' villages. The economic goal of SIP is: in 2020, the cost of hydrogen fuel for fuel cell vehicles should be the same as that of gasoline. The cost of hydrogen fuel for power generation in 2030 should be equal to the cost of natural gas. The industrial goal of SIP is to develop and cultivate the hydrogen energy industry, to achieve a domestic output value of 1 trillion yen by 2020, and to occupy a huge market share globally by 2030. Therefore, SIP is an important part of Abe's economic growth strategy. The success or failure of this project is related to the realization of the national strategy of Japan's hydrogen energy society.

2) Establishment of hydrogen production system by using renewable energy

(1) Producing hydrogen by electrolyzing water from renewable sources

Zero hydrocarbon fuel production can further expand the use of renewable energy. The method of electrolyzing water has been widely used, and hydrogen can be produced by a large scale. By using water as a raw material, hydrogen and oxygen can be generated, and electricity can be generated from nuclear power, wind power, hydropower, photovoltaics, and geothermal power. Water is inexhaustible, water is used as raw material and no harmful substances are emitted, but the efficiency is low. Hydrogen production of 1 m^3 consumes at least 3.5 kWh. Although the production of zero-hydrocarbon fuels is expected to produce hydrogen from electrolyzed water, how to improve the efficiency of electrolyzed water and reduce the cost of electrolyzed water equipment is the key issue. Therefore, in order to realize low-cost, high-efficiency, large-scale electrolysis of water to produce hydrogen, it is necessary to develop methods such as alkaline water electrolysis and high-temperature water vapor. To truly improve the efficiency of hydrogen production, Japan will focus on the hydrogen production technology of pyrolysis water and photolysis water using nuclear energy and solar heat. The focus and direction of innovative technology is to generate and separate hydrogen catalysts and separation membranes and other materials.

The cost of hydrogen production remains the primary issue in the current preparation of zero-hydrocarbon fuels. The goal of the Japanese government roadmap is to control the cost of electrolyzed water hydrogen production equipment to 50,000 yen/kW by 2020, but the use of surplus electricity from photovoltaics and wind power to produce hydrogen will be affected by climate, season and time, and the power generation will be unstable. The utilization rate of the electrolyzed water device is low. At present, the sales price of hydrogen in Japan's refueling station is 100 yen/m^3, so the cost of zero-carbon hydrogen fuel must be controlled within 100 yen/m^3, and the equipment utilization rate must be increased to more than 30% to break even. At present, hydrogen energy is still relatively high compared with traditional energy sources such as natural gas. With the increase in demand for hydrogen energy, hydrogen energy prices are likely to narrow the gap with traditional energy costs. To this end, the development goal proposed by the road map is: the hydrogen price will be 30 yen/m^3 and the power generation cost will reach 17 yen/m^3 by 2030. The hydrogen price will be 25 yen/m^3 by 2040, and the hydrogen price will reach 20 yen/m^3 by 2050.

(2) Power-to-Gas

Power-to-Gas (PtG) is a technology that converts electrical energy into fuel gas with high energy density. Japan will focus on the use of power-to-gas technology to expand the utilization and popularity of renewable energy. In recent years, with the promotion of the fixed-price acquisition system for renewable energy, Japan's renewable energy capacity has achieved rapid development, but renewable energy power generation has the characteristics of decentralization, intermittent and difficult regulation, bringing a large number of grid-connected operations great challenges, resulting in widespread abandonment of wind power and photovoltaic power due to limited grid capacity or insufficient thermal power peaking capability. The power-to-gas technology uses hydrogen as a medium to break the barrier between the traditional power system and the natural gas system, and the two-way flow of energy between the power system and the natural gas system promotes the deep integration of the gas network and power grid. Therefore, the use of power-to-gas conversion technology can maximize the revitalization of surplus power, not only improve the efficiency of renewable energy generation, but also contribute to the stability and safety of traditional power grids, providing a new way to solve the fluctuation problem of renewable energy power generation. The key issue to power-gas conversion technology is electrolyzed water hydrogen production equipment. At present, there are three main types of water electrolysis hydrogen production technologies: alkaline water electrolysis hydrogen production, proton exchange membrane water electrolysis hydrogen production and solid oxide water electrolysis technology. Japan has established the country's largest demonstration base for renewable energy hydrogen production in Fukushima. It plans to supply hydrogen energy to the Tokyo Olympic Games venues and sports villages in 2020, and use the power-to-gas technology to achieve clean hydrogen production, storage and transportation.

(3) Producing hydrogen according to local conditions

By-product hydrogen is mainly from petroleum refining, ammonia and methanol production, steel industry, etc., often used in desulfurization industrial raw materials or industrial boiler fuels in factories. For example, the gas emitted from the steel coking furnace contains 55% hydrogen, and the hydrogen produced by the entire steel industry in Japan reaches 4.5 billion cubic meters (volume at 0 °C/1 atm), but most of them are used in the factory and a limited portion is in the market. The industrial by-product from coke oven gas purification and electrolysis of caustic soda is limited, with an annual output of about 1 billion cubic meters. At present, the main source of hydrogen energy in Japan is the recovery of hydrogen-rich gas for producing hydrogen from refineries, with an annual output of about 7 billion cubic meters. The entire petrochemical industry can recover tens of billions of cubic meters of hydrogen. But the hydrogen production process emits carbon dioxide, which is not a clean energy source.

Hydrogen produced from renewable energy and carbon dioxide can synthesize methane. When methane is burned, it will emit carbon dioxide, which is in direct contact with hydrogen fuel to absorb carbon dioxide to achieve zero carbon emissions. This is the second step in the development of power-to-gas technology. Methane can be transported by means of an existing energy supply system, mixed with natural gas; or directly mixed hydrogen with natural gas at a certain proportion, and then stored and transported through natural gas pipelines. Methane as an energy carrier has great potential for heat supply. However, the problem that needs to be solved in commercialization is also to face high cost. It is necessary to reduce the cost of carbon dioxide capture and methane synthesis equipment.

Photocatalyst is a solid substance that converts solar energy into chemical energy. It is the key to the new generation of hydrogen preparation technology. Unlike solar cell silicon, the price and supply are relatively stable, but the solar energy conversion efficiency is relatively low, currently only 2%. Japan plans to increase it to 10% by 2021. Recently, Osaka University of Japan has developed a new photocatalyst that successfully achieves water decomposition and hydrogen production by using visible light and near-ultraviolet light. Kyushu University has developed a new catalyst for the integration of fuel cells and solar cells. Solar energy can be used during the day and hydrogen can be used at night.

All over Japan, based on their own circumstances, old newspapers, sewage sludge, poultry manure and other methods are used to produce hydrogen according to local conditions. Toyota Tsusho produces hydrogen from wind power generation in Hokkaido, Hokkaido, and then uses Methylcyclohexane (MCH) to store and transport it for use at local stations. Tohoku Electric Power uses solar energy to produce hydrogen in Sendai, and then uses hydrogen storage alloys for storage and transportation. Mitsubishi Chemicals Fukuoka uses sludge to produce hydrogen for use in nearby refueling stations; Toyota uses wind power to produce hydrogen in Yokohama for use by forklifts in nearby storage companies; Zhounan City, Yamaguchi Prefecture is a famous chemical city, using the by-product from caustic plant to produce hydrogen; Kushiro City, Hokkaido uses small hydropower to produce hydrogen, and supplies electricity to local agricultural production. The development and utilization of hydrogen energy in various places is innumerable.

3) Establishment of regional hydrogen energy supply system

(1) Hydrogen pipeline network construction

With the large-scale production of hydrogen and the wide application of hydrogen energy, the construction of hydrogen pipeline network is imperative. European and American countries have already experienced thousands of meters of experience, and their practice has proved their economic rationality and technical feasibility. Some residential buildings in Kitakyushu, Japan, have also been powered and supplied by hydrogen pipelines, and their hydrogen energy comes from the by-product hydrogen in the coking process of local steel mills. The most convenient way to transport regional energy supplies is to build a hydrogen pipeline. To this end, Japan's first phase is planned to build pipeline near the hydrogen energy preparation base, and it will form a hydrogen pipeline network with the refueling station; the second phase will focus on forming a regional hydrogen energy pipeline network in the coastal areas after 2030; the third phase plans to focus on the development and utilization of existing urban natural gas pipelines.

(2) Fukushima New Energy

In September 2016, in order to cultivate Fukushima, which has suffered a nuclear accident, into a demonstration base for Japan's new energy industry, the Japanese government proposed to build a "Fukushima New Energy Society Concept" plan to expand renewable energy utilization and build hydrogen energy. With the goal of building a social demonstration base and building an intelligent community, it is important to build Fukushima into the world's largest hydrogen energy society demonstration base that uses renewable energy for hydrogen production, hydrogen storage, hydrogen transportation and hydrogen utilization. The main measures are as follows: First, make use of power from renewable energy to produce hydrogen, the scale of production will reach 10,000 kW, which will put into production by 2020; second, transport the produced hydrogen to the Tokyo Olympic Games by developing and using MCH storage and transportation technology to; third, promote the utilization of hydrogen energy, improve the infrastructure such as hydrogen refueling stations, and expand the market launch of fuel cell vehicles. In addition, Japan has also established a number of hydrogen energy demonstration cities such as Fukuoka and Kitakyushu.

To sum up, two major problems must be solved from the supply side: first, how to produce hydrogen in a clean and efficient way; and second, how to achieve low-cost storage and transportation of hydrogen. Therefore, the supply side strategic goal is to develop a safe, stable, efficient, and clean hydrogen energy preparation and storage technology centered on reducing hydrogen production and hydrogen storage costs.

To focus on the mechanism and causes of metal embrittlement and develop new technologies for hydrogen energy carriers will achieve gradually the transition from black hydrogen and gray hydrogen to the green hydrogen era.

3. Demand side system

The demand side mainly refers to the application of the hydrogen energy industry chain. Hydrogen society built in Japan, from domestic cogeneration to fuel cell vehicles, from distributed fuel cell power generation systems to large-scale hydrogen fuel power plants, from chemical raw materials, energy carriers to clean energy and fuel, hydrogen energy and fuel cells are widely used in transportation industry, construction and other fields, its strategic goal is to develop hydrogen energy into the "third energy" that is in line with electricity and heat.

1) Utilization in the field of power generation

Japan's Kabushiki-gaisha Ōbayashi Gumi and Kawasaki Heavy Industries will use hydrogen as fuel to supply power to the artificial island area of Kobe City in 2018. This is the first time in the world that hydrogen fuel has been introduced to power the urban area. Hydrogen fuel power generation is the most important step in achieving a hydrogen energy society. Only by realizing hydrogen fuel power generation, the demand for hydrogen energy market can really burst out. In March 2015, the Japanese government formulated the "Research Report on Hydrogen Fuel Power Generation". In June 2016, the Japanese government formulated the latest "New Generation Thermal Power Technology Roadmap", which clearly stated the strategic goal of fully developing hydrogen fuel power generation by 2030. In the past, the development of thermal power technology in Japan focused on three key objectives, namely clean and efficient use of coal and natural gas, and carbon capture, storage and utilization technologies. The main goal is to reduce CO_2 emissions from thermal power. The introduction of hydrogen fuel power generation into the key strategic project for thermal power generation technology development has become the fourth major target of thermal power technology development in Japan. The main reason is that hydrogen fuel power generation can significantly reduce carbon dioxide emissions compared with coal-fired and gas-fired power generation. It marks Japan's formal development of large-scale hydrogen fuel power generation on the national thermal power development strategy agenda.

There are two main types of traditional thermal power units: one is the steam turbine unit and the other is the gas turbine unit. There is also a combined cycle of the two, that is, the natural gas-steam turbine combined cycle, which is a combination of a gas turbine and a steam turbine that are combined according to different thermal cycles. If you use fuel cells to generate electricity, it will form a triple combined cycle power generation, called the third coal power generation technology of IGFC. IGFC uses a solid oxide fuel cell (SOFC) for power generation, and its power generation efficiency is expected to exceed 70%. Hydrogen fuel power generation combustion technology also basically follows the traditional thermal power mode. There are two main ways. One is the mixed combustion mode of hydrogen and natural gas, which is applicable to existing natural gas power plants. At present, Japan's New Energy

Industry Technology Development Organization (NEDO) has completed the design of a 500,000-kW large-scale mixed-burning steam turbine. It plans to complete the pilot test in 2018 and put it into operation before 2020. The co-firing ratio is 20% hydrogen and 80% natural gas. The fuel temperature can reach 1700 °C, and the power generation efficiency can reach 60%. The future thermal power technology will be more efficient and clean. The second is pure hydrogen combustion. Japan plans to complete the development of hydrogen-fired steam turbines by 2025 and put it into operation after 2030. At present, Japan NEDO has completed the development of a 1000-W small-scale hydrogen-burning steam turbine and has entered the pilot stage. However, in the future, there are many technical and market challenges that need to be addressed to promote the widespread use of hydrogen fuel.

The first concern is that costs of hydrogen fuel power generation are still higher than that of coal and natural gas power generation. As shown in Fig. 5.13, Japan's hydrogen fuel generation cost target is 17 yen/kWh by 2030, but still 38% higher than coal-fired power generation (12.3 yen/kWh), 20% higher than gas-fired power generation (13.7 days) yen/kWh). Only a large-scale expansion of hydrogen energy generation can reduce costs.

The second is that the amount of nitrogen oxides emitted by hydrogen combustion is much higher than that of coal and natural gas. How to control nitrogen oxide emissions is the most important problem to be solved by hydrogen fuel power generation. There are three ways for mixed burning steam turbine. One is the premixing method, that is, the hydrogen and natural gas are mixed with the air before combustion. This method is the easiest to control the emission of nitrogen oxides, but the highest proportion of hydrogen for mixed burning is 20%. The second is the diffusion mixing method, that is, the fuel is mixed with air in the combustion process. This method

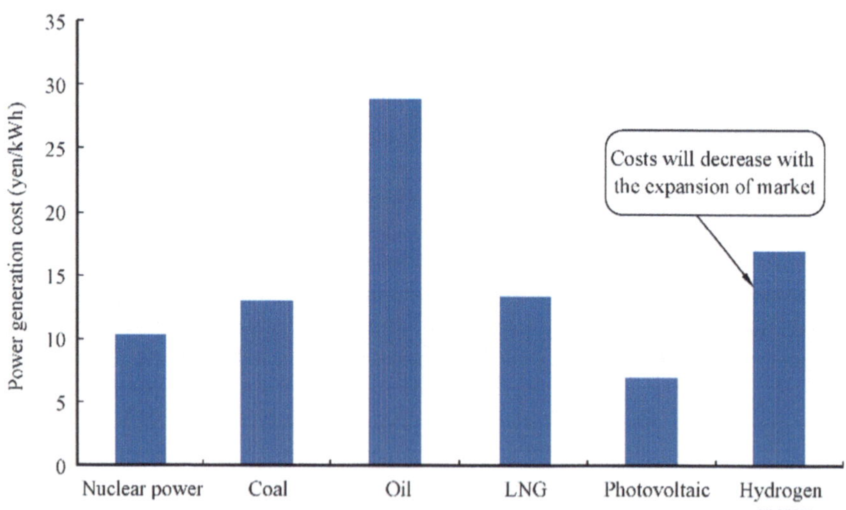

Fig. 5.13 Comparison of power generation costs in 2030

can reduce the emission of nitrogen oxides by spraying water and steam, but it will cause the combustion temperature to drop, thereby reducing the power generation efficiency. The third is the dispersive mixing method, which does not need to spray water and steam, nor the limitation of the mixing ratio, which is the main direction of future development.

The third is the breakthrough in hydrogen production technology and the establishment of a hydrogen energy supply system. Current hydrogen production mainly uses fossil fuels and by-product gas from factories. The main direction in the future is to use renewable energy to produce carbon free hydrogen fuel. Large-scale use of hydrogen fuel for power generation must establish a safe and stable hydrogen energy supply system, and build a nationwide and global hydrogen fuel supply network.

The fourth is the problem of hydrogen energy storage and transportation. At present, hydrogen energy storage and transportation methods mainly include: high pressure, liquefaction, pipelines, organic hydrides, hydrogen absorbing alloys, and the like. However, if a large amount of hydrogen fuel is imported from overseas, there are two main methods: one is to convert hydrogen directly into a liquid, and the same as the liquefied natural gas method, the hydrogen is cooled and liquefied with an ultra-low temperature of −253 °C; the second is to use methyl cyclohexane for hydrogen storage, that is, a hydrogen storage technique that utilizes a reversible reaction based on toluene and methylcyclohexane. These two methods have their own advantages and disadvantages, and they are not fully mature in technologically.

2) Utilization in the transportation field

The application of hydrogen energy in the transportation field mainly depends on the popularity of fuel cell vehicles and refueling stations. The fuel cell is the core of the hydrogen energy society. The mass production of the FCV (fuel cell vehicle) and the commercialization of the refueling station are the key. For this reason, Japan actively promotes the regulatory reform, technology development, and refueling station supporting the trinity policies of the hydrogen energy industry to accelerate the commercialization of hydrogen energy in the transportation field.

(1) Filling station

Japanese hydrogen filling stations can be divided into mobile and stationary, oil-hydrogen hybrid and single hydrogenation, non-independent preparation (offline hydrogen outsourcing) and independent preparation (on-line hydrogen production). The roadmap plans to build 100 hydrogen refueling stations by the end of 2015, but 82 were actually built. As of March 2017, 92 commercial hydrogen refueling stations in Japan has actually been built, of which 84 have been officially opened to the public. There are 19 hydrogen refueling stations capable of producing hydrogen from renewable energy, and 9 of them have been officially opened to the public. The advantages of renewable energy hydrogen refueling station are: hydrogen production to achieve a full life cycle of low carbonization, and energy self-sufficiency, can stimulate the local hydrogen market demand, and also save construction land, but also a lively textbook for coordinated development of energy and environment. However, the problems are as follows: First, the pressure specification is different, most of them are currently 35 MPa, while the commercial filling station is 82 MPa. Second, the hydrogen production capacity is low, 0.7–5 m^3/h, and the commercial filling station is 50–900 m^3/h; Third, the cost is high. To this end, the government will actively take countermeasures to increase the pressure of refueling to 70 MPa, and promote the large-scale utilization of renewable energy to improve the hydrogen production capacity and reduce costs of renewable energy.

(2) Fuel cell vehicles

As of March 2017, Japan's fuel cell vehicle sales have reached a total of 1,800 units. Toyota Motor said that the annual output of the first generation will increase from 700 units to 20,000 units and then to 30,000 units. The second generation of products will be produced after 2020, and 1,000 units are planned to be sold each month. The third generation of products will begin to target the mass consumer market in 2025. Following the successful launch of the Toyota Mirai fuel cell vehicle market, in March 2016, the Honda Clarity fuel cell vehicle was put on the market. In August of the same year, Nissan Motor Co., Ltd. released the world's first SOFC fuel cell prototype car, using e-Bio fuel cell system powered by biomass oil. In this way, all three major Japanese auto companies are competing in the market for fuel cell vehicles.

In March 2016, the Fuel Cell Commercialization Conference (FCCJ) released the "Blueprint for the Popularization of Fuel Cell Vehicles and Filling Stations". To achieve the goal of reducing greenhouse gas emissions by 80% in 2050, the following fuel cell vehicle popularization targets are proposed: 650,000 units by 2030, 3 million units by 2040, and 8 million units by 2050. By 2030, 720 refueling stations will be built (one hydrogen refueling station for every 900 fuel cell vehicles), and the goal of refueling hydrogen everywhere will be achieved after 2040.

(3) Fuel cell bus

In March 2017, two Japanese fuel cell buses were put into operation in Tokyo. The bus is equipped with 2 fuel cell stacks and 8 hydrogen tanks with a maximum output of 9 kW and a storage capacity of 235 kWh. Bus fuel batteries have advantages over lithium batteries, and large-capacity fuel cells can provide more emergency power for disaster relief. Therefore, Tokyo plans to operate 100 units of fuel cell buses by 2020. However, the price of fuel cell buses is as high as 100 million yen per unit, which is nearly five times more expensive than ordinary buses, it is difficult to expand its market. The Japanese bus market has about 230,000 units, and each fuel cell bus can consume 3,850 kg of hydrogen per year, equivalent to hydrogen used by 45 cars (86 kg). Therefore, fuel cell buses are indispensable for the utilization of hydrogen energy, which not only expands the demand for hydrogen energy market, but also helps to improve the utilization rate and economic efficiency of the filling station.

(4) Fuel cell truck

The number of commercial trucks in Japan is 3.23 million units, including 560,000 large trucks, 850,000 medium-sized trucks and 1.82 million small trucks. The hydrogen consumed by the three types of trucks is 9,192 kg, 2,894 kg and 1,103 kg per year respectively. It can be seen that the fuel cell market potential of trucks will exceed that of the bus market.

(5) Fuel cell forklifts

In November 2016, the Japanese fuel cell forklift was put into operation at Kansai International Airport. The Japanese forklift market has about 600,000 units. If 20% of them use fuel cells, there will be 120,000 fuel cell forklifts, each with an annual hydrogen consumption of 250 kg. Since the forklift is operated within a limited range, it is easy to establish and form a supporting network for the filling station. On-site preparation of hydrogen filling stations can use pipeline natural gas to decompose hydrogen or gasify the liquified storage hydrogen. The outsourcing hydrogen filling station can be supplied from the central station or by the mother and daughter station (M&D mode).

In addition, fuel cells will also be used in locomotives, ships, logistics vehicles and even aircraft. Japan has made great progress in technology research and development in various transportation fields. The promotion and application of these technologies will greatly promote the construction and development of hydrogen energy society.

3) Utilization in the industrial and construction fields

(1) Household fuel cell systems (CHP)

In the construction field, the use of hydrogen energy micro-cogeneration unit technology can greatly improve energy efficiency and promote the popularization of green energy-efficient buildings. The only country in the world where fuel cells enter thousands of households is Japan. Japanese household fuel cell system refers to a hydrogen energy micro-cogeneration unit or device that uses hydrogen from city gas and liquefied petroleum gas to generate hydrogen after chemical reaction with hydrogen in the air, and can simultaneously recover heat. The whole system consists of two parts: the fuel cell and the hot water storage tank. The fuel cell is the power generation unit, and the hot water storage tank recovers the waste heat from the power generation for heating, including boiling water, bathing and floor heating. Since the energy generated by the electrochemical reaction consumes less energy, and electricity and heat can be utilized at the same time, the overall energy efficiency is about 90%. At present, Japan's domestic fuel cells mainly include PEFC type and SOFC type. PEFC type power generation efficiency is slightly lower, about 40%, but heat recovery efficiency is higher; SOFC type power generation efficiency is higher, about 45%, but heat recovery efficiency is slightly lower. The combined efficiency of cogeneration is much higher than that of large-scale thermal power plants. The mainstream models on the market include Panasonic, Aisin Seiki and Toshiba.

Japan's domestic fuel cell system is called "ENE-FARM". Since its launch in 2009, it has sold a total of 196,000 units by the end of 2016. The price of household fuel cells for 2016 is 1.13 million yen and 1.35 million yen for PEFC and SOFC, respectively, and the payback period has been shortened to 12–13 years. In the future, the direction of technology development will focus on further improving the power generation efficiency of SOFC machines.

(2) Fuel cell distributed power generation system

Hydrogen energy can build a distributed energy network based on fuel cell technology, providing a triple supply of regional or urban electricity, heat and cold energy. The fuel cell itself has high energy conversion efficiency and low emissions.

The carbon dioxide emission coefficient of fuel cell cogeneration is 0.28–0.30 kg/kWh, which is lower than the target of average emission factor (0.37 kg/kWh) that Japanese power companies should strive to achieve by 2030, and the energy transmission distance of distributed energy systems is short and the power generation efficiency exceeds that of natural gas combined cycle power generation (GTCC). Therefore, the energy efficiency of energy systems with hydrogen fuel cells as the core is significantly higher than that of traditional energy systems. The realization of Grid parity will reduce the cost of power generation and facilitate the further popularization of the fuel cell distributed generation market.

Increasing the life of fuel cell systems to more than 90,000 h is currently the most challenging technical difficulty in developing distributed power generation of fuel cells. Japan's Panasonic has broken through this difficult problem, and its products can not only provide enterprises with large fixed fuel cells to solve the internal emergency power supply of enterprises, but also provide medium and small fixed fuel cells for emergency power supply for earthquake relief, etc. from a 5 kW small machine to a 250 kW medium machine.

In addition, Toshiba and Panasonic have also developed pure hydrogen fixed fuel cells. "H_2One" is an independent photovoltaic hydrogen energy supply system developed by Toshiba. It uses solar power to electrolyze water. The system consists of equipment, hydrogen tanks and pure hydrogen fuel cells. The hydrogen production capacity is up to 1 cubic meter per hour, and the maximum hydrogen consumption is 2.5 cubic meters per hour, the warm water supply is 75 L per hour (40 °C), the fuel cell has a maximum output of 3.5 kW, and the power reserve capacity is 350 kWh.

(3) Industrial fuels

The flame temperature of hydrogen fuel can reach 2,300 °C, and it is difficult to meet the heating or burning temperature by using electricity and natural gas. At present, high-temperature industrial sectors such as smelting, steel, and petrochemical industries use fossil fuels for heating or combustion. If hydrogen fuel is used instead, the hydrogen demand in these industries in Japan will reach 23.6 billion cubic meters.

To sum up, from the view of demand side, due to the limited demand for hydrogen energy in Japan, the scale effect of hydrogen energy is still difficult to be achieved. Only by realizing hydrogen fuel power generation can we increase market demand and reduce costs on a large scale. Therefore, the strategic goal of the demand side is to break through the bottleneck of high-efficiency hydrogen fuel power generation technology, so that the power generation efficiency can reach 70–80% or even higher, so as to deeply tap the demand of hydrogen energy market; at the same time, actively

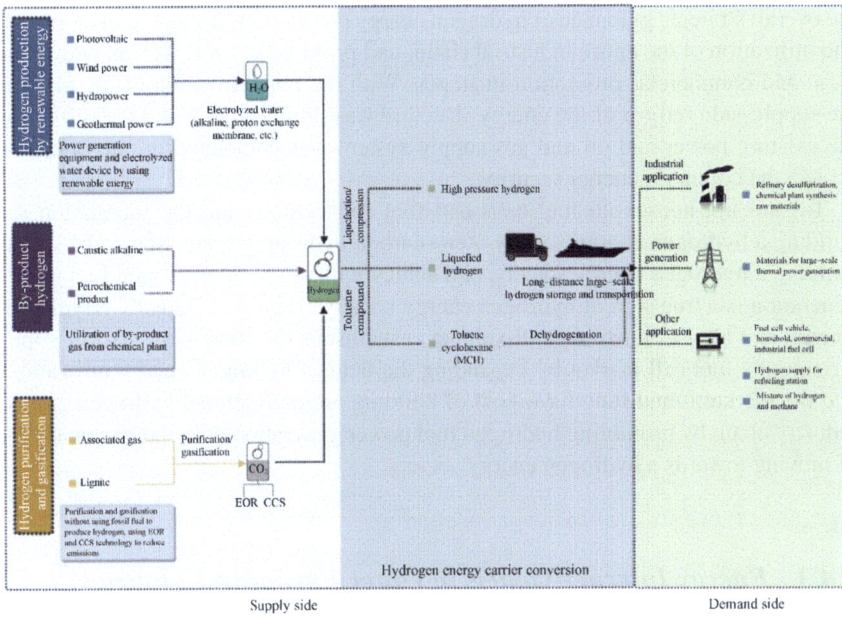

Fig. 5.14 Hydrogen energy supply side and demand side industry chain

develop high efficiency, stability and durability, low-cost, miniaturized fuel cells to expand the utilization of the hydrogen energy market (Fig. 5.14).

4. Achievement of balance between supply side and demand side

The current problems faced by Japan's energy sources are: (1) 100% of fossil fuels such as oil, coal, and natural gas are imported; (2) nuclear power is difficult to fully resume and restart, and large-scale development of thermal power units is bound to increase carbon dioxide emissions; (3) power systems are mostly centralized power generation, large-scale power outages are likely to occur in the event of an accident or disaster. Hydrogen fuels come from a wide range of sources, from fossil fuels such as oil and natural gas, bio-thermal conversion. Therefore, the promotion of hydrogen energy utilization can achieve three advantages, solving the three major problems of demand side, supply side and environmental protection currently facing hydrogen energy.

To develop hydrogen into the protagonist of future energy, we must think from the perspective of future social and economic development. If Japan does not find a way out in the direction of new energy, according to the current growth rate of thermal power generation, Japan's domestic electricity bill will be very high in 2020, production and manufacturing costs will be greatly increased, and emission reduction targets will be difficult to be met, thus making Japan lose its international competitiveness. Japan's development of a hydrogen fuel cell roadmap is to select safe, convenient, economical and clean hydrogen energy utilization technologies under the overall strategic guidance of hydrogen energy production, storage, transportation and utilization of the entire industrial chain, and promote key research and development and commercial promotion in stages. With the popularization of the market, the supply-side reform of the energy structure was implemented, and the reform of the existing power and oil and gas supply system was completed in one breath to protect the country's energy security.

To sum up, household fuel cells and fuel cell vehicles are the foundation for building a hydrogen energy society. Zero-carbon hydrogen production is the key to building a hydrogen energy society, and achieving large-scale hydrogen fuel power generation is a true sign of hydrogen energy society. Therefore, Japan's "Hydrogen Energy and Fuel Cell Strategic Roadmap" establishes the short-term goal of popularizing the fuel cell market by expanding the field of hydrogen energy utilization, and the medium- and long-term goal of building an international hydrogen energy industry chain by promoting hydrogen fuel power generation with the strategic goal of moving towards a hydrogen energy society.

5.4.3 Energy Internet Models in Other European Countries

In response to the European Commission's recommendation for a smart grid for the future, a project GRID4EU led by the European Commission with a total cost of 54 billion euros was launched in six different countries in Europe. GRID4EU is currently the largest smart grid project in Europe, it is dominated by six major energy companies and is comprised of a range of power supply and distribution companies, electrical equipment manufacturers and research institutes. The six projects are located in Cesena of Italy, Vorchrabi of Czech Republic, Nice of France, Reken of Germany, Uppsala of Sweden, and Castrion of Spain. Table 5.2 lists the scope and focus of the six demonstration projects.

It is not difficult to find through Table 5.2 that the GRID4EU Energy Internet demonstration project has the following characteristics.

(1) Utilization of the original infrastructure according to local conditions. The six pilot projects started from 2012 and ended in 2016. It is almost impossible to complete with the average construction speed of the EU countries from scratch. It can be seen that the pilot projects have utilized the original gird equipment to varying degrees, which also tells us indirectly that the energy Internet is not only the planning and construction of the new energy network architecture, but also the utilization and integration of traditional networks and resources.

Integration cannot ignore the differences between regions. The GRID4EU pilot project subtly utilizes the climatic conditions of the region and the energy modalities corresponding to this climatic condition. Germany's North Rhine-Westphalia is located in a temperate marine climate with wind and sunshine. It uses the wind turbines and solar energy of RME to provide energy for the region. Italy and France in southern Europe have unparalleled solar energy resources. The large amount

Table 5.2 Scope and focus of six demonstration projects

Demonstration project	Project focus
Cesena, Italy	Advanced system control, increasing grid carrying capacity, maximizing integration of renewable energy, and decentralized power generation units
Vorchrabi, Czech Republic	Medium and low voltage automatic operation, the combination of electric vehicle and power grid, power quality check and island operation
Nice, France	Optimizing the integration of distributed solar panels and grids through load forecasting, energy storage equipment and microgrids, while encouraging users to participate in the power market
Reichen, Germany	Grid monitoring and medium-voltage grid control based on multi-agent system
Uppsala, Sweden	Monitoring of low-voltage distribution systems based on advanced measurement systems and intelligent substations
Castrion, Spain	Realizing medium and low voltage grid automation and increasing user's participation in power grid

of utilization and management of photovoltaics is certainly the best choice; the Czech Republic is located in central Europe with a mild climate. Although wind and solar resources are scarce, the access of cogeneration equipment adds a variety of possibilities for the regulation of energy networks.

(2) Hierarchical management, clear organizational structure. The energy Internet is a big network, but it is by no means an unregulated network. When integrating the resources of the energy network and managing the network, it is necessary to first clarify the organizational structure of the network and thus manage it hierarchically. The GRID4EU project, without exception, divides the distribution network into medium-voltage networks and low-voltage networks, and gives them different roles and objectives. Medium voltage networks are more responsible for regional power flow and load regulation. The adjustment targets are mainly power load balance and voltage balance. The low-voltage network is more focused on user-side interaction, thus electric vehicles and energy-efficient buildings [including small PV and small CHP (cogeneration)] is mostly connected in low voltage networks. In addition, the low-voltage network also undertakes the task of collecting, acquiring, and feeding back user data.

(3) To generate electricity by smart utilization of energy storage like peak shaving and valley filling. The reason why energy Internet can optimize the operation of the whole network, stabilize the price of electricity, and have a high carrying capacity, energy storage equipment is indispensable. As for four projects connected to distributed energy (Germany, Italy, Czech Republic, France), Italy and France directly built lithium battery energy storage equipment; Czech projects also use EV (electric vehicle) access as small-scale storage; The only one in Germany that does not use energy storage equipment is because the capacity of RWE's wind turbines and solar

equipment is roughly equal to the maximum load in its operating area, which is basically balanced. The access of energy storage equipment helps the energy Internet to shave the peaks and fill the valley. For network operators and managers, the flow of energy is more flexible and stable. For users, the price fluctuation of energy will also be reduced accordingly.

(4) Intelligent instrumentation, demand management participated by all parties involved. For the energy Internet, the interaction of big data and information is also becoming increasingly important. The pilot project in Sweden is almost entirely focused on the use of smart meters, 15,000 users are involved in the huge information, and of course enjoy the benefits of information technology: power quality check, electricity price analysis and demand analysis of users in low-voltage distribution networks. The analysis makes virtual grid and demand-side management no longer just an energy Internet participant's talk after a meal, but a real-life engineering case that truly changes people's lives and changes people's understanding of energy Internet. Based on a large number of energy storage devices and distributed solar panels, the concept of microgrid is developed, that is, the energy is completely self-sufficient in a small area of a single or several users, which requires extensive demand side management technology at the user end. The smart meters are used to achieve energy management in a small area, and this is the guarantee of the possibility of peak shaving and valley filling in large areas. Therefore, the regulation of the user side and the collection of information are the preconditions of the entire smart grid. The smart meter, the control system of the household energy management, the battery and its management system are the hardware foundation.

(5) To guarantee power supply when disasters alike occur through supply and demand prediction. In addition to energy management at the user level, accurate data and power generation predictions can be made at the distribution network level through weather forecasting and information provided by the client. Early prevention and early preparation can minimize the impact on the grid and improve the reliability of power supply, a complete grid management system that combines big data such as climate, construction, industry and power grid is the trend of future development.

In the case of openness of the sale and distribution and the large number of related power companies, as for how to realize the resource integration and information sharing of each company, to manage and develop such a system to ensure information exchange by only one company more convenient and reliable, or the open cooperation of information making the energy Internet more dynamic based on many companies, how to choose between the two methods is still an inconclusive issue. In addition, the form that the user participates in based on the smart grid, and the kind of resource allocation can be finally achieved (whether the user can freely sell electricity, the two-way power sale agreement between the user and the power company, whether the user power information is public, the flow of funds and the direction of energy flow, etc.) are also the problems faced by the energy Internet. And the new business model derived from this (such as the power sale and distribution company rent equipment to the user and charges equipment rental fees) and how to regulate the activities, in the case of the energy Internet trend, how can more users participate in, develop corresponding policy of conforming to the actual power sales distribution structure. In Europe, the four parties of supply, distribution, sale and end user are separated, each layer of participants can be managed and monitored by analyzing the information provided by the next level of participants, or optimize resource allocation through the information provided by the upper level participants. Finally, there is a high cost problem in the realization of the energy Internet. Although the price of batteries and new energy equipment is reduced, the initial construction of a smart grid only in Nice City costs 30 million euros. To achieve this requires a pilot project like Europe, the European Commission takes the lead in cooperating with major energy companies and all walks of life, and it requires the participation of residents to complete such projects.

References

ASME B 31.12-2014. Hydrogen Piping and Pipelines[S]; Shell Hydrogen tudy-energy of the future-sustainable mobility through fuel cells and H_2, 2017.

Huang, Ming. 2013. Feasibility analysis of mixing hydrogen with natural gas pipeline. *Gas and heat*, 1000–4416.

International Hydrogen Industry Development Report. 2017.

Li, Qian and Chengzhu Yin. 2016. Electric gas conversion technology and its application in energy internet. *Electrical technology*, 1002–1388. 10.050.

PV Costs Do Not Include Grid Dispatching and Transportation Costs.

Yin, Chenhui, Dechang Yang, Guangfei Geng and Zheng Fan. 2015. Summary of German energy internet project and its Enlightenment to China *Power grid technology*, 1000–3673.

Zhang, Xiaoqiang and Qingmei Jiang. 2015. Adaptability analysis of adding hydrogen pipe in built natural gas pipeline. *Pressure vesse*, 1001–4837.

Part III
New Strategies of Traditional Energy Companies

Chapter 6
New Energy Plans of Oil Companies

The development of social civilization drives the development of energy demand in a clean, efficient and convenient direction. The advancement of science and technology drives the transformation of energy production and ultimately achieves a balanced development of human civilization and energy demand. In recent years, with the concern about climate change and its possible adverse consequences, energy consumption has been promoted to reduce the use of high-carbon fossil fuels, promoting innovation and development of energy development and application technologies, and playing an important role in ensuring climate stability and environmental improvement. Although energy exploration, development, and consumption fields such as energy mining, industry, and transportation have gradually shifted to low-carbon and new-energy sources, due to resource constraints and application technologies, the amount of carbon dioxide emission is still large, and global emissions levels are relatively high, it is 35% higher than that in 1990, and energy transformation still has a long way to go. According to current technology and development trend, it is expected that the world energy system will enter a period of zero carbon emission, and will experience two stages of low-carbon low-emission stage and carbon-free zero-emission stage, which will roughly take a hundred years (Fig. 6.1).

The first stage: low carbon low emission stage. With the energy strategy development of getting ridding of coal and the increasing application ratio of natural gas industry and low-carbon electric energy, by the middle of the 21st century, low-carbon energy in the primary energy structure accounted for more than 50%. By accelerating the development and utilization of natural gas and increasing the application of renewable energy such as hydrogen energy, new technologies of producing hydrogen from fossil fuel, renewable energy power-to-gas, power storage and transportation will be developed, the development of new electric and hybrid vehicles will be accelerated, and the heavy dependence of motor vehicles on oil will be reduced, and the transition from high-carbon energy to low-carbon energy can be achieved.

The second stage: the carbon-free zero-emission stage. Large-scale reduction of carbon dioxide emissions requires rapid development of carbon-free energy. At present, carbon-free energy such as hydrogen energy, solar energy, wind energy, biomass energy and geothermal energy has entered a stage of rapid development. In

C. Zou, *New Energy*, https://doi.org/10.1007/978-981-15-2728-9_6

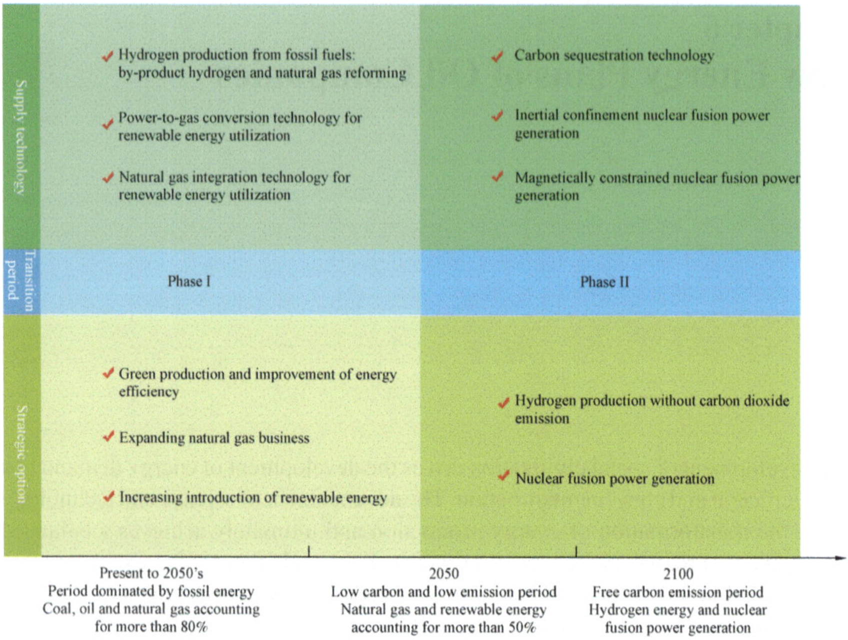

Fig. 6.1 Breakdown structure diagram of energy transformation phase

2017, the share of renewable energy in all energy sources was 4%, but the growth rate accounted for nearly one third of primary energy growth. In addition, the utilization of new energy sources such as ocean energy, tidal energy, wave energy, ocean currents and thermal convection energy has entered the exploration stage. By the end of the 21st century, with the depletion of fossil energy exploration and development potential and the development of technologies such as carbon sequestration and nuclear fusion of renewable energy, renewable energy will become the world's main energy source, accounting for up to 90% of the primary energy consumption structure, carbon emissions tend to zero.

At present, energy transition has always been one of the focuses of the whole society. As the world's largest energy consumer and producer, China has achieved remarkable results in the de-capacity of traditional energy fields such as coal and coal-fired power, the development of new energy sources such as photovoltaics and wind power, and the clean utilization of fossil energy, but the transformation of oil and gas enterprise has not yet been integrated with the international and Chinese energy transformation (Zeng 2014). The times are moving forward and technology is advancing. If oil and gas companies cannot find a favorable position in the energy transition, they will probably lose their right to speak in the process of energy replacement for oil and gas. Combined with the characteristics of oil and gas enterprises and the needs of social development, oil and gas companies should make three strategic developments: green energy, green upgrade and green technology (Zou 2018).

6.1 Green Energy Strategy

The green energy strategy aims to produce more clean energy and replace high-carbon fossil energy with cleaner and low-carbon energy. The existing energy structure is over-reliant on coal and oil, and it is necessary to gradually transform the energy structure and increase the proportion of green and clean energy in the energy structure. Despite the rapid development of renewable energy, it is currently unable to meet the global energy aggregate demand. Therefore, as a transitional energy source, natural gas plays a bridging role and realizes the replacement of energy transformation. Hydrogen energy is a kind of high-quality energy storage energy source, exerting its synergistic effect, and jointly constructing a wind-photovoltaic-hydrogen energy industrial chain system to achieve balanced development of distributed energy production and centralized consumption, and relying on basic technologies such as the Internet of Things and blockchain to create an efficient and intelligent energy Internet to realize the development of "smart sources".

6.1.1 Development of Natural Gas Business as Top Priority

The pace of transformation depends on national policies such as the feasibility of natural resources, climate change and environmental protection, and will also be affected by economic growth, technological innovation, and consumption habits. As the most realistic clean energy source, natural gas resources have great potential, and exploration and development technologies are mature, which can effectively promote energy application reform in the economic field with more energy consumption, greenhouse gas emissions and air pollution, such as power generation, industry, transportation and building heating. Oil and gas companies should take the innate advantages of natural gas energy and the golden developing period to promote the rapid development of national energy transformation (Ma 2017; Zou 2018).

1. Natural gas is an insurmountable bridge for fossil energy to new energy

1) Natural gas is the most realistic clean low-carbon energy source

Natural gas is colorless, odorless, non-toxic, easy to disperse, high in calorific value, cleanliness, people's livelihood and versatility. It can meet the most basic needs of human beings and is the best choice for meeting the needs of human ecological environment. The calorific value of 1 cubic meter of natural gas combustion is equivalent to the calorific value of 10 kWh and 1.25 kg of standard coal. The emission of natural gas, CO_2, NOx, SO_2, and dust under the same calorific value is 50% to 60%, 10%, 1/682, and 1/1479 of coal, respectively, 70% to 75%, 20%, 1/389, 1/140 of petroleum, respectively. Compared with coal and oil, natural gas is not only clean, but also with features such as ease of transportation and application. Natural gas is generally transported through pipelines and directly to households. Transportation and supply are relatively stable and convenient. It has become an energy source for residents to cook and heat, and is closely related to residents' lives. Conventional natural gas is very rich, and unconventional natural gas is 8 times that of conventional natural gas. With advances in hydraulic fracturing such as horizontal wells, it is possible to develop unconventional natural gas fields with greater potential to meet growing energy demand, generate electricity, supply heat and provide fuel for transportation, while reducing greenhouse gas emissions and improving air quality. Therefore, natural gas is an important part of energy transformation and a bridge for the transition from fossil energy to new energy. Driven by the development of social civilization and the advancement of science and technology, energy development has shifted from solid (firewood and coal), liquid (oil) to gas (natural gas), and natural gas will promote the harmonious development of human energy consumption and the ecological environment (Ma 2017).

2) New energy cannot replace fossil energy quickly

Although renewable and clean resources are developing rapidly, they are still relatively low in global energy consumption due to development and application technology constraints. Renewable energy is mainly used to generate electricity, and it only meets about one-fifth of global energy demand. Oil, natural gas and coal will continue to be the leading energy sources by 2035, natural gas growth will exceed oil and coal, and global coal consumption will peak in the next 20 years. The carbon emission growth rate is 0.6%, but the total emissions will still increase by about 13%.

There are still problems that new energy cannot meet the needs of scale applications, mainly in two aspects. First, intermittent problems caused by unstable weather (such as sunshine and wind). Since all kinds of new energy sources are ultimately converted into electricity for people's utilization. The biggest requirement for power supply is stability and reliability. At present, the ground power station that is connected to the power generation side is completely dependent on the main power grid for regulation. When the capacity of photovoltaic power generation is gradually increased, the instantaneous large-capacity fluctuation will cause a big impact on the power grid. The fluctuations allowed by China's power grid design usually do not exceed 15%, which means that if the capacity of new energy generation in a certain place exceeds the proportion of the capacity of local connected grid, the instability of power generation may cause the grid to collapse. Therefore, the inherent intermittent nature of wind power and photovoltaics is a big drawback from the perspective of power supply. As the scale of photovoltaic and wind power generation increases, the weakness will become more and more obvious, and the obstacles and restrictions on photovoltaic and wind power generation will also increase. This is a problem that must be considered in the next development of new energy sources such as photovoltaics and wind power. Second, the energy density of photovoltaic and wind power generation is low. Typically, the peak power limit per square meter of PV modules is currently below approximately 150 W, most of the time at around 100 W. This will lead to a large area of photovoltaic power generation. Therefore, large-scale ground power stations are only suitable for desert and Gobi, and it is almost impossible to build large-scale ground power stations in eastern China, especially in the southeast coast. The current 1.2 million-kilowatt generating units in thermal power plants, if calculated according to the annual power generation, the photovoltaics are far from being comparable to thermal power. The annual generating capacity of a 1.2 million-kilowatt thermal power plant is equivalent to the annual generating capacity of a 7-GW photovoltaic power plant. A photovoltaic power plant of this size needs to cover an area of 200,000 mu. Therefore, how to solve the problem of low energy density of new energy sources is also a challenge to develop and utilize new energy sources.

3) Natural gas supports new energy integration

Renewable energy sources such as wind and solar energy are critical to reducing emissions and improving air quality. According to the International Energy Agency, renewable energy generation currently accounts for about a quarter of total electricity generation and may reach 40% of total electricity generation by 2040. However, since renewable energy such as wind energy and solar energy is a variable energy source, it can only generate electricity in places with sufficient sunshine or wind, and the transportation, utilization and storage costs are large, and the natural gas power plant has the characteristics of quick and flexible starting up, which can achieve full output in minutes, respond quickly to variable energy pauses and sudden surges (Fig. 6.2). Coal-fired power plants not only produce higher levels of carbon dioxide, but often require long and expensive start-up times to preheat their boilers and steam systems. At the same time, natural gas power plants can be built near the central market due to favorable conditions such as less land occupation and less pollution, which helps to increase the speed and efficiency of power delivery. Therefore, natural gas power generation has its own advantages and is an important energy source to support the integration of new energy sources.

2. Accelerating the development of unconventional natural gas

The world's total natural gas resources are 3,922 trillion cubic meters. After nearly 100 years of exploration and development, the remaining proven natural gas recoverable reserves in 2016 is 192 trillion cubic meters, and the natural gas production in

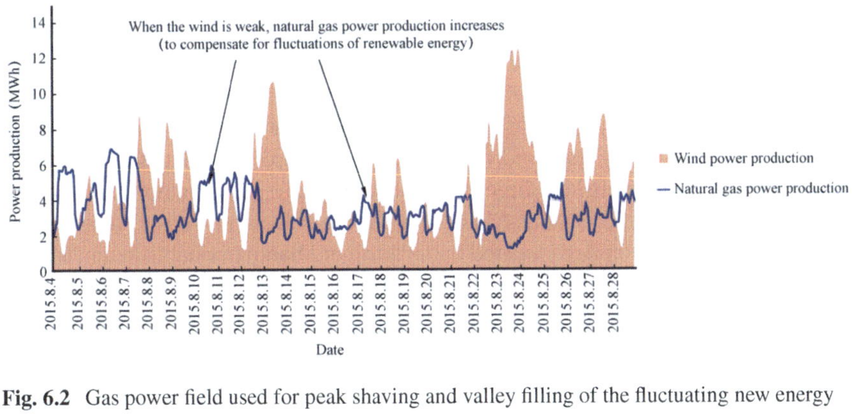

Fig. 6.2 Gas power field used for peak shaving and valley filling of the fluctuating new energy

2017 is 3.7 trillion cubic meters. The amount of unconventional natural gas recoverable resources is more than 8 times that of conventional natural gas, and the proven rate is low.

China's natural gas resources are abundant. According to the results of the dynamic resource evaluation of the Ministry of Land and Resources in 2014, the conventional natural gas geological resources are 67.4 trillion cubic meters, the tight sandstone gas geological resources are 22.9 trillion cubic meters, and the shale gas recoverable resources are 21.8 trillion cubic meters (the prospective geological resources are 121.8 trillion cubic meters), and the coalbed methane resources are 30.05 trillion cubic meters. As of 2017, the accumulated proven natural gas reserves are 14 trillion cubic meters, with a proven rate of 19%, and the annual output of natural gas ranks third in the world. Although China's natural gas reserves are growing rapidly, they still cannot meet domestic demand. In 2017, natural gas consumption was 235.2 billion cubic meters, with a growth rate of 17% and imports of 92.6 billion cubic meters. The external dependence ratio was 39.4%, which was more dependent on foreign countries than that in 2016 and increased by 2.4%. Therefore, it is necessary to vigorously carry out research on conventional and unconventional natural gas technologies, especially to increase unconventional natural gas exploration and development, and promote the rapid development of natural gas business (Ma 2017; Zou 2018).

1) Vigorously developing shale gas

The development of unconventional natural gas such as tight gas, coalbed methane and shale gas has made breakthroughs, and the output has continued to grow rapidly, which has become the main force for the growth of natural gas production (Song et al. 2009; Zhao et al. 2010). In 2017, China's unconventional natural gas production increased by 14% compared with 2016, reaching 48.2 billion cubic meters, accounting for 32.57% of total natural gas production. The production of tight sandstone gas increased rapidly in the past 10 years, and the output in 2017 reached 34.3 billion cubic meters, accounting for 23.2% of total natural gas production; coalbed methane is affected by many factors, the output growth is relatively slow. In 2017, the production output was 4.9 billion cubic meters, accounting for 3.3% of the total natural gas production. The shale gas developed rapidly in recent years, and the output in 2017 was 9 billion cubic meters, accounting for 6.1% of total natural gas production (Hou et al. 2018; Jin et al. 2018; Xie 2018).

The reserves and production of shale gas have increased rapidly in recent years and are expected to play an important role in the future natural gas industry. At present, China has established three national-level shale gas development demonstration areas, namely the Sichuan Changning-Weiyuan Demonstration Area and the Zhaotong Demonstration Area located in the north of Dianqian established in March 2012, and the Chongqing Fuling Demonstration Area established in November 2013 (Wang 2015). After years of development, the evaluation and favorable area optimization, geophysical exploration, reservoir reconstruction and development technologies of shale gas resource within 3,500 m have been quite mature, and the shale gas exploration and development deeper than 3,500 m has also made great progress

(Jia 2017; Zou et al. 2017; Cramer 2008; Warpinski et al. 2008, 2009). With the further improvement and advancement of exploration and development technology for zones deeper than 3,500 m, shale gas production is expected to increase significantly. At the same time, the geological conditions of thickness, depth, continuity and preservation of shale gas production zones in China are more complex than those in North America, and the technical requirements for exploration and development are high. Therefore, innovation should be continued in the future to strengthen the technical research on increasing and stabilizing shale gas production.

2) Actively preparing natural gas hydrates

Natural gas hydrates are unconventional natural gas resources that are present in solid state in deep-sea sediments and permafrost regions. The global resources are about 2.1 million trillion cubic meters, 99% of which are in the ocean. In recent years, the exploration of natural gas hydrates in the world has mainly focused on the Gulf of Mexico, the Yulong Basin in South Korea, the Bay of Bengal in India, the South China Sea Trough of Japan, and the northern part of the South China Sea, and has made important progress (Zou 2014).

In recent years, major breakthroughs have been made in natural gas hydrate test production, and natural gas hydrate tests conducted in the South China Sea Trough of Japan and Shenhu of the South China Sea have been successful.

In 2017, Japan implemented the second production test of marine gas hydrate. In response to technical problems such as sand production, downhole gas-water separation and long-term stable production encountered in the first marine gas hydrate production test in 2013, relevant research plans and countermeasures were formulated. The site selection of the second production test is the same as that of the first test, and the field drilling and gas production test operations are also carried out using the "Earth" ship in the selection of the operating platform. On April 7, 2017, the "Earth" ship proceeded from Shimizu Port to the Saini-Atsumi Knoll to prepare for the second marine production test; on May 4th, it began to produce gas from the natural gas hydrate zone, 350 m below the bottom of the sea with a depth of 1000 m. On May 15, due to the large amount of formation sand entering the first production well, it was decided to temporarily pause the gas production test, and the cumulative gas production for the 12-day gas production test was 35,000 cubic meters. On June 6, 2017, the second production well was switched for gas production test, it was conducted for 24 days. The cumulative gas production was about 200,000 cubic meters, and the average daily production exceeded 8,000 cubic meters.

From May to July of 2017, the China Geological Survey successfully carried out gas hydrate production test in the Shenhu area of the South China Sea by using the fluid extraction method. It is stored in silty clay and clay silt from 203 to 277 m below the seabed with a water depth of 1,266 m. The natural gas is produced in the layer; from May 10 to July 9, 2017, the continuous stable production test is carried out for 60 days, the cumulative gas production is 309,000 cubic meters, the average daily output is 5,151 cubic meters, and the methane content is up to 99.5%.

3. Plan of natural gas infrastructure and gas storage construction

China's natural gas consumption has grown rapidly and natural imports have continued to increase. In 2017, the import volume of natural gas was 92.6 billion cubic meters, of which the imported pipeline gas was 42.7 billion cubic meters, an increase of 10.9% than 2016, mainly from Turkmenistan, Myanmar, Uzbekistan and Kazakhstan; the import volume of liquefied natural gas (LNG) reached 49.9 billion cubic meters, an increase of 39.0%, mainly from Australia, Qatar, Indonesia and other countries, of which imported from Australia accounted for 46% of LNG. Taking into account the route, resources, production, and political factors of onshore natural gas imports, according to the current two pipeline import routes of Russia and Central Asia, the supply capacity in 2020, 2030, and 2050 will be 75 billion cubic meters and 120 billion cubic meters, 140 billion cubic meters; if the pipeline import route from the Middle East can be built successfully, the supply capacity in 2030 and 2050 will be expected to reach 135 billion cubic meters and 160 billion cubic meters respectively. To this end, in order to ensure the supply of natural gas, oil and gas companies need to strengthen the international route plan for natural gas import (Pan et al. 2017).

It is estimated that by 2030, China's natural gas demand will be 550 billion to 600 billion cubic meters, and domestic gas supply capacity will be 200 billion cubic meters (Fig. 6.3). The import pipeline gas supply capacity of Central Asia, China–Myanmar, China–Russia will be 158 billion cubic meters. The capacity of the LNG receiving station that has been built and under construction is 69.4 million tons. The estimated import volume is about 75 billion cubic meters based on 80% load, and the total supply capacity is about 430 billion cubic meters (Ma 2017). To this end, China's natural gas gap will reach 120 billion to 170 billion cubic meters, which needs to increase the scale of LNG imports to meet the demand for natural gas.

The import volume of natural gas is increasing, and the underground gas storage will become an emergency standby "granary" that is indispensable and irreplaceable

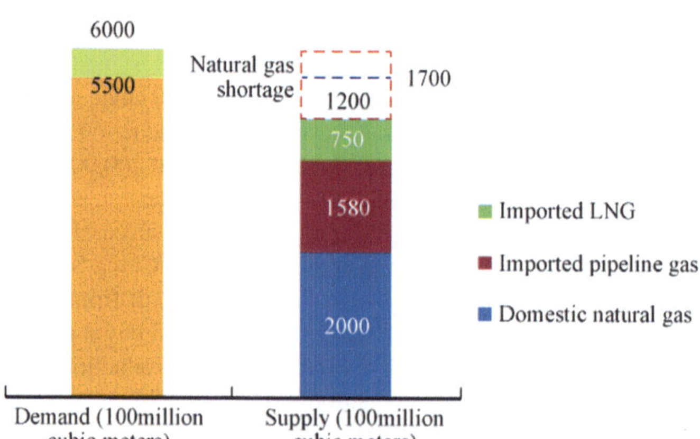

Fig. 6.3 Analysis of natural gas supply and demand in China by 2030

for gas supply under extreme weather, emergencies, strategic reserves and other conditions. The construction of China's gas storage was started in 1999. At present, 25 gas storages have been built, with a peaking capacity of 10.8 billion cubic meters and a maximum peaking capacity of 90 million cubic meters per day. With the continuous expansion of the natural gas consumption market, the function of peak shaving and stable supply of gas storage has become increasingly prominent, especially in the major cities in the north in winter. In 2016, the actual peak shaving of natural gas in China was 15.5 billion cubic meters, of which the peak shaving by gas storage was 34.2%, the peak shaving by gas field was 26.0%, the peak shaving by imported pipeline gas was 18.1%, and the peak shaving by LNG was 18.0% the peak shaving by users accounted for 3.7%; in 2017, the demand for natural gas in the Bohai Sea region was close to 190 million cubic meters, and the daily peak shaving capacity of the gas storage reached 90 million cubic meters, which can meet the maximum demand of 47%. As natural gas dependence increases, the gas storage and LNG become more urgent and important. Therefore, oil and gas companies need to strengthen research on gas storage construction planning and formulate medium and long-term plans to meet peak shaving and storage demand (Wang and Xue 2017). Recently, the underground gas storage is preferentially deployed near the import route, the pipe network hub, and the key consumer market center, initially forming the basic structure of the gas storage facility; in the medium and long term, the comprehensive peak shaving and supply system supported by the gas storage peak shaving is completed. In the system, the peak shaving ratio of gas storage will exceed 60% by 2025. In 2030, peak shaving gas storage and strategic storage are expected to be more than 50 billion cubic meters.

6.1.2 Constructing a New Energy Business Based on Geothermal

1. Large potential of geothermal resource utilization in the oil zone

Geothermal is the second largest clean energy source after solar energy. In recent years, the global geothermal industry has flourished, and more than 120 countries and regions have developed and utilized geothermal resources. It is predicted that by 2100, the global geothermal resource utilization rate will account for 30–80% of the world's total energy. In recent years, the Chinese government has actively promoted the development of the geothermal industry. In 2016, the Chinese government issued the "13th Five-Year Plan for Geothermal Energy Development and Utilization", which plans to increase the total energy consumption ratio of China's geothermal energy consumption from the current 0.5 to 1.5% by 2020, achieving an alternative standard coal of 72.1 million tons.

The oilfield is rich in geothermal resources. In addition to the deep underground hot water layer, the large amount of hot water extracted from the oil layer during oil exploitation is also an important geothermal resource. The oil layer buried several kilometers underground is 680 million cubic meters per year in the production process. The temperature of the separated sewage is generally 30–50 °C, which is an important part of the geothermal of the oilfield. It can be directly used to extract the thermal energy, calculated by extracting energy of 10°C, 1.29 million tons of standard coal can be saved per year. Therefore, the development and utilization of geothermal resources in the oil zone can not only replace the large amount of fossil energy that the oil field needs to consume, but also achieve energy conservation and emission reduction. It can also provide new ideas and new models for oilfields to cultivate new businesses and achieve sustainable development (Chen 2016; Tang 2009; Wang 2017).

2. Coordinating the development and utilization of geothermal energy

Geothermal energy is a widely-existing energy source. At present, the degree of utilization is relatively low. In order to accelerate the utilization of geothermal resources, overall planning and overall management of geothermal energy should be planned and utilized as early as possible. The first is to integrate geothermal resources into the scope of energy resources and standardize management; the second is to clarify the development and utilization of centralized management departments and organization methods as soon as possible; the third is to formulate plans for the development and utilization of geothermal energy in oil and gas enterprises, and to clarify development ideas, targets and key deployments. The fourth step is to establish a long-term communication mechanism with the safety and environmental protection departments to promote the energy conservation and consumption reduction of the oilfield as the starting point, and to promote the implementation of the geothermal heating, oil heat tracing and other pilot projects, to promote orderly geothermal energy industry development; The fifth is to let social capital and talents participate in the development and utilization of geothermal energy in the oil zone.

3. Workover of abandoned wells in oil fields for geothermal resource development

At present, the abandoned wells in the exploration area of oil and gas enterprises are basically completely discarded. The wellbores of some abandoned wells are good enough and can be converted into geothermal wells with workover. Many wells can be used for heat tracing for oil transmission, heat for separation oil gathering stations and heating and cooling in oilfield. For example, Jing 2# union station of Huabei Oilfield uses the geothermal water from two abandoned wells to heat tracing of oil transmission, saving 5 tons of self-use oil and 3,500 cubic meters of gas every day. According to the investigation of the well conditions of the abandoned wells in Daqing and Liaohe Oilfields, about 10% of the abandoned exploratory wells and 5–6% of the abandoned development wells can be recompleted for geothermal energy development. According to this ratio, more than 6,000 abandoned wells in the exploration area of oil and gas enterprises can be recompleted for geothermal energy development. The recompletion of oil well abandoned wells for geothermal energy development can save about 2/3 of development costs. According to the current geothermal industry, the average drilling cost per geothermal well is about CNY 6 million. The abandoned wells in the exploration area of oil and gas enterprises can be used for geothermal wells, which can save drilling costs of more than CNY 24 billion.

In order to speed up the utilization of geothermal resources in abandoned oil wells in the oilfield, the abandoned wells should be fully recompleted into geothermal wells as soon as possible, and the feasibility and utilization potential of geothermal wells should be analyzed to establish a foundation for the effective development and utilization of geothermal resources in the oilfields.

4. Establishment of a test base for comprehensive geothermal energy development and utilization

Geothermal energy is a valuable resource covering heat, minerals and water. Only by comprehensive utilization, cascade and recycling utilization can we make full use of it and maximize economic benefits. According to the characteristics and utilization status of regional geothermal resources, it is possible to build a pilot base for the comprehensive development and utilization of geothermal energy resources the oilfields of Huabei, Daqing and Liaohe, and promote the development of geothermal energy utilization industry. In view of the fact that Huabei Oilfield has some experience of medium and low temperature geothermal power generation, Huabei Oilfield will focus on the construction of medium and low temperature geothermal power generation and geothermal heating test base; aiming at the characteristics of good quality, shallow depth of geothermal resources and long heating time in Daqing Oilfield, it will build geothermal heating and tracing heat for oil transmission; in view of the characteristics of heavy oil and large energy consumption in Liaohe Oilfield, it will build geothermal heat for heavy oil thermal recovery and tracing heat for oil transmission pilot base.

6.1.3 Developing a Hydrogen Energy Industrial Chain

At present, major oil and gas companies in China and abroad have proposed energy transformation strategies, building a clean, low-carbon, safe and efficient new-generation energy system, and developing new clean energy is the fundamental solution to this strategy (Zhu 2010). Hydrogen energy is a kind of green and efficient secondary energy source. It can be used in water electrolysis, fossil energy conversion or cracking (Wang et al. 2017). It can be used in fuel cell power generation without pollutants, and it is more convenient for storage than electricity. The unit mass of hydrogen contains the highest heat of all fossil fuels and biofuels, reaching 143 MJ/kg, and the same mass of hydrogen is about 4.5 times that of coke, three times that of gasoline, and 3.9 times that of alcohol. Vehicles powered by hydrogen energy have the lowest carbon emissions in their entire life cycle. As a carbon free and green new energy source, hydrogen energy has the characteristics of environmental protection, safety, high energy density, high conversion efficiency and wide application range. It can realize zero emission and zero pollution from development to utilization. It is called "the ultimate energy source" by the industry.

Foreign large oil and gas companies have targeted their strategic goals in hydrogen energy and have begun planning and development and utilization. In June 2017, Total S.A. has built 10 hydrogen refueling stations in Germany and cooperated with Linde and BMW in hydrogen refueling technology. Royal Dutch Shell began to plan in the hydrogen energy field. In 2016, it cooperated with Kawasaki Heavy Industries to develop a ship for liquid hydrogen transport. In February 2017, it reached a cooperation agreement with Toyota to build seven hydrogen refueling stations in California and will increase to 100 units in 2024. BP has participated in numerous hydrogen demonstration projects around the world, including studying advanced fuel cell technology with Daimler-Chrysler and Ford, and investing $3.5 million to build China's first hydrogen refueling station in Zhongguancun, Beijing, reaching a power plant construction agreement with GE (GE), the two sides will jointly develop and build a hydrogen power plant to reduce greenhouse gas carbon dioxide emissions.

Domestic oil and gas companies have many advantages in the hydrogen energy industry, such as the use of oil and gas pipelines to transport hydrogen, the use of wind, light, electricity and hydrogen around the pipeline network to establish a hydrogen-gas joint refueling station (Zou 2018). The use of renewable energy has always been a problem of misplacement of production and consumption. With hydrogen energy as an intermediary, wind and light resources within the coverage of natural gas pipelines can be considered together and integrated into the exchange cycle chain of hydrogen energy and power, which can make full use of the abundant wind and light renewable energy in northwest China and other places.

1. Conducting demonstration of hydrogen production

Hydrogen is a clean energy carrier that enables efficient storage and utilization of renewable energy and nuclear energy. At present, hydrogen energy and fuel cell technology have been included in the important direction of future energy development and strategic emerging business development domains. Oil companies have the advantage of natural pipeline network. The participation of hydrogen energy development can effectively combine energy carriers and energy routes. After some key technologies have made breakthroughs, hydrogen energy will enter the fast lane. Therefore, it is necessary to promote the plan and demonstration of hydrogen energy in the upstream, middle stream and downstream.

(1) Demonstration of hydrogen production by wind and light energy. Hydrogen production from electrolyzed water is considered to be the most promising hydrogen production method in the development of fuel cells. The purity of hydrogen obtained by electrolyzing water is as high as 99.999%. It can be directly supplied to fuel cell vehicles without repurification. The use of wind energy and light energy to electrolyze water to produce hydrogen can realize energy recycling. The use of abandoned water, abandoned wind, and abandoned light resources to develop a low-cost, wind and light coupled hydrogen production system to improve the overall efficiency of the system

will become a key development direction through electrolyzed water technology, and oil companies can select low-cost hydrogen production tests and gradually promote demonstrations in areas where wind or light conditions are better, abandonment of wind and light are more serious, and water resources are more abundant.

(2) Demonstration of hydrogen production from underground coal gasification. Underground gasification of coal is changing physical coal mining to producing coal gas by chemical method, and one ton coal can produce low-heat gas 1,500–2,200 cubic meters, of which hydrogen accounts for 35–40%. At present, laboratory research shows that hydrogen production efficiency can be greatly improved under different catalysts and different gasification environments. On the basis of establishing the evaluation standard of coal bed for underground gasification target and find out the coal resource of the underground coal gasification target, the oil company can figure out the coal resources of the hydrogen production in the mineral rights area and optimize the favorable target for the field test in the near future to promote the field test of economical and efficient hydrogen production and gradually form a demonstration.

(3) Demonstration of joint hydrogen refueling station. Based on the advantages of the pipeline network and terminal, and existing gasoline refueling and gas refueling stations, the comprehensive refueling business of oil, gas, hydrogen and electricity will be fully developed and improved. The demonstration of joint refueling stations will be carried out to integrate the gas, hydrogen, electricity and gasoline refueling business together to develop the oil company's refueling station into a comprehensive station of medium carbon, low carbon and carbon-free energy.

2. Abundant coal, wind power and solar energy resources near gas pipeline network

The Gas Transmission System from West to East, Shanxi–Beijing Transmission Project, the China–Myanmar Natural Gas Pipeline System and the Northeast Pipeline Network System form a natural gas main pipeline network connecting east and west and running through the north and south. Coal, wind power, and solar energy are distributed near the gas pipeline network. Hydrogen can be partially mixed with natural gas and transported through the gas pipeline network (Zou 2018).

1) Abundant coal resources along the pipeline network can be used for hydrogen production

Coal is one of the largest raw materials for hydrogen production. At present, the large-scale and low-cost hydrogen production technology with coal is basically mature. Under the condition of 900–1,000 °C of isolated air, 55–60% of hydrogen and 23–27% of methane gas can be produced by coal coking, coal gasification or the like. Hydrogen production with coal has great potential for development (Table 6.1). The distribution of gas pipeline network has a good correlation with the distribution of coal resources. How to make good use of gas pipeline network is the key to achieving clean utilization of coal resources.

Table 6.1 Potential capacity of coal hydrogen production in China in 2016

Region	Coal to hydrogen production capacity		Natural gas to hydrogen production capacity	
	(100 million m^3)	(10,000 tons)	(100 million m^3)	(10,000 tons)
North China	15504	13842.9	185.2	165.4
Northeast	1021	911.6	126.4	112.9
East China	2628	2346.4	21.8	19.5
Central China	1354	1208.9	9.2	8.2
South China	36	32.1	161.6	144.3
Southwest	2647	2363.4	704	628.6
Northwest	7085	6325.9	1528	1364.3
Total	30275	27031.2	2737.2	2443.2

2) Hydrogen production from abundant wind power resources along pipeline network
China's wind energy resources are widely distributed. The richer areas are mainly concentrated in the southeast coast and nearby islands as well as in the north (Northeast, North China, Northwest) areas. There are also some areas with rich wind energy in the inland. According to the data of the China Wind Energy Association, in 2017, the national installed capacity (excluding Hong Kong, Macao and Taiwan) increased by 19.66 million kilowatts, and the cumulative installed capacity reached 188 million kilowatts, a year-on-year increase of 11.7%. In 2017, the proportion of newly installed wind power installed capacity in China's six regions was North China (25%), Central South (23%), East China (23%), Northwest (17%), Southwest (9%), and Northeast (3%). The newly installed capacity in the "Three North" area accounted for 45%, and the newly installed capacity in the Central Southeast accounted for 55%. By combining the pipeline network with regional wind power resources, hydrogen production by wind power can better solve problems such as wind power instability and energy transportation.

3) Hydrogen production from abundant solar energy resources along pipeline network
The annual average daily radiation energy in most parts of China is above 4 kWh per square meter, and the daily radiation in Tibet is up to 7 kWh per square meter. The potential for solar energy development and utilization is very large. China's solar energy distribution has the characteristics of annual total radiation in the western region higher than the eastern region and the south is lower than the north. The distribution of solar energy resources in China is divided into five categories according to the annual sunshine hours and the total annual radiation. The annual sunshine hours of the first, second and third regions are more than 2,000 h. It is a region rich or relatively rich in solar energy resources in China, and the distribution area accounts for more than 2/3 of the total area of the country. China's gas pipeline network trunk lines, especially the West to East Gas Pipeline Network, are just passing through Class I and Class II areas (Table 6.2), which is conducive to the organic integration of solar hydrogen production and natural gas pipelines.

Table 6.2 Distribution of solar energy resources in China

Region type	Annual sunshine hours (h)	Total annual radiation (MJ/m^2)	Standard coal required for equivalent heat (kg)	Regions	Remarks
Class I	3200–3300	6680–8400	225–285	Northern Ningxia, northern Gansu, southern Xinjiang, western Qinghai, western Tibet	Most abundant area of solar energy resources
Class II	3000–3200	5852–6680	200–225	Northwest Hebei, northern Shanxi, southern Inner Mongolia, southern Ningxia, central Gansu, eastern Qinghai, southeast Tibet, southern Xinjiang	Richer area
Class III	2200–3000	5016–5852	170–200	Shandong, Henan, southeast Hebei, southern Shanxi, northern Xinjiang, Jilin, Liaoning, Yunnan, northern Shaanxi, southeast Gansu, southern Guangdong	Medium area

(continued)

Table 6.2 (continued)

Region type	Annual sunshine hours (h)	Total annual radiation (MJ/m^2)	Standard coal required for equivalent heat (kg)	Regions	Remarks
Class IV	1400–2000	4180–5016	140–170	Hunan, Guangxi, Jiangxi, Zhejiang, Hubei, northern Fujian, northern Guangdong, southern Shaanxi, southern Anhui	Poor area
Class V	1000–1400	3344–4180	115–140	Most parts of Sichuan, Guizhou	Worst area

3. Business measures for oil company to develop hydrogen energy

1) Hydrogen production and hydrogen transmission by using peripheral energy along pipeline

At present, most of the hydrogen is produced from fossil fuels. Without the assistance of CCUS (carbon capture, utilization and storage technology), greenhouse gas emissions will not be reduced much during the entire life cycle. At present, as a mature energy storage technology, power-to-gas converts electric energy into gaseous fuel, which solves the problem of difficult storage of renewable energy power. There are two main methods of utilization: one is to combine hydrogen and carbon dioxide, use methanation reaction such as Sabajie reaction to convert to methane, or use bio-methanation, and then input into natural gas pipeline network; the power that can not connected to the grid is used to produce hydrogen through electrolyzed water method, and the hydrogen is incorporated into the pipeline network that transports the natural gas. Therefore, the direct integration of hydrogen into natural gas or the production of hydrogen from hydrogen and carbon dioxide and then into the gas pipeline can achieve organic integration of natural gas pipelines and hydrogen utilization, which saves costs and reduces carbon emissions.

2) New hydrogen production method and hydrogen-power conversion technology

The utilization of hydrogen energy in China is still in its infancy, and its development direction, goals and orientation of development are becoming more and more clear. In 2017, the National Development and Reform Commission and the Energy Bureau issued the "pilot scheme to promote the construction of grid-connected microgrids", clearly proposing to play the role of fuel cells in distributed energy, making it an important part of the microgrid. It can be seen that the orientation of hydrogen energy is a powerful complement to the diversification of the grid. It can rely on hydrogen fuel cell technology to establish a distributed energy network to achieve regional or urban combined supply of electricity, heat and cold energy.

Oil companies can make a difference in the production of hydrogen methods, hydrogen fuel cells, and research on the use of hydrogen energy to develop relevant business areas. For example, Ene-Farm, a family distributed energy system jointly developed by Tokyo Natural Gas Corporation of Japan and Panasonic, has been successfully promoted in Japan. The fuel cell of the system is fueled by hydrogen gas delivered by natural gas pipelines, power generation and power consumption occur at the same place, and the heat generated by the power generation is also used for heating. The energy efficiency of the entire system is as high as 95%, which is much higher than the energy efficiency of traditional natural gas power generation plus boiler heating (68.7%). According to data released by Tokyo Gas and Panasonic, Ene-Farm can reduce primary energy use by 37% and carbon dioxide emissions by 49%.

6.2 Green Upgrade Strategy

The green upgrade strategy is to convert high-carbon energy into low-carbon energy, convert high-pollution products into low-pollution products, reduce their pollution to the atmosphere; appropriately raise the standards for automotive and industrial oils, and reduce pollutant emissions levels of energy products, reduce emissions during oil and gas production and reduce emission intensity. Clean and green energy can be attained through the upgrading of production process and the products. Transition is achieved through innovation and integration to enhance the position of the company in the industry chain and value chain.

6.2.1 Producing Hydrogen by Underground Coal Gasification

1. Significance of underground coal gasification to improve coal recovery rate and reduce environmental risks

According to statistics, the coal resources of China's abandoned coal mines currently amount to 30 billion tons. Underground gasification of the coal resources of abandoned mines can not only revitalize the existing fixed assets of abandoned mines, but also recover coal resources discarded by traditional coal mining methods to convert non-clean energy into clean energy. From the perspective of coal gasification costs, underground gasification production costs are significantly lower than surface gasification (Table 6.3). The gasification value of underground gasification is greater than

Table 6.3 Economic and technical analysis of surface and underground gasification

Project	Surface gasification	Underground gasification
Investment (CNY/m^3)	350–450	120–150
Cost (CNY/m^3)	0.4–0.6	0.15–0.25
Production process	Coal preparation, sorting, gasification	Direct gasification of underground coal

12.6 MJ/m^3 and the hydrogen content is above 50%, which can be used to extract pure hydrogen (purity of 99.9%).

From the perspective of environmental benefits, underground coal gasification can avoid many environmental problems caused by open pit mining and underground mining (Zou 2018). The ash, oxides, radiation and coal gangue after underground gasification are left underground, achieving slag-free accumulation on the ground and greatly reducing the probability of surface subsidence. At the same time, in the underground gasification process, the sulfur in the coal is basically converted into hydrogen sulfide, which can be concentrated and purified on the surface; the carbon dioxide in the gas can be washed with the waste alkali liquor, and then fixed in the high calcium fly ash, and filled in the underground gasification burning area by grouting. In addition, underground gasification does not pollute surface water. The pollution of groundwater by organic substances such as phenols in coal tar is also greatly reduced by the adsorption of coal ash and the natural regeneration ability of groundwater.

2. Mature technology in screening underground coal gasification area

The main influencing factors of screening geological settings of coal underground gasification include buried depth, thickness, pinch, geological structure and so on. Buried depth is mainly related to air tightness and environmental protection, and the increase of the depth and pressure is conducive to the reaction of carbon and hydrogen to generate more methane; the thickness of coal bed is too thin and with too many barriers, which will affect gasification stability, and result in low thermal efficiency; Keep away from faults to avoid leaks, water cross flow and environmental problems. Under the current technical and economic conditions, the shallow depth of 1,500 m is the favorable depth range for underground coal gasification of drilling type.

3. Major advantages of oil companies in developing underground coal gasification

At present, underground coal gasification technology is basically mature, the natural gas market is expanding, and the external environment for the development of underground coal gasification industry is becoming more mature. Oil companies have five major advantages in developing underground coal gasification.

1) Considerable resources

In general, coal bed mining with a depth of less than 1200 m is mature and economical. The buried depth of coal resources in China's oil and gas mineral areas is generally more than 1,200 m. It does not have the economic and technical conditions for mining, but it is very beneficial for underground coal gasification. According to preliminary evaluations, coal resources suitable for gasification along the West to East Gas Pipeline are abundant. If these coals are fully gasified, 160 trillion cubic meters of gas can be produced.

2) Available technology

The key links of site selection, furnace construction (horizontal well drilling), gas injection and ignition in the favorable area of underground gasification can be combined with geological evaluation, horizontal wells, in situ combustion, oil shale in situ retorting and other oil and gas exploration and development technologies; Surface construction, purification and synthesis, and multi-component utilization can utilize existing natural gas processing and transportation technologies.

3) Market available

First, relying on the existing pipe network, the terminal sales can be directly realized in the mature technology and market areas; second, the oilfield can use it itself to realize the substitution of heat tracing and power generation, such as low purity and low calorific value of gasification gas taking the place of natural gas for power generation in the oilfield, and the natural gas replaced by gasification gas can be used for terminal sales.

4) Environmentally controllable

The oil and gas geological evaluation technology can find out the coal and coal quality characteristics, caprock characteristics and groundwater distribution characteristics of coal beds. In the underground gasification, the existing technologies can control in situ combustion; geomechanics monitoring, pressure control, water zone monitoring in the process of gasification, as well as waste water and gas treatment after gasification, can effectively monitor and control environmental risks by means of technical means and treatment processes formed by oil companies' long-term natural gas development practices.

5) Synergy achievable

The integration development of oil company's entire industrial chain and coal underground gasification has a natural advantage. The underground coal gasification in the oilfield, in addition to harvesting gas products, is also of great significance to the healthy development of the oil and gas industry, which is manifested in the following aspects: the gas can be used to improve the efficiency of oil and gas production, such as enhancing oil recovery of heavy oil and oil sands by thermal recovery and shale oil by gas injection, etc.; coal gas can be used as raw material for refining and chemical industry, high value-added heavy hydrocarbons and hydrogen in gas composition can be used as important raw materials for refining and chemical industry; promote energy conservation and emission reduction, replace natural gas with coal gas. At the same time, it can carry out carbon capture and storage; carry out waste furnace utilization, use waste gasification furnace to transform gas storage, and hope to solve the urgent problem of difficulty in location selection, high cost and long cycle of current gas storage. The relevant technologies used are mature, and the future synergy effect is worth looking forward to.

6.2.2 Improving the Quality of Refined Oil by Adding Hydrogen

The General Administration of Quality Supervision, Inspection and Quarantine (AQSIQ) issued a more stringent national standard for motor gasoline, GB17930-2006, "Auto Gasoline" (National III), which was implemented nationwide on January 1, 2010. The standard has made increasingly strict regulations on the sulfur content, benzene content, aromatic hydrocarbon content and olefin content in gasoline. Among them, more stringent restrictions have been placed on reducing the sulfur content in gasoline.

On May 7, 2015, the National Development and Reform Commission and other seven departments jointly issued the "Work Plan for Accelerating the Quality Upgrade of Refined Oil Products", stating that accelerating the upgrading of refined oil quality is an important action that is conducive to improving the environment and managing smog. Promoting green development and increasing people's livelihood and well-being will also help expand investment and consumption and promote industrial restructuring and upgrading. The National Development and Reform Commission, the National Energy Administration, the Ministry of Finance, the Ministry of Environmental Protection and other relevant departments jointly organized and implemented special programs to strengthen inter-ministerial coordination and close cooperation.

In the face of the national oil product quality upgrade requirements, oil companies can actively introduce advanced technology to implement technical improvements and transformations in gasoline hydrogenation processes and facilities (Wang et al. 2018). For hydrogenation process technology, domestic and foreign are based on the characteristics of sulfur and olefin in the light and heavy fractions of catalytic gasoline. Firstly, the raw materials are pre-fractionated or pretreated to meet the requirements of producing low-sulfur products and reducing octane loss. A catalytic gasoline desulfurization production that meets the requirements for producing low sulfur while minimizing octane loss is selected.

6.2.3 Constructing Four-in-One Station to Supply Oil, Gas, Hydrogen and Electricity

1. Rapid development of hydrogen refueling station construction

According to the report data released by the foreign website H_2stations.org in February 2017, there were 92 new hydrogen refueling stations in the world in 2016, with a growth rate of 70%. The construction of the hydrogen refueling station entered a stage of rapid development.

As of January 2018, there are 327 hydrogen stations operating worldwide, including 139 in Europe and 118 in Asia. Japan, the United States, Germany, and China are the four countries with the largest number of hydrogen refueling stations, 91, 40, 45, and 12 respectively. In addition, there are 19 hydrogen refueling stations under construction in China.

2. Large-scale development of the joint construction mode of hydrogen refueling stations and gas stations

The United Kingdom, Germany, Japan and other countries have already started the joint construction model. On March 27, 2017, ITM Power (Energy Storage and Clean Fuels) and Shell's official website announced that the UK's first gasoline-hydrogen mixing station was open and will be fully owned and operated by ITM Power. The gasoline-hydrogen mixing station is located in the Beaconsfield service area of the UK M40 highway. In addition, Shell has more hydrogen sites in California, with the goal of establishing a network of up to 400 hydrogen sites by 2023.

On October 2017, Sinopec and Guangdong Yunfu carried out the groundbreaking ceremony for the construction of the hydrogen refueling station-gas station; in March 2018, Shaanxi Coal Chemical Industry Group and Guangdong Yunfu carried out the groundbreaking ceremony for the construction of the hydrogen refueling station-gas station.

3. Four-in-one construction of hydrogen, gasoline, gas refueling and charging station

By the end of 2015, there were 96,800 gasoline stations in China, including 20,700 gasoline stations of CNPC, 30,600 gasoline stations of Sinopec, 45,500 gasoline stations of CNOOC and private stations. How to make full use of the good refueling stations will have a great impact on the construction of comprehensive energy service companies for oil companies.

Figure 6.4 is a schematic diagram of the operation mode of the hydrogen refueling station. The hydrogen refueling station mainly includes a high-pressure hydrogen storage device, a compression device, a refueling device, and a station control system. Regardless of the cost of land, the total investment cost of building a hydrogen refueling station is estimated to be CNY 13 million to 15 million. Oil companies only need to make full use of the resources of existing refueling stations, which will save a lot of land and personnel management costs.

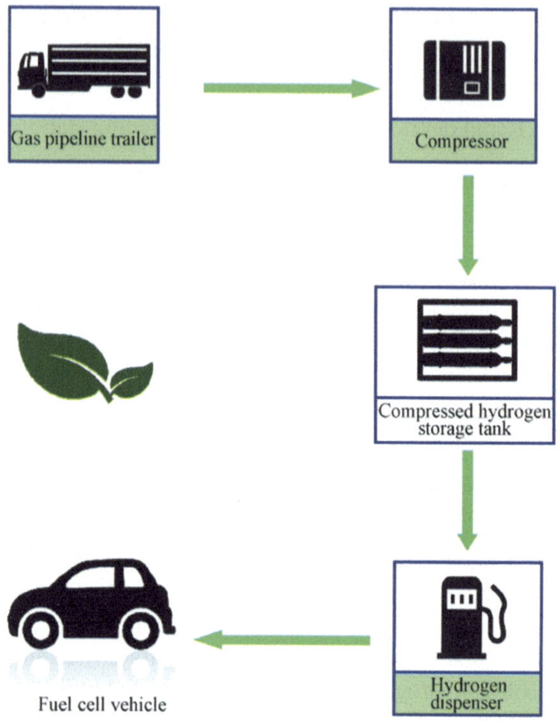

Fig. 6.4 Operation mode of the hydrogenation station

6.2.4 Biomass Energy Utilization

On September 13, 2017, 15 ministries including the National Development and Reform Commission, the National Energy Administration, and the Ministry of Finance jointly issued the "Implementation Plan for Expanding Biofuel Ethanol Production and Promoting the Use of Ethanol Gasoline for Vehicles". According to the program requirements, the use of ethanol gasoline for vehicles will be promoted nationwide by 2020.

Ethanol gasoline is a new type of energy formed by blending fuel ethanol and ordinary gasoline in a certain proportion. At present, the ethanol gasoline promoted in China is made up of 90% of ordinary gasoline and 10% of fuel ethanol, which is E10 ethanol gasoline. Adding fuel ethanol to gasoline can effectively reduce carbon emissions in automobile exhaust, fine particulate matter such as PM2.5, and other toxic substances, thereby improving air quality. Unlike traditional fossil fuels, ethanol is a renewable energy source. Therefore, adding a certain amount of ethanol to the gasoline can reduce the use of non-renewable energy such as gasoline, and effectively utilize the straw and a large amount of expired and degraded reserve grain to convert the waste into usable fuel (Wang 2015; Wei 2013).

According to the 13th Five-Year Plan for Biomass Energy Development issued by the National Energy Administration in December 2016, the annual production of fuel ethanol in China is about 2.1 million tons by 2015; it is estimated that China's fuel ethanol production will reach 4 million tons by 2020.

6.3 Green Technology Strategy

The goal is to achieve the green energy strategy by leading the transformation and upgrading of the industry with scientific and technological innovation, giving full play to the support and leading role of technological progress in industrial transformation, and realizing the low-carbon clean development of oil and gas enterprises. "Internet plus" smart energy is already the direction of energy development promoted by the state. Shared energy and distributed energy need to use the Internet to drive the further development of traditional industries (Zou 2018). The Internet also needs the participation of traditional industries to promote new forms of production. Conduct detailed research on key scientific and technological issues in green development, such as low-carbon technologies, environmental technologies, and strategic emerging technologies. Taking the technological innovation system and the green technology guarantee system construction as the strategic focus of future development, it will lead a new technological revolution and industrial revolution characterized by green, intelligent and sustainable, and accelerate the formation of new industrial products.

6.3.1 Constructing Energy Internet and Distributed Energy System

The energy Internet is an important measure to promote the energy production and consumption revolution. It covers the coordination between primary energy and secondary energy, and the overall planning between the power grid and oil and gas networks. Oil and gas is an effective support for the energy source to realize the "source-network-load-storage" operation mode and the coordination and optimization of the entire energy system. The energy Internet will have an impact on the oil and gas industry from the diversification of energy production and supply models, the intelligent use of users, the interconnection of transmission and trading systems, and the clean access to large scale energy. Oil and gas enterprises should change the traditional operation mode and management mode, actively realize the deep integration of automation and information, connect seamlessly the upstream chains, improve the construction of storage facilities, and shape the "source-network-load-storage" coordination optimization mode of midstream of oil and gas.

The "source-network-load-storage" coordinated optimization model will be further extended to the entire energy Internet field, which is the coordinated and optimized operation mode of the entire energy system. Among them, "source" includes oil, natural gas, electric power and other energy resources; "network" includes various resource networks such as oil pipeline network, heating network and power grid; "load" includes power load and multiple energy needs of users; "storage" mainly refers to a variety of storage facilities and storage methods for energy resources. First, the "transverse multi-source complementarity" in the energy Internet refers to the mutual complementation and coordination of various energy resources such as electricity, oil, hydro and natural gas, highlighting the substitutability between various types of energy, and users can not only choose any kind of energy they like, but also choose the way of utilizing energy resources. The second is the "longitudinal source-network-load-storage coordination" in the energy Internet. Through various energy conversion technologies and information flow and energy flow interaction technologies, the mutual coordination between the development and utilization of energy resources and resource transportation networks and among energy transmission networks are realized. To unify the user's multiple energy needs into one makes the power demand management further expand into a comprehensive energy management in the whole energy field, and regards the user's comprehensive energy demand and the user's energy storage equipment as controllable resources, to realize the integration of information flow and energy flow between energy demand, energy storage and energy supply, and forms a highly regulated energy supply and demand system, thereby improving the operational efficiency and safety and stability of the entire energy system.

The effective development of the energy Internet will enable the "Internet plus" thinking to penetrate the oil and gas industry in depth. Oil and gas companies will change their traditional management methods and management models, and prepare for the energy Internet era in every aspect of the industry's upstream, middle stream and downstream.

1. Connecting the upstream "nodes" of oil and gas

Actively respond to the technical driving connotation of the energy Internet, and achieve deep integration in the field of automation and information technology. Through the application of corresponding technologies and information management methods in oil and gas upstream enterprises, the "nodes" of production, management and control will be connected seamlessly, the allocation of enterprise resources will be optimized, and the competitiveness of enterprises will be improved. Specifically, it includes three aspects: First, build a smart research platform. Based on the problems of 2-dimension regional distribution of oil and gas enterprises, single source of information channels, and scattered information, build an enterprise-level public information consultation and service platform, build a scientific data sharing platform and standardization system, and promote the integration and sharing of scientific research resources. Second, build intelligent oil and gas exploration and intelligent oil fields. The information technology such as sensors and intelligent devices are applied to oil and gas exploration and oil well production remote monitoring, and the physical conditions and production data of oil wells and oil fields are collected and processed, thereby achieving efficient oil and gas development, safe production, reducing labor intensity and improving work efficiency and management level, etc. Third, build a smart energy-saving emission reduction system. Based on Internet means such as cloud computing, integrate servers to promote efficient use of equipment, use information technology to conduct real-time energy consumption monitoring, intelligent analysis of energy consumption, establish an effective energy plan at the group level, and build an efficient, energy-saving, green and environmentally friendly production work environment.

2. Shaping the "capillary " in the midstream of oil and gas

The characteristics of open, interconnected, equal, and shared energy Internet have put forward new requirements for infrastructure construction of energy supply. For the oil and gas field, it is necessary to first improve the infrastructure construction of the midstream of the industry, and to shape the "capillary" for transportation and storage on a wider range of users to ensure the smooth flow of energy. The first is to build a smart gas pipeline network. Similar to the current construction of power transmission and distribution network, expand and improve the extension and application scope of gas pipeline network. With the advancement of urbanization construction, each household becomes a gas terminal, the phenomenon of lack of supply, shortage of supply and no supply will be reduced. At the same time, with intelligent technology, installation of application facilities and construction of smart gas pipeline network, achieve forecasting and early warning, safety supervision and emergency disposal, etc. and ensure safety of energy utilization. The second is to promote the development of natural gas storage capacity. On the basis of the existing transportation pipeline network, the construction of facilities related to natural gas storage capacity will be promoted. Based on the construction concept dominated by distributed type and assisted by centralized type, the natural gas flexible adjustment characteristics are fully utilized. In the construction of microgrids or industrial parks, natural gas storage capacity is used as a soft tray for energy supply, and the overall flexibility and openness of the energy Internet, and intelligent collaboration are realized, which ensure energy security, cleanliness and efficiency.

3. Activating the "organs" of downstream of oil and gas

With the advancement of energy Internet construction, the continuous development of China's oil and gas industry will be promoted in the future. As the sales and energy utilization of the "organs" in the downstream industry, it is necessary to integrate market and innovative thinking, activate and pay attention to maintain the overall vigorous life of the oil and gas industry. The first is to develop hardware manufacturing of smart gas. According to the existing base of users and gas supply experience, the integration of energy Internet concept, manufacturing technology innovation, cost-efficient smart gas meter, smart gas appliances and other smart gas hardware, it is applicable to realize a series of functions such as real-time monitoring, automatic reporting, gas consumption analysis, early warning beyond benchmark and remote cut off through mobile Internet, and to obtain user-related information, gas consumption data of system, user-side access point, and certain financial resources. The solidified pattern of gas market will be changed through the product innovation, while more user needs can be created and more revenue can be attained by meeting user needs. The second is to set up an integrated energy service provider. Pilots will be conducted preferentially in areas where oil and gas resources are abundant and concentrated, and the market is mature. Starting from the actual situation of the pilot project, aiming at the user's highest energy efficiency and lowest cost, the company provides comprehensive solutions including information technology and energy technology through data provided by intelligent hardware to help terminal users realize a variety of energy options and switching such as gas or electricity. The operation of integrated energy service providers needs to reflect economic and social benefits.

6.3.2 Technical Preparation of Carbon Capture and Underground Gas Storage

Coal plays an important role in China's energy structure and it is necessary to develop and prepare the technology of CCUS. Firstly, the development of carbon capture technology can provide technical means for the green development of energy; secondly, the reserve technology will serve the clean utilization of coal resources in China. Compressed air energy storage technology is the second largest energy storage technology. Oil companies should take advantage of their own advantages, utilize the technology and resource advantages of gas storage construction and operation, and combine with wind energy and solar power technology to store renewable energy. It is also possible to convert electrical energy into hydrogen energy to supplement natural gas resources.

1. Carbon capture technology

Carbon capture is the capture of carbon dioxide released into the atmosphere, which will be compressed, and then injected back into depleted oil or gas fields or other safe underground sites. Carbon capture is an important way for China to reduce carbon emissions in the future. In addition, the captured carbon dioxide can be recycled, improve oil recovery, and even use high-purity carbon dioxide as a food additive supplying to food manufacturers such as beer and soda plants for free.

Carbon capture and storage can collect carbon dioxide from large power plants, steel plants and other sources, and then store them in various ways and bury them underground. It includes three steps of carbon dioxide capture, transportation and storage, which can reduce carbon emissions by 85–90% per power unit. The cost of carbon from capture to storage is relatively high. In particular, the technology of storage is difficult and the reliability is still being explored. There are three ways for carbon sequestration in the world: the first is the storage method of utilizing oil and gas fields, which can improve the recovery factor of oil and natural gas and realize carbon sequestration, and the technology is basically mature. The second type is deep coal bed storage, which is used to improve the recovery factor of coalbed methane. It is currently in the preliminary exploration and testing stage. The third is deep salt water zone storage. The amount of carbon dioxide sealed by the storage method is much higher than the previous two methods, and the technology is becoming more mature, but there is currently no industrial device for commercial operation.

Oil companies should step up research on carbon sequestration technology. China's geological structure is various, and the formation with potential for injection can store 1.5 trillion tons of carbon dioxide, which can meet the needs of geological storage of carbon dioxide for hundreds of years. By using carbon dioxide flooding technology, oil recovery can be greatly improved, and carbon dioxide can be stored in the reservoir after displacing crude oil for a long time, realizing scale reduction of carbon emission. Due to the high cost of carbon capture and storage, the government has introduced a corresponding subsidy policy or raised the carbon tax to offset the cost of emissions, and adopted carbon emission limits and trading permits to obtain taxes, prompting enterprises to improve production and reduce pollution.

2. Underground gas storage technology

1) Gas storage construction

With the continuous expansion of the natural gas consumption market, the peak shaving and stable supply of gas storage has become increasingly prominent, especially in the major cities in the north in winter. In 2016, the actual peak shaving amount of natural gas in China was 15.5 billion cubic meters, of which the gas storage accounted for 34.2%. The higher the dependence of natural gas on foreign countries, the more urgent and important the construction of gas storage.

2) Professional team

The construction and operation of gas storage is obviously different from the development of conventional gas reservoirs. Gas storage is with multiple cycles, high-speed huff and puff, the speed of injection and production is 20–30 times that of gas reservoirs, which make the encroachment of water and gas channeling serious, fluid distribution and space utilization complicated, and its construction standards, operational requirements and risks are much higher than conventional gas reservoirs. There is a general lack of professional technology in China's gas storage, and the gap in the future of gas storage professionals will be further expanded. Strengthening the building of professional talent team is the key to ensure the rapid and healthy development of the future gas storage.

3) Foreign operation experience

Since the depth of China's gas storage is generally higher than that of foreign countries, the operating pressure and the consequent risk are much higher than those of foreign countries. China's gas storage is mostly located in densely populated areas, and the consequences of accidents are unimaginable. China has built gas storage for less than 20 years, and the history of foreign gas storage has accumulated rich experience and lessons. Although the oil company's construction technology of gas storage has reached advanced level, it is still necessary to learn from the mature experience of foreign countries about the operation and safety and maintain the status of "zero accidents".

4) Speeding up marketization

The driving force for the rapid development of China's gas storage construction is mainly the strategic needs of the national economy and the people's livelihood. Gas storage includes system value, financial value and insurance value. The system value of the gas storage is the physical effect of the gas storage on the pipe network. The existence of gas storage can reduce the pipe diameter of pipeline construction to a certain extent and play a role in optimizing the cost of pipeline network. The financial value is to the use of time or seasonal spread arbitrage by gas storage operators. The value of insurance is the commonly mentioned stable supply function. In 2014, the National Development and Reform Commission issued the "Guiding Opinions on Accelerating the Construction of Gas Storage Facilities" to encourage various ownership economies to participate in the construction and operation of gas storage facilities, and to increase financing support for investment enterprises in gas storage facilities. In the oil and gas reform plan announced in 2017, emphasis was placed on establishing a natural gas peak shaving policy and a grading reserve peak shaving mechanism, clarifying that gas supply enterprises and pipeline enterprises should shoulder seasonal peak shaving and emergency responsibility, and local governments should be responsible for coordinating the implementation of the daily peak shaving responsibility. Although the relevant policies promulgated by the state are actively guiding the gas storage, they have not made specific enforceable regulations on gas storage financing, price adjustment and operational subsidies.

6.3.3 Top Level Plan

1. Top Level Design

The work scopes are to strengthen the top-level design, clarify workflow, make long-term plan, and choose the opportunity for arrangement, determine the vision of energy transition and development. The environment is the scarcest resource, and ecology is the most valuable asset. The future green development will be the pursuit of enterprise development. Green and low carbon is the basic development strategy of oil companies. This aim can be achieved through strengthening the top-level design and persisting in the two directions of the cleanliness of producing process and products. Realize the transformation from oil and gas production and suppliers to energy service providers, and cultivate green and sustainable development into the new core competitiveness of oil companies.

(1) Develop green energy and promote the transformation of social energy to low carbon. Enhance the development of green and low-carbon energy, develop vigorously natural gas to assist in the optimization and adjustment of energy structure. Promote the upgrading of oil quality and actively explore the development and utilization of new oil products such as ethanol gasoline and biodiesel.
(2) Cultivate emerging energy and realize the transformation of enterprise development. Based on the oil fields, develop vigorously geothermal and hydrogen energy businesses, and use natural gas and electricity as a link to steadily develop energy storage, wind, light and electricity, and actively explore other emerging businesses.
(3) Persist in clean production and vigorously implement energy conservation and emission reduction. Improve the level of on-site clean production, control pollution and protect the ecology from the origin.
(4) Adhere to scientific and technological innovation and lead the green development of enterprises. In the areas of energy cleanliness, energy efficiency improvement, hydrogen energy development, and new energy storage materials, focus on the objectives, clarify the issues, and give full play to the strategic support of basic research.

2. Deployment of Foundation and Innovation Platform

Scientific and technological innovation is inseparable from laboratories. The laboratory has played a role in the scientific and technological innovation and the transformation of scientific research achievements. The laboratory is the cradle of science, the base of scientific research, the source of scientific and technological development, and plays a very important role in the development of science and technology. It is necessary to establish a key enterprise laboratory with a strategic vision, develop application foundations, cutting-edge technologies and common technology research and development, and establish joint laboratories with universities and institutes within the enterprise to carry out basic and cutting-edge innovation research around the construction of independent innovation capabilities.

For energy conservation, environmental protection and new energy, select key common technologies and major application demonstrations to strengthen the supporting role of basic research in energy transformation and development.

(1) In terms of energy conservation, environmental protection and new energy, consider basic research on clean and efficient use of coal and new energy-saving technologies, renewable energy and hydrogen energy, smart grid, strengthen clean conversion of carbon-containing energy, and the basically supporting and leading research of "source-network-load" synergy mechanism.
(2) In the aspect of transformation and upgrading, focus on basic research of network collaborative manufacturing, key infrastructure materials, cloud computing and big data, high-performance computing, broadband communications and new networks, new energy vehicles, functional molecular materials and devices deployment, solve the key technical basic issues of industrial commonality, provide scientific support for cultivating strategic emerging industries.

(3) Strengthen energy Internet and blockchain research in energy sharing and smart production. The de-centering of energy interconnection and blockchain is highly compatible with the distributed development trend of the energy system. Through the decentralized trust system and intelligent contract execution system, it is basic guarantee of the effective connection to realize energy Internet energy flow, information flow and capital flow.

In the face of global energy transformation, international oil companies have responded positively, and their development strategies have both commonalities and characteristics. In general, adhere to the core strategy, continue to create economic benefits, practice social responsibility, cooperate with the society, guarantee oil and natural gas supply, and continue to play a strong role in low-carbon energy by exploring new business models and technologies.

As the world becomes more and more oriented towards low-carbon energy, international oil companies will take advantage of their inherent advantages to provide cleaner energy solutions, such as the combination of natural gas and renewable energy, and the creation of new energy businesses and investment in new areas of energy, including wind, solar, hydrogen, and new materials. Explore various low-carbon energy development pathways and realize the transformation from international oil companies to international comprehensive energy companies.

6.4 Shell's Energy Transition

In 2018, Shell released an energy transformation report that outlines the company's new insights into the transition to a low-carbon energy system, and outlines how to drive business strategies in the context of climate-related risks and opportunities. Shell has set three strategic goals that provide a solid foundation for managing the risks and opportunities associated with transitioning to low-carbon energy systems. The first goal is to provide a world-class investment case, which means that Shell is the leading company in the industry in terms of total return, enabling it to invest in areas where growth is foreseen, and to bear oil price volatility and downstream manufacturing and marketing risks associated with fluctuations in profit margins. The second goal is to transform and thrive into a low-carbon energy source by meeting the needs of society for more and cleaner energy. With the development of the energy system, we provide our customers with the product portfolio they need. The third goal is to maintain Shell's social business license and make a real contribution to people's lives (Zhu 2010; Hamburg: Shell Deutschland Oil GmbH 2017; London: Shell Group of Companies 2017).

If the goal of the Paris Agreement is to be achieved, it will be necessary to stop increasing greenhouse gas emissions such as carbon dioxide into the atmosphere by 2070. To achieve these goals, the amount of greenhouse gases generated per unit of energy consumed by 2050 must be halved. Shell's plan to achieve the goals of the Paris Agreement means that Shell needs to reduce the net carbon footprint of energy products by about half by the middle of the 21st century, with major measures including reducing emissions from Shell's own operations and changing emissions of product selling to customers etc. The specific approach is to use more renewable energy, biofuels and electric vehicle charging stations, and provide more natural gas for electricity, industry and transportation, develop underground safe mining and carbon storage technologies, and develop natural carbon sinks such as forests and wetlands to compensate for unavoidable emissions by society.

6.4.1 *Initiatives of Transition*

In order to promote Shell's continued success and adaptability to competition, it actively participates in many growth areas, manages emissions including methane, expands existing businesses in areas such as LNG, investing in new fuels, charging electric vehicles and providing new services such as electricity to households.

1. Reducing pollution emissions during operation

A series of actions boost Shell's emissions performance, including improved equipment maintenance programs, installation of more energy efficient equipment, management of operational emissions, provision of low carbon products to customers, and development of technologies such as carbon capture and storage (CCS).

Greenhouse gas and energy management plans must include greenhouse gas emissions sources and at least 10 years of expected emissions projections.

To assess the adaptability of new projects, Shell considers the potential costs associated with greenhouse gas emissions when evaluating all new investments. This means that if greenhouse gas emissions are expected to be too high, or if design changes are needed to reduce greenhouse gas emissions, the project may stop at an early planning stage.

In terms of integrated gas, the Pearl gas-liquid conversion plant in Qatar uses heavy paraffin synthetic waste gas as fuel to power the plant, reducing CO_2 emissions by 700,000 tons per year.

Actively expand the deployment of solar photovoltaic (PV) in the business. In California, Shell is using a photovoltaic project to provide solar power to the Stockton fuel distribution terminal for on-site service. It also develops a solar power plant at the Moerdijk chemical base in the Netherlands, which will provide a peak capacity of approximately 20 MW of renewable energy.

Shell Oil has a 34% share of the, which is building a 1,000 MW solar thermal power plant that will generate 6,000 tons of steam per day for production of Petroleum Development Oman in daily operations with the use of sunlight, reducing carbon emissions.

2. Providing customers with lower emission products

Shell produces and sells a variety of low-emission energy products, technologies and services that help reduce overall emissions from energy systems, and sells more natural gas for electricity, transportation and industry.

Shell is the world's largest independent oil and gas company, LNG distributor, with offices in Japan, Korea, Taiwan and China. Currently, natural gas accounts for about 50% of Shell's product portfolio.

Develop electricity business. As energy consumption shifts from high carbon to low carbon energy, electricity (including electricity produced through renewable energy) will be an important part of Shell's future development and hopes to become the fourth largest role after oil, gas and chemicals business. pillar. Shell plans to apply the oil and gas market model to electricity, including providing power to end users, buying and selling electricity, and producing electricity.

Shell has decades of experience in electricity trading and can be wholesaled to energy retailers. Shell currently generates electricity through renewable energy projects in the US and Europe. In the US, Shell has been the top three wholesalers of electricity for the past 10 years. More than 10,000 MW of electricity are managed in North America, more than one-third of which comes from renewable energy producers. In 2017, MP2 energy trading expanded its ability to directly supply electricity to the business market, giving Shell direct access to large industrial and commercial markets in Texas and the Northeastern of United States. In 2017, Shell agreed to acquire First Utility, a leading independent energy supplier in the UK, and power 825,000 homes.

Shell also wants to improve electricity supply in developing countries. It will provide reliable access to people who lack electricity by providing access to companies that invest in small grids and through secure pay-as-you-go systems. In early 2018, Shell invested in Husk, based in India, to supply electricity to rural areas in Asia and Africa.

3. Investing in electric vehicles

In order to achieve the goal of selling more and cleaner energy, Shell is expanding its power supply for transportation. In 2017, Shell acquired New Motion, a Dutch-based company that has one of Europe's largest electric vehicle (EV) charging stations networks.

New Motion has more than 30,000 private electricity toll collection points of homes and offices in the Netherlands, Germany, France and the UK, and more than 50,000 public charging points for 100,000 registered toll card users in 25 countries in Europe.

Shell has introduced electric charging stations in residential areas in the UK, and there were 10 stations in 2017. A network of fast charging points was established at the Shell retail station in the Netherlands. Shell also signed an agreement with high-performance charging network operator Ionity to provide faster charging points for 10 European countries.

Shell also provides customers with smart charging technology, which helps electric vehicles to concentrate on the grid when the total electricity demand is low, which is beneficial to the power grid and to customers, making their vehicles much cheaper to charge. The system was piloted in three pilot projects in the UK, Germany and the US and is looking for a larger opportunity to use this expertise.

4. Developing advanced biofuels

Shell is actively developing advanced biofuels using biomass such as plant inedible waste and cellulose.

In 2015, Raízen opened its first advanced biofuels plant at the Costa Pinto plant in Brazil. In 2017, the plant produced 10 million liters of cellulosic ethanol from sugarcane residues and is expected to produce 40 million liters of cellulosic ethanol per year.

Shell has built a demonstration plant in Bangalore, India, to convert waste, including food, cardboard, plastics and paper, into gasoline or diesel that can drive cars. It is still in the experimental stage, and it is still a few days away from commercialization. The process was developed by the Gas Technology Institute, a research center in the United States, called IH2.

The Shell IH2 process uses heat, hydrogen and catalyst to convert large molecules in the waste into smaller pieces, removing oxygen and other contaminants to produce both hydrogen and carbon. The two are then combined to form hydrocarbon molecules such as gasoline, diesel and jet fuel.

Shell is also one of the world's largest biofuel blenders and distributors. In 2017, Shell used approximately 9 billion liters of biofuels in gasoline and diesel, and sold biofuels in many countries including the United States, Brazil, the European Union, Turkey, Thailand, Malaysia, the Philippines, Argentina and Canada.

5. Expanding carbon capture and storage

Carbon capture and storage (CCS) is a key technology to help large industrial facilities and power plants reduce CO_2 emissions. Carbon dioxide is captured and buried underground to prevent its release into the atmosphere. In the process of carbon capture, utilization and storage (CCUS), the captured carbon dioxide is used directly rather than stored. For example, in the oil and gas industry, captured carbon dioxide can be injected into later-stage oil fields to enhance oil recovery and as a raw material in industrial production processes.

There are currently 21 large-scale carbon capture and storage (CCS) projects in operation or under construction around the world that have the combined capacity to capture approximately 40 million tons of carbon dioxide per year.

In cooperation with the Canadian government and Shell's joint venture partners, Shell operated the Quest CCS project in Alberta, capturing and sequestering carbon dioxide from Scotford Upgrader. Upgrader uses hydrogen to convert bitumen into synthetic crude oil. In less than two years, the Quest project has captured and safely stored more than 2 million tons of carbon dioxide.

Carbon Capture and Storage (CCS) technology was developed by the Shell subsidiary—Shell Cansolv for the Border Dam power station in Saskatchewan, Canada. It is the largest coal-fired power station in Saskatchewan and the largest electricity producer in the region. The use of carbon capture and storage (CCS) technology captures and sequesters most of the sulphur dioxide and carbon dioxide produced by the Border Dam power station, making it the world's first commercial thermal power plant capable of capturing its own carbon dioxide emissions.

At the Mongstad Technology Center (TCM), Shell and the Norwegian state-owned Statoil and Sasol companies research and develop carbon capture and storage (CCS) technology to help reduce costs. In 2017, Shell reaffirmed its commitment to participate in continuous testing of the Mongstad Technology Center (TCM) by 2020.

Shell is also participating in the Gorgon CO_2 Gas Injection Project in Australia, which will become the world's largest carbon capture and storage (CCS) project upon completion. The Gorgon Carbon Capture and Storage (CCS) project will separate and reinject 3.4 million to 4 million tons of CO_2 storage annually. About 100 million tons of carbon dioxide are expected to be captured and sequestered throughout the life of the project.

6. Increasing investment in science and technology

Shell continues to invest in research and development to improve the efficiency of its products, processes and operations, and to commercialize technologies that transform to low-carbon energy in the future. In 2017, Shell invested $922 million in research and development.

Shell operates a global network technology center with major hubs in Houston of the United States, Amsterdam of Netherlands, and Bangalore of India. Thousands of employees in the Technology Center network are engaged in R&D projects, including R&D projects such as seeking to convert natural gas into more efficient and cleaner fuels, developing oil from formations which are several thousand meters below sea level, and reducing Shell's net carbon footprint.

R&D projects often involve collaboration with public or private entities, including universities, government laboratories, technology startups, and incubation base. For example, in 2017, Shell agreed to support research by the American Institute of Energy Biosciences, using biochemical processes to store or transport energy, or to synthesize high-value chemicals.

Shell's open innovation programs include:

(1) Shell Game Changer. The program works with start-ups and unconfirmed early creative companies having the potential to influence the future of energy. Shell provides support, expertise and seed funding to companies while keeping them independent so they can make their own decisions.
(2) Shell Technology Venture Capital. This is the corporate venture capital department of Shell. The companies it invests in are developing promising technologies that are expected to complement Shell's business—primarily oil and gas, new energy and information technology.
(3) Shell Tech Works. Shell Tech Works, based in Massachusetts, USA, accelerates the adoption of proven technologies from other industries and applies them to the oil and gas industry. Launched in 2013, the program works with companies, universities, research institutions, and start-ups to help them develop and deploy technology quickly, cost-effectively.

6.4.2 Future Transition Strategy

1. Medium-term Plan (before 2030)

The transition to low-carbon energy brings opportunities and risks to Shell. Shell's strategy, portfolio and strong financial framework form the company's transformation path by 2030.

1) Diversification of business areas

Energy systems will evolve in different countries and economic sectors, and business risks and opportunities will vary widely. Shell's diversified business helps reduce the risk of unexpected changes in any industry or country, and enables companies to acquire and divest assets and operations based on growth prospects.
Shell has seven strategic themes: conventional oil and gas, deep water, shale, integrated gas, petroleum products, chemicals, and the recently created new energy business. Shell's development and adaptability has been enhanced, with a focus on electricity and new fuels.

The business scope includes the extraction, processing and sale of energy, enabling Shell to manage risk and return flexibly as the energy system evolves. In the past three years, downstream businesses (including chemicals, marketing, refining and trading) have generated good returns. This helps offset the impact of falling oil and gas prices on upstream and integrated gas businesses, indicating that every part of the energy system may be affected differently by changes in demand, supply and commodity prices.

2) Geographical diversity

Shell operates in more than 70 countries around the world and is spread across different economic systems and development stages.

In 2017, operations in 19 countries accounted for approximately 80% of Shell's operating cash flow, including Australia, Brazil, Canada, Nigeria, Qatar and the United States. Shell expects a similar distribution of operating cash flows over the next 10 years.

During this period, Shell predicted that some countries' oil and gas demand will decrease, while other countries will grow rapidly. The sky scenario predicts that oil demand will begin to decline globally after 2025, but oil demand in countries such as India and China will continue to grow, and its growth will continue into the middle of the 21st century.

3) Active investment portfolio

Shell reshapes its investment portfolio by developing integrated gas, chemical and new energy businesses. As the world's energy consumption shifts to low-carbon energy, Shell's business in integrated gas, chemical and new energy will continue to grow in the next 10 years. In 2016, Shell acquired the BG Group and took an important step in restructuring its portfolio. The acquisition accelerates the growth of Shell's integrated gas business and also enhances Shell's position in oil and gas exploration in the deepwater sector.

Shell is expanding the electricity market. It is expected that the power of the energy system will increase in the coming decades, and the power industry will turn to the use of natural gas and renewable energy with lower carbon dioxide emissions to generate electricity. Shell has invested in wind power in the Netherlands and in supplying electricity to UK retail customers. The increase in investment has leveraged existing gas and electricity trading capabilities while building new business models for future growth. Shell continues to adjust its business to meet the changing needs of different countries. For example, in addition to liquefied natural gas (LNG) and biofuels, hydrogen and electric vehicle charging are also offered in European markets such as Germany for faster transition to low-carbon energy.

Shell plans to increase the number of retail outlets in countries such as China, Mexico, India, Indonesia and Russia, where demand for petroleum products is growing in the next 10 years.

Shell's ongoing portfolio changes have reduced the risk of non-economic asset operations and reduced the economical oil and gas reserves due to changes in demand or changes in carbon dioxide regulation.

4) Continuing upstream core business

The upstream business covers three strategic themes: conventional oil and gas, deepwater oil and gas, and shale oil and gas.

By 2030, oil and gas demand is expected to be higher than today in three scenarios, including mountains, oceans and skies (Shell's projected high, low, and low). To meet this demand, Shell is expected to continue investing in the search for and production of oil and gas. At present, Shell has proven reserves available for development for approximately 8.8 years, and 2P reserves available for development for approximately 13 years, and has a resource for development for 20 to 26 years (2P 2C). Therefore, Shell believes it is possible to extend upstream business to the 2030s.

Shell will continue to strengthen oil and gas exploration and discover more oil and gas resources to meet the oil demand expected in the sky scenario in the 2030s. Shell will continue to expand and develop oil and gas exploration and development projects, and remain competitive in terms of oil and gas costs. Even if oil and gas supply is excessive and oil prices are low, it will continue to develop.

Since 2015, Shell's upstream costs have fallen by more than 20% and oil production has increased by 20%. Under the current oil price conditions, the upstream is still creating a large amount of cash flow for Shell, enough to pay for the taxes and reinvestment. Even if the oil price falls to $40 per barrel, there will still be operating cash flow in upstream.

In the Vito deepwater project in the Gulf of Mexico, overall capital investment costs were reduced by 70% compared to the original concept. In the Permian Basin in the United States, direct on-site expenditures for shale operations decreased by 33% in 2014 and have decreased by 60% since 2015.

When investing, Shell considers the following factors to improve its adaptive ability: (1) short-term investment and flexibility to enable production to increase or decrease as demand or price changes (shale oil and gas projects); (2) focusing on projects that generate positive cash flow in the short term (for example, adding new wells to existing deepwater oil and gas fields); (3) increasing capital efficiency to lower breakeven prices; (4) considering the specific implementation criteria of carbon dioxide emission intensity of various assets when investing in new assets; (5) optimization techniques to further enhance resilience, including the use of carbon capture and storage (CCS) and renewable energy technologies in upstream assets; (6) adoption of greenhouse gas and energy management to reduce carbon intensity and potential cost of operating assets.

5) Integrated natural gas business

Shell's integrated natural gas business manages LNG activities and gas-to-liquids (GTL) fuels and other products, selling and trading natural gas, liquefied natural gas, crude oil and electricity.

In 2017, the total demand for LNG in the world increased by 29 million tons to 293 million tons. According to current demand forecasts, supply shortages are expected in mid-2020 unless companies are soon investing in new LNG production projects.

Liquefied natural gas is expected to be a transportation fuel with strong demand. Replacing heavy vehicle fuels with liquefied natural gas can help reduce air pollution. For example, during transportation, natural gas reduces nitrogen oxides from combustion by 80% compared to heavy fuel oil, and almost no sulfur dioxide is emitted from natural gas combustion.

The current market for LNG as a transportation fuel is about 14 million tons per year. If the entire ocean shipping and heavy cargo sector is converted to LNG, it will require 22 million tons of LNG fuel per year.

Shell has a global portfolio of 13 operational LNG plants and nearly 70 different customers in more than 25 countries, enabling the integrated gas business to adapt to changes in regional demand.

Since the acquisition of the BG Group, Shell has reduced its basic operating expenses for integrated gas from 15% to approximately 11%, while sales have increased by 15%. Therefore, the LNG forward delivery cost of the supply is lower than the expected pricing level.

In addition to existing assets, Shell is increasing the cost competitiveness of future supply projects. In the past few years, Shell has reduced the unit supply cost of its portfolio. Shell's goal is to reduce costs and enable all projects to be executed to produce and deliver LNG at a price that is competitive in the natural gas market. This is also a necessary condition for further investment in LNG supply projects.

Shell has increasingly incorporated third-party natural gas and liquefied natural gas into its portfolio to reduce its reliance on equity gas reserves.

6) Strengthening new energy business

Shell expects that by 2020, the average annual capital investment in new energy will reach $1 billion to $2 billion. Most of the investment is expected to be used for electricity and for power generation through solar, wind and natural gas.

The decline in solar and wind power generation costs and the electrification of energy systems make renewable energy development increasingly attractive to society and an attractive investment opportunity for Shell.

Regulatory uncertainties in some electricity markets can lead to uncertainty in long-term income. To avoid this, Shell is looking to invest in today's commercially viable projects. In addition, based on a rapidly evolving technology environment, the best technology options will be determined for each project. For power projects, Shell strives for a return on investment of 8–12%.

The electricity business gives the investment portfolio the ability to withstand oil price volatility, because although electricity prices are tied to natural gas, coal and carbon dioxide prices, most of the electricity prices are decoupled from oil prices. Higher carbon dioxide prices and higher electricity prices can also generate higher revenues for renewable energy projects that sell electricity.

Shell focused on expanding its electricity market with existing customer bases in specific markets in North America and Europe, and is expected to provide new opportunities for trading and selling natural gas, renewable energy and warehousing. Shell is also focusing on economic powers and the fast-growing renewable energy sector.

The new energy business is exploring new transportation fuels, ranging from the development of advanced biofuels made from waste and non-food plants to the start-up of hydrogen refueling stations and electric vehicle (EV) charging operations.

2. Long-term plan (after 2030)

By 2050, Shell will halve the net carbon footprint of its energy products, which means reducing emissions from its operations, most of which will come from the change of product combination. Ways to achieve this include selling more natural gas, biofuels, electricity, developing more carbon capture and storage (CCS) capacity, and adopting natural solutions such as planting forests or recovering wetlands for carbon sinks.

As the world transitions to low-carbon energy, an important approach adopted by Shell is to reduce the net carbon footprint of energy products. The goal includes not only the emissions that it operates, but also the emissions that customers generate when they use energy products sold by Shell.

1) Reduce the net carbon footprint

Shell plans to synchronize with society to reduce the net carbon footprint of energy products. Shell uses a unique net carbon footprint approach, using Sky Scenario Analysis and the International Energy Agency's energy technology outlook 2017 as a hypothetical condition. A reasonable opportunity has been identified to reduce the net carbon footprint of the energy system by limiting global warming to well below 2 °C.

To achieve the stated goals of the Paris Agreement, human society will achieve a net zero emissions of greenhouse gases such as carbon dioxide by 2070. It is required that by 2050, the amount of carbon dioxide consumed per unit of energy in the world will be reduced from about 74 g/MJ to about 43 g. By 2050, Shell plans to match the net carbon footprint of the global energy system. To achieve this goal, Shell needs to go faster than society. In the middle of the 21st century, the net carbon footprint of the energy portfolio needs to be cut in half, from about 83 g of carbon dioxide per megajoule to about 43 g of carbon dioxide per megajoule. As a temporary measure, Shell plans to reduce its net carbon footprint by around 20% by 2035.

Shell's plans cover the full energy lifecycle of the product, making it unique in the energy industry. It includes emissions from the production of energy products, as well as emissions from its energy products, of which about 85% are related to Shell's energy products.

2) Improve emissions from Shell's energy products

Reducing the net carbon footprint will require Shell to reduce emissions from its operations. But most of the reductions will come from changing the combination of products to provide customers with more products that produce lower emissions. Shell will do this in a way that is commercially viable in response to changing consumer demands and in sync with social progress.

6.4.3 Background Information

1. Mountains scenario

The Mountains scenario assumes a steady global economic development and policy guidelines play an important role in building a global energy system and environmental protection. Burning cleaner natural gas will be the backbone of the global energy system, replacing coal as a fuel for power generation in many areas and beginning to be widely used in transportation.

Around 2035, the profound changes in the transportation industry have brought the global demand for oil to a peak. By the end of the 21st century, most cars and trucks on the road will be powered by electricity and hydrogen. Technologies that capture carbon dioxide from power plants, refineries, and other industrial facilities will be widely used, making carbon dioxide emissions in the energy industry close to zero by 2060. Nuclear energy will grow in total global power generation, and by 2060, nuclear power will increase its market share by about 25%.

With the changes in the global energy system, greenhouse gas emissions began to decline after 2030. However, total greenhouse gas emissions are still higher than emissions targets that can control global warming within 2 °C.

2. Oceans scenario

The world outlined in the ocean scenario is prosperous but unstable, and the energy outlook is largely determined by the market and civil society, not by government policy. The resistance from the public and the slow implementation of related policies and technologies will limit nuclear energy development and natural gas development outside North America. At least until the middle of the 21st century, coal is still widely used for power generation.

Without the strong support of the decision-making levels, carbon dioxide capture and storage technology will develop slowly. By the middle of the 21st century, capture and storage accounted for only 10% of global carbon dioxide emissions. By 2075, the ratio may reach 25%. Therefore, the ocean scenario achieves zero emissions of electricity generation 30 years later than the mountain scenario.

High energy prices will promote the development of hard-to-find oil resources and drive the production of biofuels. In the 2020s and 2030s, oil demand will continue to increase until it reaches a steady level in 2040. By the middle of the 21st century, 70% of land travellers will still use liquid fuel.

The high price of fuel will also increase energy efficiency and stimulate the development of solar energy. By 2070, solar photovoltaic panels will become the world's most important source of energy. The public's opposition to the installation of large wind turbines will lead to slow development of wind energy. Due to increased demand for coal and oil, carbon dioxide capture and storage technologies are not supported, and natural gas development outside of North America is slow, greenhouse gas emissions from the ocean scenario will be 25% higher than that from the high mountain scenario.

6.5 BP's Energy Transition

In 2018, BP published a report on Promoting Energy Transformation. In this report, BP examines how the energy world is changing rapidly, articulates its low-carbon goals, and shows how it can help drive energy transformation.

BP believes that the world is growing at an unprecedented rate, creating opportunities for billions of people. All of this growth is inseparable from energy. While the world needs more energy, it also requires new ways of producing and delivering energy and reducing emissions. In response to this double challenge, BP has always focused on the future, adapting to change, and meeting such challenges (London: BP 2016).

6.5.1 Development Goals

Twenty years ago, BP was one of the first energy companies to address the threat of climate change, taking the lead in developing alternative energy sources such as wind, solar and biofuels. In order to make renewable energy a true alternative energy source, BP has invested billions of dollars. In order to significantly reduce emissions, each type of energy needs to be better and cleaner, and it is not enough to compete on renewable energy alone. This is the main reason why BP has to make bold changes to the entire business.

1. Energy structure changes

Human demand for energy continues to grow, mainly due to increased incomes in emerging economies and a rise in the global population to 9 billion by 2040. At the same time, with the advancement of technology, the shift in consumer preferences, and the evolution of policy measures, the energy mix is changing.

Today, renewable energy is the fastest growing energy source ever, and BP estimates that by 2040, renewable energy will account for at least 14% of all energy consumption. In 2040, oil and gas will meet at least 40% of the world's energy needs, even with the goal of limiting the global warming within 2 °C of the Paris Agreement.

Natural gas provides a cleaner alternative to coal power generation and can reduce emissions on a large scale. It also provides valuable backup support for the intermittent nature of renewable energy, providing the high-temperature heating required by industry and increasingly used for transportation.

Oil is currently the main fuel for transportation. BP expects oil's share of the overall energy mix to decline gradually as energy efficiency increases in traditional engines, more use of biofuels and natural gas, and an increase in the number of pure electric and hybrid vehicles over the next few years.

Given the high demand for oil and gas in the coming years, it is crucial to take steps to reduce emissions in production and use.

2. Transformation outlook

In order to achieve the goal of low carbon, society is seeking more energy and providing energy solutions in new and better ways. In response to this double challenge, BP has designed its own strategy.

Although BP can't predict the future, some of the ideas from Energy Outlook and Technology Outlook help shape BP's strategic thinking. BP considers how policy, consumer behavior and technological advances over the next few decades will affect the pace of energy transformation and how it will be produced and used.

In all of BP's forecasts, renewable energy is expected to grow rapidly, but it also shows that oil and gas still dominate in the next 20 years. As a result, BP's portfolio balances the advantages of oil and gas, competitive downstream industries, trade in all forms of energy, and a broad range of low-carbon businesses.

In making strategic decisions, BP considered a variety of potential medium-term supply and demand scenarios, including a faster transition to low-carbon energy. To determine whether an investment potential is commercially significant, BP will test a range of oil, gas and carbon prices to prepare for uncertainty and opportunity.

Regardless of the speed and path that the world will choose, BP will actively plan how to participate in the energy transition and how to maintain a competitive edge in it, which will bring BP a resilience. To reinforce this belief, BP also included part of the long-term compensation of the company's executives in the implementation of the strategy.

3. Low carbon transformation

The growing global demand for energy is a real opportunity for expanding BP's business and generating higher returns for investors. As the business grows, BP's net operational emissions will not increase, and will help others curb their emissions.

BP will fulfill this commitment by reducing emissions from operations, improving products and services, and creating low-carbon businesses. This is just the latest step in BP's 20-year journey, but it is a significant step and a step BP plans to take in the coming years.

6.5.2 Transition Strategy

1. Reducing operating emissions

1) Zero net growth in operating emissions as BP's target

According to estimates by the International Energy Agency, energy efficiency can contribute about 40% of the required emission reductions in order to meet the global warming below 2 °C. BP is reducing greenhouse gas (GHG) emissions by improving the company's existing operational efficiency and designing new major projects.

BP set a sustainable emission reduction target of 3.5 million tons by 2025. Operating companies will achieve this by improving energy efficiency, reducing methane emissions, and reducing flaring, all of which will result in permanent, quantifiable greenhouse gas (GHG) emissions reductions.

As part of the World Bank Initiative, the goal is to achieve zero flaring by 2030. In addition, in order to ensure that carbon emissions will not increase as the business grows, emissions that are not included in BP's sustainable emissions reduction activities will be compensated for any increase in emissions at the 2015 level.

2) Responding to methane challenge

According to data from the Intergovernmental Panel on Climate Change, methane accounts for around 20% of man-made greenhouse gas (GHG) emissions. Since methane is a major component of natural gas, BP is committed to playing a leading role in addressing the methane challenge.

In the atmosphere, methane has a shorter life span than carbon dioxide, but it has a stronger potential for global warming. Therefore, BP's goal is to achieve a 0.2%

methane emission intensity and keep it below 0.3%. This includes methane emissions from operations, where natural gas is marketed as a percentage of gas components.

To manage methane emissions, BP uses technologies such as infrared cameras to identify and help prevent small leaks from becoming more dangerous leaks. It is estimated that 13 of the 22 major projects delivered by 2021 will be natural gas projects, and BP is designing these projects to reduce methane emissions from the outset.

In response to this challenge, BP is working with industry peers to share best practices and invest in potentially breakthrough technologies. For example, BP is actively involved in the Oil and Gas Climate Initiative, whose members produce more than 25% of the world's oil and natural gas. The organization aims to bring methane emissions from the gas value chain close to zero. In addition, BP supports research such as Princeton University to increase scientific understanding of methane and its contribution to global warming.

Focus on Khazzan: BP's Khazzan project will provide Oman with a new major source of natural gas, which is expected to reach around 40% of Oman's total natural gas supply. From the outset, BP has designed the Khazzan project to be characterized by its own efficiency and low emissions. The project has a central processing facility, so there is no need to set up processing facilities at each well site. Fewer treatment sites reduce the likelihood of emissions.

Focus on Lower 48: BP is one of the largest natural gas producers in the United States. BP's Lower 48 business is responsible for approximately half of BP's total methane emissions, so the company has prioritized methane reduction.

2. Improving the product

1) Producing more natural gas

When natural gas is used for power generation, the amount of carbon dioxide (CO_2) produced is about half that of coal. This means that natural gas can have a major impact, as is the case in the United States, where the full use of shale gas has helped reduce carbon dioxide emissions in the United States to the levels of 1990s.

Natural gas is an ideal complement to renewable energy because of the variability of wind, solar and hydropower, which can be a low-carbon, cost-effective backup energy source.

Focus on the Southern Corridor: As one of the largest projects in BP's portfolio, the Southern Gas Corridor will deliver Caspian natural gas directly to Europe for the first time. Natural gas will start in the Shah Deniz gas field in the Caspian Sea and deliver to Italy via five countries, with a total length of 3,500 km. BP plans to supply natural gas to Turkey in 2018 and natural gas to the European market in 2020.

Once peak production is reached, the project will provide enough natural gas to meet more than twice the demand in the capital cities along the Southern Corridor.

Focus on Tangguh: In the operation of Tangguh, Indonesia, converting natural gas into liquid, making transportation in Indonesia and other countries more practical and commercially viable. In addition to meeting the needs of Indonesia, these LNG will also be exported to China and South Korea.

BP's latest expansion project in Tangguh will increase LNG production by 50%, a large part of which will be used in Indonesia.

2) Helping consumers reduce carbon emissions

80–90% of carbon dioxide emissions from oil and gas products comes from the transportation, power generation, industrial and construction industries. Therefore, one of BP's greatest contributions to driving energy transformation is to provide consumers in these industries with products and services that help reduce their carbon footprint.

Many of BP's products and services have been recognized by the Low Carbon Advancement Program. These products include transportation fuels, heating and lighting energy, lubricants that keep the engine running, and petrochemicals used to make everyday items such as paints, clothing and packaging materials. More than 20 carbon neutral products and services through the use of advanced technology and BP's compensation program were developed and opportunities were provided to customers to compensate for their own carbon emissions.

3. Creating a low carbon business

1) Expanding BP's renewable energy business

BP has been operating renewable energy for more than 20 years and is one of the largest operators in the industry and is expanding its business.

Biofuels: BP uses sugar cane to produce ethanol in Brazil, which emits 70% less greenhouse gas emissions over its life cycle than traditional transportation fuels. In addition, BP's joint venture with partner Copersucar to operate a large ethanol repository will help further expand into the bulk fuel market in Brazil.

BP is working with DuPont to develop Butamax technology, which converts corn glucose into bioisobutanol. Compared to ethanol, this biofuel has higher heat and can be blended with higher concentrations of gasoline and delivered through existing fuel lines and infrastructure.

Bio-power generation: Bio-electricity is generated by burning the fiber left behind by crushing sugarcane straw, that is, bagasse, and about 70% of the bio-energy is output to the local grid. This is a low-carbon power generation energy that absorbs CO_2 during the growth of sugar cane, offsetting the CO_2 emitted during the burning of bagasse.

Solar: BP is working with Lightsource, Europe's largest solar energy development company, which focuses on acquisitions and long-term management of large-scale solar projects.

Wind: BP is one of the largest wind energy producers in the United States, operating 13 wind farms in seven states and holding an interest in another power generation facility in Hawaii.

2) Investment in low carbon enterprises

Innovation can be disruptive and will have a huge impact on future development. For example, if concrete carbon reduction technology is globally promoted, it will reduce the emissions of artificial greenhouse gases by 1%.

Carbon management: Oil and gas remain the world's major energy sources in the coming decades, and they may still account for 40% of all energy sources by 2040. BP is investing in a variety of ways to reduce carbon dioxide emissions into the atmosphere.

Enable carbon offset: BP is one of the world's largest carbon traders and is investing in emissions reduction projects to help companies and other organizations compensate for their carbon footprint.

Focusing carbon reduction on concrete: Carbon emissions from cement production account for 5–7% of global carbon emissions. BP has invested in Solidia to reduce its carbon footprint by up to 70% by using lightweight concrete production technology.

BP invests approximately $200 million annually to help incubate and grow low-carbon solutions. It plans to allocate at least $500 million a year to low-carbon activities, including renewable energy businesses and acquisitions. BP regards these activities as the core of the strategy and is likely to make a real contribution to the future.

Advanced transportation: by 2040, more than 30% of the passenger car's mileage may be driven by electricity. In addition, BP believes that more and more people will use shared rides and carpools to travel.

Charging piles for electric vehicles: BP has partnered with Free Wire, which has developed a smart battery system for fast charging of electric vehicles. In addition, from the United States to Europe and New Zealand, BP is piloting charging piles at retail locations.

Digitally connected fleet: BP is investing in Peloton, whose technology enables two or more trucks to travel closely and safely together. This reduces aerodynamic resistance, saves fuel, and reduces carbon emissions.

Biological and low-carbon products: demand for low-carbon fuels, industrial materials and other products is increasing. In the aviation industry, air travel is expected to grow and its emissions will be halved by 2050.

Aviation fuel from waste: BP's partner Fulcrum BioEnergy has developed a jet fuel made from domestic waste with a carbon footprint of only 20% compared to conventional equivalent fuels. BP will distribute and supply bio-aviation fuel to aircraft at major air hubs in North America.

Durable building materials: BP is working with Tricoya to produce lower carbon alternatives for concrete, metals and plastics. By using the acetylation process to change the chemical properties of the wood, weatherable building materials that do not expand or shrink can be produced.

Digital Transformation: artificial intelligence, faster data processing and other digital technologies have great potential for improving efficiency and reducing emissions.

Low-carbon electricity and storage: Nearly two-thirds of the projected world energy growth demand in the coming decades will be in the form of electricity.

BP is conducting research of how to meet customer demand for power and storage, such as developing advanced battery technologies.

3) Persistence in low carbon activities

To encourage each of BP's divisions to seek low-carbon emissions opportunities, BP has designed a new low-carbon advancement certification program to provide a framework for companies to focus on activities that show better low-carbon outcomes.

Qualification activities range from BP's operational emissions reduction to carbon-neutral products, from investments in low-carbon technologies to renewable energy businesses. BP conducts these activities through its own business and in collaboration with others.

Deloitte has evaluated BP's plans and standards and secured these activities and their greenhouse gas (GHG) emissions and compensation.

BP's low-carbon propulsion program highlights many of its initiatives in the low-carbon sector. The activities identified in the first year of the program include the following aspects.

(1) Reduce greenhouse gas (GHG) emissions. In BP's operations, continuous greenhouse gas reductions are made to increase energy efficiency and reduce methane emissions and flaring. Mainly reflected in: producing less carbon than competitors or industry standards; using recycled edible oil to make jet fuel; using more energy-efficient new ships as tankers; Castrol low-viscosity lubricants help improve fuel economy.
(2) Provide renewable energy. For example, biofuels and bioelectric power and wind energy in Brazil.
(3) Compensate for the carbon produced. Castrol Magnatec Bio-synthetic carbon neutral oil made from 25% plant-derived oil compounds; Castrol Optigear carbon neutral lubricant for the wind industry; Castrol GTX ECO, an engine oil, can reduce CO_2 emissions over the life cycle of a product compared to Castrol GTX Diesel 15W-40.
(4) Participate in various cleaning activities. For example, the Climate and Clean Air Coalition's Oil and Gas Methane Partnership, the Oil and Gas Climate Initiative, and the World Bank's Global Gas Flaring Reduction Partnership and Zero Flaring Initiative 2030.
(5) Enable BP or other interested parties to meet their low carbon targets. BP's global environmental products business, investing in forestry projects to reduce emissions and generate carbon credits; BP Target Neutral, developing carbon neutral products and services; encouraging EKO Green Carbon Fund to invest in carbon offset forestry projects.

4) Use carbon offset to support low carbon targets

Through carbon offsets, a reduction in local greenhouse gas emissions offsets emissions elsewhere. BP is the leader in establishing and using compensation programs, and it will use compensation to support low carbon targets. Carbon offsets are critical to achieving the goals of the Paris Agreement, which BP believes is a valuable addition to its own emissions reduction activities.

A carbon offset mechanism was created by investing in activities that reduce greenhouse gas (GHG) emissions or absorb carbon dioxide (CO_2). This can be seen as an initiative to provide low-carbon alternatives, such as renewable energy or stoves that can replace open flames, or natural resources that protect or enhance carbon dioxide absorption in the atmosphere, such as land and forest projects.

BP's Target Neutral program provides individuals and organizations with the means to reduce their carbon footprint through compensation.

Over the past 10 years, BP has accumulated a wealth of expertise in carbon management projects around the world and has helped customers compensate for more than three million tons of CO_2.

BP plans to offset the increase in operating emissions above the emissions level of 2015, which is not included in sustainable emissions reduction activities. This means that as of 2025, even if BP's output will increase, there will be no net increase in carbon footprint.

BP currently offers customers more than 20 carbon neutral products and services, using the Target Neutral program to offset emissions.

In addition, BP is helping to grow the carbon credit market by selling and buying carbon credits and by increasing its overall supply. BP can leverage its strong market insight and innovation platform to help companies achieve their emissions reduction commitments while providing revenue to project operators. In 2017 alone, subsidized low-carbon projects have resulted in CO_2 emissions reductions of more than 12 million tons.

5) Incentives

1. Carbon pricing.

BP believes that carbon pricing is the most effective way to motivate energy producers and consumers to play their role in reducing emissions. It makes energy efficiency more attractive and makes low-carbon solutions such as renewable energy and carbon capture, utilization and storage more cost-competitive.

It is estimated that by 2020, countries accounting for about two-thirds of BP's direct emissions will introduce emissions and low-carbon policies. BP has played an important role in helping governments design their trading systems, and has been an active player in the system since the World Emissions Trading System began operations.

Carbon pricing adds cost to industrial production and products, but it also benefits the energy industry by providing a roadmap for future investments and providing an environment where all energy is fairly competitive.

2. Internal carbon pricing.

To help predict more regulatory requirements that affect greenhouse gas (GHG) emissions, BP uses carbon costs to evaluate new large-scale project plans and project plans that may have high emission costs. In industrialized countries, BP's current internal carbon price is \$40 per ton of carbon dioxide and stress testing is conducted at a carbon price of \$80 per ton.

3. Carbon capture, utilization and storage (CCUS).

BP believes that carbon capture, utilization and storage (CCUS) plays a vital role in achieving the goals of the Paris Agreement. It can achieve deep reductions in existing power infrastructure and energy-intensive industries that rely on fossil fuels.

The technology has been in use for more than 20 years, but government support is still needed to accelerate its deployment through carbon prices and other policy measures. Through the Oil and Gas Climate Initiative, BP is working to identify policy mechanisms to best facilitate the deployment of carbon capture, utilization and storage (CCUS) on a regional basis.

BP is exploring opportunities to deploy carbon capture, utilization and storage (CCUS) in its operations, projects and products. For example, as a participant in a joint venture in the United Arab Emirates, BP is using carbon dioxide from industrial production processes to improve oil recovery.

References

BP. 2016. BP Sustainability Report 2016.

Chen, Conglei. 2016. The utilization of geothermal resources in oilfield surface engineering. *Sino-Global Energy* 12: 28–31.

Cramer, D.D. 2008. Stimulating unconventional reservoirs: lessons learned, successful practices, areas for improvement. SPE Unconventional Reservoirs Conference.

Hou, Qijun, Haiqing He, Li Jianzhong, et al. 2018. Recent progress and prospect of oil and gas exploration by PetroChina Company Limited. *China Petroleum Exploration* 32 (1): 1–13.

Jia, Chengzao. 2017. Breakthrough and significance of unconventional oil and gas to classical petroleum geological theory. *Petroleum Exploration and Development* 44 (1): 1–11.

Jin, Zhijun, Xunyu Cai, Jinlian Liu, et al. 2018. The recent exploration progress and resource development strategy of China Petroleum and Chemical Corporation. *China Petroleum Exploration* 23 (1): 14–25.

Ma, Xinhua. 2017. Natural gas and energy revolution: A case study of Sichuan-Chongqing gas province. *Natural Gas Industry* 37 (1): 1–8.

Pan, Jiping, Lili Yang, Luxin Wang, et al. 2017. Strategy on the development of natural gas resources in China under the new situation. *International Petroleum Economy* 25 (6): 12–18.

Shell Deutschland Oil GmbH. 2017. New energy of the future.

Shell Group of Companies. 2017. World energy model.

Song, Yan, Shaobo Liu, Mengjun Zhao, et al. 2009. Coalbed gas reservoirs: Boundary types, main controlling factors of gas pooling, and forecast of gas-rich areas. *Natural Gas Industry* 29 (10): 5–9.

Tang, Zhiwei. 2009. Development and utilization of waste water resources in oilfields. *Chemical Industry and Engineering Progress* 28: 423.

Wang, Shaoyong. 2015a. Some thoughts on new energy development by oil companies. *Sino-Global Energy* 20 (8): 18–22.

Wang, Zhigang. 2015b. Breakthrough of Fuling shale gas exploration and development and its inspiration. *Oil and Gas Geology* 36 (1): 1–6.

Wang, Jiyang. 2017. Advancement and developmental trend in the geothermics of oil fields in China. *Earth Science Frontiers* 24 (3): 2–10.

Wang, Xiangzeng. 2018. Advances and prospects in oil and gas exploration and development of Shanxi Yanchang Petroleum (Group) Co., Ltd. *China Petroleum Exploration* 23 (1): 36–43.

Wang, Zhen, and Qing Xue. 2017. To fully exert the important role of natural gas in building a modern energy security system in China: An understanding of China's National 13th Five-Year Plan for Natural Gas Development. *Natural Gas Industry* 37 (3): 1–8.
Wang, Geng, Jinyang Zheng, Lijun Jiang, et al. 2017. The development of hydrogen energy in China. *Science and Technology Review* 35 (22): 105–110.
Warpinski, N.R., M.J. Mayerhofer, M.C. Vincent et al. 2008. Stimulating unconventional reservoirs: Maximizing network growth while optimizing fracture conductivity. SPE 114173.
Warpinski, N.R., M.J. Mayerhofer, M.C. Vincent, et al. 2009. Stimulating unconventional reservoirs: Maximizing network growth while optimizing fracture conductivity. *Journal of Canadian Petroleum Technology* 48 (10): 39–51.
Wei, Wei. 2013. The overview and prospect in biomass energy exploitation and utilization in China. *Journal of Agricultural Mechanization Research* 3: 7–10.
Xie, Yuhong. 2018. New progress and prospect of oil and gas exploration of China National Offshore Oil Corporation. *China Petroleum Exploration* 23 (1): 26–35.
Zeng, Zheng. 2014. A study on the development of oil industry under energy conservation and emission reduction policy. *Ecological Economy* 30 (2): 131–134.
Zhao, Wenzhi, Hongjun Wang, Xu Chunchun, et al. 2010. Reservoir-forming mechanism and enrichment conditions of the extensive Xujiahe Formation gas reservoirs, central Sichuan Basin. *Petroleum Exploration and Development* 37 (2): 146–156.
Zhu, Kai. 2010. Research on new energy business of large foreign oil companies. In *Proceedings of the first China Academy of Engineering/Energy Forum of the National Energy Administration*.
Zou, Caineng. 2014. *Unconventional oil and gas geology*. Beijing: Geological Publishing House.
Zou, Caineng. 2018. Energy revolution and oil company transformation strategy of new era. Journal of Beijing Petroleum Managers Training Institute 25(4): 3–15.
Zou, Caineng, Yunhong Ding, Lu Yongjun, et al. 2017. Concept, technology and practice of "man-made reservoirs" development. *Petroleum Exploration and Development* 44 (1): 144–154.

Chapter 7
New Structure of Coal Power Companies

In the 21st century, China's economy has begun to enter a new normal, the economic structure has been continuously optimized and upgraded, and it has developed in the core direction of innovation and environmental friendliness. Energy transformation is a key factor in economic development. China's energy industry is undergoing energy restructuring and deepening reforms, and is in a period of strategic transformation opportunities to accelerate the construction of a clean, low-carbon, safe and efficient energy system (Zou, 2018). The position of coal companies and power companies in China's energy supply system is very important. Under the current situation, we should proceed from our own characteristics, achieve strategic transformation, and lay the foundation for sustainable development.

7.1 Coal Companies

Coal plays an important role in China's energy consumption. The clean utilization of coal is an inevitable trend and a must for coal companies to develop. Under the new situation, the coal company launched a coal cleaning strategy, an industry chain extension strategy and a new energy business strategy to continuously optimize the company's energy structure and promote sustainable social and environmental development.

7.1.1 Clean Production and Utilization of Coal

1. Green production

In 2016, the "13th Five-Year Plan for Coal Industry Development" issued by the National Energy Administration clearly stated that the clean development and utilization of coal will be arranged from two aspects: clean production and efficient

C. Zou, *New Energy*, https://doi.org/10.1007/978-981-15-2728-9_7

utilization. At the same time, it will clarify the development of mining area circular economy and coordinate the development of mining areas and related industry construction layout to improve the construction level of the circular economy park. Support coal enterprises to build power plants of comprehensive utilization of washing coal gangue, coal and mud according to the principle of "equal capacity replacement", develop coal gangue and coal ash building materials industry, promote the industrial utilization of mine drainage, and improve the comprehensive utilization level of resources.

The in situ conversion of coal can maximize the economic benefits and environmental benefits, which is an indisputable fact in the industry. For example, using supercritical units to generate electricity, low coal consumption, strong economy, energy saving and environmental protection, the desulfurization rate of more than 98%, almost zero emissions can be achieved.

The combination of coal production and power generation can build a green production cycle. The raw coal produced by the coal mine is washed by the coal washing plant, and the clean coal is partially transported by the railway special line. The low calorific value coal that is screened out directly enters the pit power plant and the comprehensive utilization power plant to generate electricity, or is used in the coal chemical project to produce methanol; The discharged fly ash, desulfurized gypsum and slag are used as raw materials of the cement plant; the coal mine water and the domestic sewage of the nearby enterprises enter the sewage treatment plant, and are used for power plant cooling, downhole spraying, coal washing and greening in the park after treatment. In addition, the waste generated in coal mining, coal gangue, can also be used to refine kaolin or to form sintered bricks. It not only solves the environmental problems of coal gangue stacking and landfill disposal, but also increases economic benefits.

2. Green utilization

Under the background of China's accelerated construction of ecological civilization, strict control of the total amount of coal consumption and ensuring continuous reduction of coal consumption are important tasks for promoting energy production and consumption revolution and solving smog and other prominent environmental problems (Zou 2018). Improving coal energy efficiency and in-depth analysis of the contribution of energy conservation to controlling total coal consumption will help to formulate more scientific and rational coal control targets, identify key industries and priority areas for coal control, and achieve coal consumption control and energy structural optimization and environmental quality improvement goals in the most economical and effective way.

(1) Improve efficiency

Optimizing the coal gasification system is an important means of green utilization of coal and improvement of efficiency.

The first is to increase the pressure of coal gasification. Not only can the efficiency of the gasification unit itself be increased, the single furnace capacity can be expanded, the equipment investment can be reduced, the device scale and energy consumption of the subsequent system can be effectively reduced, and isostatic synthesis can be realized for coal-to-methanol.

The second is to increase the size of the unit, mainly to improve the size of coal gasifiers, the scale of synthetic systems, the scale of air separation units. It is also conducive to improving efficiency, reducing overall equipment investment, and facilitating production operations and operation management.

The third is efficient recycling of by-product steam. It mainly recovers a large amount of medium-pressure or low-pressure steam, which is by-produced by the gasification unit, a large amount of low-pressure steam, which is by-produced by the conversion unit, and a large amount of medium-pressure or low-pressure steam, which is produced as a by-product of the synthesis unit.

For the coal gasification unit, it is necessary to actively adopt the waste heat boiler process or the waste heat boiler chilling combined process to recover the coal gasification high-grade heat energy as much as possible. The coal gasification high-temperature gas cooling adopts the waste heat boiler process, which can recover the high-grade heat in the flue gas to a large extent, and produce high-pressure and superheated steam for power generation or work, which can improve the energy efficiency of the coal chemical project by 1 to 2%.

(2) Clean burning

The cleanliness of energy is not only related to the characteristics of the energy itself, but also closely related to the energy utilization technology. From the source point of view, the gas produced by coal gasification as a fuel can effectively reduce the excess air coefficient, thereby improving its energy use efficiency, and with the low-cost purification of gas, the pollution emission index of the flue gas can be greatly reduced, and the combustion pollution is super low emissions, thereby improving the economy and convenience of clean and efficient combustion of coal. In the combustion process, the use of bias flames combustion, staged combustion and flue gas circulation can also reduce the amount of nitrogen oxides generated; at the end of the combustion process, fine dedusting offlue gas, desulfurization, denitrification and removing heavy metals can achieve clean and efficient combustion of coal. The high-efficiency pulverized coal boiler technology with independent intellectual property rights has been widely promoted in China. The boiler fuel burnout rate is 98%, which is 28% higher than that of ordinary coal-fired boilers. The pollution emission indicators such as soot, sulfur dioxide and nitrogen oxides are equivalent to standards of natural gas boilers. Based on the demonstration of more than 10 provinces and regions such as Shanxi, Inner Mongolia, Shandong, Gansu, etc., quick industrial development was realized quickly. At the same time, the promotion of technologies such as coal-water slurry, briquette, and low-rank coal grading and utilization have also improved the level of clean and efficient use of coal in China.

7.1.2 Extension of Industrial Chain

1. Coal gasification

Coal gasification refers to the thermochemical process of converting the combustible part of solid coal or coal coke into gaseous fuel by chemical reaction at a certain temperature and pressure by means of coal or coal coke as raw material, oxygen (air, oxygen-rich or pure oxygen), water vapor or hydrogen as gasification agent. Coal is gasified to produce syngas (mainly carbon monoxide and hydrogen) for use in synthetic ammonia, coal-to-fuel, coal-based chemicals, etc., which can increase the value of coal and ancillary products and improve the comprehensive utilization rate of coal (Zou et al. 2016; Zou 2018). In addition, high-sulfur coal, which accounts for about 20% of the country's total coal reserves, must be gasified by a gasifier to obtain efficient and clean utilization. It is worth noting that in addition to coal-to-olefins, coal-to-oil, coal-to-gas and other large-scale coal chemical projects that maintain a large demand for syngas, the traditional synthetic ammonia enterprises have upgraded their coal gasification and refining and chemical projects to support the construction of coal-to-hydrogen plants. The demand for syngas is also a highlight of the growth of the coal gasification market in recent years.

The added value of coal-to-hydrogen production is much higher than that of direct sales of coal. Coal enterprises attach importance to the development of coal hydrogen production technology and technology to promote the transformation and upgrading of the coal industry. Large-scale coal gasification hydrogen production not only has cost advantages, but also optimizes the material balance of the refinery. It is an important direction for hydrogen production in Chinese refineries. Gasification is the core technology of hydrogen production from coal. The hydrogen production unit requires efficient and stable operation of the gasifier. As of 2017, hundreds of coal gasifiers with more than 20 gasification technologies in China's refinery hydrogen and coal chemical industries are in operation. The 15 refining and chemical integration projects and hydrogen production scales to be under construction in China are shown in Table 7.1. Advanced coal gasification technologies are moving towards high energy efficiency, large scale and wide coal type adaptability.

Shenhua Group has vigorously developed coal gasification hydrogen production technology. The cost of hydrogen production from coal has been lower than the cost of hydrogen production from natural gas, and the basic framework of new energy industry based on hydrogen energy has been established. Based on coal-based hydrogen production, solar energy and wind energy hydrogen production, we have established operational capabilities and industrial bases in the fields of hydrogen production, hydrogen storage, hydrogen transportation, and sales, promoted the development of the coal-to-hydrogen industry, and used hydrogen fuel cells for automobiles.

2. Coal chemical industry

In recent years, the strong demand for transportation fuels and petrochemical products in China has led to the rapid development of modern coal chemical industry such as coal-to-liquids, coal-to-olefins, coal-to-natural gas, coal-based ethylene glycol, etc.

Table 7.1 Summary of China's planned refining and chemical integration project and hydrogen production scale in 2017

Company	Refining capacity	PSA scale (m^3/hr)	Supplementary hydrogen production scale (m^3/hr)	Hydrogen process
Hengli Petrochemical	20 million tons/year, integrated refining and chemical	730,000	500,000	Coal to H_2
CNOOC Huizhou Refinery	22 million/year, refining expansion and 1 million tons/year of ethylene project	–	278,000	Coal to H_2
Zhejiang Petrochemical	40 million tons/year, integrated refining and chemical	480,000	145,000	Coal to H_2
Shenghong Refinery	16 million tons/year, integrated refining and chemical	440,000	230,000	Coal to H_2
Sinopec Hainan Refinery	1 million tons/year of ethylene project and refining expansion	90,000	96,000	Coal to H_2
China Ordance Industry Group	Fine chemicals and 15 million tons/year of raw material engineering	–	183,000	Coal to H_2
Sinochem Quanzhou	1 million tons/year of ethylene projectand refining expansion	–	140,000	Coal to H_2
China-Venezuela Joint Oil Refinery	20 million tons/year heavy oil processing project	130,000	120,000	Coal to H_2

(continued)

Table 7.1 (continued)

Company	Refining capacity	PSA scale (m^3/hr)	Supplementary hydrogen production scale (m^3/hr)	Hydrogen process
North Huajin Chemical Industry Group	Optimization and upgrading of existing material on the base	–	100,000	Coal to H_2
Sinopec Yanshan Petrochemical	12 million tons/year, integrated refining and chemical	–	100,000	Coal to H_2
Zhongke Joint Venture-Guangdong	15 million tons/year, integrated refining and chemical	80,000	–	Coal to H_2
PetroChina Yunnan Petrochemical	13 million tons/year of refining	110,000	170,000	Natural gas to H_2
Luoyang Petrochemical	18 million tons/year of refining expansion project	80,000	–	–
Tangshan Xuyang	15 million tons/year, integrated refining and chemical	180,000	50,000	–
Hebei Xinhua United Petrochemical	20 million tons/year, integrated refining and chemical	–	–	–

Note The information comes from Asia chemical consulting's "China hydrogen industry annual report 2017"

Modern coal chemical industry has entered the industrial production from upgrading demonstration and the period of large-scale capacity expansion, and gradually have an increasingly large impact on petrochemicals.

(1) Coal to olefins

Coal-to-olefins have become the mainstay of China's olefin production capacity growth, and its competition with petroleum olefin production has begun to show.

It should be noted that coal-to-olefin products are generally ordinary polyethylene and polypropylene products. The existing products have serious homogenization problems, lack of high-end special materials, and the market competitiveness of them is relatively weak. If large-scale development, enterprises face more and more competitive pressure.

(2) Coal to ethylene glycol

Petroleum ethylene routes and ethylene glycol production are constrained by the source of raw materials, resulting in low operating rate of device. The new capacity of the coal-to-liquids project is mainly derived from coal-based ethylene glycol. At present, the domestic coal-to-ethylene glycol technology is not fully mature, and the stability of the operation of the device is poor. Therefore, to accelerate the development of the coal-based ethylene glycol industry, it is necessary to further improve the technology to reduce the investment risk of new projects.

(3) Coal-to-gas

In recent years, China's natural gas production growth rate is significantly lower than the growth rate of consumption. The shortage of production is partly supplemented by imported pipeline gas, imported liquefied natural gas (LNG) and coal-based natural gas. At present, the state has begun to adjust the price of natural gas, and the problem of reversed price of natural gas is expected to be gradually solved, providing opportunities for the development of coal-based natural gas (Zou 2018). Core technologies such as China's methanation catalysts are still underway, and there are risks in terms of technical reliability, equipment size, and long-term operation.

(4) Coal to liquids

From the perspective of coal-to-liquids technology itself, direct liquefaction technology still needs to be improved. The indirect liquefaction currently has a small scale of operation equipment, and projects with a scale of more than one million tons are under construction, and the technology has not been further verified. Coal-to-liquids projects have a large investment, and new projects are facing financial pressures and environmental pressures.

Under the pressure of China's increasing dependence on foreign oil, the development of modern coal chemical industry is a useful supplement to petrochemical industry. It is an important way to develop China's comparative advantages in coal resources, reduce foreign dependence on oil, and ensure China's energy security. At this stage, the development of China's coal chemical industry is still being explored. It is necessary to conduct in-depth research, steadily pilot, and firmly grasp the autonomy of modern coal chemical technology. Through continuous efforts, we will explore sustainable development path to improve energy efficiency and reduce pollution emissions.

7.1.3 Developing New Energy

In a long period of time in the future, new energy cannot be the main energy source to support industrial restructuring and economic development. For most coal companies, on the one hand, the clean and efficient use of coal should be the key development direction to improve the quality of terminal energy products; on the other hand, actively participate in the industrialization and large-scale development of non-fossil energy such as wind, light, biomass and hydrogen, and increase the proportion of new energy supply.

The National Energy Group, which was formed by the merger of Guodian Group and Shenhua Group, is not only the world's largest coal company, but also the world's largest installed capacity of wind power. After the establishment of the National Energy Group, it was proposed to focus on the development of coal, new energy, transportation and other businesses in 2018, of which power investment accounted for 68% of the total investment, and renewable energy investment such as hydropower and wind power accounted for 31% of the main investment. The newly added coal production capacity was 17 million tons, the newly installed thermal power installed capacity was 1.67 million kW, the wind power installed capacity was 3.52 million kW, and the hydropower installed capacity was 1.45 million kW. The intensity of reducing coal-fired power generation should be strengthened and a shift from speed-scale to quality-benefit will be achieved.

Since 2005, Shenhua Group, which is mainly engaged in coal production, has been involved in large-scale renewable energy business mainly based on wind power and solar photovoltaic power generation, and has insisted on intensive cultivation and stable operation. By the end of 2016, Shenhua Group's wind power grid-connected installed capacity had reached 7.36 million kW, and photovoltaic grid-connected installed capacity was 330,000 kW. At the same time, we are constantly exploring in the strategic emerging industries, and actively carry out research and development in the field of renewable energy such as thin film solar cells, energy storage technology, and photothermal; focusing on the development of offshore wind power, low wind speed wind power generation, ground photovoltaic power generation, the micro-grid technology of integrating renewable energy, energy storage and other technologies; signed a strategic cooperation agreement with major nuclear power companies in China, and conducted business exchanges with Tara Energy Corporation of the United States; completed the basic research on the development of Shenhua hydrogen energy industry.

Tongmei Group is the largest coal and electric power enterprise in Shanxi Province, with a power installed capacity of 17 million kW, of which new energy power generation installed capacity is 1.543 million kW, accounting for nearly 10%. In 2011, the construction of the Zhinuquan Wind Farm was started. The first phase of the power station was connected to the grid in October 2012. The second and third phases of the project were also put into operation. The fourth phase is about to start construction with a planned capacity of 100 MW. In 2014, as a demonstration project of the photovoltaic group of Tongmei Group, the Tashan Photovoltaic Power Plant of 20,000 kW unit was "rooted" in the Tashan Industrial Park of Tongmei Group and was the first photovoltaic power generation project of Tongmei Group. Not only that, Tongmei Group has adopted the coal mining subsidence area to develop photovoltaics according to local conditions. The coal-fired Jiajiagou photovoltaic base has a total installed capacity of 1 million kW in the Datong coal mining subsidence area in Shanxi, of which the Tongmei group photovoltaic project accounts for 100,000 kW. The project is an important project for the transformation of Datong and Tongmei Group, and is a successful practice for the transformation of underground primary energy into green energy on the ground. The utilization of abandoned land by planting and treating achieved clean and green development, brought economic benefits to surrounding villagers, especially poor households. In the future, the development of the new energy sector of Tongmei Group will focus on distributed wind power and low wind speed wind power. In the centralized photovoltaic field, it will focus on super runners and front runners, as well as distributed photovoltaic and photovoltaic poverty alleviation projects, renewable biomass power generation and biomass coupling projects, low gas production power generation projects, energy storage frequency modulation projects and energy storage peak shaving projects.

In recent years, Jinneng Group has vigorously developed new energy power plants based on photovoltaics and wind power to provide clean energy for social and economic development. In 2013, the installed capacity of photovoltaic and wind power generation operations was 170,000 kW. Two wind power projects were approved, with an installed capacity of 99,000 kW, four photovoltaics, and the installed capacity of 130,000 kW; four wind power and photovoltaic projects with installed capacity of 470,000 kW were approved. At the same time, two photovoltaic power plants were put into operation at the same time in Shanxi Province, achieving the country's first wind-solar photovoltaic complementary power station. Pinglu Photovoltaic Power Plant is located at the north side of the Bohubao Wind Farm Boosting Station in Pinglu District, Shuozhou City. The total planned construction scale is 10 MW, and the scale of the first phase project is 5 MW. Youyu Photovoltaic Power Station is located inside Xiaowutai Wind Farm of Youyu County, Shuozhou City. The total planned construction scale is 20 MW. The scale of the first phase project is 10 MW, including the fixed installation angle of crystalline silicon and the 100 kW crystal silicon dual-axis tracking system, 100 kW thin film battery fixed inclination installation method, 100 kW low magnification concentrating power generation. Daliuzhuang project of 60 MW photovoltaic power station in Yingxian successfully realized grid-connected power generation.

7.2 Power Companies

The current energy Internet is mainly based on the energy grid of the power grid. The power company has a large number of basic advantages such as power grids and has innate conditions for energy transformation. In the process of transformation, the combination with the resource advantages of regional renewable energy, persistence in the power structure clean transformation strategy and integrated energy service strategy will promote the cleanliness of energy (State Grid Energy Research Institute Co., Ltd. 2017).

7.2.1 Power Structure Transition

With the continuous advancement of energy of low carbon transformation, energy structure adjustment is more obvious in the power sector, because the main utilization of alternative energy sources such as wind and solar energy is converted into electricity. According to the "13th Five-Year Plan for Energy Development" and the "13th Five-Year Plan for Power Development" promulgated at the end of 2016, the proportion of thermal power installed capacity in the country will be controlled within 55% by 2020. From the current point of view, solar power generation has achieved the "13th Five-Year Plan" development goal three years ahead of schedule. It is expected that the power supply structure will be further optimized in 2020 (Beijing: National Development and Reform Commission 2016, 2017).

Under the background of the continuous growth of electricity demand, China's power industry is developing rapidly and the overall scale is growing, but the imbalance between supply and demand is still the same, the growth rate of social electricity consumption is accelerating, the power structure is more optimized; and the market pattern is under new adjustment, the overall transformation trend is mainly manifested in the following aspects.

1. Continuous optimization of power structure and gradual increase in the proportion of new energy power generation

In the global energy transformation, China is an active advocate, practitioner and leader. With the low-carbon as the main feature of the current round of energy transformation, wind power, solar power and other low-carbon, carbon-free energy sources are rapidly emerging, power is more prominent in the center of the energy system, and the structure of installed capacity is constantly changing. In 2017, among the nation's power installed capacity, hydropower installed capacity was 341 million kW, accounting for 19.20%, down about 1% year-over-year; thermal power installed capacity was 1.106 billion kW, accounting for 62.24%, down 2% year-over-year; nuclear power installed capacity was 35.82 million kW, accounting for 2.01%, basically the same as last year; wind power installed capacity was 164 million kW, accounting for 9.21%, up 0.2% year-over-year; solar power installed capacity was 130 million kW, accounting for 7.33%, up 2.6% year-over-year.

According to the 10-year historical data (Table 7.2), the trend of power supply structure optimization and adjustment is more obvious, showing that the proportion of traditional fossil energy power generation installed capacity has dropped significantly, and the proportion of new energy installed capacity has increased significantly. The data shows that the proportion of hydropower and nuclear power installed capacity has not changed much in recent years, while the proportion of thermal power and

Table 7.2 National power installation structure from 2008 to 2017. Unit: 10,000 kW

Power structure	2008 年	2009 年	2010 年	2011 年	2012 年	2013 年	2014 年	2015 年	2016 年	2017 年
Hydropower	17,260	19,629	21,606	23,298	24,947	28,044	30,486	31,954	33,207	34,119
Thermal power	60,286	65,108	70,967	76,834	81,968	87,009	92,363	100,554	106,094	110,604
Nuclear power	885	908	1082	1257	1257	1466	2008	2717	3364	3582
Wind power	839	1760	3258	4923	6142	7652	9657	12,796	14,807	16,418
Photovoltaic power	–	–	–	212	341	1589	2486	4318	7631	13,025

Note The data from 2017 comes from the information released by the National Energy Administration, and the other is from the China power industry association annual report "China Power Industry Annual Development Report"

new energy power generation has undergone major changes. The proportion of thermal power installed capacity has decreased by 15.38% compared with 2007, and the proportion of new energy power generation installed capacity has increased by 15.3% (Figs. 7.1 and 7.2). At present, China's new energy power generation installed capacity ranks first in the world, and hydropower, wind power, solar power and other single installed capacity is also the highest in the world.

Although the national installed capacity of power generation increased by 134 million kW, a record high, but due to the different structure of new installed capacity, the available usage time of new energy is low, and the power supply grows non-linearly. As of the end of 2017, the cumulative installed capacity of solar power generation was 132.5 million kW, an increase of 53.94 million kW, a year-over-year increase of 71%, and exceeded the scale of thermal power growth for the first time, becoming the fastest growing single power generation category.

Wind power, which was once as "promising" as solar power, has apparently taken a different path of expansion in recent years from solar power (Beijing: National Energy Administration 2017). In 2009, the scale of newly built wind power was hundreds of times more than the newly added solar power generation. However, the

Fig. 7.1 National power supply structure in 2007

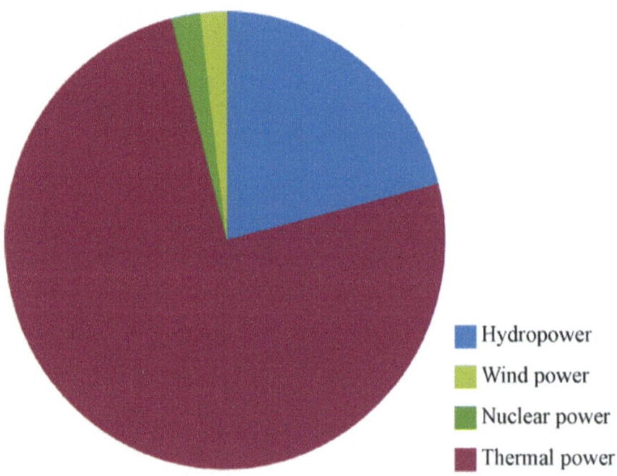

Fig. 7.2 National power supply structure in 2017

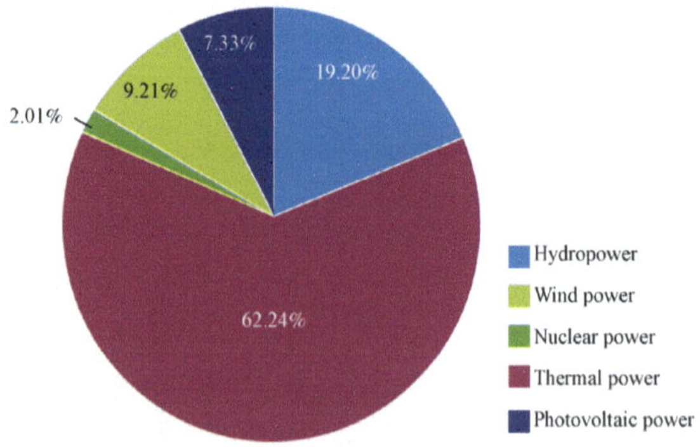

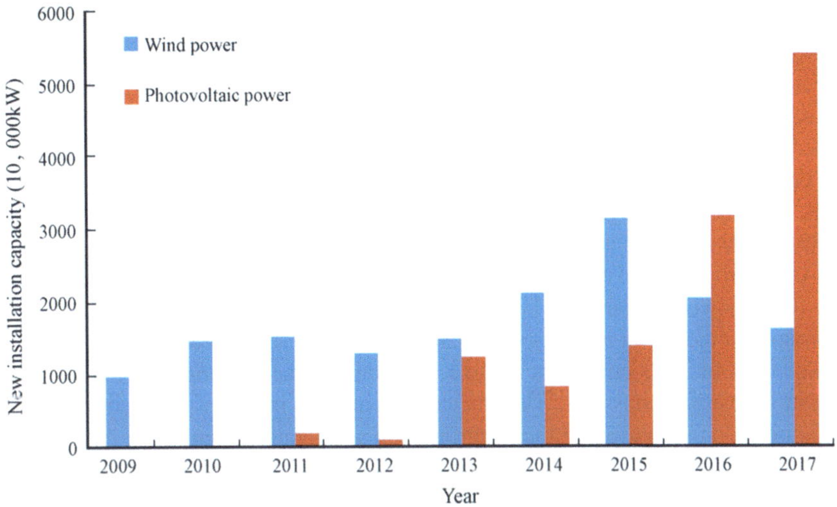

Fig. 7.3 New installation capacity of wind and photovoltaic power generation from 2009 to 2017

newly installed capacity of wind power reached a historical peak of 31.39 million kW in 2015, and suddenly reduced by more than 10 million kW, lower than that of solar power generation. the new installed capacity in 2017 was reduced by 4 million kW, less than one-third of the increase in solar power generation in the same year (Fig. 7.3).

The utilization of new energy has improved, and the problem of abandoning wind and abandoning photoelectric has been alleviated. In 2017, the national abandoned wind power was 41.9 billion kWh, a year-over-year decrease of 7.8 billion kWh, and the wind power abandonment rate was 12%, down 5.2% year-over-year, achieving a "double down" of abandoned wind power and abandoned wind power rate. The national abandoned photoelectric output was 7.3 billion kWh, and the photoelectric abandonment rate was 6%, down 4.3% year-over-year. Mainly in Xinjiang and Gansu, there was a serious photoelectric abandonment phenomenon. The abandoned photoelectric output was 2.82 billion kWh and 1.85 billion kWh respectively. The photoelectric abandonment rate exceeded 20%, but it reduced by more than 9% year-over-year, and it tends to be under benign development.

Onshore wind power and photovoltaic power generation will be the fastest-growing power sources. The total installed capacity in 2030 and 2050 is expected to exceed 30 and 50%. Gas power is constrained by cost factors and has limited growth space. The optimized installed capacity in 2030 and 2050 will reach 180 million kW and 370 million kW respectively. Nuclear power development is affected by factors such as site location and construction speed. It is estimated that the installed capacity in 2030 and 2050 will reach 120 million kW and 220 million kW respectively. The potential for hydropower development is relatively limited. It is estimated that the installed capacity in 2030 and 2050 will reach 460 million kW and 540 million kW respectively. Coal power will gradually be transformed into a capacity-supported power supply with deep adjustment capability. The installed capacity in 2030 and 2050 is expected to reach 1.2 billion kW and 700 million kW. In addition to wind power and photovoltaic power generation, although the installed capacity growth of other powers is limited, they will play an important role in the power system. In the case of large-scale development of volatile power supply, the power balance and peak shaving flexibility of the power system are guaranteed by all of them, and all kinds of power supply shows a coordinated development trend.

As the proportion of clean energy generation increases, the carbon emission intensity of the power industry will be greatly reduced. It is expected that the total carbon emissions of the power industry will peak around 2025, with a peak level of about 4.2 billion tons. In 2050, emissions fell to 1.4 billion tons, accounting for less than 30% of national carbon emissions. The unit carbon emission intensity of the conventional transformation scenario and the electrification acceleration scenario will reduce to 400 g/kWh and 363 g/kWh in 2030, respectively, and will drop to 114 g/kWh and 96 g/kWh in 2050, one-fifth below the current level.

2. The power supply layout continues to tilt towards resource-rich areas, and interconnected power grids and energy storage will play an important role

Recently, due to multiple factors such as the consumption situation and subsidy policy, the power supply layout should put the same emphasis on both the load center in the east and the resource-rich areas in the west and north. Support of power installation capacity in the western and northern areas with better resource conditions is a more scientific energy transformation program. Among them, the layout of coal-fired power installations will be mainly in North China and Northwest China. Due to the high density of coal-fired power units and the severe situation of environmental emission reduction in the consumption area of East China, the exit trend of coal-fired power is obvious. The gas-electric installation is mainly in the Central China and East China in the near or middle term. The policy-supported cogeneration and the industrial park's distributed triple supply unit have potential for development, and the northwest region with advantages of gas prices in the long term is more competitive; the wind power layout will still be dominated by the "Three North" region, and the decentralized wind power in the eastern and central regions is meeting the period of development opportunities. With the abandoned wind power problem alleviated, the development of wind power in high-quality resource areas is a more economical optimization plan; photovoltaic power generation should be both centralized and distributed, and the eastern and central distributed PVs have larger growth space, but limited by the available roof area and lighting resources, distributed photovoltaics are difficult to meet the power demand of the load center, and can not be the main part of China's vigorous development of clean energy, centralized development in high-quality resource areas will still account for a large share.

The national interconnected grid will play a more important role in the new generation of power systems. The national conditions of reversed distribution of China's energy resources and load demand determine the objective needs of optimizing the allocation of resources nationwide. A high proportion of new energy sources puts higher demands on power system flexibility and energy storage. By adopting a more flexible and optimized operation mode, the cross-reginal interconnected power grid can realize the dynamic balance of power supply and demand with the cooperation of energy storage systems, which will effectively promote the consumption and utilization of high proportion of new energy. The overall planning results of the power system show that the optimized capacity of China's trans-regional transmission channels will reach 300 million kW and 500 million kW respectively in 2030 and 2050, and the scale of "West-to-East Power Transmission" will gradually expand and will be dominated by the transportation of clean energy. As a basic platform for large-scale and high-efficiency allocation of energy resources, the importance of the power grid will become more and more prominent. In the new generation of power systems, there will be more coordinated interactions between various types of power sources, power grids, demand side resources and energy storage, and jointly promote the optimal operation of power systems in a flexible and efficient manner, which will effectively promote new energy consumption. As the penetration rate of new energy sources continues to increase, flexible power sources such as gas to electricity, hydropower, and photothermal will play an important role in peak shaving. Coal-fired power will also undertake more peak-shaving tasks. Only some high-parameter large-capacity coal-fired power units continue to bear the base load. Trans-regional transmission lines will operate more flexibly, effectively supporting clean energy to achieve full consumption on a larger scale. The operation mode of emerging flexible resources such as demand response and energy storage is optimized and adjusted with the output of wind and photovoltaic power, and supports the optimal operation of the system (Figs. 7.4, 7.5 and 7.6).

3. Development of large-scale natural gas combined cycle power generation

Gas power generation usually adopts gas-steam combined cycle mode. The combined cycle consists of Brayton cycle and Rankine cycle. The system inlet temperature can be up to 1,300 °C, the exhaust gas temperature is 500–600 °C, and the simple cycle thermal efficiency is up to 45–50%. The waste heat boiler is used to further recover waste heat and improve thermal efficiency, generally a double pressure or triple pressure system. The thermal efficiency of contemporary large-scale 9F gas-steam combined cycle power generation is as high as 58 to 60%, which is much higher

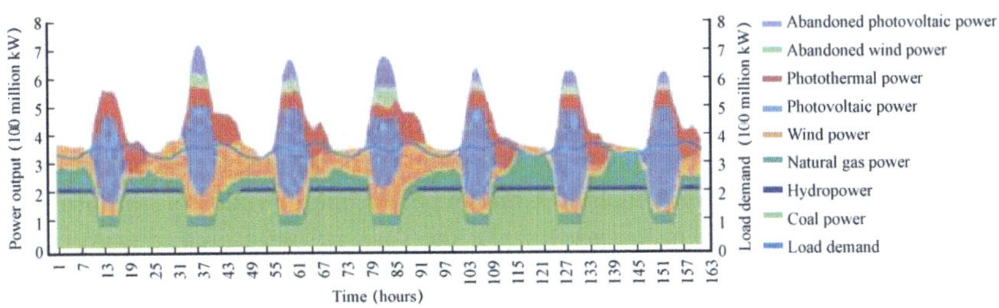

Fig. 7.4 Optimization results of various power supply output in a typical winter week in Northwest China in 2050 under electrification acceleration scenario

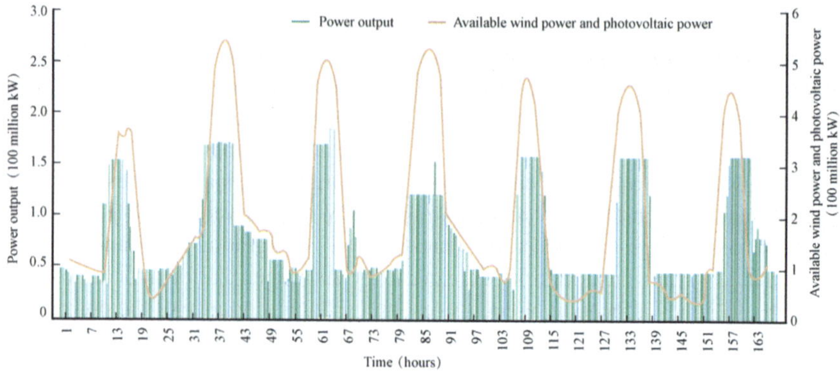

Fig. 7.5 Optimization results of trans-regional transmission in a typical winter week in Northwest China in the 2050 under electrification acceleration scenario

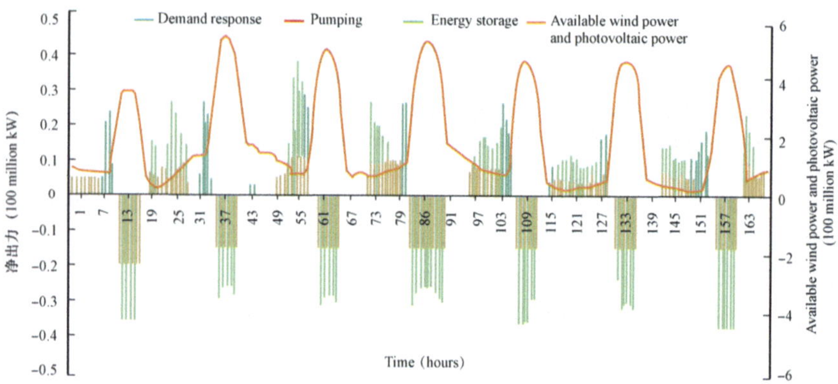

Fig. 7.6 Optimization results of demand response and energy storage output in a typical winter week in Northwest China in 2050

than the thermal efficiency of coal-fired power generation. The thermal efficiency of coal-fired generating units is generally 46 to 48% even for ultra-supercritical 600 and 1,000 MW units. The thermal efficiency of the two types of units differs by 10 to 20%. After converting power generation to standard coal consumption: gas power generation is 205 to 213 g/kWh, coal-fired power generation is 260 to 280 g/kWh, the difference between the two is 55 to 60 g/kWh.

The actual situation of China's resource endowment is rich in coal but insufficient in oil and gas, and the production of resources and users are misplaced. It is difficult to change the status of fossil fuels in China's power energy industry within 30 to 50 years. Therefore, it is necessary to take the road of clean coal, develop coal chemical industry, IGCC (integrated gasification combined cycle power generation) and develop and utilize unconventional oil and gas resources, deepwater oil and gas and other oil and gas production to ensure energy security. In addition to coal power, nuclear power is operating at full capacity; hydropower is affected by regional resources and seasonal dry and flood changes, and power generation is uneven; renewable energy power such as wind power and solar power generation has random and intermittent instability characteristics. It is not suitable for base-load power generation; the cost of fuel power generation is relatively high. In order to cope with global climate change and China's haze weather, to ensure the demand of clean energy of people's lives and production, to adjust and optimize the power energy structure, gas power generation has become the main alternative to coal-fired power generation (Editorial Board of China Electric Power Encyclopedia 2014).

In the 21st century, China's natural gas power generation has developed rapidly. By the end of 2013, the installed capacity of gas power generation was 42.5 million kW, accounting for 3.4% of the national power generation capacity. The installed capacity of coal power is 78.621 million kW, accounting for 63% of the total installed capacity. China's natural gas power generation is mainly distributed in the Yangtze River Delta, the southeast coast, and the Beijing-Tianjin region. The installed capacity of gas power plants in Guangdong, Fujian and Hainan provinces reached 17.5 million kW, accounting for 34% of the total installed capacity of gas-fired power generation in the country. The gas-fired power plants in Jiangsu, Zhejiang and Shanghai accounted for about 32%, and the Beijing-Tianjin region accounted for about 23%. In recent years, gas power plants have also been put into operation in Shanxi, Ningxia and Chongqing.

At present, China's natural gas power generation operations are mainly divided into three categories: the first category is state-owned large-scale power generation central enterprises, such as Huadian Group, Huaneng Group, Datang Group, China Power Investment Group, etc.; the second category is local power investment group and energy groups invested by local governments, such as Shenneng Group, Zhejiang Energy Group, Guoxin Group and Jingneng Group; the third category is oil and natural gas production and supply companies, such as China Petrochemical Corporation and China National Offshore Oil and Gas Group. In order to take the advantage of their respective advantages and complement each other, gas-fired power plants are mostly joint ventures.

The industrial chain of China's natural gas power generation industry is mainly composed of three types of entities. Upstream is a natural gas supplier, including domestic oil companies and city gas companies. Power generation companies are responsible for investing in gas-fired power plants and converting natural gas to electricity through buying gas from upstream gas suppliers. Oil and gas companies can supply natural gas through LNG or pipeline directly to power plants, or from city gas companies; power generated by power plants is supplied to downstream grid companies at on-grid tariffs. The price of the on-grid electricity of a natural gas power plant in China varies from one factory to another and from one unit to another. It is determined by the local price management department and submitted to the National Development and Reform Commission for approval. The main pricing methods are pricing consisting of two parts and single pricing. (1) The two-part pricing consists of electricity price and capacity price. Shanghai implements two-part electricity price, the electricity price (on-grid price) is CNY 0.504 per kWh, the capacity electricity price is arranged by 2,500 h for the whole year, and the electricity price compensation standard is CNY 0.22 per kWh to compensate for the gas power plant in the power grid for the peak shaving power generation. For the smaller 9E unit series, the on-grid electricity price within 500 h of annual power generation is CNY 0.544 per kWh. (2) Except for Shanghai, the gas turbine power plants in other regions implement a single price. The gas source is the same, the gas price is similar, and the on-grid electricity price is similar. For example, the on-grid price of the peak-shaving power plant for gas supply from the West to East Gas Pipeline in Jiangsu Province is CNY 0.581 per kWh, and the on-grid price of gas supply from the West to East Gas Pipeline in Henan Province is CNY 0.553 per kWh, and the on-grid price of thermoelectric power is CNY 0.605–0.656 per kWh. After the gas price reform, some of the on-grid tariffs were raised, such as Zhejiang Banshan Power Plant, which was CNY 0.606 per kWh. There are more gas-fired power plants in Guangdong Province, with diversified gas sources, and large differences in gas prices. The pricing method of one price for one power plan, or even one price for one power unit is adopted, with a minimum of CNY 0.553 per kWh and a maximum of CNY 1.1 per kWh. The development of on-grid tariffs is broadly divided into three categories. First, the provision of temporary feed-in tariffs based on the cost-plus method, mainly refers to 9E units which uses Australia LNG imported by Guangdong Dapeng, the implementation of unified on-grid tariffs of CNY 0.553 per kWh; Second, the approved temporary feed-in tariff is CNY 0.72 per kWh, the implementation of this pricing is mainly the state-approved gas power units; Third, the pricing is to use the coal-fired unit benchmark price and subsidy. At present, some 9E units in Guangdong Province have no formal government approval price, only temporary settlement of electricity prices, grid companies make settlement based on coal-fired benchmark price of CNY 0.5042 per kWh, the government subsidized the gap.

China's power installed capacity and power generation rank first in the world, but the degree of development of gas-fired power generation lags far behind developed countries. At the end of 2013, China's total installed capacity of electric power reached 1.25 billion kW, of which thermal power was 860 million kW, accounting for 69%, while gas-fired installed capacity was 42.5 million kW, accounting for only 3.4%, and power generation accounted for 2.5%; With natural gas consumption of more than 150 billion cubic meters, power generation gas accounts for 17.20% of the total natural gas consumption, which is too small. At present, China's gas power generation is mainly divided into two categories: thermal power plants and peak shaving power plants. Thermal power plants are mainly based on heat supply and supplemented by power generation. From the perspective of heat load, the north is dominated by heat load for winter heating, while the south is dominated by industrial heat load. The peaking shaving power plant generally operates at peak load and intermediate load. Because the gas peak and the electric peak coincide in time, it is difficult for two types of gas generating units to obtain sufficient gas source in winter, which cannot meet the peak shaving effect of peak power generation, and the thermal power plant cannot guarantee the heating quality, and also reduces the amount of power generation of heating, reduces energy conservation and economic benefits. Most of the gas-fired power plants built in China are supporting projects for natural gas pipelines and LNG receiving stations. For example, the first line of the West to East Gas Pipeline Project has built a number of gas-fired power plants in Jiangsu and Henan, and the CNOOC Gas-Electric Power Group has built supporting power stations for Guangdong Dapeng and Fujian Putian LNG receiving stations. These gas-fired power plants undertake the task of peak shaving for the natural gas pipeline network. When the gas supply is tight, especially in the winter, the gas supplier will reduce or even stop the gas supply, and give priority to ensuring the use of gas by other users such as residents' living and heating. From the perspective of power demand, winter and summer are the peaks of electricity consumption for one year. Because gas-fired power plants do not have sufficient gas sources, they cannot play the role of peak-shaving. For thermal power units, heating and industrial heat loads cannot be interrupted. The negative impact of source supply interruption is greater. Moreover, the utilizing hours of the thermal power unit is reduced due to interruption of gas supply, the power generation and heating are affected, the cost of heat and electricity is increased, and the economic benefit of the power plant is reduced.

The heating price of natural gas thermal power plants far exceeds that of coal-fired thermal power plants and even exceeds that of the heating boiler room, resulting in a lack of competitiveness in the heating market. In order to increase market share, gas-fired thermal power plants must significantly reduce the price of heating, which is equal to or slightly higher than the price of coal-fired thermal power plants, resulting in the more heat supply, the greater the loss. At present, the main reason for the loss of natural gas power generation enterprises is that the gas price and electricity price system have not yet been marketized. The various power generation fuels do not reflect the real costs including external costs such as resource scarcity and environmental factors. The value of environmental benefits and peak shaving benefits of natural gas power generation enterprises has not been fully reflected.

The cost of gas-fired power generation is higher than the cost of coal-fired power generation, which will last for a long term. This is because natural gas is much higher than coal under the same calorific value. Gas-fired power plant fuel costs account for 70–80% of electricity cost. Natural gas prices are not market-oriented. Gas price reform will be further deepened in the future. The integration of stock gas and incremental gas prices has made the survival environment of domestic gas-fired power plants more disadvantageous. In October 2013, the National Development and Reform Commission issued a document deciding to appropriately solve gas-fired power price contradictions in some regions and improve on-grid price of natural gas power generation in eight provinces and cities such as Shanghai, Jiangsu, Zhejiang and Guangdong, while maintaining the same level of sales electricity prices. The rise of on-grid price is used to solve the increase in power generation costs due to the adjustment of the stock natural gas price. Zhejiang Province has raised the on-grid price of gas-fired power plants by about 20% (up to CNY 0.16 per kWh), and achieved a certain degree of gas-electricity linkage mechanism. Shanghai also raised the feed-in tariff by CNY 0.05 per kWh, and several areas in Jiangsu province raised heating steam prices to improve the economics of natural gas thermal power plant projects.

In combination with the new 13th Five-Year Plan for urbanization construction and urban and rural natural gas pipeline layout planning and construction, fully consider the comprehensive benefits of triple supply of heating, electricity and cooling for natural gas units, priority should be given to the development of distributed energy systems, and large-scale natural gas power stations should be developed according to local conditions. In the southern region, the demand for heating and cooling is solved in principle. The needs of small and medium-sized heating and cooling of users should be solved in the northern region, and the cascade utilization of energy through the combination of cooling, heating and electricity is realized. In areas where large-scale development of new energy such as wind power and system peak shaving capacity is seriously insufficient, natural gas generator sets are used to undertake peak shaving and frequency-modulating tasks, improve system operation flexibility and reliability, and reduce the abandonment of wind power, hydropower and photoelectricity. In combination with the access of the West to East Gas Pipeline and the external pipelines and the import of LNG, it is appropriate to develop large-scale combined cycles power generation and heating system in the receiving areas and cities, according to the needs of heat supply, heating and environmental protection, and to improve the haze weather in accordance with local conditions and so on.

7.2.2 Integrated Energy Service

1. Multi-energy complementary service

China is the country that uses the most electricity in the world and has the largest electricity market in the world. According to the National Energy Administration, the national electricity consumption in 2016 was 59,198 million kWh. According to data from the National Development and Reform Commission, in 2016, the marketized trading power of the country including direct transactions exceeded 1 trillion kWh, accounting for about 19% of the national electricity consumption. At present, most provinces in the country have carried out reforms in the electricity sales side, established a preliminary electricity market, and implemented medium and long-term electricity trading. The country is vigorously developing renewable energy power generation, UHV transmission, etc. With the development of new energy vehicles, electrified railways, electric energy replacement, energy storage and new energy technologies, electricity will become the main form of energy in the future. With the gradual advancement of economic development and power marketization, the country's electricity consumption has increased year by year, the amount of electricity involved in market transactions has gradually expanded, and the corresponding power trading, incremental distribution network and auxiliary service market have great prospects, which will share the trillions of electricity trading market.

Multi-energy complementarity is the complement of multiple energy sources, in order to comprehensively use energy and improve energy output and utilization efficiency. The first batch of demonstration projects of multi-energy complementary systems for wind, light, hydro, thermal, and storage was confirmed in July 2016 by the National Development and Reform Commission and the National Energy Administration to jointly issue the "implementation opinions on promoting the construction of multi-energy complementary integrated optimization demonstration projects". The implementation of this will help improve the utilization efficiency of renewable energy systems, and it is also an important means to solve the long-standing problem of abandonment of wind, light, and hydro for new energy sources such as wind and light in China.

At present, China's energy consumption supply, energy structure transformation, and energy system form show a new development trend. Integrated energy service is a new type of energy service that meets the diversified energy production and consumption of end customers, covering energy planning and design, engineering investment and construction, multi-energy operation services, and investment and financing services. With the acceleration of Internet information technology, renewable energy technologies and power reform, the development of integrated energy services has become an important development direction for improving energy efficiency, reducing energy costs, and promoting competition and cooperation. In order to adapt to the new situation of energy development, the power company tries to enhance the company's new development capabilities and expand the company's development path through comprehensive energy service business.

In order to implement the energy "four revolutions" and the national energy conservation, clean and safe energy development strategy, the power company actively adapts to the new requirements of energy supply side reform and power system reform, and takes energy Internet, smart energy and multi-energy complementarity as the development direction. With smart grid, big cloud and mobile services as the supporting means, we will build a new energy consumption market with electricity as the center and smart application, providing customers with diversified comprehensive energy services and promoting the healthy and sustainable development of the company. The power company has traditional advantages in terms of brand, tech nology and customer resources. With energy-saving service companies as the main body of implementation, it can provide a variety of services such as energy diagnosis, energy monitoring, energy-saving transformation, electric energy replacement, distributed new energy generation, and combined triple supply of heating, cooling and power and energy storage. It can also provide comprehensive energy services for water, electricity, gas, heating (cooling) and other energy sources for public buildings, industrial enterprises and parks. Carrying out comprehensive energy services is conducive to the implementation of national energy development strategies, promoting energy conservation and emission reduction in the whole society; conducive to consolidating the company's electricity sales market, expanding business scope, and enhancing new capabilities of customer service; facilitating the replacement of electric energy, driving the development of the company's related industries, and cultivating new market formats, adding new benefits growth points. The power company actively promotes the development of integrated energy service business, making the integrated energy service business stronger and bigger, transforming from electric energy suppliers to integrated energy service providers, creating new profit growth points and enhancing market competitiveness.

At the same time, the power company actively promotes the development of "Internet plus" smart energy, builds a coordinated development of energy production, transportation, utilization and energy storage systems, integrates complementary energy Internet, and promotes demonstration projects of multi-energy complementary systems such as wind, light, water, thermal and storage. The biggest significance of the power system reform for the market is that the market is open to a large number of power users. The power sales company can generate direct business relationships with each power user, including the procurement and operation and maintenance of power equipment, energy management systems, distributed photovoltaics, and multi-energy complementary, energy storage, energy-saving renovation, power engineering, etc. It can be said that the expansion from the aspects of equipment, engineering, services and information technology to other forms of energy is integrated energy services.

Taking large-scale energy storage equipment as an example, the significance of energy storage equipment for the power industry and power system is epoch-making. It is an ideal equipment for realizing flexible power grids and automatic power generation control and demand side response. There are two main reasons why it cannot be widely promoted in the market at present: first, the economy is not ideal; second, the energy density is not enough, and the application scenarios are limited due to large floor area. As the power spot market and the auxiliary service market are liberalized, the electricity prices at different time periods are reflected by the supply and demand relationship at that time. The market-based income from the demand side response, the income of the large-scale energy storage equipment and household energy storage will be greatly improved.

Power operators must provide basic power trading services to widely distributed power users through professional information tools, especially for spot transactions, which must collect user's power information. Operators also provide users with a variety of deeper integrated power services, such as energy management systems, distributed photovoltaic power generation management systems, energy storage operating systems, operating systems of charging station and charging pile, demand side management systems, power operation and maintenance platforms, even smart home systems, etc., users' electricity information and even energy information naturally access the Internet. Auxiliary service providers and power generation companies can use information tools to implement auxiliary services such as intelligent power generation and automatic power generation control, and interconnect power generation side information through the electricity market.

2. Building a new business model

Under the influence of China's new power reform and the energy Internet wave, traditional energy companies have taken measures such as transforming development methods, integrating energy supply systems, developing distributed energy, and innovating business models, driven by new economic normalities and energy restructuring. Promote the transformation from traditional single energy industry enterprises to integrated energy suppliers by integrating power generation, electricity sales, heating, cooling and water supply, and simultaneously build service models, business models and core capabilities that match the integrated energy system.

Power companies have explored the transformation from traditional energy providers to comprehensive clean energy production experts, integrated energy allocation experts and integrated energy users, and actively promoted the cleaning of traditional energy, the scale production of clean energy, and the industrialization of energy conservation and environmental protection. Under the new situation, the power company actively adapts to the adjustment of energy structure and the development of the power market. It is guided by market demand, persistence in precise management, pragmatic and efficient operation, steady progress, innovative development, and builds clean and efficient coal-fired power as the main body, four new business developments, e.g. distributed (multiple supply) energy, renewable energy, distribution of electricity business, carbon resources development as a supplement to the new pattern of energy development. Among them, in the aspect of distributed energy, we will build a comprehensive energy system with coordinated development, integration and complementation energy system covering "source, network, load, storage and control", and to provide a comprehensive energy integration solution based on distributed energy-led and multi-energy complementary and energy Internet has become the direction of the power company.

Vigorously develop distributed energy. Distributed energy is an effective way to control air pollution. It can realize the cascade utilization of energy through the combined supply of heating, electricity, cooling and gas, and greatly improve the comprehensive utilization efficiency of energy. It will become an effective supplement to large-scale energy systems, and it is an important way of efficient and clean utilization of energy. According to different resources and utilization objects of energy, multi-energy complementary integration optimization realizes that various distributed energy sources such as coal-based, natural gas, wind and light are complemented and coordinated to meet the energy demand of users. To promote the construction of multi-energy complementary integrated optimization system is to meet the needs of end users for electricity, heating, cooling, gas and other energy needs, adapt to local conditions, coordinate development, complement and utilize traditional energy and new energy, optimize layout and build integrated energy supply infrastructure, achieve multi-energy collaborative supply and energy integrated cascade utilization through coal-based, natural gas triple supply of heating, power and cooling, renewable energy distribution and energy intelligent micro-grid.

The power company-led energy Internet is based on the power system as the core and the link, and the energy network of multiple energy sources is interconnected. The energy comprehensive utilization efficiency is greatly improved through multi-energy synergy; the integrated energy system and information physics system are deeply integrated to help transform energy; the user-centered, innovative energy power operation business model and service format provide users with convenient energy and information integrated services.

"Source, network, load, storage, control" coordinated development, integrated and complementary comprehensive energy system works with functional areas as a unit, using distributed energy technology to provide multi-energy complementary, "Internet" smart energy (energy Internet) integrated energy solution, forming a cross-connection and synergistic supply between the multiple "electric, heating, cooling, gas and water" energy sources, and the coordinated development and integrated complementation of the vertical "source-network-load-storage-control" energy supply chain, establishing a new ecological integrated energy system with high integration of energy and information. Among them, "source" refers to various primary energy sources such as coal, natural gas, solar energy, geothermal energy, wind energy, biomass energy, and secondary energy such as electricity; "network" covers energy transmission such as power grid, gas grid, heating grid, and cooling grid; "load" and "storage" represent various energy needs and storage facilities; "control" is the use of intelligent data acquisition and analysis system to achieve multi-energy coupling, comprehensive control.

The integrated energy solution is divided into two modes. One is energy Internet multi-energy integration, multi-service areal integrated energy solution based on coal or natural gas distributed energy and new energy system (including solar energy, geothermal energy, wind energy, biomass energy, etc.), and "Source, network, load, storage, control", which is applicable to urban industrial parks, large communities, etc.; the other is a complementary, integrated and optimized urban complex energy solution based on natural gas distributed energy and new energy systems (including rooftop photovoltaics, geothermal energy, wind energy, biomass energy, etc.) for factories, schools, hospitals, single buildings and other users.

The basic business model of integrated energy services is based on distributed energy and energy Internet. Through energy transmission network, information physics system, integrated energy management platform and information and value-added services, energy flow, information flow and value flow exchange and interaction are realized. The multi-objective requirements of ecological, safe and reliable, high-quality economy are met. In the implementation of distributed energy projects, many entities such as government, industrial parks, electricity, gas, power generation and users will be involved. Conventionally, government, industrial parks, and electricity belong to the passive side. Gas, power generation, and users are active parties. However, in the future, passive parties may become active parties and implement the main driving force for consideration of performance pressure, industrial environment, and grid security. With the national energy industry regulatory reform, energy technology innovation and distributed energy development, traditional energy industry participants (including power companies, power grid companies, gas companies, equipment vendors, system integrators and design institutes) are planning the transition of comprehensive energy service industries. Distributed energy spans multiple industries such as electricity, natural gas, heating, and cooling. It not only changes the form of electricity and supply, but also eliminates policy and technical barriers between industries; the convergence trend between new energy and traditional energy appears in the energy industry, which has played a role in reducing costs and diversifying risks, improving the efficiency of comprehensive energy utilization, reducing environmental pollution, optimizing the energy structure, and achieving mutual benefit and win-win results.

The breakthroughs in key technologies for distributed energy and energy Internet are important "catalysts" for the commercial model of integrated energy supply. The business model itself will also play an important role in promoting the maturity of technology. Considering the continuous development of the integrated energy service business, the requirements for core capacity building should be gradually improved in terms of key technology. Coal-based distributed energy is a total energy solution for high-efficiency and low-emission coal-based distributed system by integrating advanced and mature high-efficiency pulverized coal furnace technology, energy power center multi-connection technology, coal-fired boiler tail gas emission reduction technology and intelligent control technology. In the implementation of coal-based distributed energy, it is necessary to consider factors such as policy planning, thermal (cooling) load, thermal efficiency, fuel supply and price, operating hours, investment and operating costs, and environmental indicators. Natural gas distributed energy is the link for efficient energy conversion of multiple energy sources and can serve as a combination point and hub for integrated energy supply. In the implementation process of natural gas distributed energy, factors such as setting power on the basis of heat, under-matching, grid connection, running time, gas price and investment interface need to be considered. In the actual project, under the same configuration conditions, factors such as power generation efficiency, gas price, waste heat utilization, electricity price, operating time, and operating years have a greater impact on energy prices.

3. Promoting power generation of renewable energy

According to the "Guidelines on the Implementation of the 13th Five-Year Plan for the Development of Renewable Energy", by the end of 2020, the installed capacity of new energy power generation in the country will reach 390 million kW, accounting for 29% of the total installed power, of which the installed capacity of wind power is more than 210 million kW, solar power installed capacity is more than 160 million kW. At present, the installed capacity of wind power has reached 164 million kW, and the installed capacity of solar energy has reached 130 million kW. In the future, wind power still has a large room for growth, and the focus will be on offshore wind power. If solar energy installations maintain the growth trend of the previous two years, they will reach and even exceed the expected total installed capacity by 2020, while encouraging photothermal power generation to explore multi-mode development.

Wind power development continued to shift to the eastern, central and southern regions. By 2020, the cumulative installed capacity in the eastern and central regions reached 74.7 million kW, accounting for 36% of the national wind power installed capacity, while accounting for 25% in 2017. By 2020, the scale of offshore wind power construction will reach 10 million kW, and the cumulative installed capacity will reach 5 million kW. Focus on promoting offshore wind power construction in Jiangsu, Zhejiang, Fujian, Guangdong and other provinces, with installed capacity of 3 million kW, 300,000 kW, 900,000 kW and 300,000 kW respectively.

In the future, we will follow the idea of combining centralized and distributed solar power generation development. In the "Three North" area, the solar photovoltaic power generation base will be built in an orderly manner. In 2020, the installed capacity of photovoltaic power generation base will exceed 71 million kW; the comprehensive construction of distributed photovoltaic power generation will be promoted. In 2020, the national distributed PV installed capacity will reach 60 million kW or more. The central and southern regions account for more than 70%.

Various types of solar power projects will be promoted in the future. (1) Photovoltaic front-runner plan: Combining with coal mining subsidence area, desertification land management, unified planning and construction of photovoltaic power generation technology base will be conducted. From 2017 to 2020, 8 million kW of leading base projects will be arranged each year. (2) Photovoltaic poverty alleviation project: With the goal of solving poverty-stricken filing card households with non-working capacity, we will vigorously promote distributed PV poverty alleviation, and preferentially adopt village-level power stations (including household systems) photovoltaic poverty alleviation mode in the central and eastern regions. (3) "Photovoltaic" comprehensive utilization project: Encourage the combination of barren hills and wasteland and comprehensive utilization along the beach, protected agriculture, fishery and other methods, and carry out various "photovoltaic" application projects according to local conditions to promote the organic integration of photovoltaic power generation with other industries. (4) Photothermal power generation: In 2020, the installed capacity of solar thermal power generation will reach 5 million kW or more. In Qinghai, Xinjiang, Gansu and other places, we will explore the establishment of large-scale renewable energy power generation bases with complementary utilization of photothermal power generation, photovoltaic power generation and wind power, and controllable and adjustable power generation.

References

Editorial Board of China Electric Power Encyclopedia. 2014. *China Power Encyclopedia (third edition)*, 177–178. Beijing: China Electric Power Press.

National Development and Reform Commission. 2016. The 13th five-year plan for power development.

National Development and Reform Commission. 2017. The 13th five-year plan for power development.

National Energy Administration. 2017. Guidance on the implementation of the "13th Five-Year Plan" for renewable energy development.

State Grid Energy Research Institute Co., Ltd. 2017. *Comparative analysis report of the world's top 500 power companies*, 116–117. Beijing: China Electric Power Press.

Zou, Caineng, Qun Zhao, Guosheng Zhang, et al. 2016. Energy revolution: From a fossil energy era to a new energy era. *Natural Gas Industry* 36 (1): 1–10.

Zou, Caineng. 2018. Energy revolution and oil company transformation strategy of new era. *Journal of Beijing Petroleum Managers Training Institute* 25 (4): 3–15.

Part IV
Theories of New Energy

Chapter 8
Hydrogen Energy

Hydrogen is considered to be an important green energy source in the 21st century. Hydrogen has storable properties and, with hydrogen-electrical interchange, will be one of the most important storage energy sources of the future. With the gradual maturity of hydrogen energy application technology and the continuous increase of global pressure on climate change, the development of hydrogen energy industry has attracted much attention. Japan, the United States, the European Union and other developed countries or regions have upgraded the development of hydrogen energy industry to the height of national energy strategy. It has taken substantial steps in the application of hydrogen energy industrialization. Both the government and industrial capital are actively deploying and promoting the development of the hydrogen energy industry. Especially in the field of energy transportation, hydrogen energy is playing an increasingly important role, gradually starting a revolution in energy transformation in the fields of transportation. Global hydrogen is in the best period of development (Wang et al. 2017).

In 2017, at the UN Climate Change Conference in Bonn, the Hydrogen Council, a consortium of 18 companies in the automotive, oil, gas, industrial LPG and machinery manufacturing sectors, proposed that hydrogen energy can be the driving force behind the transformation of renewable energy systems and clean energy carriers in a wide range of applications. At the beginning of 2017, the Hydrogen Energy Commission, established at the World Economic Forum in Davos, was the first global CEO organization to promote the role of hydrogen energy technology in the global energy transition.

Current Hydrogen Energy Committee members include 3M, Air Liquide Group, Alstom, Anglo American, Audi, BMW Group, China Energy, Daimler, ENGIE, General Motors, Great Wall Motor, Honda, Hyundai Motor, Iwatani, JXTG New Japan Petroleum and Energy Corporation, Kawasaki, Plasticomnium, Royal Dutch Shell, Norwegian National Oil Company, Bosch Group, Linde Group, Total, Toyota and Weichai, and 15 dynamic companies from the entire value chain (Balard, France Bay Industries, Faurecia, First Element Fuel, Gore, Hexagon Composites, Hydrogenics, Marubeni, McPhy, Mitsubishi Corporation, Mitsui & Co., Nel Hydrogen, Plug Power, Toyota Tsusho and Royal Vopak).

C. Zou, *New Energy*, https://doi.org/10.1007/978-981-15-2728-9_8

8.1 Current Status of Hydrogen Energy Development

8.1.1 Global Hydrogen Market

1. Global industrial hydrogen market is beginning to take shape

In recent years, the hydrogen energy market has developed rapidly. In 2011, the global industrial hydrogen market was 187.82 billion US dollars. In 2017, the global industrial hydrogen market reached 251.493 billion US dollars (Fig. 8.1). The growth rate reached 34.4% in the past 7 years.

The United States is the country with the largest hydrogen imports. Between 2012 and 2016, US imports totaled US$248 million; Belgium ranked second with imports totaling US$229 million; France, the Netherlands and Germany imported US$98 million, US$50 million and US$34 million, respectively.

Netherlands is the country with the largest export of hydrogen. Between 2012 and 2016, the Netherlands exported a total of 342 million US dollars; Canada ranked second with a total export value of US$246 million US dollars; Belgium, the United States and Germany exported a total of 159 million US dollars, 48 million US dollars, 36 million US dollars.

2. Hydrogen production is regional

In terms of regional distribution, the Asia-Pacific region ranks first in the world in hydrogen consumption and hydrogen production, followed by North America. The rapid economic growth in China, India and other Asia-Pacific countries is spurring strong growth in clean energy demand in the Asia Pacific region (Fig. 8.2).

In 2017, the industrial hydrogen market in North America was US$55.08 billion, the industrial hydrogen market in Europe was US$51.757 billion, and the industrial hydrogen market in the Asia Pacific region was US$107.136 billion.

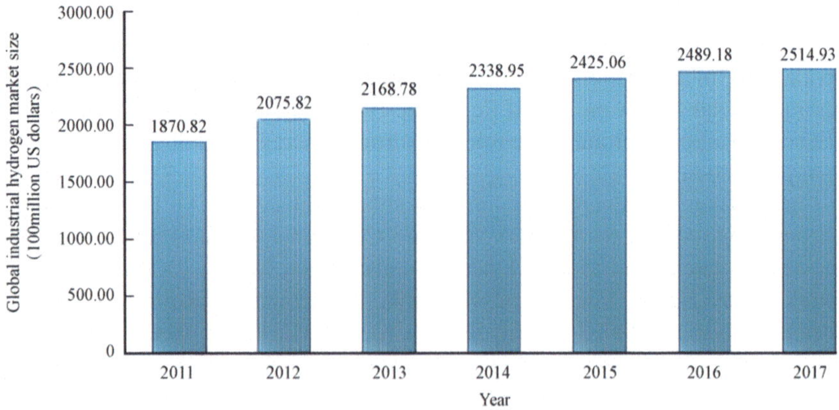

Fig. 8.1 Trends in the global industrial hydrogen market from 2011 to 2017

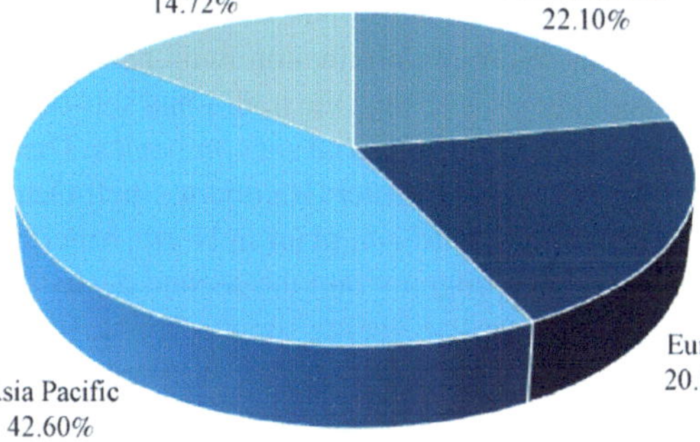

Fig. 8.2 Global industrial hydrogen market status

Table 8.1 Major liquid hydrogen producers in Western Europe in 2017

Company name	Device capacity (tons/day)
L'AirLiquide SA	10.5
LindeGasAG	4.4
AirProductsNederlandBV	5.4
Total	20.3

Most of the hydrogen production in Western Europe is for the self-use of the equipment. Commercial hydrogen accounts for only about 2% of the total. The transportation of cylinders or tank trucks accounts for only a small part of the hydrogen consumption, and most of the hydrogen is transported through pipelines. In 2001, the production capacity of hydrogen in gaseous commodities in Western Europe was about 700,000 tons/year. The main equipments were concentrated in Denmark, France, Germany, Italy, Portugal and the United Kingdom. The hydrogen production capacity of self-use hydrogen plants in Western European refineries and petrochemical enterprises is about 1.06 million tons/year (excluding the hydrogen production capacity of ammonia and methanol plants). The main producers are shown in Table 8.1.

The hydrogen energy economy in Germany is taking a top-down development path. First, it focuses on the development of the power-to-gas conversion mode, in order to provide a convenient infrastructure for downstream hydrogen energy applications as soon as possible, thereby activate downstream application scenarios. In terms of the layout of hydrogen energy infrastructure, the German government has taken the core city as the center and gradually extended outside based on the complete natural gas pipeline system.

The United States is the world's leading producer and consumer of hydrogen, with a large supply of commercial hydrogen. Most refineries and petrochemical plants in the United States have hydrogen production facilities. The hydrogen production capacity of US hydrogen production units totals 2.7 million tons/year (which does not include the hydrogen production capacity of ammonia and methanol plants. Because hydrogen is not the target product in ammonia and methanol plants, the hydrogen production capacity of these hydrogen units is not counted). Due to the improvement of oil quality indicators and the increase of hydrogen consumption, most refinery hydrogen production units have higher operating rates. At the same time, some refineries change the traditional self-built hydrogen production facilities, which are supplied by industrial gas companies. Therefore, the growth of commercial hydrogen production is faster (Hamburg: Shell Deutschland Oil GmbH 2017).

The first liquid hydrogen plant in North America was built in the United States for the US space program, and the second device was established in Canada using a relatively inexpensive energy supply. So far, four companies in the United States control the liquid hydrogen market in North America, namely Air Products and Chemicals, Praxair, BOC, and Air Liquide.

The hydrogen energy roadmap for the central and western United States has been developed jointly by the Renewable Hydrogen Fuel Cell Collaborative (RHFCC) and the Midwest Hydrogen Center of Excellence (MHCOE). The RHFCC was established to promote the Midwestern United States as a national leader in the use of hydrogen fuel cell vehicles through education, advocacy and research.

The use of renewable energy in the United States has enormous potential. Preliminary calculations, if using wind, solar and other biomass energy, can produce 1 billion tons (equivalent to 11.2 trillion cubic meters) of hydrogen per year.

Japan's major liquid hydrogen producers include Iwatani Industrial Gases and Pacific Hydrogen with production capacities of 12.4 tons/day and 17.5 tons/day respectively. The production capacity of gaseous commercial hydrogen is about 45,610 m^3/h (equivalent to 33,000 tons/year).

In recent years, China's industrial hydrogen consumption and output have risen simultaneously, maintaining a balance between supply and demand for many years, with both demand and output ranking first in the world. China is the world's largest hydrogen-utilizing country and has maintained its number one position in the world for nine consecutive years since it first exceeded 10 million tons in 2009 (Fig. 8.3). At present, there are nearly 30 major manufacturers of hydrogen in China (Table 8.2).

3. Hydrogen production from fossil energy dominates

Globally, in 2017, more than 96% of the world's major hydrogen production materials were derived from thermochemical reforming of traditional energy sources, of which 48% came from natural gas reforming, 30% from alcohol reforming, and 18% from coke oven gas. In addition, about 4% is derived from electrolyzed water (Fig. 8.4). However, different countries have their own characteristics, and there are also significant differences in the main raw materials and methods for hydrogen production.

Natural gas and coal are the main raw materials for hydrogen production in China. In China's hydrogen production, natural gas produces 61.7% of hydrogen, and coal produces 18.7% of hydrogen, other raw materials produce hydrogen (including refinery dry gas, coke oven gas, methanol, electrolyzed water and other by-products) produces about 19.6%.

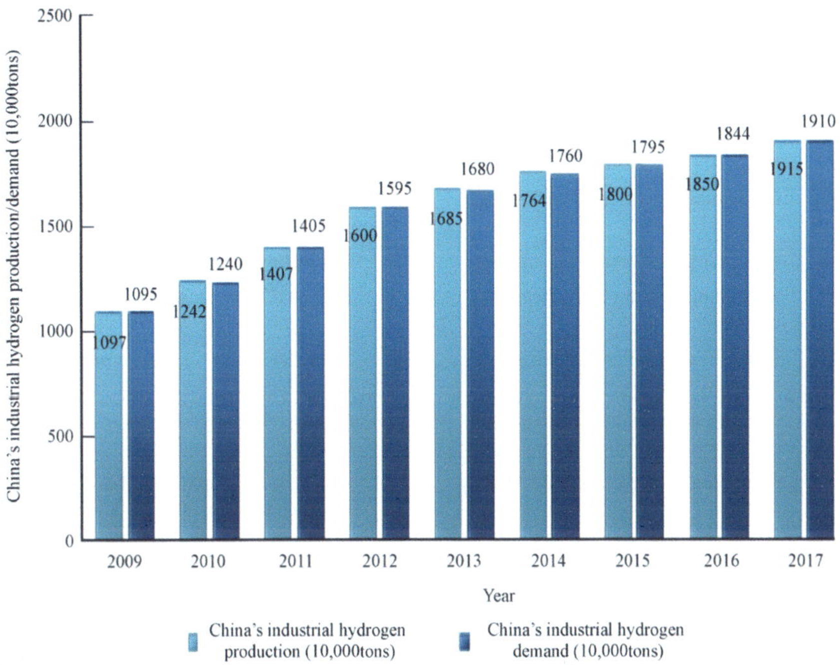

Fig. 8.3 Relationship between industrial hydrogen production and demand in China

Electrolytic hydrogen production has a special position in the manufacture of hydrogen in Japan. The capacity of brine electrolysis accounts for 63% of all hydrogen production capacity, natural gas produces 8% hydrogen, ethylene produces 7% hydrogen, coke oven gas produces 6% and methanol produces 6% (Fig. 8.5).

The contradiction between supply and demand of natural gas in China is outstanding. It is neither economical nor guaranteed to use natural gas as a raw material for hydrogen production. If hydrogen is produced by natural gas, it will not be able to meet the huge demand for hydrogen in the future hydrogen energy society. For a long period of time in the future, one of the most realistic sources of large-scale hydrogen should be hydrogen produced by electrolyzed water, followed by coal hydrogen production and hydrogen extraction from steel mills, emissions of exhaust gas of refining plants, and large-scale, low-cost ultimate hydrogen production technology should be solar hydrogen production (photocatalysis, photothermalpyrolysis) and nuclear energy hydrogen production.

At present, most of the hydrogen is produced from petroleum, natural gas and coal. The production of hydrogen from fossil fuels emits a large amount of greenhouse gases, which adversely affects the environment. However, compared with other hydrogen production methods, the process of producing hydrogen from fossil fuels is relatively mature, and the prices of raw materials are relatively low. China's coal reserves are relatively abundant, and the use of coal to produce hydrogen is one of the routes for China to achieve large-scale hydrogen production in the future.

Table 8.2 China's major commodity hydrogen production plants

No.	Manufacturer	No.	Manufacturer
1	Beijing Praxair	14	Yunnan Kunming Electrochemical Factory
2	Changshu Chemical Plant	15	Nantong Pesticide Factory
3	Chongqing Tianyuan Chemical General Factory	16	Zhejiang Ningbo Fertilizer Factory
4	Chunwang Industrial Gas Company	17	Henan Puyangchlor-alkali plant
5	PetroChina Fushun Petrochemical Company	18	Shanghai Bioxi Gas Co., Ltd.
6	PetroChina Jilin Petrochemical Company	19	Shanghai Chlor-Alkali Chemical Group Co., Ltd.
7	Fushun Organic Chemical Factory	20	Shanghai Zhongyuan Chemical Group Co., Ltd.
8	Guangdong Haotian Chemical Group Co., Ltd.	21	Shanghai Baosteel Group Bao Hydrogen Company
9	Jiangsu Pesticide Chemical Plant	22	Sinopec Baling Petrochemical Company
10	Jiangsu Jinlong Group Corporation	23	Sinopec Tianjin Petrochemical Industry Co., Ltd.
11	Jiangsu Lianyungang Chemical Plant	24	Sinopec Yangzi Petrochemical Co., Ltd.
12	Jiangsu Meilan Chemical Group Co., Ltd.	25	Wuxi Chemical Group Corporation
13	Sichuan Zigong Honghe Chemical Factory	26	Xinjiang Chlor-alkali Plant

4. Pressure swing adsorption purification of hydrogen is large in scale and economical

Pressure Swing Adsorption (PSA) is a new type of gas adsorption separation technology, which has features such as high product purity (it can work at room temperature and low pressure, no heating during bed regeneration), energy saving and cost efficient, simple equipment, continuous cycle operation and can fully achieve the advantages of automation. Therefore, when the new technology came out, it was concerned by the industrial circles of various countries, competing for development and research, developing rapidly and becoming more mature.

From the method of obtaining hydrogen, the pressure swing adsorption purification hydrogen has become an important production mode for hydrogen production due to its large scale and good economy. In recent years, with the expansion of domestic methanol production scale, hydrogen production converted from methanol steam has developed rapidly. At present, the self-produced pressure swing adsorption (PSA) single-unit hydrogen extraction unit can reach a scale of 300,000 m^3/h, and can obtain hydrogen with a purity of 99.999% or more in one step. PSA technology

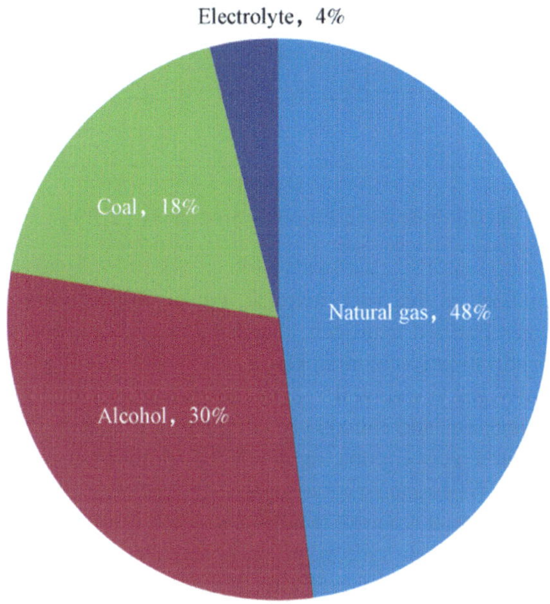

Fig. 8.4 Proportion of major sources of hydrogen raw materials in the world in 2017

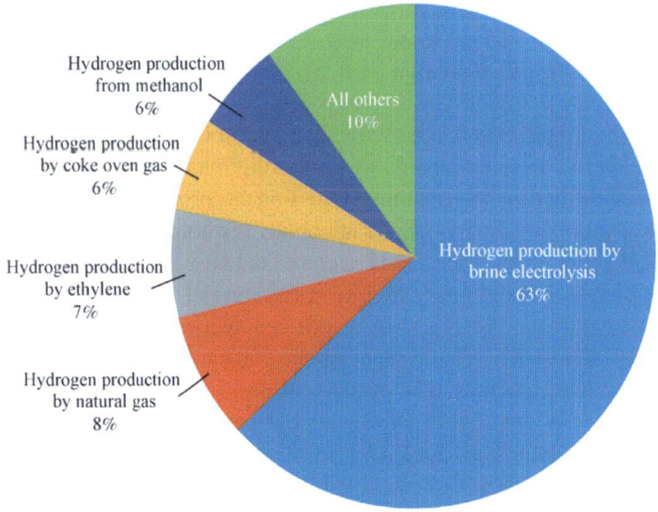

Fig. 8.5 Proportion of major sources of hydrogen raw materials in Japan in 2017

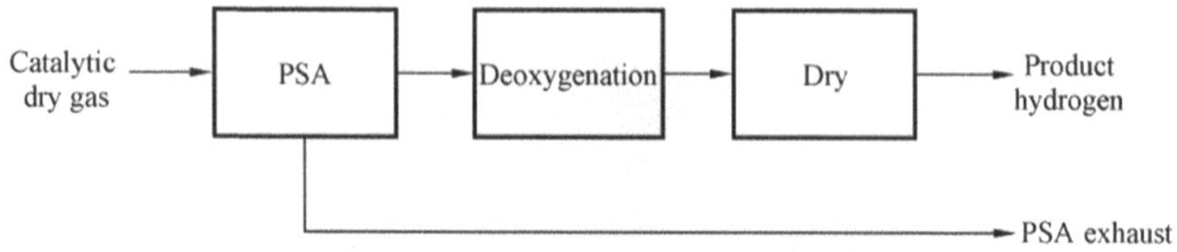

Fig. 8.6 Diagram of catalytic hydrogen extraction process of dry gas PSA

has been industrialized in the petrochemical industry for reforming gas, catalytic dry gas, shifting gas-purified hydrogen, catalytic dry gas recovery of ethylene, and recovery of olefins from polyolefin tail gas.

The catalytic dry gas contains 0.5–1.2% oxygen in addition to hydrogen and hydrocarbon components, and can also be purified by PSA technology. In order to obtain a high hydrogen recovery rate, the product contains a trace amount of oxygen in the hydrogen gas. When the oxygen content of the desired product hydrogen is low, the catalytic oxidation reaction is required to add a drying process (Fig. 8.6). The catalytic dry gas has a low hydrogen content and a large amount of hydrocarbon impurity components. In order to increase the hydrogen recovery rate, an evacuation desorption process is often employed.

5. The development potential of coal gasification hydrogen production is large

With the development of coal-based syngas and coal-to-liquids industry, coal-to-hydrogen production has increased year by year, with large scale and low cost. When the coal price is appropriate and the process flow and gasifier type are selected correctly, the hydrogen cost can be reduced to about CNY 20 per kg. At the same time, the number of devices capable of recovering hydrogen with a purity greater than 99% from the hydrogen-containing emission gas produced in the production of chemical products (including synthetic ammonia, methanol, etc.) is increasing.

Coal gasification refers to the conversion of coal into a gaseous product by reaction with a gasifying agent under high temperature and normal pressure or raised pressure. The gasifying agent is water vapor or oxygen (air), and the gas product contains components such as hydrogen, and the content thereof varies depending on different gasification methods. The purpose of gasification is to produce chemical raw materials or city gas. Large industrial coal gasifiers such as Lurgigasifer are fixed bed gasifiers, which produce 37–39% hydrogen (volume fraction), 17–18% carbon monoxide, and 32% carbon dioxide. Methane accounts for 8–10%. China has a large Lurgi furnace with a capacity of up to 100,000 m^3/h per furnace. Another new type of furnace is an entrained flow coal gasification furnace, called Texaco coal gasification furnace. The coal water slurry is used as raw material. China began research and development in the 1960s. At present, industrial production facilities have been built to produce synthetic ammonia and synthetic methanol raw gas, with a gas composition of 35–36% of hydrogen, 44–51% of carbon monoxide, 13–18% of carbon dioxide, and 0.1% of methane by volume.

As of the end of 2017, the domestically developed airflow bed clean coal gasification technologies mainly include: fixed bed, fluidized bed, coal water slurry gas flow bed and dry powder air flow bed (Table 8.3). It covers coal-water slurry refractory

Table 8.3 Status of major coal gasification technologies and applications in China

Technical title	Technology development department	Technology type	Application status
Crushed coal pressurized gasification technology	Saiding Engineering Company	Fixed bed	With good application performance
Yunmei furnace gasification technology	Yunmei Group	Fixed bed	13 gasifiers were put into production in total
ICC ash fusion gasification technology	Shanxi Coal Chemical Institute, Chinese Academy of Sciences	Fluidized bed	9 gasifiers put into production or under construction
Multi-nozzle coal slurry gasification technology	East China University of Science and Technology and Yankuang Group	Coal water slurry entrained flow bed	92 gasifiers of 33 projects put into production or under construction
HT-L furnace gasification technology	China Aerospace Science and Technology Group	Dry powder entrained flow bed	More than 30 gasifiers of 20 projects put into production or under construction
Multi-component slurry gasification technology	Northwest Research Institute of Chemical Industry	Coal water slurry entrained flow bed	About 100 gasifiers of 40 projects put into production or under construction
Two -stage furnace gasification technology	Xi'an Thermal Engineering Research Institute	Dry powder entrained flow bed	12 gasifiers for 10 projects signed (2 units put into production)
SE Oriental furnace gasification technology	Sinopec Ningbo Engineering Company, Ningbo Technology Research Institute and East China University of Science and Tehcnology	Dry powder entrained flow bed	8 gasifiers for 2 projects signed
Ningmei furnace gasification technology	ShenhuaNingmei Group and China Wuhan Engineering Company	Dry powder entrained flow bed	–

brick gasification technology, dry coal powder water wall gasification technology and coal-water slurry water wall gasification technology, there are both chilling process and waste boiler process, among which, coal-water slurry water wall gasification technology, that is to say, the second generation of Tsinghua furnace pioneered the world's first set of coal-water slurry water-cooling gasification industrial plants, making domestic clean coal gasification technology at the forefront of the world.

Coal underground gasification methods have attracted people's attention in recent decades. Underground gasification technology has the advantages of high utilization of coal resources and reduction or avoidance of surface environmental damage. However, the technology of coal underground gasification hydrogen production is still in the exploration stage, and there is still a considerable distance from commercial development.

It is an ideal and realistic solution to use renewable energy such as abandoned wind, abandoned light, and abandoned electricity to produce hydrogen by electrolyzing water to form green secondary energy. The elimination of affluence and abandonment of electricity is a rational allocation of energy and will drive the balanced development of various industries.

8.1.2 *Storage Methods of Hydrogen Energy*

Hydrogen storage has low-temperature liquid hydrogen storage, high-pressure gaseous hydrogen storage, solid-state hydrogen storage and organic liquid hydrogen storage. Several hydrogen storage methods have their own advantages and disadvantages. The hydrogen delivery state can be divided into gaseous hydrogen transport, liquid hydrogen transport, and solid hydrogen transport.

Different hydrogen storage methods have different hydrogen storage densities. As can be seen from Fig. 8.7, the gaseous hydrogen density is the smallest, the metal hydride has the highest hydrogen storage density, and the liquid and solid hydrogen storage is the second.

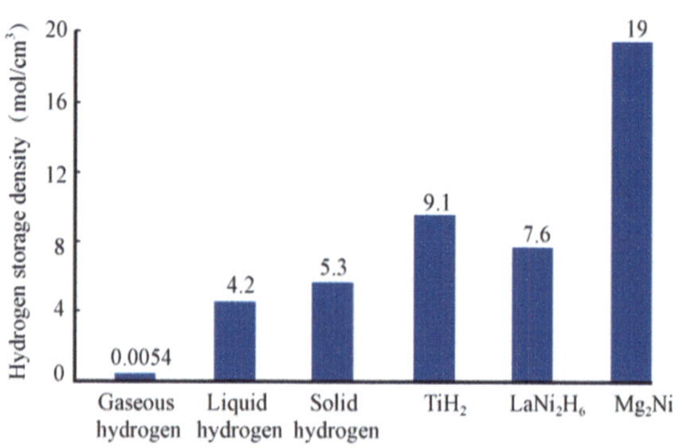

Fig. 8.7 Hydrogen storage density of different hydrogen storage technologies

1. Cost of low temperature liquid hydrogen storage is high

The density of liquid hydrogen is 845 times that of gaseous hydrogen. The volumetric energy density of liquid hydrogen is several times higher than that of hydrogen under compression. If hydrogen can be stored and transported in liquid form at a low price, its replacement of traditional energy sources will be just around the corner.

Hydrogen liquefaction is expensive, and liquefaction consumes 4–10 kWh/kg, which is almost equivalent to the cost of one-third of liquid hydrogen. Liquid hydrogen storage vessels require excellent adiabatic devices to insulate them from boiling vaporization.

For the needs of human space research programs, liquid hydrogen storage containers tend to be larger. Large liquid hydrogen insulated storage tanks with a capacity of more than 1,000 m^3 have been built. For example, the US Atomic Energy Commission (now the Department of Energy) has a large liquid hydrogen tank of 1,893 m^3 built in the test base in Nevada. Long-distance transportation of liquid hydrogen is usually carried out by rail tanker. The liquid hydrogen capacity of the tank is 36–107 m^3. A specially designed high-vacuum low-temperature insulation system is adopted to meet the requirements that the absolute pressure of the vacuum insulation layer is less than 0.133 Pa (1 μm mercury column), heat transfer coefficient is 0.1163 W/(m^2 K). In addition to the high vacuum jacket, the insulation system also includes a plurality of layers of radiation shielding. A well-designed liquid hydrogen tank can store liquid hydrogen at a temperature of 20 K for a longer period of time. The daily evaporation rate of liquid hydrogen is less than 0.3%. Due to the low density of liquid hydrogen (70 kg/m^3), a 107 m^3 liquid hydrogen tank with a load of less than 7.5 tons can be transported by tank truck. Due to the large gas-liquid volume ratio of hydrogen, liquid hydrogen storage tanks can also be used as working storage tanks in the process requiring a large amount of hydrogen.

2. High-pressure gaseous hydrogen storage technology is mature, but the capacity is small

High-pressure gaseous hydrogen storage is currently the most commonly used and mature hydrogen storage technology. It is stored by high pressure to compress hydrogen into a high-pressure vessel. The high-pressure gaseous hydrogen storage device has a fixed hydrogen storage tank, a long tube gas cylinder and a long tube bundle, a cylinder and a cylinder group, and a vehicle hydrogen storage bottle.

At present, the most commonly used high-pressure gaseous hydrogen storage container is a steel cylinder, which has the advantages of simple structure, low energy consumption for preparation of compressed hydrogen, fast filling and discharging speed, and disadvantages of leakage and explosion hazard, poor safety performance and low volume specific capacity.

Long tube cylinders and long tube trailers have also been successfully manufactured in China and have been installed and operated in some hydrogen production plants, hydrogen users, and hydrogen refueling stations. At present, hydrogen storage stations have been built and under construction in China, and this kind of hydrogen storage equipment is generally used.

3. Solid-state hydrogen storage density is high and technology is not yet mature

The solid-state hydrogen storage method can effectively overcome the shortage of high-pressure gas and low-temperature liquid hydrogen storage methods, and has high hydrogen storage bulk density, easy operation, convenient transportation, low cost, safety, etc., and is particularly suitable for occasions with strict volume requirements, such as the application of fuel cell vehicles is one of the most promising hydrogen storage methods.

There are many types of hydrogen storage materials, which can be mainly divided into physical adsorption hydrogen storage and chemical hydride storage. Among them, physical adsorption hydrogen storage can be divided into metal organic framework and nanostructure carbon material; chemical hydride hydrogen storage can be divided into metal hydride and non-metal hydride (including borohydride and organic hydride).

The metal hydride hydrogen storage process is simple, in sharp contrast to compressed gas and low temperature liquefaction. By selecting a suitable metal hydride, hydrogen can be stored in the metal hydride at room temperature and less high pressure. The outstanding advantage of hydrogen storage with metal hydrides is safety. Hydrogen is present at low pressure in combination with another substance (hydrogen storage alloy) to form a quasi-compound state. High pressure and low temperature are not required.

Metal hydride hydrogen storage has a high hydrogen storage density and high purity (hydrogen released from the heating of hydride has a very high purity, usually up to 99.999%).

However, the current rare use of metal hydride hydrogen storage for large-scale industrial production is mainly due to four reasons: (1) hydrogen storage alloy is too expensive; (2) complex structure. Due to a large amount of heat release during hydrogen storage, heat exchange equipment must be added to the storage to remove the heat released. (3) the hydride itself is very unstable and susceptible to harmful impurity components. After repeated use, the performance is significantly reduced; (4) although hydrogen storage density is high, the hydrogen storage mass ratio is too low, that is, only 2–4% of hydrogen can be stored in terms of mass fraction.

Metal hydride hydrogen storage is still in the experimental research stage and has not yet entered commercial applications. At present, the research and development of hydrogen related with automobile engine is mainly carried out. For example, Suzuki Motor Co., Ltd. has developed a Ti–Zr hydrogen storage composite using magnesium as a binder for hydrogen-burning vehicles, which has undergone 1,000 cycles without degradation. The hydrogen storage device (approx. 280 kg) with a hydrogen storage capacity of 37 m^3 drives the vehicle for 200 km.

4. Organic liquid hydrogen storage has received much attention

The organic liquid hydrogen storage technology realizes hydrogen storage by reversible hydrogenation and dehydrogenation of unsaturated liquid organic substances. In theory, olefins, alkynes, and certain unsaturated aromatic hydrocarbons and their corresponding hydrides, such as benzene-cyclohexane, methylbenzene-methylcyclohexane, can be hydrogenated and desorbed without destroying the carbocyclic backbone structure, and the reaction is reversible.

Organic liquids have high mass and volumetric hydrogen storage density. Currently used materials (such as cyclohexane, methylcyclohexane, decalin, etc.) can meet the specified standards; cyclohexane and methylcyclohexane are at room temperature. It is liquid under normal pressure and similar to gasoline. It can be stored and transported by existing pipeline equipment. It is safe and convenient, and can be transported over long distances. The catalytic hydrogenation and dehydrogenation reactions are reversible, and the hydrogen storage medium can be recycled. It can be stored for a long time, solving the problem of energy shortage to a certain extent.

There are also many shortcomings in the storage of hydrogen in organic liquids: the technical operating conditions are more stringent, and the equipment configuration of catalytic hydrogenation and dehydrogenation is higher, resulting in higher cost; the dehydrogenation reaction needs to be transmitted under low pressure and high temperature heterogeneous conditions. Restricted by heat and mass transfer and reaction equilibrium limit, dehydrogenation reaction efficiency is low, and the side reaction is easy to occur, so that the released hydrogen is impure, and the pore structure of the dehydrogenation catalyst is easily destroyed under high temperature conditions, resulting in coking inactivation.

8.1.3 Hydrogen Transportation by Pipeline

Hydrogen delivery methods include high pressure gaseous hydrogen, liquid hydrogen, and pipeline transportation. Nearly ten hydrogen refueling stations currently operated in China are transported by high-pressure gas tanks; short-distance hydrogen pipeline transportation is generally used in refineries and chemical plants; liquid hydrogen transportation mainly uses liquid hydrogen storage tanks through trailers and tank trucks.

Hydrogen long-distance pipelines have been constructed and operated in some areas in recent years. It is understood that the hydrogen pipelines of 50 km at 2.0–4.0 MPa have been in operation, and the inner diameter of the pipeline has reached 400 mm. However, it is still necessary to carry out systematic basic research work on the compatibility of pipeline steel with high-pressure hydrogen and the operation management of pipelines, and create conditions for the construction of long-distance, high-pressure scale hydrogen pipelines.

8.1.4 Construction of Hydrogen Refueling Station

With the continuous improvement of the hydrogen energy industry chain, the hydrogen energy market continues to expand. Especially in the rapid development of hydrogen fuel vehicles in the past two years, the demand for hydrogen has increased, and the construction of hydrogen refueling stations is also accelerating.

1. Status of global hydrogen refueling station construction

As of January 2018, there were 327 hydrogen refueling stations in operation worldwide. Among them, there are 139 in Europe, 118 in Asia, 68 in North America, 1 in South America and 1 in Australia.

Japan: Nissan, Honda and Toyota jointly signed a memorandum to cooperate in the construction of a hydrogen refueling station. According to the previous Japanese government's "Hydrogen Fuel Cell Strategic Plan", 160 hydrogen refueling stations will be completed by 2020.

South Korea: There are currently 165 hydrogen fuel cell vehicles, of which 140 are non-commercial vehicles and 25 are for commercial use. Twelve hydrogen refueling stations have been established, six of which are for commercial use and six for research. In 2018, 28 hydrogen refueling stations will be established.

Germany: As of January 2018, there were 45 hydrogen refueling stations in Germany. The number of German hydrogen refueling stations has surpassed that of the United States, second only to Japan, and has become the second largest country in the world with hydrogenation infrastructure.

2. Status of China's hydrogen refueling station construction

In October 2016, the "China Hydrogen Energy Industry Infrastructure Development Blue Book (2016)" issued by the Resources and Environment Branch of China National Institute of Standardization and China Electrical Equipment Industry Association first proposed the development roadmap of China's hydrogen energy industry (Beijing: National Development and Reform Commission 2016; Beijing: National Development and Reform Commission 2017; Beijing: National Energy Administration 2017). Hydrogen refueling stations and fuel cell vehicle development goals have been planned. By 2020, the number of hydrogen refueling stations is expected to reach 100, fuel cell vehicles will reach 10,000, and hydrogen rail vehicles will reach 50. By 2030, the number of hydrogen refueling stations is expected to reach 1,000, and the number of fuel cell vehicles will reach 2 million. By 2050, the network of hydrogen refueling stations is expected to be completed, and the number of fuel cell vehicles will reach 10 million.

As of February 2018, there are 31 hydrogen refueling stations built and under construction in China, and 12 are in operation, located in 8 provinces and cities including Beijing, Shanghai, Guangdong, Jiangsu, Henan, Hubei, Liaoning and Sichuan. The total hydrogenation capacity is 4,000 kg/day (Table 8.4) (Chen et al. 2018).

The 12 hydrogen refueling stations in operation are small in scale and relatively centralized in distribution. From the point of view of hydrogen refueling capacity,

Table 8.4 Statistics of domestic hydrogen refueling stations in February 2018

No.	Province	City	Name	Refueling amount (kg/day)	Total refueling amount (kg/day)	Remarks
1	Beijing		Beijing Yongfeng hydrogen station	200	200	Operating
2	Shanghai		Shanghai anting hydrogen station	200	700	
3			Shanghai electric drive hydrogen station	500		
4	Guangdong	Zhongshan	ZhongshanShalang hydrogen station	1000	1400	
5		Yunfu	Silao hydrogen station	200		
6		Foshan	RuihuiFoshan hydrogen Station	200		
7	Jiangsu	Rugao	Nantong Baiying hydrogen station	200	200	
8		Changshu	Toyota hydrogen station	–		
9	Henan	Zhengzhou	Zhengzhou Yutong hydrogen station	200	200	
10	Hubei	Shiyan	Shiyan hydrogen station	500	500	
11	Liaoning	Dalian	TongjiXinyuan Dalian hydrogen station	400	400	
12	Sichuan	Chengdu	Pidu district hydrogen station	400	400	
Total in China				4000		
1	Shanghai		Shanghai Jinshan hydrogen station	500	2750	Under construction
2			SongjiangWanxiang hydrogen station	500		
3			QingpuYunda hydrogen sation	1000		
4			Jiangqiaojia hydrogen industry	750		
5	Guangdong	Zhongshan	Guzhen hydrogen station	1000	3400	
6		Foshan	GuonengLiansheng hydrogen station	1000		

(continued)

Table 8.4 (continued)

No.	Province	City	Name	Refueling amount (kg/day)	Total refueling amount (kg/day)	Remarks
7		Yunfu	Yunfu Sinopec	1000		
8		Yunfu	Luoding No. 1 hydrogen station	400		
9	Jiangsu	Zhangjiagang	Qinfeng hydrogen station	1000	1500	
10		Yancheng	Auxin auto hydrogen station	500		
11	Zhejiang	Taizhou	Hydrogen energy town hydrogen station	500	700	
12		Jiashan	Aideman hydrogen station	200		
13	Hubei	Wuhan	Qingxiong hydrogen station	500	1000	
14		Xiangyang	Test site hydrogen station	500		
15	Shandong	Liaocheng	Zhongtong bus hydrogen station	200	200	
16		Binzhou	Binhua hydrogen station	–		
17	Anhui	Liu'an	Tomorrow hydrogen	400	400	
18	Shanxi	Xi'an	Youth bus	500	500	
19	Hebei	Zhangjiakou	Zhangjiakou hydrogen Station	–		
Total in China				10,450		

the daily filling capacity of hydrogen refueling stations is generally small, and the filling capacity of a single hydrogen refueling station is mostly 200–500 kg/day, and the maximum is only 1,000 kg/day, and there is only one of this kind. From the distribution of hydrogen refueling stations, mainly distributed in the central and eastern coastal areas, four hydrogen refueling stations have been put into operation in the Yangtze River Delta, and eight are under construction. Three hydrogen refueling stations in the Pearl River Delta region have been put into operation and four are under construction. It is worth mentioning that in Hubei, one hydrogen refueling station has been put into operation, two are under construction, and the filling capacity of a single hydrogen refueling station is 500 kg/day, which is a large-scale hydrogen refueling stations that have been built and are under construction.

3. The cost of construction of hydrogen refueling station remains high

At present, the construction cost of a new hydrogen refueling station is between 2 million and 5 million US dollars. The cost of building a medium-sized hydrogen station (300 m^3/h) in Japan is 5 million to 5.5 million US dollars; the construction of a medium-sized hydrogen refueling station in the United States requires about 2.8 million to 3.5 million US dollars.

Compared with foreign countries, the construction cost of China's hydrogen refueling stations is relatively low. The cost of building a hydrogen refueling station (35 MPa) in China is between 2 and 2.5 million US dollars. With the increase in the number of hydrogen refueling stations, scale effects will inevitably occur, and the construction cost of hydrogen refueling stations will be expected to decline (Wang et al. 2017).

The main equipment of the hydrogen refueling station includes hydrogen storage device, compression equipment, filling equipment, station control system and so on. The compressor costs the most, accounting for about 30% of the total cost (Fig. 8.8). China needs to accelerate the localization of hydrogen compressors, reduce the construction cost of hydrogen refueling stations, and promote the development of the hydrogen energy industry chain.

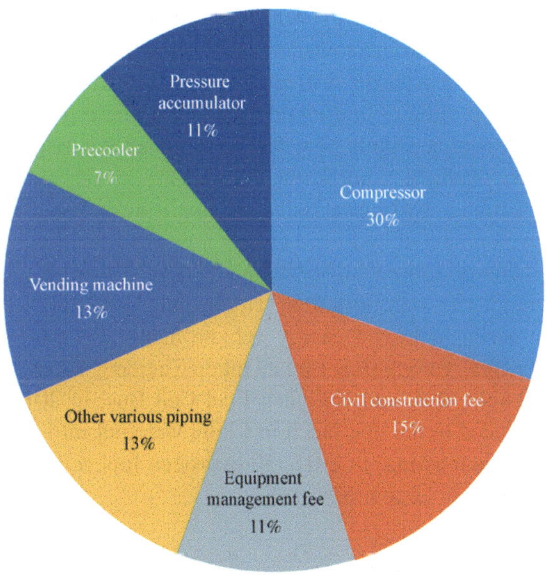

Fig. 8.8 Analysis of construction cost of China's hydrogen refueling stations in 2017

8.1.5 Fuel Cell Vehicle Technology

With the rapid development of new industries such as new energy, new materials and environmental protection, the application of industrial hydrogen has been greatly expanded. The demand for food and beverage and medical and health industries is rising, and the development of industrial hydrogen in China is on a new platform. From a global perspective, hydrogen fuel cell vehicles, which are important technological innovations, are gradually becoming an important domain for large-scale commercial application of hydrogen energy.

According to the US Department of Energy, in 2016, the US hydrogen industry created about 16,000 jobs and more than 3,500 hydrogen fuel cell vehicles. By the end of 2017, Japan had more than 2,000 hydrogen fuel cell vehicles, making it the world's largest hydrogen fuel cell vehicle market. It is estimated that by 2020, the number of hydrogen fuel cell vehicles that can be used in Japan will reach 40,000, reach 200,000 by 2025, and reach 800,000 by 2030. 900 widely distributed hydrogen refueling stations will provide hydrogen refilling services for these vehicles.

In 2017, 794,000 new energy vehicles were sold in China. In 2020, the sales target of new energy vehicles will be 2 million units, and in 2030 it is expected to reach 10 million units.

Hydrogen energy vehicles are still in their infancy in the world. Although the upstream of entire hydrogen energy industry chain (hydrogen production—hydrogen storage—hydrogen transport—hydrogen utilization) is developing fast, the hydrogen application in the downstream of the industrial chain is still in its infancy. In particular, the proportion of household hydrogen vehicles is negligible. People have a gradual process of accepting new things, and it takes a long time for the full-scale hydrogen energy vehicle to break out.

China's hydrogen fuel cell application has begun to take shape with the development of hydrogen energy logistics vehicles and buses (large passenger cars).

The development of China's hydrogen energy logistics vehicles and hydrogen energy city transportation vehicles has been fully launched. At present, the main domestic hydrogen energy vehicle manufacturers include Guangdong FoshanFeichi Bus, Yutong Bus, FutianOuhui, Zhejiang Youth Auto, Jiangsu Aoxin Company, Chengdu Bus, and Keliyuan Fuel Cell Bus. In 2017, 28 hydrogen fuel cell buses produced in the first batch in China were put into trial operation in Yunfu and Foshan, Guangdong. In 2018, 50 hydrogen fuel cell logistics vehicles in Jiangsu were operated in large urban areas.

8.2 Development Prospect of Hydrogen Energy

Hydrogen is recognized as a clean energy source and is regarded as the most promising secondary energy source in the 21st century. Hydrogen energy plays an important role in solving the energy crisis, global warming and environmental pollution, and its development and utilization has received worldwide attention. Hydrogen has played an important role in many industrial processes for many years. Almost half of the world's hydrogen production is used in the chemical industry to produce ammonia and methanol. The remaining large amounts of hydrogen are also used in the metal processing, glass manufacturing, electronics and food industries to reduce emissions from desulfurized gasoline and diesel in internal combustion engines and refineries. Hydrogen is widely used in industry, including hydrogenated shortening, electronics industry, metal heat treatment, float glass, petroleum and refining, hydrogen fuel cells, hydrogen refueling stations (hydrogen vehicles), glass surface polishing, heat treatment quenching and heat treatment annealing, brazing and heat treatment powder metallurgy and so on (Guan et al. 2012).

It is predicted that refining industry, new energy vehicles and clean energy power generation will be the largest end markets for hydrogen, with the global refining industry accounting for about 90% of the world's total hydrogen consumption. As fuel specifications become more stringent, refinery hydrotreating requires more hydrogen to produce low-sulfur clean fuels, which will greatly stimulate the rapid growth of hydrogen demand.

8.2.1 Complementarity of Hydrogen Energy and Electricity

Hydrogen and electric energy are important secondary energy sources, and they are also one of the main green clean energy sources in the future. They have the characteristics of long-distance transportation, storage and hydrogen-electricity interchange.

Hydrogen and electric energy are widely used in various fields such as industry, agriculture, electronics, steel, and civil use. However, each type of energy has a peak-to-valley problem at different times and in different periods. Each energy source may have an affluence or shortage of about 10% during peaks and valleys. Hydrogen-electricity swap is one of the effective means to solve the fluctuation of energy peaks and valleys. How to effectively utilize the exchange advantages of hydrogen energy and electric energy, and to play intellectual interconnection and complementarity of energy is the key to improving energy efficiency.

Power by hydrogen combustion: There may be a short-term surplus of hydrogen in use. This excess hydrogen can be stored in a high-pressure gas storage or a gas storage tank, or it can be converted into electric energy by hydrogen combustion to alleviate the shortage of electricity.

Hydrogen production from electrolyzed water: Electricity fluctuates in different seasons, daytime, etc., especially wind energy, solar power generation, etc., with the weather changes, daytime, etc., which will fluctuate greatly. How to effectively use this part of electricity has been a world problem. The storage capacity of electric energy through a battery or the like is limited, and it is unable to satisfy the storage challenge of a large amount of electric energy. Hydrogen production from electrolyzed water is one of the effective means to accept rich electricity. In particular, hydrogen produced by electrolyzed water can make good use of discontinuous wind and photovoltaic energy.

8.2.2 Hydrogen Vehicles

1. Hydrogen can lead the transformation of new energy vehicles

Hydrogen fuel cell vehicle industry kicked off the commercialization of hydrogen energy. Hydrogen has been used on a small scale for large logistics vehicles, urban transportation vehicles, and home vehicles. Even some have been used in trains, bicycles, model airplanes, drones, etc. Hydrogen energy will set off a small-scale revolution in the field of energy and transportation, especially in the field of new energy vehicles.

Hydrogen energy passenger cars based on high-pressure hydrogen technology in Japan and South Korea have been mass-produced. Among them, Honda Clarity can run 750 km after refilling hydrogen for 3 min; Model 2019 Hyundai NEXO can run 609 km after refilling hydrogen for 5 min. According to the Asahi Shimbun website, Japan's large chain convenience store and Toyota Motor Corporation have cooperated to launch trucks that use hydrogen fuel cell (FC)-driven vehicles to transport goods and establish a zero-emission logistics system in the spring of 2019.

In March 2017, Germany successfully tested the world's first "hydrogen train", a zero-emission hydrogen-powered train. Germany and Alstom of France signed an agreement to start construction of a series of hydrogen-powered trains in 2021. Alstom announced that they will build 14 non-emission trains with a full hydrogen fuel tank for a cruising range of 1,000 km with a top speed of 140 km/h.

In December 13, 2017, the French "European Times" reported that the first batch of hydrogen energy electric bicycles in the city of Saint-Louis in the Lower Normandy region of France was officially put into use. The bike weighs 25 kg, which is about the same as the normal electric bicycle, but the normal electric car takes 3 h to charge, while the hydrogen energy electric bike takes 2 min to charge and twice the charging amount of the former.

In the past few years, China has focused on the development of electric vehicles. In 2018, the Chinese Prime Minister's visit to Japan became a turning point in the use of hydrogen fuel cell vehicles in China. In February 2018, the China Hydrogen Energy and Fuel Cell Industry Innovation Alliance led by the National Energy Investment Group was officially established in Beijing. At present, more than 30 listed companies have been involved. In addition to Geely, FAW, Great Wall and other car companies, energy companies such as the National Energy Investment Group and China Guangdong Nuclear Power Corporation are also beginning to deploy hydrogen fuel vehicles. Due to the relatively late start, the basic research of Chinese fuel cell vehicle manufacturers lags far behind Japan, South Korea, Europe and the United States.

2. Hydrogen energy consumption will account for 20% in 2050

It is predicted that in 2050, the proportion of natural gas in energy consumption will exceed that of oil and coal for the first time, entering the era of "four major components" of natural gas, oil, coal and new energy. Among them, hydrogen energy will account for more than 20% of energy consumption, and the world will enter a clean energy era (Beijing: China Quality Inspection Press, China Standard Press 2016).

According to a report released by the Hydrogen Energy Council in November 2017, hydrogen energy will account for about 20%, reducing carbon dioxide emissions by 6 billion tons per year, and the annual output value of the hydrogen energy industry chain will reach 2.5 trillion US dollars.

In 2100, new energy will occupy the first place in energy consumption. The total ratio of hydrogen energy and natural gas over total energy consumption will exceed 60%, and the proportion of hydrogen energy will reach 30%. Humans will enter a true era of green energy, and hydrogen society will emerge in the early 22nd century.

Three of the world's top five coal producers are in the Asia Pacific region. China is the mainstay of global coal consumption, accounting for 50.6% of global coal consumption. China has formulated a strict emission reduction plan. China's policy and efforts of developing new energy vehicles are the largest in the world and the development is fastest. China's hydrogen energy development started later than Japan, the United States and the European Union. However, since 2016, China has formulated hydrogen energy development targets and a clear technology roadmap, China's hydrogen energy economy will catch up from behind (Wang 2015).

The major countries in the world have set a timetable for the withdrawal of fuel vehicles from the historical stage (Table 8.5). China is in the process of setting a timetable for stopping the production and sale of traditional energy vehicles. Around 2050, global hydrogen vehicles are expected to reach around 20% of market share, and will reach 50% by 2060. In 2100, the global energy system will completely enter the era of secondary energy based on hydrogen and electricity.

Table 8.5 Time table for sales of fuel vehicles in major countries of the world

Country	Sales starting	Sales ending	Remarks
California, US	August 2015	2030	Ban on the sale of traditional fuel vehicles
Netherlands	April 2016	2025	Ban on the sale of fuel vehicles
Norway	May 2016	2025	Ban on the sale of fuel vehicles
Germany	October 2016	After 2030	Ban on the sale of traditional internal combustion vehicles
India	June 2017	2030	Nationwide ban on the sale of fuel vehicles
France	July 2017	2040	Nationwide ban on the sale of fuel vehicles
United Kindom	July 2017	2040	Nationwide ban on the sale of fuel vehicles

8.2.3 Roadmap of Hydrogen Energy Development in Various Countries

Developed countries have successively upgraded the development of hydrogen energy industries to the height of national energy strategies, and many countries have set clear development goals and detailed road maps for action.

1. Roadmap of Hydrogen Energy Development in US

The United States is a pioneer in the development of hydrogen energy and a leading country in the development of hydrogen fuel cells in the world. In 1970, the development of hydrogen energy technology was started. In November 2002, the US Department of Energy issued the "National Roadmap for Hydrogen Energy Development", on the development goals of hydrogen energy in the United States, various factors affecting the development of hydrogen energy, and the status of hydrogen energy technology in detail. The challenges faced and the future development path have been carefully designed, which indicates that the US "hydrogen economy" concept has begun to move from the concept stage to the action stage (London: Shell Group of Companies 2017).

Since 2007, the southern part of California has implemented a tax-free policy on hydrogen fuel cell production and research equipment; Ohio has a tax-free policy for fuel cell systems below 250 kW. In 2014, the United States launched a new hydrogen energy program at the national level, promulgating the "Comprehensive Energy Strategy", re-confirming the leading role of hydrogen energy in transportation transformation, and announced in 2017 that it will continue to support the construction of 30 hydrogen energy projects to promote the rapid development of the hydrogen energy industry. According to data of the US Department of Energy, in 2016, the US hydrogen industry created about 16,000 jobs, with more than 3,500 hydrogen fuel cell vehicles and 60 hydrogen refueling stations. The hydrogen energy economy is expected to be fully realized between 2030 and 2040.

2. Roadmap of Hydrogen Energy Development in Japan

Japan focused on building a hydrogen energy society. In April 2014, the Fourth Energy Basic Plan was formulated to define the strategic direction for accelerating the construction and development of a hydrogen energy society. In 2015, it was designated as "the first year of hydrogen energy", and in 2020 it was defined as "the first year of hydrogen energy Olympics". In 2025, it was defined as "the first year of hydrogen energy going aboard", and in 2030 it was defined as "the first year of hydrogen fuel power generation".

By 2030, Japan's hydrogen-related industries will reach a scale of 1 trillion yen (about 8.8 billion US dollars). Hydrogen energy cars developed by Japan's leading vehicle company Toyota have been on the market in Japan at the end of 2014, and Honda and Nissan are also developing their own hydrogen energy cars. The Japanese government has proposed to build a hydrogen energy society in the future, and it will also make the 2020 Tokyo Olympics a "hydrogen energy event". Athletes and spectators will travel to and from the Olympic venues to take the latest hydrogen energy buses. It is expected that Japan will build a nationwide hydrogen energy supply network by 2040.

Japan proposes three paths to promote low-cost, clean hydrogen technologies: (1) promote the establishment of overseas hydrogen energy supply systems; (2) use low-cost lignite from overseas to achieve low-cost hydrogen production; (3) use overseas renewable energy to produce hydrogen from electrolytic water. In 2030, the hydrogen supply will reach 300,000 tons/year, and in 2050, it will reach 5 million to 10 million tons/year.

3. Roadmap of Hydrogen Energy Development in Germany

In 2016, Germany revised its strategic plan for hydrogen energy transportation, and there were three main initiatives in the newly announced hydrogen transport strategy.

The first is to increase investment. The German federal government has launched a second national innovation program for hydrogen and fuel cell technology with the goal of ensuring continuity of research and development and building the competitiveness of hydrogen and fuel cell vehicles in the market. With the development of electric vehicles, autonomous driving and car networking technology, people are facing the biggest revolution since the invention of the car. Fuel cells are one of the core technologies of this revolution.

The second is to promote cooperation. According to the plan, 400 hydrogen refueling stations will be built in Germany in 2023. In 2015, 6 companies jointly established the German Hydrogen Transportation Company, which is responsible for the construction of a hydrogen energy transportation infrastructure network in Germany with a total investment of 350 million Euros.

The third is to encourage innovation. The German Ministry of Transport is preparing a series of preferential measures to encourage innovation and activate the market. For example, it will focus on supporting autonomous vehicles for logistics and autonomous power supply technologies for critical or off-grid infrastructure and equipment.

At the end of 2017, there were 56 hydrogen refueling stations in Germany, of which 45 were public hydrogen refueling stations. It is estimated that by 2019, there will be 100 hydrogen refueling stations in German, surpassing the United States, second only to Japan, and become the second country with the world's second largest hydrogen refueling infrastructure.

4. Roadmap of Hydrogen Energy Development in UK

In 2018, the Green Pioneer Energy Test at the University of Keele in the UK will conduct the first test of a hybrid network of hydrogen and natural gas. This is the first test in the UK for a mixed network of domestic hydrogen and natural gas.

The trial will be officially launched in the near future, led by the gas network Cadent and in collaboration with Northern Gas Networks and a coalition of technical experts, known as the HyDeploy project.

The HyDeploy project is supported by the OFGEM Network Innovation Competition and aims to establish a hybrid supply network of hydrogen (up to 20%)

with conventional natural gas. Beginning in 2019, Keele University will conduct a one-year real-time trial of the school's private gas network.

As early as the 1970s, the natural gas used in the UK natural gas network consisted of up to 60% hydrogen, but it was not universal. In the HyDeploy project, users will use hydrogen as their energy source for the first time at home.

5. Roadmap of Hydrogen Energy Development in France

According to the report of China hydrogen energy website, the French Hydrogen Energy Program states that from 2019, the French Environment and Energy Agency (ADEME) will invest 100 million euros in the deployment of hydrogen in the industrial, transportation and energy sectors, the main road map for action is as follows.

1) Creating a carbon-free industry

Target: a complete hydrogen traceability system will be built in 2020, and the use of hydrogen in the industry will reach 10% in 2023 and 40% in 2028.

2) Developing renewable energy storage capacity

Through the electrolysis unit, hydrogen service is provided in real time for areas with hydrogen demand; the technical and economic conditions for creating a hydrogen network are created. The production of electrical energy from renewable energy sources to produce hydrogen to achieve power-to-gas conversion, gas transporters and distributors determine the corresponding technical and economic conditions, and make an interim report at the end of the year.

3) Achieving zero traffic emissions

A large-scale hydrogen transportation system, with 5,000 light commercial vehicles, 200 heavy-duty vehicles and 100 hydrogen refueling stations will be further deployed in 2023; in 2028, the scale of light commercial vehicles will be expanded to 20,000–50,000 vehicles, heavy vehicles (buses, trucks and ships, etc.) will be expanded from 200 in 2023 to 800–2,000. In 2028, the hydrogen refueling station will be further increased to 400–1,000. The further work is as follows: supporting the development of a series of heavy road vehicles, developing innovative applications with existing tools; further improving the deployment and management of hydrogen ecosystems, and promoting project development; further clarifying safety and risk precautionary provisions on the basis of existing work.

6. Roadmap of Hydrogen Energy Development in China

China is the country with the most hydrogen utilization. Since it exceeded 10 million tons for the first time in 2009, it has maintained its number one position in the world for nine consecutive years. In recent years, China has attached great importance to the development of hydrogen energy industry and has introduced a series of related industrial policies that are conducive to the development of hydrogen energy. In

April 2016, the National Development and Reform Commission and the National Energy Administration jointly issued the "Energy Technology Revolution Innovation Action Plan (2016–2030)", which proposed a roadmap for key innovation actions in the energy technology revolution. The roadmap contains three strategic directions and six innovative actions.

1) Three strategic directions

(1) Hydrogen production, storage and transportation and hydrogen refueling stations. Focus on R&D and technical innovative in large-scale hydrogen production, distributed hydrogen production, hydrogen storage and transportation materials and technologies, and hydrogen refueling stations.
(2) Advanced fuel cells. Focus on R&D and technical innovative in hydrogen/air polymer electrolyte membrane fuel cells, methanol/air polymer electrolyte membrane fuel cells.
(3) Distributed power generation of fuel cells. Focus on R&D and technical innovative on proton exchange membrane fuel cells, solid oxide fuel cells, metal air fuel cells, and integrated design and system integration of distributed hydrogen and fuel cells.

2) Six innovative actions

(1) Large-scale hydrogen production technology. Conducting research on low-cost hydrogen production technology based on renewable energy and advanced nuclear energy, focusing on breaking through key technologies such as solar photolysis hydrogen production and thermal decomposition hydrogen production, building demonstration systems; breaking through high-temperature iodine-sulfur cycle water decomposition hydrogen production and high-temperature electrochemical hydrogen production, completing the design of commercial high-temperature nuclear energy water decomposition hydrogen production program. Develop a new generation of coal catalytic gasification hydrogen production and methane reforming/partial oxidation hydrogen production technology
(2) Distributed hydrogen production technology. Study renewable energy power generation and proton exchange membrane/solid oxidation type battery electrolysis water hydrogen integration technology, break through the core technology of materials and components such as high-efficiency catalysts, polymer membranes, membrane electrodes and bipolar plates, and master the rapid variable load of renewable energy high-efficiency medium-pressure electrolysis hydrogen-making electrolysis cell technology, conduct demonstration and promotion of hydrogen production from renewable energy electrolyzed water; perform research on high-efficiency catalytic hydrogen production technology for traditional energy and chemical products such as distributed natural gas, ammonia, methanol and liquid hydrocarbons and the process, with high efficiency and low cost membrane reactor hydrogen production and hydrogen purification technology, form a standardized hydrogenation station on-site hydrogen production mode and demonstration applications.

(3) Hydrogen storage and transportation technology. Develop 70 MPa grade carbon fiber composite materials and hydrogen storage tank equipment technology, hydrogen storage station with high pressure and liquid hydrogen storage technology; study light-weight hydrogen storage materials of low development cost, good cycle stability, use temperature close to the fuel cell operating temperature such as nitrogen-based, boron-based, aluminum-based, magnesium-based and carbon-based; develop long-distance, large-scale hydrogen storage and transportation technology using liquid compounds and ammonia as hydrogen storage media, design and develop high activity, high stability and low cost hydrogenation/dehydrogenation catalyst.
(4) Hydrogen/air polymer electrolyte membrane fuel cell technology. Aiming at the major demand for clean and efficient new energy power sources, we will focus on low-cost long-life electrocatalysts, polymer electrolyte membranes, orderly membrane electrodes, high-consistent stack bipolar plates for hydrogen/air polymer electrolyte membrane fuel cells. Key technologies such as modular system integration, intelligent process detection control, and hydrogen source technology to solve key problems such as hydrogen/air polymer electrolyte membrane fuel cell performance, life, cost, etc., and realize hydrogen/air polymer electrolyte membrane fuel cell electric vehicle demonstration operation and promotion.
(5) Methanol/air polymer electrolyte membrane fuel cell technology. Aiming at the major demand for clean and efficient new energy power sources, we will focus on core technologies such as the high temperature and long life electrocatalysts of methanol/air polymer electrolyte membrane fuel cells, new high temperature resistant polymer electrolyte membranes, orderly membrane electrodes, integrated organic fuel reforming, integrated optimization and intelligent control of stack system under high temperature conditions, and demonstration operation and promotion of methanol/air polymer electrolyte membrane fuel cells for electric vehicles (no hydrogen production, hydrogen storage, hydrogen refueling station).
(6) Fuel cell distributed generation technology. Focus on the development of proton exchange membrane fuel cell hydrogen source technology, solid oxide fuel cell technology, and metal air fuel cell technology. Under the conditions of decentralized power stations, we will break through key technologies such as proton exchange membrane fuel cells, solid oxide fuel cells, key materials for metal air fuel cells, core components, system integration and mass balance management, and establish a distributed power generation industrialization platform, to realize the promotion and application of the proton exchange membrane fuel cell system of magnitude of kilowatts to 100 kW in the fields of communication base stations and decentralized power stations; to realize the demonstration application of the distributed energy system for power generation of 100 kW to MW solid oxide fuel cells, the power generation efficiency is over 60% and to develop distributed power plants suitable for remote cities, industrial and mining enterprises, etc.; to realize demonstration operation or scale application of

metal air fuel cell systems in the fields of intelligent microgrid, communication base stations and emergency rescue.

Hydrogen energy and fuel cell technology innovations indicate that the hydrogen energy industry has been incorporated into the national energy strategy; in June 2016, the National Development and Reform Commission, the National Energy Administration, and the Ministry of Industry and Information Technology jointly issued the "Made in China 2025—Energy Equipment Implementation Plan". The fuel cell and hydrogen energy development and utilization as a special chapter, listed as the disruptive technology leading the industry change by the National Innovation Driven Development Strategy Outline (2016); in 2017, the National Energy Administration approved a number of hydrogen production and energy storage project of abandoned winds and abandoned light.; in January 2018, the Ministry of Science and Technology has included renewable energy and hydrogen energy technologies as key special projects and actively supported research. On February 11, 2018, an interdisciplinary, cross-industry, and inter-departmental national-level industry alliance led by the National Energy Group and the State Grid Corporation of China, China Hydrogen Energy and Fuel Cell Industry Innovation Strategic Alliance (referred to as the "China Hydrogen Energy Alliance") was established in Beijing, and many state-owned enterprises took the lead in participating, marking the opening of China's large-scale commercialization of hydrogen energy.

Driven by national strategic orientation and market development, the investment scale of China's hydrogen fuel cell industry is growing rapidly. According to statistics, in 2017 alone, the hydrogen fuel cell project (in production in 2020) will have an investment of more than CNY 100 billion and a capacity of 170,000 hydrogen fuel cell engines.

3) The "Hydrogen Energy China" Strategy

China is a large country in the world of hydrogen energy utilization. It has formed seven hydrogen energy industrial clusters such as Beijing-Tianjin-Hebei, Yangtze River Delta, Pearl River Delta, Central China, Northwest China, Southwest China and Northeast China. By consuming surplus new energy such as abandoned hydro, abandoned wind, and abandoned light, reducing utilization of fossil fuels such as coal, oil and natural gas, and the clean and efficient utilization of coal will gradually lower costs and steadily improve safety. Through the "three major development stages", we will establish a support system that is conducive to the development of the hydrogen energy industry, and build a nationwide network of hydrogen energy supply and utilization infrastructure.

(1) In the near term (to 2030), a major breakthrough will be made in the development of fossil-based hydrogen energy industry represented by coal-to-gas, and the industrial chain demonstration was initially completed. At present, it is necessary to speed up the evaluation of China's coal underground gasification hydrogen production resources, the series of technical research and field tests of economical and efficient hydrogen production, and the ultra-deep, supercritical

water gasification hydrogen production technology preparation, especially to deepen the research on the mechanism of high-efficiency hydrogen production. Research and development of geological evaluation, engineering technology, monitoring and control, safety and environmental protection, as well as major equipment such as high temperature and high pressure downhole tools and high-strength anti-corrosion pipes. The potential of underground coal gasification in China is huge. Only the coal resources with a depth of 1,000–2,000 m in the Ordos Basin will reach 1.3 trillion tons. It is conservatively estimated that the commercial industrial hydrogen producible through gasification will be about 10 trillion cubic meters (equivalent to 0.9 billion tons). China's development of gasification industry of coal underground should be planned according to shallow layers (formation pressure below 10 MPa), mid-deep layers (formation pressures between 10 and 22 MPa) and ultra-deep, supercritical water (formation pressures over 22 MPa), and the basins such as Erdos, Erlian and Junggar are chosen and carried out for on-site experimental research.

(2) In the medium term (from 2035 to 2050), the hydrogen energy industry will become an important part of China's new economic growth point and new energy strategy; creating new materials, energy storage and hydrogen energy industry chains; strengthening the preparation of new materials such as graphene and nano-super in advance. Through research and development and large-scale production of high-standard, high-performance automotive and marine batteries through independent, cooperative, technology purchase, and high-quality enterprise mergers and acquisitions, we will cooperate with major auto manufacturers or participate in the equity for the promotion of standardized battery utilization in the transportation field. Give full play to the advantages of enterprise gas station layout, build a large warehouse-type charging center, quickly establish a network of high-efficiency charging stations for new energy vehicles, and seize the opportunity of the energy revolution in the transportation sector. Taking the full advantages of oil pipeline layout, develop the low-cost electrolytic hydrogen production from abandoned wind, abandoned hydro and abandoned light, transport hydrogen through natural gas pipeline network, and develop hydrogen-rich natural gas, liquefied hydrogen, hydrogen refueling station and other services.

(3) In the longterm (from 2050 to 2100), hydrogen energy will become an important part of China's energy consumption structure, relying on new energy sources to achieve national "energy independence". China's "energy independence" concept refers to the realization of basic self-sufficiency and consumption safety of energy production through China's new energy production revolution. As for China's primary energy production in 2017, coal accounted for 70%, oil accounted for 8%, natural gas accounted for 5%, and new energy accounted for 17%. The innate conditions of China's abundant coal resources but relatively insufficient oil and gas have determined that energy production and consumption must have Chinese characteristics and constitute China's energy structure of "one major component and three non-major components" (coal is the major component and oil/natural gas/new energy are non-major components). The growth rate of solar energy and wind energy production is the fastest, and the

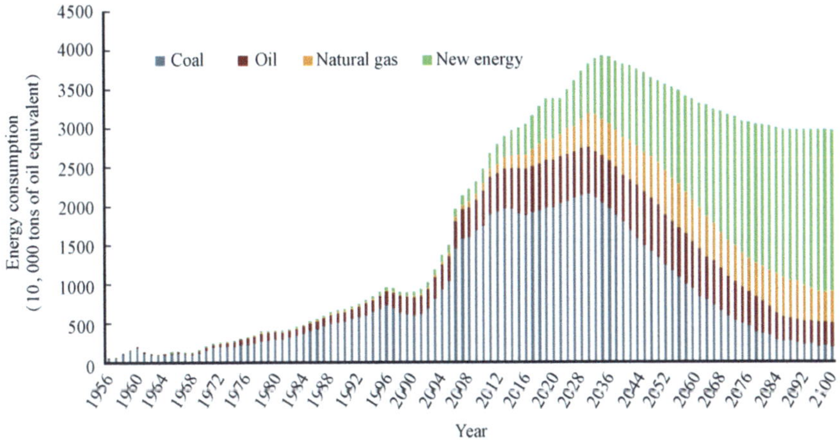

Fig. 8.9 Primary energy consumption in China from 1965 to 2100

proportion of hydropower and nuclear power production is the highest. Hydrogen energy, energy storage, new materials and new energy are the most disruptive. The clean coal utilization and the early arrival of "two large scales" of new energy should be accelerated to reduce the time span and safety pressure of oil and gas in China's energy utilization path. China's "energy independence" may be achieved when the new energy takes the leading position (Zou 2018).

China needs to plan to accelerate the realization of the conventional-unconventional oil and gas production revolution, the clean revolution of coal development and the speed revolution of new energy development, and strives to achieve the energy structure transformation revolution of "three major components" such as coal, oil and gas, and new energy from "one major component and three non-major components" around 2050. At that time, coal accounts for about 40% of primary energy consumption, oil and gas accounts for 30%, and new energy accounts for 30%.

By around 2100, it is possible to rely on new energy sources to achieve national "energy independence". The proportion of fossil energy in the primary energy consumption structure drops to 30%, and non-fossil energy accounts for 70%, realizing the position change of the two roles (Fig. 8.9).

8.2.4 Development of Hydrogen Energy

1. Global hydrogen refueling station construction enters the corner overtaking stage

According to the report data released by the website H2stations.org in February 2017, there were 92 new hydrogen refueling stations in the world in 2016, 38 more hydrogen refueling stations than in 2015, with a growth rate of 70%, a record high growth rate. The construction of hydrogen refueling stations has entered a stage of rapid development (BP. BP World Energy Statistical Yearbook 2017).

In August 2017, there were 286 hydrogen refueling stations operating worldwide, including 112 in Europe and 109 in Asia. Japan, the United States, Germany, and China are the four countries with the largest number of hydrogen refueling stations, 91, 60, 31, and 12 respectively. The total hydrogen refueling stations built and under construction in China are 31, of which 12 are in operation.

In the next five years, the global hydrogen refueling station will enter a rapid development stage. By 2020, there will be more than 435 global hydrogen refueling stations, and more than 1,000 in 2025. Japan, Germany, the United States and China will have 320, 400, 100 and 100 respectively.

With the continuous advancement of the hydrogen energy economy, the hydrogen energy society has gradually become clear. In the future, hydrogen refueling of new energy vehicles is no longer a problem. Especially in China's coastal developed areas, a hydrogen station for every 100 km will become a reality, which can meet the needs of hydrogen vehicles. The construction plans for hydrogen refueling stations in major countries of the world are shown in Table 8.6.

By 2030, the number of China's hydrogen refueling stations is expected to reach 1,000, and the number of fuel cell vehicles will reach 2 million. By 2050, the construction of China's hydrogen refueling station network will be completed, and the number of fuel cell vehicles will reach 10 million.

2. Hydrogen production from electrolysis water runs through the development of the hydrogen energy industry chain

With the rise of the global hydrogen economy, the demand for hydrogen has increased dramatically, and hydrogen production technology is changing with each passing day. Although hydrogen production from coal produces a lot of carbon dioxide, it is still the best way to achieve hydrogen production with large scale and low-cost due to abundant coal resources and relatively low price. New paths such as carbon dioxide flooding and storage can be found to effectively solve the problem of carbon dioxide, so hydrogen production from coal will continue until 2050. Blast furnace flue gas, chemical tail gas, etc. contain a large amount of hydrogen, which can achieve low-cost recovery through PSA (pressure swing adsorption) technology, and is also an important component of future hydrogen sources. Solar (photocatalytic, photothermal) hydrogen production technology is the ideal hydrogen production

Table 8.6 Construction plan of hydrogen refueling stations in major countries of the world

Year	2020	2025	2030	2050
Japan	160	320	–	–
South Korea	80	210	520	–
Germany	–	400	–	–
United States	–	100	–	–
China	100	–	1000	Hydrogen station network completed

technology in the future, but due to problems such as conversion efficiency and cost, it may be difficult to produce in large scale before 2030 (Hamburg: Shell Deutschland Oil GmbH 2017).

Hydrogen production from electrolyzed water will continue throughout the wave of hydrogen energy industry chain development (Table 8.7 and Fig. 8.10), which is the main source of hydrogen in hydrogen energy society. Hydrogen produced by electrolyzed water can effectively absorb wind power, photovoltaic power generation and other peak and valley power. As technology continues to evolve, the cost of hydrogen produced by electrolyzed water will gradually decrease to a level of commercial acceptance (Xin 2015).

Hydrogen production from electrolyzed water is also the best way to achieve distributed hydrogen production. A single gas station can build a small electrolysis water hydrogen production unit, or a centralized hydrogen production unit can be built. A medium-sized electrolysis water hydrogen production station can meet the

Table 8.7 Forecast schedule of major sources of hydrogen in 2050

Year	2018	2020	2030	2050	2100
Electrolyzed water hydrogen production	4	8	20	40	50
Natural gas, methanol hydrogen production	78	60	35	10	0
Coal hydrogen production	18	20	30	35	30
Industrial exhaust gas hydrogen production	0	5	10	10	10
Solar energy hydrogen production	0	0	0.1	5	10
All others	0	7	5	0	0

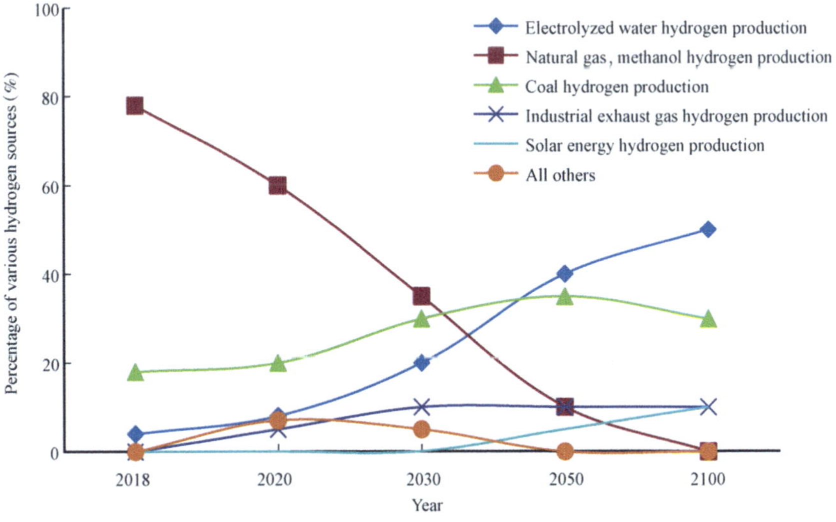

Fig. 8.10 Forecast trend of main sources of hydrogen in 2050

Table 8.8 Forecast schedule of hydrogen storage mode in 2050

Year	2020	2030	2050	2100
High pressure gaseous hydrogen storage (%)	90	75	40	25
Low temperature liquid hydrogen storage (%)	9	20	45	55
Organic liquid hydrogen storage (%)	1	5	10	15
Solid state hydrogen storage (%)	0	0	5	5

hydrogen consumption of dozens of hydrogen refueling stations in the nearest area to realize the intelligent interconnection of hydrogen energy.

On April 16, 2018, the H2FUTURE project consisting of VAI, Siemens, Verbund and the Austrian power grid, together with research partners K1-MET and ECN, reached an agreement to build the world's largest "green" electrolytic hydrogen production pilot plant with capacity of 6 MW.

3. Liquid hydrogen will be the main form of hydrogen storage in 2050

At present, hydrogen is mainly stored in a high-pressure gaseous state, and a small portion is in a liquid state. Liquid hydrogen is mainly used in the aerospace industry due to cost issues. As technology continues to mature, liquid hydrogen is expected to be the primary form of hydrogen storage in 2050 (Tables 8.8).

There are two main ways of transporting hydrogen: the first one is the utilization of dedicated hydrogen pipelines. Due to the development of electrolyzed water hydrogen production technology, blockchain distributed hydrogen production will become the mainstream, and hydrogen produced by hydrogen production stations is directly sent to the hydrogen refueling station after being compressed through the compressor, and the pipeline transportation of hydrogen is mainly within close distance; the second one is to use filling and transportation, and the hydrogen produced by the hydrogen production station is processed into a liquid state by low temperature, and is transported to the hydrogen refueling station by a special liquid hydrogen transportation tanker.

References

BP. 2017. BP World Energy Statistical Yearbook.

Chen, Shuoyi, Weidong Zhu, Li Zhang, et al. 2018. Development status and trend of hydrogen fuel cell technology. *Science and Technology of China* 5: 11–13.

China National Institute of Standardization, National Hydrogen Standardization Technical Committee. China Hydrogen Industry Infrastructure Development Blue Book (2016). Beijing: China Quality Inspection Press, China Standard Press. 2016.

Guan, Genzhi, Xiaoqiong Zuo, and Jianpin Jiag. 2012. Nuclear power technology. *Hydropower and New Energy* 1: 7–9.

National Development and Reform Commission, National Energy Administration. Energy Technology Revolution Innovation Action Plan (2016–2030). Beijing: National Development and Reform Commission. 2016.

National Development and Reform Commission. The 13th Five-Year Plan for Power Development. 2017.

National Energy Administration. Guidance on the implementation of the "13th Five-Year Plan" for renewable energy development. 2017.

Wang, Shaoyong. 2015. Some thoughts on new energy development by oil companies. *Sino-Global Energy* 20(8): 18–22.

Shell Deutschland Oil GmbH. 2017. New energy of the future.

Shell Group of Companies. 2017. World energy model.

Wang, Geng, Jinyang Zheng, Lijun Jiang, et al. 2017. The development of hydrogen energy in China. *Science and Technology Review* 35 (22): 105–110.

Xin, Peiyu. 2015. *Comprehensive evaluation and application prospects of solar power generation technology*, 5–21. Beijing: North China Electric Power University.

Zou, Caineng. 2018. Energy revolution and oil company transformation strategy of new era. *Journal of Beijing Petroleum Managers Training Institute* 25 (4): 3–15.

Chapter 9
Energy Storage and New Materials

Energy storage technology is the key to achieve sustainable energy development and can be used in power, transportation, and industrial production. Large-scale energy storage systems are a key part of smart grid construction. To a certain extent, the application of energy storage technology determines the development level of renewable energy and determines the success or failure of energy Internet. At the same time, the discovery, invention and application of new materials are inseparable from the technological revolution and industrial transformation. New materials will play an irreplaceable foundation and support role in the smart grid. Accelerating the development of new materials is of strategic importance for promoting technological innovation, supporting industrial upgrading, and building a strong manufacturing country (Qi et al. 2007; Wu et al. 2017).

9.1 Energy Storage

Energy storage technology plays a very important role in energy transformation and is an important means to meet large-scale access to renewable energy. Traditionally, fossil energy power station systems will generate, transmit and distribute electricity according to their electricity demand. For new renewable energy sources (such as wind and solar), power generation systems are intermittent and fluctuating due to natural conditions, direct grid connection will cause great security risks to the grid. The energy storage technology can effectively solve the volatility and intermittent problems of renewable energy generation such as solar energy and wind energy, smooth the power station output, reduce fluctuations, improve the reliability and safety of power grid operation, and enable efficient utilization of energy such as wind energy and light energy. Energy storage is also an important part of the distributed energy system and electric vehicle industry, and plays an important role in the energy Internet. As more and more countries put energy storage technology into strategic planning, the investment scale of the energy storage market continues to increase, and the bottleneck of research and development of emerging technology continues

C. Zou, *New Energy*, https://doi.org/10.1007/978-981-15-2728-9_9

to break through. The industrial chain of the energy storage industry and its business model will gradually mature.

9.1.1 Current Status of Energy Storage Development

Energy storage refers to the technology of utilize certain media to store energy such as electric energy in a certain form, and then release it into power generation when there is market demand. Energy storage technology can be mainly divided into three categories, physical energy storage (such as pumped storage, compressed air energy storage, flywheel energy storage, etc.), chemical energy storage (such as lead-acid batteries, redox flow batteries, sodium-sulfur batteries, lithium-ion batteries, etc.) and electromagnetic energy storage (such as superconducting magnetic energy storage, supercapacitor energy storage, etc.). Physical energy storage is the most mature, lowest cost, and largest scale energy storage method. Chemical energy storage is the most widely used and most potential energy storage technology (Yan et al. 2017).

The application of energy storage technology mainly includes five major areas: (1) power generation and auxiliary services, involving auxiliary dynamic operation, replacing or delaying new units, frequency modulation, voltage regulation, peak shaving, and spare capacity; (2) renewable energy grid connection, mainly involving peak shaving and valley filling, tracking plan output, smoothing output, etc.; (3) user-side electricity fee management, power quality, distributed power generation and micro-grid applications; (4) power transmission and distribution, such as reactive power support, mitigation line blocking, delay expansion and upgrade, substation DC power supply, etc.; (5) electric vehicles, such as orderly charging, power battery cascade utilization.

The stages of development and application of different energy storage technologies are currently different. From the energy storage technology maturity curve drawn by the Australian Renewable Energy Agency (Fig. 9.1), pumped storage is the most mature technology, and the application of energy storage technology in the global grid connection, the proportion of pumped storage capacity reaches 99%. Secondly, air compression technology, sodium-sulfur batteries and lithium-ion batteries are in the application stage, and their future costs are expected to be further reduced. The types of lithium-ion batteries that have been applied or are in the research and development stage are also different, including lithium cobalt oxide batteries, lithium nickel cobalt manganese oxide batteries, lithium manganese oxide batteries, lithium titanate batteries, lithium iron phosphate batteries, and lithium polymer batteries, etc. Compared with other energy storage technologies in the application stage, lithium-ion battery energy storage technology has a faster response time and can supply power to the grid in a timely and continuous manner to ensure the stability of the power grid. At present, the research hotspot of lithium-ion battery technology is to increase the energy density, thereby reducing the cost and meeting the economic

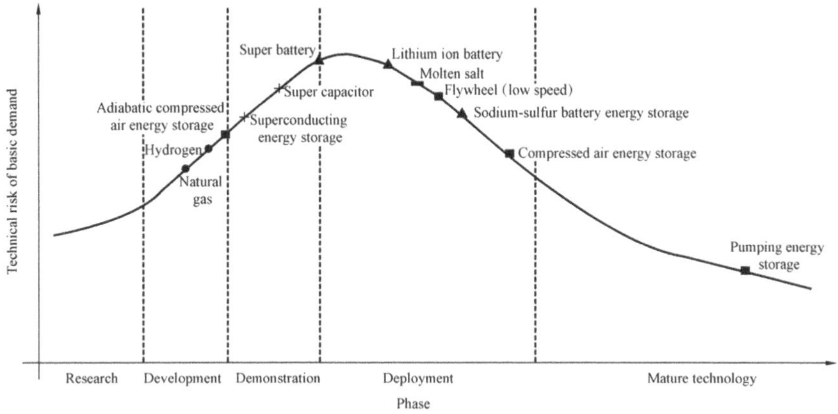

Fig. 9.1 Energy storage technology maturity curve

requirements of commercial applications. Lead-acid battery manufacturers are optimizing their technology to develop advanced lead-acid batteries by merging carbon in the electrodes, combining supercapacitors or other means.

As of the end of 2015, the installed capacity of energy storage projects in the world has reached 146.1 GW. China's accumulated energy storage projects have an installed capacity of 23.7 GW, ranking second in the world. Chile is the only South American country to enter the top 10 global energy storage capacity, and Germany is the country with the largest installed capacity in Europe.

Pumped storage applications are the largest. Pumped storage is the world's largest installed energy storage technology, accounting for 98% of the world's total energy storage capacity, Japan, China, the United States installed the world's top three installed capacity. The scale of pumped storage has reached 300 MW, which is the most mature energy storage technology. As of December 2015, the total installed capacity in the world reached 142.1 GW. The world average of pumped storage units in a country's total installed capacity is around 3%, and some countries have more than 10%, such as Austria (16.3%), Switzerland (12%), Italy (11%), Japan (10%), France (13%) and Germany (11.2%). By the end of 2015, there were 327 energy storage projects in operation worldwide, and the installed capacity increased from 50 MW in 2005 to 950 MW in 2015, and 180 projects under plan and under construction.

Compressed air energy storage technology is moving towards industrialization. As the current storage technology of compressed air storage in addition to pumped storage, the largest capacity and most mature energy storage technology has attracted much attention from the industry. The international approach to isothermal compressed air energy storage technology has made breakthroughs, and small air compressors are in a small-scale trial stage. Compressed air energy storage has now been scaled up in commercial applications in Germany. In terms of new compressed air energy storage, only the Institute of Engineering Thermophysics of the Chinese

Academy of Sciences (1.5 MW supercritical compressed air energy storage, 10 MW advanced compressed air energy storage), and General Compression Company of the United States (2 MW regenerative type compressed air storage), US SutainX (1.5 MW isothermal compressed air energy storage) and UK High View Power (MW scaleliquid gas storage) have a megawatt production design capability. The Institute of Engineering Thermophysics of the Chinese Academy of Sciences has successfully developed the first 1.5 MW supercritical compressed air energy storage system with independent intellectual property rights in China, which is more than 10% more efficient than the traditional compressed air energy storage system. It has opened up new space for the development of the energy storage application of China's power grid.

Electrochemical energy storage, mechanical energy storage and thermal energy storage are also developing rapidly. In 2015, the installed capacity of thermal storage accounted for 43% of the above three energy storage methods. The cumulative installed capacity of Spain, the United States and South Africa was more prominent, accounting for 97% of the global thermal energy storage. The global electrochemical energy storage project has an installed capacity of 910 MW. The cumulative installed capacity of electrochemical energy storage in the United States, Japan and China ranks among the top three in the world, accounting for 43%, 33% and 11% respectively. All vanadium redox flow batteries have made significant progress in the design and integration of key materials, stacks, battery systems, and the industry chain has gradually improved. The overall industry has entered the initial stage of marketization, and has gradually started to replace lead-acid batteries in Japan, Canada, the United States, Australia and other countries; lithium-ion batteries are still a hot research topic in the field of energy storage, and electric vehicles have become an important factor driving the development of lithium-ion battery technology. At present, for lithium ion batteries, the positive electrode materials lithium iron phosphate and nickel cobalt manganese ternary materials are the research focus, and the negative electrode materials nano silicon and graphene are research hotspots. The heat storage technology has developed rapidly, and some thermal energy storage technologies have been very mature, especially sensible heat storage, but the market size is still small, mainly due to the high cost of heat storage, and the society lacks sufficient attention to heat storage. It is estimated that heat storage systems can save 30–40% of the world's energy.

The scale of hydrogen fuel cell applications is gradually expanding. Hydrogen fuel cells are still the mainstream of fuel cell development, and related technologies have basically met industrialization requirements, and are used in small-scale transportation tools such as trains, buses, logistics vehicles, forklifts, drones and other transportation tools.

Energy storage technology has been regarded as an important part of the power system "power generation-transmission-power substation-distribution-utilization". Based on the energy Internet architecture, there are two application modes of energy storage in the energy Internet: (1) Wide-area energy network applications. In the backbone network, large-scale energy storage technology is used to coordinate centralized energy production, participate in wide-area energy management, provide energy buffer for large-scale energy production and transmission, provide support for system wide-area energy dispatch, and maintain system supply and demand balance. The operating entities of large-capacity energy storage directly participate in the energy trading market, flexibly purchase or sell energy according to changes in energy market prices, or provide adjustment services. (2) Local energy network application. In the local energy network, the energy storage and energy conversion devices cooperate to maintain the economical and efficient operation of the system. The local energy network management system combines the energy storage information with the energy price information to make decisions on production and consumption of energy in the local area network, and buy or sell energy from the energy market. In virtual energy station (VEP) applications, it is difficult to predict the behavior of distributed producers. Therefore, it is difficult to manage distributed power and electric vehicles, so it is of great significance to introduce energy storage to the management and operation of VEP. Under the influence of energy storage, dispersed energy producers have more credible energy supply capabilities, which enable them to participate in energy market transactions.

9.1.2 Industrial Policy of Energy Storage

Energy storage may be the basis for realizing future energy system changes. Developed countries or regions such as the United States, Japan, and Europe have been conducting R&D from the key technical areas of energy storage at the national level. Relevant technology research and development and demonstration activities are progressing rapidly. Utility companies such as power grid operators and large energy equipment manufacturers and some small and medium-sized technology companies are optimistic about the market prospects of the energy storage industry, and have entered the field.

1. International Energy Agency (IEA)

In 2014, the International Energy Agency (IEA) released the "Energy Storage Technology Roadmap", which includes: investigating and classifying the advantages of energy storage technologies in energy systems; exploring new methods to make use of its energy storage advantages, reduce costs, identify barriers to deployment and conduct competitive analysis of other technologies. The main research techniques include electricity storage (mechanical conversion, chemical conversion, etc.) and thermal storage (water/ice thermal storage, thermochemical storage).

2. The United States

The US energy storage industry has developed rapidly and the relevant supporting policies are relatively complete. Since 2009, the US government has gradually introduced various policies directly related to energy storage. For example, the Energy Storage Plan of US Department of Energy from 2011 to 2015 investigates R&D,

demonstration projects and commercialization and sets corresponding short-term and long-term goals. Act No. AB2514 defines the important role of energy storage systems in peak shaving of power grids, accessing renewable energy, reducing power costs and reducing greenhouse gas emissions; the Federal Energy Management Commission (FERC) No. 745 compensates for the demand response resources for the wholesale market for electricity supply in the United States; Federal Energy Regulatory Commission (FERC) Order 755 requires regional transmission organizations (RTOs) and independent system operators (ISOs) to compensate suppliers that can provide rapid and accurate FM services, not just paying at the base price; the Self-Generation Incentive Program-California (SGIP-CA) subsidizes distributed power generation technology installed at the client; Renewable and Green Energy Storage Technology Act of 2009 (S.1091) standardizes investment tax cuts for grid-side and user-side energy storage equipment; American Recovery and Reinvestment Act (ARRA) funded 16 energy storage technology demonstration projects with 185 million dollars.

In September 2014, the Electricity Advisory Committee adopted the 2014 Energy Storage Plan Evaluation Report, which reviewed and summarized the US Department of Energy's strategy and actions on energy storage plans, and proposed relevant suggestions for the Department of Energy to develop and implement its energy storage projects. It is recommended to focus on the research or evaluation of the following aspects: storage connection in the transport layer; pumped storage and compressed air energy storage technology; power electronics costs; regulation and market design and its impact on energy storage.

3. Japan

Japan is an energy-poor country. It has developed renewable energy and energy storage earlier, and has long supported the development of energy storage technology. Its battery energy storage technology has been commercialized.

In 1974, the Japanese government promulgated the "Daylight Plan" to start the development of new energy technologies. In 1978, the "Moonlight Plan" was introduced, and a large amount of research and development funds were invested in the development of energy storage technology. In 2008, the Ministry of Economy, Trade and Industry of Japan proposed a high-performance energy storage technology roadmap under the framework of "Cool Planet—Energy Technology Innovation Project". The technology roadmap focuses on energy transmission issues, including an efficient high-conductivity transmission technology roadmap, a high-performance power storage technology roadmap, and a hydrogen energy production, transportation and storage technology roadmap. For high-performance power storage technology, through the development of new concept batteries (such as metal air batteries), enhanced lithium batteries, sodium-sulfur batteries, redox flow batteries, nickel metal hydride batteries, enhanced nickel-hydrogen batteries, new concept capacitors, which greatly improves performance and reduces costs. The main associated technologies include residential energy management systems, building energy management systems, and local-level energy management systems. For the hydrogen production, transportation and storage technology route in Japan, hydrogen production from hydrolysis, fossil fuel hydrogen production to renewable energy hydrogen production and photocatalytic hydrogen production can greatly save costs. Hydrogen transportation technology is transformed from compressed hydrogen transportation, liquid hydrogen transportation, and pipeline transportation to high-pressure transportation and liquid transportation, which can greatly improve transportation efficiency and safety. The hydrogen storage technology consists of ultra-high pressure vessels, liquid hydrogen vessels to complexes and organometallic structures, which can greatly save costs, extend storage time and improve safety. The main associated technologies include: hydrogen supply technology (building small hydrogen refueling stations, local and national-scale hydrogen fuel supply systems), fuel cell electric vehicles and fuel cell facilities.

In May 2014, the Ministry of Economy, Trade and Industry launched a new round of subsidy programs for lithium-ion battery energy storage systems, with the aim of increasing the utilization ratio of renewable energy, effectively managing peak loads, improving power stability, and helping the government measure the impact of large-scale production on battery costs. In April 2016, the "Energy and Environmental Technology Innovation Strategy 2050" issued by the Japanese government included energy storage in it, pointing out that it is researching low-cost, safe and reliable advanced battery technology for rapid charge and discharge, its energy density reaches seven times of the existing lithium-ion battery, the cost is reduced to 1/10, and the small-scale electric vehicle will run more than 700 km with this technology. The technology will also be used for renewable energy to achieve a larger scale of renewable energy grid connection.

4. German

In 2011, after the German government decided to abandon the nuclear, the pace of development of renewable energy, mainly wind and solar energy, continued to accelerate, and energy storage also received more attention. The government issued the "latest on-grid tariff policy (self-consumption) tax, which stipulated all photovoltaic systems that are in operation in 2011 will face a 13% subsidy reduction, and the household self-consumption tax on the installation of energy storage equipment has dropped significantly. In addition, the Renewable Energy Act further increases the share of renewable energy generation in the power generation structure and plans to replace three-quarters of traditional energy through green technology in a few decades. In September 2012, a new energy storage agency, the German Energy Storage Association (BVES), was established, claiming that the German energy storage technology roadmap was the top priority. In 2013, KFW and the German Federal Ministry of Environment, Nature Conservation and Nuclear Reactor Safety (BMU) introduced a distributed photovoltaic energy storage subsidy policy, which subsidizes energy storage facilities for photovoltaic power generation, marking Germany's distributed photovoltaic policy from the subsidized power generation unit alone to the energy storage unit that guarantee photovoltaic power generation.

5. China

Although the energy storage industry started late in China, the development speed in recent years is remarkable with the support of the government. According to the survey conducted by the Center for Information and Consultation of the Chinese Academy of Sciences (CICCAS), as of the end of December 2015, the cumulative installed capacity of power storage (excluding pumped storage, compressed air and thermal storage) in the mainland reached 105.2 MW, involving 136 projects (including built, completed, put into operation and under construction).

Since 2015, domestic support policies for the energy storage industry have been intensively introduced. Energy storage was included in the 100 major projects of "Thirteenth Five-Year Plan", and officially entered the national development plan for the first time. A new round of power reforms, especially the electricity price reform and the electricity sector's open policy, will start new growth points for energy storage.

The National Development and Reform Commission and the National Energy Administration jointly issued the "Energy Technology Revolution Innovation Action Plan (from 2016 to 2030)" and the "Energy Technology Revolution Key Innovation Action Roadmap", which require research on high temperature thermal storage of solar photothermal efficient use, distributed energy system large-capacity heat storage (cold) technology. Research on energy storage technologies for power grid peak shaving and efficiency improvement, physical energy storage technology for regional energy supply applications, renewable energy grid-connected, distributed and micro-grid, electric vehicle energy storage technologies, master key technologies of energy storage technology, complete demonstration verification, the overall technology has reached the international leading level, guiding the international energy storage technology and industrial development.

In June 2016, the National Energy Administration issued the Notice on Promoting the Participation of Electric Power Storage in the pilot work for the Electric Power Auxiliary Service Compensation (Market) Mechanism in the "Three North" Area (Guo Neng Supervision [2016] No. 164). The notice clarifies that the electric energy storage facilities built on the power generation side can participate in the peak shaving and frequency modulation jointly with the units, or participate in the auxiliary service market transactions as independent entities; the electric energy storage facilities built on the user side can be used as independent market entities or with power generation. Enterprises jointly participate in auxiliary services such as frequency modulation, deep peak shaving, and start and stop peaking.

6. Other Countries

In addition, the United Kingdom, France, Spain, Italy, South Korea and other countries have also formulated energy storage plans or corresponding electricity price mechanisms, directly and indirectly support the application and development of energy storage by adopting peak and valley electricity prices and renewable energy feed-in tariffs. For example, in the United Kingdom, the Future Low Carbon Center in 2011 published the "Path to Energy Storage in the UK", which has developed a path for energy storage deployment from three aspects: user-led, distributed, and centralized. Although South Korea's energy storage technology started late, the Korean government has also integrated the development of energy storage into energy development and power development policies. The Korean government has issued the "Strategic Plan for Energy Storage R&D and Industrialization" to increase the construction of energy storage system projects, focusing on research and development of energy storage technologies and industries in Korea; the green energy strategy roadmap for 2011 aims to increase the market share of green energy in South Korea from the current 1.2–18%.

9.1.3 Development Prospects of Energy Storage

1. Development Bottleneck

Energy storage is an indispensable support technology for smart grid, renewable energy access, distributed power generation, microgrid and electric vehicle development. Its application runs through the power generation, transmission and distribution, and power consumption of power systems. The development of China's energy storage market has just started, and there are still the following problems in technology research and development and industrial application.

(1) Technical aspects. Key materials, manufacturing processes and energy conversion efficiencies are common challenges for various energy storage technologies, and further stability, reliability and durability issues must be addressed in large scale applications. Some major technical bottlenecks have yet to be resolved, such as high-load compressor technology in compressed air energy storage, system development is still in the demonstration stage; key technologies such as high-speed motors, high-speed bearings and high-strength composite materials for flywheel energy storage have not yet broken through; the preparation and batching/scale technology of key materials in chemical battery energy storage, especially electrolytes, ion exchange membranes, electrodes, module packaging and sealing, etc., still have significant gaps with international advanced level; high-performance materials and high-power modular technology in supercapacitors and superconducting energy storage medium and high temperature superconducting materials and superconducting current limiting technology have not broken through. At present, all kinds of energy storage technologies have advantages and defects, and there is no technology that is dominant in terms of cost, safety, stability and applicability, and has not yet formed a clear technical route. Although China has formed a pattern of coexistence of multiple technical routes, it lacks application practices. Therefore, it is urgent to carry out research on technological development routes and provide guidance for the development of energy storage industry.

(2) Promotion and application. The application of energy storage in the power system mainly includes peak-shaving and valley-filling, peak-shaving and frequency modulation and spare capacity, alleviating peak power supply tension, delaying investment in new units, investment in transmission and distribution lines, improving power supply quality and reliability, reducing user electricity consumption costs and achieving renewable energy access. However, most of the energy storage projects are still demonstration applications, with short running time, high cost, and lack of clear application direction. It is still unable to carry out perfect economic analysis. Although energy storage systems can achieve multiple applications, due to the complexity of the application scenario, the quantification of multiple benefits is currently difficult to define, increasing the difficulty of commercial promotion.

(3) Policy making. The United States, Japan, Europe and other countries have more mature peak-to-valley tariff policies, which effectively reflect the value of energy storage applications. In addition, the US government has introduced "Pay for Performance" and "Self-Generation Incentive Program" (SGIP) policies. In order to enable energy storage services to participate in the US power market, especially the operation of FM services, commercialization of energy storage in some areas has been realized. In China, government departments such as the Development and Reform Commission, the Ministry of Science and Technology, the Ministry of Industry and Information Technology, and the National Energy Administration have been paying attention to the development of the energy storage industry, and generally define energy storage as a key technical field. However, China's competitive power-assisted service market has not yet formed, affecting the commercial operation and promotion speed of energy storage technology.

(4) Industrial development. In the past 20 years, the United States, Japan and other countries have established mechanisms for the development of energy storage industry according to local conditions. Japan is a country with a lack of resources. Since the sodium-sulfur battery technology of Japan NAGAKI (NGK) can extract sodium and sulfur from seawater without resources constraints, it has received attention. Due to its abundant resources and high degree of marketization of electricity, the United States has increased its support from the application field, and developed a variety of technologies and emphasize complementarity. In particular, power-assisted services and distributed power generation applications have become the focus of policy support. China's energy storage industrialization is starting and the number of demonstration projects is small, the scale is limited, the application time is short and the application scenarios are not rich enough, and the demonstration of energy storage economy is lacking. In addition, China's power system reform has not yet been completed. Under the current power system, it is difficult to define how much energy storage will benefit the participants in the application of power generation, transmission and distribution, and power consumption. Therefore, it is impossible to determine who will bear the cost of the energy storage system. Although China

has already announced some demonstration projects for distributed energy storage, it mainly uses photovoltaic power generation services for industry and commerce and residents, and solving the problem of power supply of resident without electricity, border defense and special operation. Most of the projects are still under planning. There is still a certain gap between the construction and the scale of industrialization.

(5) Standard specifications. Standards are the basis for industrialization of technology and an important factor supporting the healthy development of the industry. Energy storage is an emerging industry. The standards for energy storage at home and abroad are still in the exploration stage. The number of standards is very small, and the establishment of the standard system has just started. At present, many countries are actively formulating energy storage standards. China should also speed up the formulation of energy storage standards, keep pace with international standards, strive for more voices in international standards, and strive to incorporate Chinese technology into international standards and to avoid the phenomenon that the standard lags behind the market.

2. Prospects

Energy storage technology has a great impetus to achieve energy-saving emission reduction targets and optimize energy structure. The development prospects are broad, mainly reflected in: energy storage technology can promote the clean and efficient use of traditional energy, promote energy-saving emission reduction; energy storage technology can promote the application of new energy to optimize energy structure; energy storage technology can promote the development of electric vehicles and green transportation, reduce fuel consumption, optimize energy structure; energy storage technology can promote smart grid construction and promote energy structure optimization.

Under the background of the global promotion of clean energy, the development of energy storage technologies with higher energy density, longer cycle life, lower system cost and better safety performance has become an important direction of research support programs in various countries. Driven by the rapid development of the renewable energy industry, the electric vehicle industry and the energy Internet industry, the energy storage industry is expected to show explosive growth. Renewable energy power storage costs will continue to decrease, and energy storage system performance and technology costs will enter a new phase of virtuous cycle development. The current battery energy storage cost and energy density are still some distance away from people's expectations. From the current research results, battery technology is expected to achieve a major breakthrough and the market prospect is broad.

Energy storage technologies, especially high-capacity energy storage technologies, are critical to the development of the global energy Internet. In the development of the global energy Internet, the core is to persist in the two basic principles of clean development and global allocation. This means that clean energy such as wind power and solar energy will become the main source of electricity. However, wind power and solar power generation are random and intermittent etc., its access to the grid will bring voltage, frequency fluctuations and other power quality problems, and may even affect the safe and stable operation of the grid. Energy storage technology can increase the energy storage link in the power system, making the real-time balanced rigid power system more flexible, thus improving the safety, economy and flexibility of the grid operation.

Forecasts of future global and China's energy storage market scales by major institutions around the world show that the energy storage market has great potential for development: According to estimates by Navigant Research, global commercial and industrial storage will reach 9.1 GW in 2025, while industrial income will reach \$10.8 billion; McKinsey and Company expects the market value of energy storage to reach \$0.1 trillion to \$0.6 trillion by 2025; market research firm IHS predicts that global capacity of grid-connected energy storage system will surge to 21 million kilowatts by 2025, of which lithium-ion batteries will account for more than 80% of the global energy storage market; the "Energy Storage Industry Research White Paper 2016" released by Energy Storage International Summit in 2016 proposes that the maximum installed capacity of China's energy storage will reach 24.2 GW by 2020. In 2020, the market size will reach CNY 100 billion.

China's energy storage industry is still in the initial stage of development, with demonstration and application as the mainstay. According to the International Energy Agency (IEA), the global installed capacity of energy storage will reach 800 GW or more by 2050, and the proportion of total installed capacity will increase to 10–15%. Among them, China's install energy storage capacity will reach 200 GW by 2050, and China's demand for energy storage is huge and urgent.

China's energy storage technology is still maintaining a diversified development pattern. China's energy storage industry has accelerated its pace of development in terms of project planning, policy support and capacity distribution. The energy storage industry has gradually revealed its spring and is poised for growth. The development of China's pumped storage industry is relatively slow, while the growth rate of the electrochemical energy storage market is significantly higher than that of the global market. Photothermal energy storage is still in its infancy. Thanks to technological advances and cost reductions, in the current unsubsidized situation, energy storage has achieved conditional commercial operation in peak-valley spread arbitrage, ancillary services market and renewable energy rationing solutions. According to the statistics of the China Energy Storage Alliance (CNESA) project library, in 2016, a number of large-scale projects were planned or put into operation. The scale of China's newly-invested energy storage projects was 28.5 MW, and the storage capacity scale continued to grow rapidly. At the same time, energy policies have been intensively introduced, and energy storage has gradually become a key area of planning and layout. Local governments have also arranged energy storage projects and demonstrations to boost local industrial transformation and upgrading. In the next few years, with the rapid development of the renewable energy industry, the energy storage market will also achieve rapid growth.

Domestic energy storage applications have not been widely recognized, and different application fields have different technical requirements for energy storage, resulting in different application scales and potentials of various energy storage technologies in different application fields. At present, lead-acid battery technology is the energy storage technology with the widest application range and the highest technical maturity. Sodium-sulfur batteries have the highest proportion of applications in power grid peak shaving, load transfer and spare capacity (rotary standby, etc.) and renewable energy grid-connection. In addition to accounting for a considerable proportion of these fields, lithium-ion battery technology has the most outstanding performance in grid frequency regulation and is the main technical choice for power system frequency modulation. Flywheel energy storage and flow batteries also have the ability to respond instantaneously. There are also some applications in the FM field, but due to their production and cost constraints, their applications are small. On the user side, lead-acid batteries are still regarded as mainstream energy storage technologies, mainly related to the application history and cost advantages of lead-acid batteries, but with the rise of household energy storage, the capacity and performance advantages of lithium-ion batteries will gradually be reflected.

In the large-capacity large-scale energy storage technology, the pumped storage and compressed air technologies are relatively mature, and are suitable for energy storage systems with a scale of more than 100 MW. Sodium-sulfur batteries, vanadium batteries, lithium batteries, lead-acid batteries and flywheel energy storage have begun to be used in megawatt-level grid-connected renewable energy, distributed micro-grid and off-grid projects, as well as charging, discharging, storage, and conversion all in one power stations and so on. China has launched a number of renewable energy grid-connected projects, such as the Zhangbei wind-photovoltaic storage and transportation project, the Dunhuang wind-photovoltaic energy storage demonstration project, the Heishan Tangfang wind farm project in Jinzhou City, Liaoning Province, and the Longyuan Faku Woniushi wind farm project. As China is constantly adjusting its energy mix and developing a proactive renewable energy development goal, the application of future energy storage projects in this area will maintain a rapid development trend.

Driven by policies, renewable energy distributed generation is also a recent development focus. During 2012, China constructed and planned more energy storage projects for distributed microgrid and off-grid, such as Xinjiang Turpan Microgrid Demonstration Project, Nanji Island wind-photovoltaic-diesel complex system, Luxi Island microgrid demonstration project, Chenbarhu Qi distributed power/energy storage and micro-grid experimental research projects, Chang Island solar photovoltaic power generation and energy storage projects in Yantai and Sansha Yongxing Island multi-energy complementary microgrid in Hainan and other projects.

Energy storage projects also have some applications in China's power transmission and distribution sector, such as the South China 10 MW battery energy storage station project, the Baiyin superconducting energy storage substation in Gansu Province, the Anxi mobile lithium battery energy storage power station, and the Meizhou Island energy storage power station in Putian. However, compared with grid-connected renewable energy and distributed power generation energy storage, the number of energy storage projects on the transmission and distribution side is currently small, and the installed capacity is not large.

With the increasing scale of electric vehicles, China has also built a number of integrated power station projects for charging, discharging, storage and replacement, such as Gao'antun electric vehicle charging and replacement station, Xi'an electric vehicle charging station, and Xuejiadao electric vehicle intelligent charging-replacement-storage-discharge integrated demonstration power station in Qingdao.

9.2 New Materials

New materials refer to emerging materials with excellent properties or special functions, or materials with improved performance or new functions after improvements in traditional materials. The new materials are national strategic emerging industries, covering a wide range of industries such as rare earths, magnetic materials, diamond materials, new energy materials, special ceramic materials, optoelectronics, information materials, smart materials and biomedical materials. According to the properties of materials, it can be divided into metal materials, inorganic non-metallic materials (such as ceramics, gallium arsenide semiconductors, etc.), organic polymer materials, advanced composite materials. According to the performance, it can be divided into structural materials and functional materials. The structural materials mainly utilize the mechanical and physical and chemical properties of the materials to meet the high performance requirements of strength, stiffness, hardness, heat resistance, wear resistance, corrosion resistance and radiation resistance; functional materials mainly use the electrical, magnetic, acoustic, optical, thermal and other effects of materials to achieve certain functions, such as electrochemical materials, semiconductor materials, magnetic materials, photosensitive materials, heat sensitive materials, stealth materials and nuclear materials used to manufacturing atomic bombs and hydrogen bombs. The use of new materials is of great significance to the country's construction. For example, the successful development of ultra-pure silicon and gallium arsenide has led to the birth of large-scale and ultra-large-scale integrated circuits, which has increased the computing speed of computers from hundreds of thousands of times per second to more than one billion times per second, and the operating temperature of aero-engine materials increased by 100 °C, the thrust can be increased by 24%, etc. (Zhou 2014; Zhu et al. 2014).

9.2.1 Current Status of New Materials Development

The scale of the global new materials industry continues to expand, and regional differences are increasingly evident. In 2010, the global new materials market exceeded US$400 billion, and by 2016 it was close to US$2.15 trillion, with an average annual growth rate of over 10%. With the rapid growth of the global high-tech industry and the continuous upgrading of the manufacturing industry, as well as the continuous advancement of sustainable development, the demand for new materials will be more vigorous. The products, technologies and models of new materials will be continuously updated and iterated, the market will be broader, and the industry will continue rapid growth. The major innovators of the new materials industry are developed countries and regions such as the United States, Japan and Europe. It has an absolute advantage in economic strength, core technology, research and development capabilities, market share and other aspects, occupying a monopoly position in the global market. Among them, the leading country is the United States, Japan has advantages in nanomaterials, electronic information materials, etc. Europe has obvious advantages in structural materials, optics and optoelectronic materials. China, South Korea and Russia are closely behind them and currently belong to the second tier in the world. China has advantages in semiconductor lighting, rare earth permanent magnet materials, artificial crystal materials, Korea in the display materials, storage materials, Russia in aerospace materials and so on.

Since the "Twelfth Five-Year Plan", the development of China's new materials industry has made great progress. The total output value of the new materials industry has increased from CNY 0.65 trillion in 2010 to nearly CNY 2 trillion in 2015. China has established about 138 new material industry bases. There are three types of distribution of China's new material industry bases: (1) Technology-driven regions. For example, in the Beijing-Tianjin-Hebei region, these regions are centered on the capital Beijing, represented by Zhongguancun, and gathered China's cutting-edge scientific and technological strength, and have produced a number of independent intellectual property rights. This model is the key to independent innovation in the new materials industry. (2) Market-oriented areas. Such as the Yangtze River Delta and the Pearl River Delta, these regions have gained a large amount of foreign investment and international enterprises transfer with their superior geographical advantages, resulting in the export-oriented economy of these regions, thus promoting the development of new materials industry. (3) Areas where resources are supported. This resource based type is mainly concentrated in the central and western regions with abundant resources. The development of these areas is based on local resources. Firstly, develop the basic processing links in the upstream of the industry, secondly, stabilize the initial processing, and finally, upgrade the industry, complete the downstream industries of related materials, that is, further improve the added value of products through the refining of the processed products., which makes it an industrial advantage, and no longer the original resource-based advantage.

New materials will play an irreplaceable foundation and support role in the smart grid. In the field of smart grids, new energy-saving materials, new electrical insulating materials, new smart materials and new energy materials will be continuously developed and applied. There is still a big gap between China and developed countries. (1) Sumitomo Electric and Furukawa Electric of Japan, Bierli of Italy, Nexon of France, and General Cable Corporation of US have developed a series of world-class electrical aluminum and aluminum alloy conductor materials and wire products, and large-scale application and promotion on the transmission line abroad have been conducted. China State Grid Corporation has carried out scientific and technological research from 2010 to 2015, and carried out research and application technology on 61% IACS standard heat-resistant aluminum alloy wire, 59.2% IACS standard medium-strength all-aluminum alloy wire and 63% IACS standard hard aluminum wire and has obtained certain technological breakthroughs; application technology of 0.27 mm oriented silicon steel in high voltage/UHV transformer, 0.18 mm oriented silicon steel R&D technology, high voltage grade transformer oriented silicon steel evaluation technology, application technology of amorphous alloy in relevant research of distribution transformer has been carried out in other fields and is at the international advanced level. (2) Foreign countries are in a leading position in the research of electrical insulation materials and application technologies. They are mastering core technologies in high-grade electrical materials such as epoxy, polyethylene, and liquid silicone rubber, and monopolizing the global market in the field of high-end electrical raw materials. (3) The technical level of Bonner, Honeywell of American, Mitsubishi Group of Japan, Shike of Germany and other well-known sensor companies in the development of intelligent sensing monitoring devices and control systems and their applications in the safe operation of the power grid, intelligent regulation and state monitoring and early warning, etc. is in the leading position in the world and related technology products provide good technical support for the construction of smart grids abroad and intelligent and safe operation. China's intelligent materials for sensors in power grid-related fields are still dominated by traditional functional materials such as silicon, quartz, lead titanate, lead zirconate titanate, zinc oxide, shape memory alloys, lead telluride, and antimony telluride. (4) Foreign countries have carried out a lot of research work on hydrogen storage and fuel cells, and developed countries such as Japan and the United States are in the forefront of the world. Representative units of domestic hydrogen storage materials include the Iron and Steel Research Institute, Zhejiang University, Beijing University of Science and Technology, and the Chinese Academy of Sciences, and the Institute of Applied Chemistry, but the practical process of hydrogen storage new materials is far from the foreign countries. In the case of solid oxide fuel cells, core materials such as $LaMnO_3$, YSZ electrode materials and $ZrO_2–Y_2O_3$ electrolyte materials need to be solved urgently. Domestic research on chemical storage electrode materials such as supercapacitors started late, but the follow-up speed is relatively fast, and a large number of basic research and development and application technology research work needs to be carried out urgently.

New materials technologies are moving towards the development of smaller, smarter, more versatile, environmentally friendly and customizable products and components. In the 1990s, the global gradual upsurge in nanomaterials research began. Since nanotechnology fundamentally changes the manufacturing methods of materials and devices, nanomaterials exhibit many characteristics that are not available in conventional materials in terms of magnetic, optical and electrical sensitivities, and have broad application prospects in many fields. Experts predict that the research and development of nanomaterials will be a technological revolution, which will lead to another industrial revolution in the 21st century.

In 2004, the University of Manchester successfully produced graphene through graphite. People's attention to this magical graphite single-layer two-dimensional material has increased year by year, and a large number of theoretical and applied researches have been carried out. Because graphene has excellent mechanics, thermoelectric performance, which can be used in electric vehicles, aerospace, military equipment, solar cells, touch panels, super-strong structural materials, optoelectronic devices, heat-dissipating films and other fields with a wide range of applications. According to the 2016 global graphene industry research report, although graphene has developed rapidly, it is still in its infancy and has not yet formed a complete industrial chain.

Superconducting materials have a huge market in the fields of electric motors, transformers and maglev trains. For example, the manufacture of electric motors with superconducting materials can increase the ultimate output by 20 times and reduce the mass by 90%. The key to the development of superconducting materials is to increase the critical temperature of the material. If the problem is solved, it will cause major changes in many fields. In 2017, scientists discovered new superconducting materials with critical temperature updates, which made humans take a big step in developing room temperature superconducting materials. It has been found in Japan that magnesium diboride can be a superconductor at −234 °C. This is the metal compound superconductor with the highest critical temperature found so far. The magnesium diboride superconductor is easy to synthesize and easy to process, and it is easy to form a film or a wire. In recent years, American scientists have made significant progress in developing more practical superconducting materials and have begun to enter the practical stage. The Freis by Power Station in Detroit, USA, has laid more than 360 m of superconducting cable underground. The 123 kg wire in the cable is made of oxide superconducting porcelain containing bismuth, strontium, calcium and copper. This is the first practical superconducting transmission line in the world. China has also made major breakthroughs in high-temperature superconducting industrialization technology. At present, production lines of high-temperature superconducting wire have been put into production. It is estimated that by 2017, the scale of superconducting products will be 130 to 160 billion US dollars. However, it should be pointed out that in addition to superconducting materials, there are many supporting technologies that need to be solved, as well as the continued development of high-temperature superconductors, such as room temperature superconducting materials.

High-performance structural materials have the advantages of high temperature strength, wear resistance and corrosion resistance. The high-temperature structural ceramic materials currently under development include silicon carbide, silicon oxide, silicon nitride, boride, toughened zirconia ceramics and fiber-reinforced inorganic synthetic materials. For example, replacing metals with ceramics in an internal combustion engine can reduce fuel consumption by 30% and increase thermal efficiency by 50%. High-performance structural materials can be designed according to requirements to make materials, so that the material can enhance its strengths and avoid its weaknesses. Current research focuses on fiber reinforced plastics, carbon/carbon composites, ceramic matrix composites and metal matrix composites. Polymer functional materials are the fastest growing organic synthetic materials in recent years, increasing at a rate of 14% per year. In addition, American scientists have discovered a compound that binds to glass. This silane compound can adhere to the surface of phosphate glass to form a single molecular layer and multi-molecular layer, which can protect the glass surface and minimize corrosion. This finding is important for improving the corrosion resistance of glass.

With the advancement of science and technology, the scope of new materials continues to expand, and new materials will develop in a higher and newer direction. The chemical industry produces a large number of new chemical materials to provide technical support for the development of new materials. At the same time, the development of new materials can also promote the technological advancement of the chemical industry and the changes in industrial structure.

9.2.2 Industrial Policy of New Materials

In view of the strategic and fundamental role of new materials, new materials technology has become one of the hotspots of competition among countries. To this end, major countries around the world have developed corresponding new material development strategies and research plans.

1. National Manufacturing Innovation Network Strategic Planning and Materials Genome Project in the United States

The United States' leading position in the world of technology benefits from the long-term focus and continued support for new materials research. For a long time, the leading direction of scientific research in the United States has been to serve the defense field. The research and development of materials are mainly concentrated in the fields of national defense and nuclear energy, which have led to the rapid development of related materials applications in the aerospace, computer and information technology industries in the United States. In 1991, the United States proposed to improve the quality of life of citizens, strengthen national security, increase industrial productivity, and promote economic growth by improving material manufacturing methods and improving material performance. The United States listed new materials as one of the areas proposed for national economic prosperity and national security, and placed new material development at the national strategic level.

In order to promote the development of clean energy, the US Department of Energy released the Key Materials Strategy in December 2010 to address safety issues due to origin, weak supply chains and the lack of suitable alternative materials.

In February 2012, the United States issued the National Strategic Plan for Advanced Manufacturing, creating a federal government portfolio of four areas, including advanced materials, to promote the development of advanced materials. At the same time, the National Nano-Plan was released, and eight major support areas such as nanomaterials and nanofabrication were identified.

In 2014, the United States released the Strategic Planning of the Materials Genome Project, which mainly includes 63 directions in 9 fields covering biomaterials, catalysts, photovoltaic materials, energy storage systems, lightweight structural materials, and organic electronic materials.

In 2016, the United States issued the "National Manufacturing Innovation Network Strategic Plan", and established the Light Modern Metal Manufacturing Innovation Institute and the Composite Materials Manufacturing Innovation Institute, focusing on the development of key materials such as advanced alloys, emerging semiconductors, and carbon fiber composite materials.

2. Horizon 2020 Plan and the European Metallurgical Plan of EU

In September 2009, the European Commission published the document "Preparing for Our Future: Overall Strategy of Developing Key Enabling Technologies in Europe", which will include nanotechnology, microelectronics and semiconductors, optoelectronics, biotechnology and advanced materials, which are recognized as a key enabling technology.

In November 2011, the European Commission announced a 7-year "Horizon 2020" proposal with a cost of €80 billion, proposing special support for ICT, nanotechnology, microelectronics, optoelectronics, advanced materials, advanced manufacturing processes, and biology technology, space technology and multi-discipline research of these technologies.

In 2011, the European Union launched the "Accelerative Metallurgy" (ACCMET) project under the EU's Seventh Framework Programme, driven by demand for high-performance alloy materials. In 2014, the European Union proposed the "Graphene Flagship Plan", investing 1 billion euros to support 13 aspects of graphene preparation and application, and launched "Nano Science, Nanotechnology/Materials and New Manufacturing Technology" (NMP) project and "Research Network Plan" to accelerate the development of high performance alloys and new generation materials.

3. "Industry 4.0" of Germany

In June 2012, Germany launched a long-term research project on "Nanomaterial Safety". In November 2012, Germany launched the "Raw Materials Economic Strategy" research project. In order to encourage various social forces to participate in the research and development of new materials, Germany has successively promulgated 3 plans, e.g. the "Material Research MatFo" (1984–1993), the "Material Technology MaTech" (as of 2003) and the "Material Innovation WING for Industry and Society" (beginning in 2004).

In April 2013, Germany issued a white paper on the "Recommendations for Implementing the Industry 4.0 Strategy". Germany then incorporated the "Industry 4.0 Project" into the 10 future projects of "High Technology Strategy 2020", promoting a new industrial revolution characterized by smart manufacturing, the Internet, new energy, new materials and modern living. The German business community generally believes that ensuring and expanding its leading position in materials research and development is the key to its success in international competition.

4. Science and Technology Basic Plan of Japan

In 1996, Japan implemented the first "Science and Technology Basic Plan". In the fourth issue of "Science and Technology Basic Plan (2011–2015)", special emphasis was placed on the important position of high-tech materials such as materials in the national development strategy, and the important development direction of the material industry is determined.

5. "Materials and Technology Development Strategy 2030" of Russia

Russia has always regarded the new material-related technology industry as the national strategy and the leading industry of the national economy. The "Materials and Technology Development Strategy 2030" released in April 2012 listed 18 key material strategies as development directions, including smart materials, intermetallic compounds, nanomaterials and coatings, single crystal heat resistant super alloys, and Nb-containing composite materials, etc., and also developed a development strategy for the main application areas of the new materials industry.

The Russian Academy of Sciences released the "Technology Development Forecast to 2030" in 2015, which mainly includes seven priority areas for science and technology development, namely information and communication technology, biotechnology, medical and security, new materials and nanotechnology, rational use of natural resources, transportation and aerospace systems, energy efficiency and energy conservation.

6. Guide to New Material Industry Development in China

In order to guide the healthy and orderly development of the new materials industry, the Ministry of Industry and Information Technology of the People's Republic of China has formulated guidelines for the development of new materials industries in accordance with the "13th Five-Year Plan" and the "Made in China 2025". By 2020, the development trend of large-scale and agglomeration of new materials industry will basically take shape, breaking through the technical equipment constraints in the fields of metal materials, composite materials, advanced semiconductor materials, etc., in the fields of carbon fiber composite materials, high-quality special steel, advanced light alloy materials, etc. More than 70 kinds of key new materials are industrialized and applied, and a technological equipment support system matching the development level of China's new materials industry is established.

9.2.3 Development Prospects of New Materials

The development of materials has a major impact on the social form and the quality of human life. The history of mankind has proven that the discovery of every new material can make a huge difference in human society. The discovery of chemical fiber has made the garment industry a new bright spot in the world. The legend of people's colorful clothing and even the exchange of shirts for aircraft is derived from the industrial production of chemical fibers. New materials play an important role in developing high-tech technologies, transforming and upgrading traditional industries, and enhancing overall national strength and national defense strength.

Materials, energy, information technology and biotechnology all together constitute the four pillars of a civilized society. The development level of new materials reflects the level of scientific and technological development and comprehensive national strength of a country. It is the forerunner of the development of the material industry, the key guarantee of the development of high-end manufacturing and the development of the national defense industry etc., the important support and guarantee for national economic development, national defense modernization, and energy security, and the focus of strategic competition among countries. Since the middle of the 20th century, countries around the world have paid more and more attention to the research and development of new materials. In the history of mankind, a new era of material revolution has emerged. The access to new materials is different from that of traditional materials. It is the result of multi-discipline, interactive, comprehensive research, has high-tech performance, meets the needs of cutting-edge technology and equipment manufacturing. The development of new materials is more closely related to its specific application, and it pays more attention to the coordination of ecological environment and resources compared with traditional materials. In the world, the wave of new material research is on the rise, and the competition for new materials industrialization is also intensifying. The development of new materials has entered a golden age.

At present, the developed countries in the world attach great importance to the development of new materials technology, and regard the development of new materials technology as an important part of the technology development strategy. When formulating national science and technology and industrial development plans, the new material technology is listed as the key to the development of the 21st century, which is the technology to focus on for maintaining its leading position in the economy and technology. The combination of new materials technology and information technology and energy technology constitutes the three pillars of the current world new technological revolution and is the three most important and most promising areas of the world in the 21st century. The development of new materials has benefited from technological innovation and accelerated transformation of research achievements. Breakthroughs in cutting-edge technologies have led to the emergence of new materials industries, while new materials are increasingly integrated with information, energy, healthcare, transportation, and construction industries. The cross-cutting and scale of materials science engineering and other disciplines are expanding. Governments around the world attach great importance to the development of new materials industries, and have formulated relevant plans to promote the development of new materials industries and science and technology, and have given strong support in funding, thereby promoting the improvement and development of technological innovation capabilities in this field, especially new materials technologies and the combination of energy technologies will have a disruptive impact on the future energy sector, and the energy materials we use today are likely to be replaced by new energy sources in the near future.

In today's world, the scientific and technological revolution is developing rapidly, new materials and products are changing with each passing day, and industrial upgrading and material replacement are accelerating. The new material technology is integrated with nanotechnology, biotechnology and information technology. The trend of structural function integration and intelligent functional materials is obvious. The environmentally friendly characteristics of materials such as low carbon, green and renewable recycling have attracted much attention. The new materials show the following trends.

(1) Continue to pay attention to high-performance new metal materials. The new metal materials are still the leading materials in the 21st century. By adopting high-tech and new processes (reasonable physical metallurgical design of alloy components, addition and control of trace elements, etc.), the material properties can be greatly improved, and the metal materials have high strength, high toughness, high temperature resistance, low temperature resistance, corrosion resistance, radiation resistance and other properties.
(2) Conduct the synthesis and functionalization of structural materials. In order to meet the comprehensive requirements of high-tech for materials performance, it is necessary to adopt a new type of composite material, which can obtain better properties than monomer materials or properties not possessed by the original materials, so the composite and functionalization of different materials become an important trend in the development of structural materials.
(3) Expand the application of low-dimensional materials continuously. Low-dimensional materials refer to zero-dimensional (ultra-fine) materials, one-dimensional (fiber) materials, two-dimensional (thin film) materials, etc. These materials are a new type of material that has developed rapidly in recent years and can be used as structural materials and functional materials.

   Zero-dimensional materials, such as nano-materials, refer to materials with nano-scale (0.1–100 nm) grains in powders or materials. Due to the small grain size, the proportion of interfaces and surface atoms increases (up to 50%). The surface atoms are highly active, which can make these materials exhibit new properties in physicochemical properties such as sintering, diffusion, strength, etc. The size of the nanomaterials has reached the de Broglie wavelength of electrons, and the movement of electrons in the material must be considered. Its volatility and size effect will inevitably lead to new properties such as electricity, magnetism and light.

   The most prominent one-dimensional material is optical fiber, which can be used as a communication engineering material. Fibrous structural materials are equally important and are the major reinforcing components in composite materials, which determine the critical properties of the composite. Carbon fiber, organic polymer fiber and ceramic fiber have broad application prospects.

   Two-dimensional materials are developing rapidly. With the miniaturization of electronic devices, various thin film insulation, semiconductor, dielectric and magnetic materials are required. Diamond film can be used in microchips for high-speed electronic computers. Polymer separation membranes have been developed for ion exchange membranes, dialysis membranes, microporous membranes, gas separation membranes, etc. High-temperature superconducting membranes will open up new fields of superconducting technology.

(4) Functional materials are developing towards multi-functional integration, intelligence, and structural functional integration. Functional materials are the material basis for new energy technologies, information technology, computer technology, and biomedical technologies. Optoelectronic materials will be the fastest growing and most promising information material. New energy materials are key materials for the development of green secondary batteries, hydrogen storage materials, fuel cells, solar cells and nuclear energy. There are three research and development directions for biomedical materials: the first is to simulate the composition, structure and function of human soft and hard tissues, organs and blood, to carry out bionic or functional design and preparation; the second is to give materials excellent biocompatibility and biology activity or life activity, to achieve the replacement of human organs to the repair of organs; the third is the biological simulation in industrial production. The development trend of nanomaterials is to carry out basic research in the fields of nano-processing, nanoelectronics, nano-medicine and robotics, which can form emerging leading industries in the future, and transform and upgrade existing information high-tech industries and traditional industries. The development trend of superconducting materials is to continuously explore higher temperature superconductors and realize the application of high temperature superconducting materials industrialization technology in the fields of energy, electric power, mobile communication and national defense. Intelligent materials are one of the important directions for the development of high-tech in the 21st century. In important projects and cutting-edge technologies, there are important application prospects such as bridges, dams, buildings, aerospace, high-speed train safety monitoring, active shape control, reduction of noise and vibration, self-healing and improvement in the compatibility of biomedical materials.

(5) Materials used in special environments. Under low temperature, high pressure, vacuum, high temperature and radiation conditions, the structure and composition of the material will change, and thus the performance changes. Studying these external conditions is conducive to creating and improving materials. For example, in structural materials under high pressure, the material will be converted from an insulator to an electrical conductor due to the shortened distance between atoms, and the superconductor materials such as Nb_3Sn, Nb_3Ge and Nb_3Si are all completed under high pressure.

(6) Requirement of low carbon, environmental protection and recyclable material. The new materials will reduce the use of carbon emissions and toxic metals, reduce pollution to the atmosphere and the environment, reduce material costs and waste of resources through recycling, and move toward green, environmentally friendly and recyclable direction.

High-performance electrochemical materials have also made great progress in recent years, and the discovery of lithium ion intercalation materials directly contributed to the invention of lithium ion batteries. The new battery materials have many advantages such as high capacity, long life, safety and reliability, and high power, attracting energy-related manufacturers to join the lithium-ion battery industry. Now, the lithium-ion battery related industry has grown into a huge field in the world. The application of lithium-ion batteries in power and energy storage is accelerating the transformation of the world's energy structure. In the future energy structure, it is likely that distributed energy and high-performance batteries will lead the way.

At present, new energy materials technology has developed into a stage, and more advanced materials research has entered a bottleneck period. The performance of lithium-ion battery materials has been unable to meet the increasing user demand. People are not satisfied with the use of small batteries for the supply of portable electronic products, but also hope that small batteries can solve various complex problems such as large-scale energy storage, power of sports car-class electric vehicles. New materials will revolutionize the future of energy.

References

Qi, Baosen, Gang Zhang, and Daocheng Luan. 2007. *New materials and applications*. Harbin: Harbin Institute of Technology Press.

Wu, Qisheng, Xia Zhang, and Zhenhua Dai. 2017. *New energy materials*. Shanghai: East China University of Science and Technology Press.

Yan, Xiujing, Kuiren Liu, and Qing Han. 2017. *New energy technology*. Beijing: Chemical Industry Press.

Zhou, Ji. 2014. *New materials industry*. Shanghai: Shanghai Science and Technology Literature Publishing House.

Zhu, Jiping, Paifeng Luo, and Chenxi Xu. 2014. *New energy materials technology*. Beijing: Chemical Industry Press.

Chapter 10
Geothermal

Geothermal resources refer to hot water fields of various temperatures and depths that exist beneath the surface of the Earth. They are the second largest clean energy source after solar energy. Water vapor or high-temperature hot water can be extracted into the surface by drilling for use, such as power generation, direct utilization, and heating and cooling. Since geothermal resources are heat generated by the Earth's core, even in geothermal areas that rely on hot water fields, the amount of water produced can be reinjected, making it a sustainable energy source. Through proper hot water reservoir management, the heat recovery rate is balanced with the natural heat exchange rate, which ensures the effective and sustainable utilization of geothermal resources.

10.1 Current Status of Geothermal Development and Utilization

Geothermal energy is a kind of green, low-carbon, recyclable renewable energy. It has the characteristics of large reserves, wide distribution, clean and environmental protection, stable and reliable. Geothermal resources have been developed in more than 120 countries and regions around the world, including countries such as the United States, Japan, the Philippines, New Zealand, Indonesia, and Iceland, where the level of geothermal resources utilization and development is high.

10.1.1 Global Geothermal Resources Distribution

It is estimated that the global shallow geothermal resource less than 5 km is about 4900 trillion tons of standard coal (3,430 trillion tons of crude oil), and the geothermal resource less than 10 km is about 43,000 trillion tons of standard coal (30,100 trillion tons of crude oil). It is more than 30,000 times the total amount of conventional oil

C. Zou, *New Energy*, https://doi.org/10.1007/978-981-15-2728-9_10

and gas resources, and the annual recoverable geothermal resources is 170 trillion tons of standard coal.

Geothermal resources can be divided into high temperature, medium temperature and low temperature according to temperature. High-temperature geothermal system refers to those in the form of steam at a temperature greater than 150 °C; medium-temperature geothermal system refers to those in the form of a mixture of water and steam at 90–150 °C; low-temperature geothermal system refers to those at a temperature between 25 and 90 °C, in the form of warm water (25–40 °C), warm and hot water (40–60 °C), and hot water (60–90 °C). Global high-temperature geothermal resources are concentrated at the edge of the tectonic plate, developing four major thermal regions around the world, namely the Pacific Rim geothermal belt, the Mediterranean-Himalayan geothermal belt, the Atlantic Ridge geothermal belt, the Red Sea-Gulf of Aden-East Africa Rift Valley geothermal belt. Global low-temperature geothermal resources are mainly distributed in the interior of the plate near the boundary.

Pacific Rim geothermal belt: Located at the border of the Eurasian, Indian and American and the Pacific Plate, it is characterized by significant high heat flow, young orogenic movements and active volcanic activity. Its distribution includes the Aleutian Islands, Kamchatka Peninsula, Kuril Islands, Japan, Taiwan, the Philippines, Indonesia, New Zealand, Chile, Mexico and the western United States. There are many world-famous geothermal fields in the Pacific Rim geothermal belt, such as the US Long Valley, Roosevelt, Japan's Songchuan, Dayue, and Taiwan's Macao and so on.

Red Sea—Gulf of Aden—East Africa Rift Valley geothermal belt: Located at the border of the Arabian Plate and the African Plate, it is exposed to the oceanic ridges between the Arabian Plate and the African Plate and the East African Rift Valley Belt, with features of high heat flow, active volcanoes and fault tectonic activities. The famous geothermal fields include Djibouti, Dalor in Ethiopia, and Orkaria in Kenya and so on.

Mediterranean—Himalayan thermal belt: Located on the border between the Eurasian plate and the African plate and the Indian Ocean plate, stretching from Italy on the north coast of Mediterranean in the west, through Turkey, Pakistan in southeast to the southwestern part of the Ali region in China, through the Yarlung Zangbo River to the Nujiang River in the east, and then connecting geothermal activity belt of western region of Sichuan province and western region of Yunnan province. World-famous geothermal fields such as Ladrillo in Italy, Kyzyldale in Turkey, Yangbajing in Tibet, and Tengchong in Yunnan are all distributed in the geothermal belt.

Atlantic Ridge geothermal belt: Located in the Mid-Atlantic Ridge Expansion Zone, along the Americas and Eurasia, Africa and other plate boundary, characterized by high heat flow, active volcanic and fault tectonic activity. Famous geothermal fields in the geothermal belt are Crafrain Iceland, Azores Islands and so on.

10.1.2 Geothermal Exploration and Development Technology

The geothermal resources located in the deep underground need to be identified by means of geological survey, geophysics, geochemistry, geothermal drilling, deliverability test analysis and dynamic monitoring to determine their distribution, resource quantity, quality and development and utilization conditions.

1. Geophysical exploration

Geophysical exploration is an important means of deep geothermal geological exploration and is a work that needs to be carried out before deep geothermal drilling. Geophysical exploration uses geophysical instruments to detect differences in physical properties (gravity, magnetism, electrical, etc.) in different layers below the surface, to divide the strata, determine the depth of thermal storage and make judgments on geological structures, and provide the design data for geothermal drilling. The geophysical methods commonly used in geophysical exploration are: electric method, magnetic method, gravity method, artificial earthquake, and temperature measurement method and so on (Wang et al. 2011).

The magnetic method is based on the magnetic data to judge the location and orientation of the fault tectonic belt, and to delineate the extent of the intrusive rock mass. Combined with the geological data, the approximate geological time and lithology of the intrusion formation can be inferred, thereby predicting the degree of influence of the intrusive rock mass on the formation of the geothermal field. Magnetic data can also help people understand the extent of sedimentary basins and the basal fluctuations in the basin. According to the speculative results of the magnetic survey data, combined with the relevant geological data, it is possible to predict the prospective area where geothermal heat may be formed and reduce the exploration area of the geothermal field.

The electric (magnetic) sounding method is one of the most commonly used geophysical methods in geothermal exploration. The electric (magnetic) methods commonly used in geothermal exploration include symmetric quadrupole sounding, frequency sounding, and transient electromagnetic (TEM), magnetotelluric sounding method (MT), audio magnetotelluric sounding method (AMT), controllable source audio magnetotelluric sounding method (CSAMT), and EH-4 electromagnetic sounding method combined with controllable source and natural source. The resistivity of the heat storage zone is a comprehensive reflection of the resistivity of various media in the volume range of the heat storage zone, including formation lithology, formation temperature, water ion type and concentration, formation porosity, permeability, fragmentation degree of formation rock and so on. In addition, the solubility of hot water is relatively high. As the temperature of groundwater increases, the solubility of groundwater increases, the density and viscosity of water decrease, and the salinity of groundwater increases, the activity of ions increases, and the resistivity decreases, so a low resistivity anomaly characteristic of the heat storage structure zone occurs. However, high-temperature steam geothermal systems exhibit relatively high-impedance anomaly characteristics because steam is similar to a high-resistance medium filled with pores in the formation.

Seismic method can reflect the distribution and occurrence of underground rock layers, structures, fluids, gases, etc. through imaging, inversion and so on through the reflection, refraction and transmission of seismic waves. Micro-motion is a kind of seismic method. It is the daily small vibration of the earth's surface. It is different from the micro-seismic with specific earthquakes and earthquake events, and can be observed at any time and place. The micro-motion survey method is based on the stationary stochastic process, extracting the surface wave (Rayleigh wave) dispersion curve from the micro-motion signal, and obtaining the geophysical prospecting of the S-wave velocity structure of the subsurface medium by inversion of the dispersion curve method. The related research and experimental results show that the concealed fault and fracture zone has obvious low-velocity anomaly on the micro-motion S-wave velocity profile, which becomes an important indicator for the micro-motion profile to explain the hidden geothermal structure, and provides an important basis for geothermal well site selection.

Temperature measurement methods include geothermal flow measurement, formation temperature measurement, and formation temperature gradient measurement. The thermal conductivity of the rock formation and the geothermal heat flow are very important for the formation of the geothermal field. The magnitude of the geothermal flow value reflects the amount of heat transferred from the deep part. The higher the heat flow value, the more obvious the high temperature anomaly in this area. The heat flow value is the only parameter that can be measured on the earth's surface about releasing heat inside the earth. It is an important parameter to predict whether a region can form a geothermal field. The higher the heat flow value, the more favorable it is to form a geothermal field.

Radon gas measurement is a convenient and effective radioactive detection technology, which has been widely used in many fields. Radioactive elements are widely found in nature, and many daughters are produced during their decay. Among them, Radon (222Rn) is the only gaseous body that can migrate from the subsurface to the surface to reflect the deep underground information.

2. Geothermal drilling

The medium and low temperature geothermal well drilling technology is mature, and a series of special drilling and completion technologies have been formed in the well structure design, low density drilling fluid drilling, logging, cementing, determination of basic performance parameters of the liner and sand control.

The problem facing high temperature geothermal well drilling is how to choose the right drilling fluid, cementing cement slurry, logging instrument, casing and so on under high temperature conditions. At present, the geothermal field water temperature is generally lower than 200 °C. China's temperature-resistant 200 °C drilling fluid, cementing cement slurry, logging instrument and technology are mature. It can be said that domestic drilling technology can basically meet the demand of high-temperature geothermal drilling. The bottom hole temperature of the GR1 well drilled in China's Gonghe Basin in Qinghai is as high as 236 °C. Its high temperature water-based drilling fluid system has stable performance at 236 °C, no obvious thickening and thinning phenomenon, and excellent anti-high temperature rheology.

The performance, suspension and carrying ability are strong, and the wall protection effect is remarkable under high temperature conditions. At the same time, we should pay attention to drilling fluid treatment, noise pollution and safe drilling (blowout prevention, cooling) and other issues to protect the ecological environment.

3. Deliverability test

The flow of heat storage and oil and gas reservoirs is similar, and the well testing analysis and productivity prediction methods of oil and gas reservoirs can be used for reference. However, geothermal wells generally measure wellhead pressure, flow rate and other data through surface blowdown test. Because there is no flow control in the blow out test method, the flow rate changes with time, which is equivalent to variable production, and does not meet the requirements of constant flow rate for conventional well test analysis. If the bottom hole pressure under constant production conditions can be calculated from the discharge test data, mature well test analysis and productivity prediction methods can be used for production dynamic prediction.

Productivity forecasting is divided into steady-state productivity forecasting and transient productivity forecasting. The steady-state deliverability prediction regards the formation flow as a steady-state seepage, ignoring the compressibility of the fluid, and the calculation is relatively simple. The transient deliverability prediction estimates the formation flow according to transient seepage, and can calculate the change of production over time, which is more realistic. When the production time is longer, the calculated output of steady-state seepage and transient seepage is closer.

4. Supporting technology

(1) Technology of converting abandoned well to geothermal well

Before the recompletion of abandoned wells, it is necessary to comprehensively analyze and integrate the existing geological, geophysical and geochemical exploration, drilling and exploration results, and focus on strengthening the comprehensive analysis of well logging, mud logging, cuttings and other data of oil and gas wells, and identify the available thermal reservoirs and provide reliable information for the next step of recompletion (Wang 2017). To pay attention to the well spacing between the injector and producer when selecting well pairs will prevent the heat connection from occurring.

(2) Insulation technology of geothermal well

Geothermal wells are similar to oil wells. The closer to the wellhead, the greater the temperature drop of geothermal water and the greater the heat loss. The use of enhancing fluid rate technology can significantly reduce the heat loss of the wellbore and greatly increase the temperature of the geothermal wellhead (Wang 2015).

(3) Geothermal recharge technology

After the geothermal fluid is utilized, the tail water is reinjected into the produced thermal reservoir by natural or pressurized means. It is the most effective method to solve the problem of excessive under groundwater level continuous decline and waste water discharge pollution caused by excessive geothermal fluid exploitation. At present, the geothermal waste water recharge in China is still in its infancy, and the theoretical model for geothermal waste water recharge technology and scheme optimization design is still not mature.

10.1.3 Geothermal Utilization Technology

1. Direct utilization technology

The direct application of shallow geothermal heat is mainly to use heat pump technology to provide heat to the main utilizing unit. The use of deep geothermal heat includes the use of heat pumps, indirect heat transfer and downhole heat transfer to provide heat to the main utilizing unit.

The heat pump technology converts low-level heat energy that cannot be directly used into high-level heat energy by consuming a certain high-level energy. It is mainly used for cooling and heating. The heat pump is divided into two types: compression type and absorption type (Table 10.1).

The indirect heat exchange is to heat the radiator of the user terminal after the hot water is exchanged by the heat exchange device, and the geothermal water after the heat exchange is cooled, and discharged or reinjected after comprehensive utilization.

Downhole heat transfer is to use the downhole heat exchange device to achieve direct heat transfer downhole. It takes heat without discharging water, with high heat exchange efficiency, no need for recharge, low cost. This method includes coaxial casing heat transfer injection and superconducting heat exchanger form.

2. Power generation technology

Geothermal power generation is a power generation technology that uses underground hot water and steam as a power source. The process of geothermal power generation is the process of converting underground thermal energy into mechanical energy and then converting mechanical energy into electrical energy. According to the type of heat carrier, temperature, pressure and other characteristics, geothermal power generation technology can be divided into dry steam power generation technology, flash steam power generation technology, dual cycle (dual working media) power generation technology and full flow power generation technology (Fig. 10.1).

(1) Dry steam geothermal power generation technology

The dry steam geothermal power generation technology is suitable for high temperature, using steam turbines to generate electricity, and steam directly enters the steam turbine to drive the generator to generate electricity. The system is simple, the investment is small, and the payback period is short, but the requirements for steam quality, anti-corrosion and anti-scale ability of the unit equipment are high.

Table 10.1 Classification and comparison of compression and absorption heat pumps

Project	Compression heat pump	Absorption heat pump
Principle	The medium uses electrical energy or mechanical energy to produce a phase change to work	The medium uses thermal energy to produce changes in phase and concentration to work
Compensation method	Compensation by consuming mechanical energy or electrical energy	Using heat energy as a compensation method
Heat transfer medium	For various types of refrigerants, such as R22, R134, etc., it is harmful to the environment, affected by the nature of the refrigerant, and the operating temperature range is limited	It is an environmentally friendly lithium bromide solution, and it is non-toxic, and the change in equipment output does not affect performance
Energy efficiency ratio	COP is 4–5 or even higher	When heating, the COP is 1.4–2.0 When cooling, the COP is 1.0–1.2
Temperature	The temperature of the heating water is 45–60 °C, very few can reach higher, and the cooling temperature is 7–12 °C	The water temperature can reach 85 °C when heating, the temperature is related to the heat source temperature, and the cooling temperature is 7–12 °C
Capacity	Single unit is small and has a maximum capacity of 1,500 kW	Up to 30 MW, but cooling of the same machine is only 1/3 of the heating
Scope of application	Wide range of applications, suitable for single rooms or several rooms, both for summer cooling and winter heating, more suitable for areas with low electricity prices	Suitable for large and medium-sized projects such as central heating and refrigeration, suitable for areas with rich waste heat and low price of heat energy

(2) Flash geothermal power generation technology

The flash geothermal power generation technology is suitable for medium and high temperature, and the flash steam from depressurized hot fluid drives the steam turbine to drive the generator to work. The advantage is that the investment is small and the investment recovery period is short, but the on-site installation and construction is complicated and the setting up is difficult.

(3) Dual cycle geothermal power generation technology

The dual-cycle geothermal power generation technology is suitable for medium and low temperature, and the heat generated by the geothermal fluid can cause the low-boiling medium fluid to flash to form steam, which drives the steam turbine to generate electricity. The advantage is that the equipment system is compact, easy to install,

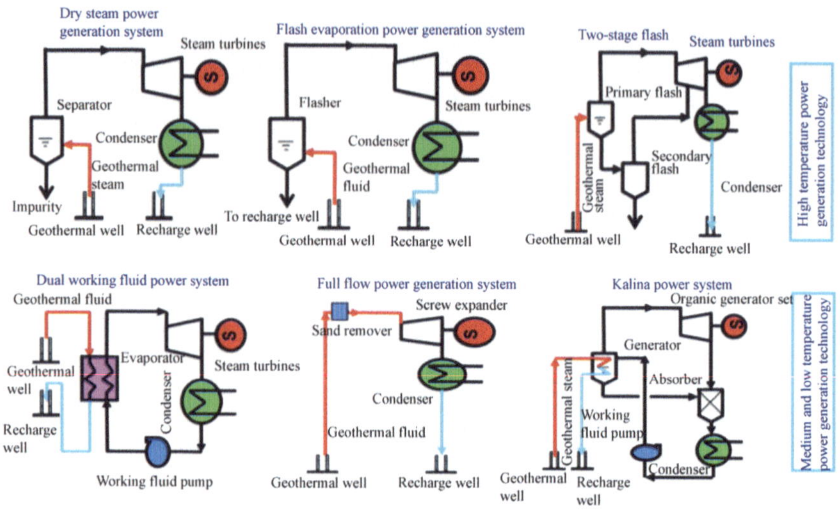

Fig. 10.1 Diagram of geothermal power generation system

and the wide applicability of geothermal water, but there are problems such as high engineering price and risk of leakage.

(4) Full-flow power generation technology

The full-flow power generation technology is applicable to a wide temperature range, and uses the full flow of hydrothermal fluid to enter the screw space for power generation. It has the advantages of high efficiency, low operating cost, simple system and convenient installation, but the installed capacity is small.

Manufacturers of high-temperature power generation equipment include Mitsubishi, Fujitsu of Japan, and General Electric of the United States. The market for medium and low-temperature power generation equipment is mainly occupied by UTC of the United States and ORMAT of Israel, etc.

In recent years, there have been some breakthroughs in hot dry rock power generation technology. Compared with the conventional use of dry steam power generation and geothermal water power generation, Enhanced Geothermal System (EGS) has its unique advantages, mainly in: huge, widely distributed hot dry rock resources; no waste gas and other fluid or solid waste, almost zero emissions; safe development system; good thermal continuity; good economical result and so on. Since the feasibility study of the technology in the United States in 1993, some breakthroughs have been made in recent years, and hot dry rock power generation has entered a small-scale commercial development stage from the early experimental stage. Currently, the countries that use dry hot rock to generate electricity mainly include France, Germany, Australia, the United States, and Japan. Among them, the world's largest EGS project under development is located in the Australian Cooper Basin currently. According to preliminary statistics, the Cooper Basin has a potential of 5,000–10,000 MW of power generation.

10.1.4 Current Status of Development and Utilization of Geothermal Resources in Major Countries Around the World

1. Global geothermal resource is mainly used directly, and high-temperature geothermal power generation is growing rapidly

At present, more than 100 countries around the world have developed and utilized geothermal energy, and they are growing at an annual rate of 18%. In 2015, the global geothermal installed capacity reached 83 GW (direct use of 70.3 GW, geothermal power generation of 12.7 GW).

The direct use of geothermal energy in the world has developed rapidly. The ground source heat pump technology has broadened the direct utilization of geothermal energy. In 2015, the direct utilization capacity reached 70.3 GW, and its installed capacity accounted for 70% of the world's direct use of geothermal energy, with the amount of utilization of 163.2 billion kWh (66.5 million tons of standard coal), the utilization of geothermal power generation in 2015 was 73.6 billion kWh (29.73 million tons of standard coal). The geothermal power generation utilization coefficient is higher, more stable, and has a smaller footprint than other renewable energy sources. According to the measured data, the geothermal energy utilization coefficient can reach 72–76%, which is 3–4 times that of wind power generation, and 4–5 times that of solar power generation (according to the statistics of the World Energy Council); The occupied area of installed capacity per unit is at least one order of magnitude lower than that of wind power and photovoltaic. Wind power and photovoltaics are at least one order of magnitude lower. Since 2000, global high-temperature geothermal power generation has grown rapidly. The United States, the Philippines, and Indonesia have the world's top three installed capacity, accounting for 54.3%; geothermal power generation in Kenya, Iceland and other countries accounts for more than 20% of domestic power generation. Hot dry rock power generation has high utilization rate and stable system. Developed countries have invested in the development of hot dry rock. A number of experimental projects have been built, and only the United States, Germany and France have realized commercial operation. The medium-low temperature geothermal power generation technology is still immature, and there is no large-scale promotion, which needs further research.

2. Geothermal utilization with its own characteristics in typical national

(1) The United States ranks first in geothermal power generation capacity in the world

The United States is the most active and most developed country at utilizing geothermal resources in the world. The United States has abundant geothermal resources, mainly distributed in the west, with an annual utilization of about 16.01 million tons of standard coal. In 2015, the direct utilization of geothermal energy was 17.4 GW, and the annual utilization was equivalent to 8.48 million tons of standard coal, of which ground source heat pumps accounted for 88% of direct utilization of installed capacity. The potential for geothermal power generation in the United States exceeds 3 million MW. After 60 years of development from the first geothermal power station built in 1960, the installed capacity geothermal power generation reached 3.45 GW in 2015, accounting for 27.2% of the world, and the annual power generation is 16.6 billion kWh.

(2) Iceland's geothermal utilization system is the most perfect

Iceland is located at the junction of the Eurasian plate and the American plate. The boundary between the two plates runs through the island from the southwest to the northeast. Active crust activity and complex geomorphology make Iceland one of the most abundant countries in geothermal resources. There are 250 geothermal areas in the country, and the thermal energy is huge, and the per capita utilization is the highest in the world. In 2015, Iceland's geothermal installed capacity reached 2,698 MW, of which direct utilization capacity was 2033 MW.

Since the 20th century, Icelanders have begun to make large-scale use of geothermal resources for heating and power generation, 85% of Iceland's population uses geothermal heating. The capital, Reykjavik, has all used geothermal heating as the cleanest city in the world, and its geothermal technology has been promoted by the United Nations as a model.

Iceland is rich in high-temperature geothermal resources. Geothermal power generation began in 1969. At present, seven large-scale geothermal power stations have been built (four in the capital, three in the north, with a maximum installed capacity of 303 MW), and three thermal power stations around the capital are providing hot water and electricity 15 millions of Icelanders. In 2015, the installed capacity of geothermal power generation was 665 MW, and the annual power generation was 5 billion kWh, accounting for 26.6% of the national power generation.

Iceland's geothermal utilization system is the most complete. In addition to heating and power generation, geothermal energy is widely used in greenhouse farming, factory drying, snow melting, bathing and tourism services and so on.

(3) Rapid growth of geothermal power generation in Kenya

Kenya is rich in high-temperature geothermal resources, and the use of geothermal resources is mainly used for power generation except for a small amount of direct utilization. At the end of 2006, Kenya began to build a geothermal power generation project. Geothermal power generation has shown a rapid development trend, and it is the country with the largest increase in global geothermal power generation in recent years. In 2017, Kenya's geothermal power generation capacity was 1,074 MW, with an annual power generation of 2.68 billion kWh, accounting for 43.8% of the country's electricity generation, making it the largest country in Africa to use geothermal power. Kenya plans to build six 4 MW geothermal power plants in 2019. It is estimated that the installed capacity of geothermal power will reach 5,000 MW in 2030.

10.1.5 Current Status of Geothermal Development and Utilization in China

1. Distribution of geothermal resources in China

China's geothermal resources are abundant, and the amount of geothermal resources less than 5,000 m is about 200 trillion tons of standard coal, accounting for about 1/6 of the global geothermal resources. The geothermal resources less than 10,000 m are about 857 trillion tons of standard coal (600 trillion tons of crude oil) (Yan et al. 2013).

(1) Distribution of shallow geothermal resources

China's shallow geothermal resources are mainly distributed in the central and eastern regions, of which 143 cities above the prefecture level are the most suitable areas for the development and utilization of shallow geothermal energy. The annual recoverable amount is equivalent to 460 million tons of standard coal, which can realize the heating and cooling for a building area of 21 billion square meters and basically meet the needs of heating and cooling. The annual recoverable amount of shallow geothermal energy resources in 336 prefecture-level cities across the country is equivalent to 700 million tons of standard coal, and the heating and cooling building area can reach 32 billion square meters (Ma 2015).

(2) Distribution of medium and deep geothermal resources

China's middle and deep hydrothermal geothermal resources are mainly with medium and low temperature, supplemented by high temperature. The medium-low temperature geothermal energy resources with a depth of 200–4,000 m in China can be divided into three types: high-temperature convective geothermal resources, medium-low temperature convective geothermal resources, and medium-low temperature conductive geothermal resources. The high-temperature convective geothermal resources are mainly distributed in Yunnan, Tibet and Taiwan. The high-temperature geothermal resources suitable for power generation are mainly distributed in southern Tibet, western Sichuan and western Yunnan, with an installed potential of about 6 million kilowatts (Bai et al. 2006). The medium-low temperature convective geothermal resources are mainly distributed in the southeast coastal areas, including Guangdong, Hainan, Guangxi, Jiangxi, Hunan and Zhejiang; medium and low temperature conduction geothermal resources are mainly distributed in large and medium sedimentary basins such as North China, Songliao, North Jiangsu, Sichuan, Erdos, etc. this type of geothermal resources is the most abundant. China's middle-deep hydrothermal geothermal resources are mainly distributed in the eastern oil and gas bearing sedimentary basins. The mineral rights area of petroleum enterprise contains a large number of hydrothermal geothermal resources. Among them, the amount of geothermal resources in China's petroleum exploration area is 590 billion tons of standard coal, accounting for 47.2% of the total hydrothermal geothermal resources in China; the geothermal resources in Shengli Oilfield of Sinopec is about 23.5 billion tons of standard coal.

2. Status of development and utilization of geothermal resources in China

The utilization of geothermal resources in China is mainly direct use, and the proportion of power generation is relatively low. The direct use of geothermal resources in China includes the use of ground source heat pump technology for heating/cooling and direct utilization.

(1) Shallow geothermal heating/cooling

China introduced ground source heat pump technology in the 1990s. Shallow geothermal heating/cooling has developed rapidly, and the application area has expanded to 31 provinces, municipalities and autonomous regions across the country. In 2000, the shallow geothermal heating/cooling construction area was only 100,000 m^2. In 2004, the shallow geothermal heating/cooling construction area reached 7.67 million square meters. After 2010, it grew at an average annual rate of 28%. By the end of 2017, China's ground source heat pump installed capacity reached 20,000 MW, ranking first in the world. The annual use of shallow geothermal energy is equivalent to 19 million tons of standard coal, achieving a heating/cooling area of more than 500 million square meters, mainly distributed in Beijing, Tianjin, Hebei, Liaoning, Shandong, Hubei, Jiangsu, Shanghai and other provinces and cities, of which Beijing-Tianjin-Hebei has the largest development and utilization.

(2) Direct utilization of medium and deep geothermal energy

Since the 1990s, the direct utilization of geothermal in China has developed rapidly and the utilization mode of geothermal heating in Tianjin and Hebei has been adopted. The geothermal utilization model of hot spring tourism and recuperation has been formed in Beijing and the southeastern coastal areas. In the past 10 years, China's middle and deep hydrothermal geothermal utilization has grown at an average annual rate of 10%. By the end of 2015, the medium and deep geothermal heating area reached 102 million square meters, of which Tianjin's medium and deep geothermal heating area was 21 million square meters, ranking first in China's cities, accounting for 6% of the city's central heating; middle and deep geothermal energy in Hebei Xiongxian was with a heating area of 4.5 million square meters, it met 95% of the county's heating needs and created China's first smoke-free city. According to incomplete statistics, as of the end of 2017, China's medium and deep geothermal heating area reached 150 million square meters, with Shandong, Henan and Hebei growing rapidly.

(3) Geothermal power generation

In the 1970s, China built seven medium-low temperature geothermal power stations in Fengshun of Guangdong, Huailai of Hebei, Yichun of Jiangxi, Huitang of Hunan, Xiongyue of Liaoning, Xiangzhou of Guangxi and Zhaoyuan of Shandong, and built medium and high temperature geothermal power station in Yangbajing, Tibet. Due to power generation efficiency, economic benefits, etc., only geothermal power stations in Fengshun of Guangdong and Yangbajing of Tibet are still generating electricity. At the end of 2017, China's geothermal power generation capacity was 27.88 MW.

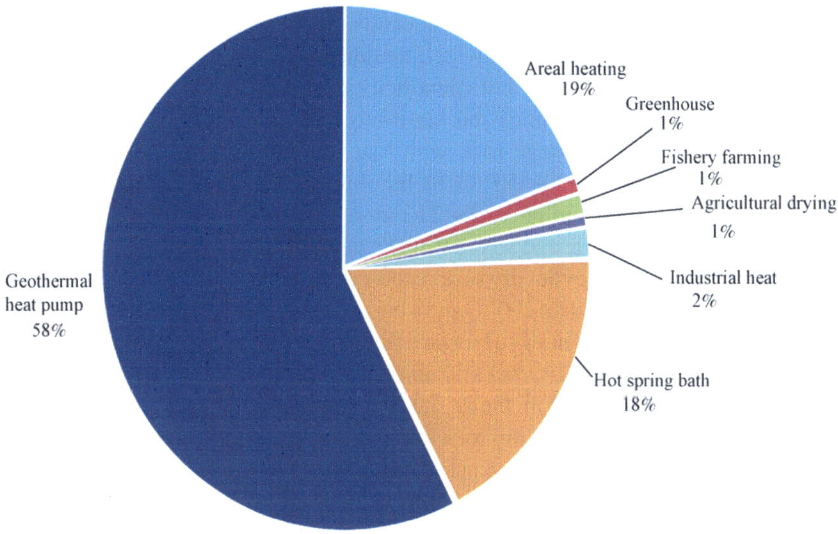

Fig. 10.2 Percentages in terms of geothermal utilization in China in 2014

In July 2017, the power generation test of the 10 MW geothermal power station in Ruili, Yunnan Province was successful. At present, there are already four 400 kW units, which have a net power generation capacity of 1.2 MW.

In 2017, China's geothermal utilization was close to 25 million tons of standard coal, accounting for 0.55% of China's total energy consumption. With the introduction of ground source heat pump technology, the direct utilization of geothermal energy in China has maintained a growth rate of more than 10% in the past 10 years. At present, geothermal utilization in China is mainly concentrated in ground source heat pumps, geothermal heating and hot spring baths (Fig. 10.2).

10.2 Development Prospects of Hot Dry Rock

10.2.1 Hot Dry Rock Overview

Hot Dry Rock (HDR) is a medium for the occurrence of rock thermal energy inside the Earth. Since the concept of dry hot rock geothermal energy was proposed by the Los Alamos National Laboratory in the United States in the 1970s, the definition of hot dry rock has been continuously developed. The hot dry rock currently widely recognized is a kind of water-free and steam-free dense thermal rock, with a depth of 3–10 km and temperature of 150–650 °C. In addition, considering its objectivity, scientificity, feasibility and economy, the basic meaning of hot dry rock can be divided into two types: hot dry rock in general meaning and hot dry rock in narrow meaning. The hot dry rock in general meaning mainly considers its objectivity and scientificity, and is considered to be a heat storage rock with little fluid content and temperature of 150–400 °C. In addition to objectivity and scientificity, the hot dry rock in narrow meaning must also consider the economics and feasibility of geothermal power generation, mainly referring to heat storage rock with low fluid content, buried depth of 3–8 km, and temperature of 200–350 °C. Its lithology is mainly composed of various metamorphic rocks or crystalline rocks. The more common hot dry rocks are biotite gneiss, granite, granodiorite and so on. China's newly released geothermal resource terminology in 2017 defines hot dry rock resources as: high temperature rock masses with a temperature greater than 200 °C and a buried depth of 3–10 km without internal fluids or only a small amount of subsurface fluids.

The distribution of hot dry rocks is almost universal and is widespread in the depths of the earth. Judging from the dry hot rock resources that have been studied in the world, the main areas of hot dry rock resources with economic and technological exploitation are concentrated in the periphery of modern volcanoes, high heat flow granite development areas and tectonic activity areas.

According to the current geological exploration results, there are four types of hot dry rock types in China: high-radiation heat-generating type, sedimentary basin type, modern volcanic type, and strong tectonic activity belt type. High-radiation heat-generating hot dry rocks are mainly concentrated in the southeastern coastal areas, including Guangdong, Fujian, Jiangxi, Hainan, and parts of Guangxi. The large-scale acidic rocks formed in the Yanshanian period are used as deposits to form hot dry rock resource-rich areas; The basin-type dry hot rocks are mainly distributed in the lower part of the Mesozoic-Cenozoic fault basins in Guanzhong, Guide, and Northeast. Due to the high geothermal gradient of the sedimentary cover, they are usually symbiotic with the hydrothermal geothermal field; the modern volcanic hot dry rock is distributed in Tengchong in Yunnan, Changbai Mountain in Jilin, Wudalianchi in Heilongjiang, and other places. Its heat source characteristics are closely related to the history of bottom magmatism and the characteristics of magmatism. The hot dry rocks of strong tectonic belt type are distributed in Tibet, and are squeezed by the Eurasian plate and the Indian Ocean plate. Since the Cenozoic, the Qinghai-Tibet Plateau has been gradually uplifted, and local magma underflow has existed, which provides good conditions for the formation of dry hot rock resources.

1. Highly radioactive heat-producing hot dry rock enrichment area in the southeast coast

During the Mesozoic period in the southeastern coastal areas of China, large-scale granite intrusion occurred, with a large area and a wide range, which is rare in the world. According to previous studies, the granitoid emplacement events in the Mesozoic in the southeastern coastal areas have obvious spatial and temporal distribution patterns. According to the geological age, the most recent Triassic granitoid magmatism in the Mesozoic Era occurred mainly in the inland areas away from the Eurasian-Pacific plate contact zone, and the Jurassic granitoid magmatism was closer to the plate contact zone than the Triassic. The Cretaceous granitoid magmatism is distributed along the southeastern coast of China, relatively close to the plate contact zone. Overall, the granitic magmatism in the southeastern coastal areas has an overall trend from west to east, and the age of rock mass emplacement has changed from old to new. Granite magmatism, which is widely occurring in the Mesozoic, provides important conditions for the formation of hot dry rock resources.

The geodetic heat flow value is one of the important geothermal geological indexes of the distribution of dry and hot rock resources. The soil heat flux value of the southeast coastal area of China is obviously affected by the geotectonic background: The overall tectonic activity of the basin area on the northwest side is stable, which is characterized by the lower geothermal flow value, while the south-southeast coast near the plate contact zone is characterized by a higher geothermal flow value.

Under the trend that the overall heat flow value from west to east is gradually increasing, the local deep thermal structure controls the level of geothermal flow. The thermal flow values of Fuzhou and Zhangzhou in Fujian and Yangjiang-Maoming in Guangdong Province reaches the highest, and the heat flow value reaches 95 $\mu W/m^2$ or more. Radioactive heat generation is one of the main sources of heat in the lithosphere. U (uranium), Th (thorium) and natural radioisotope ^{40}K (potassium) are the main heat-generating elements. According to the existing research results, most areas in the southeast coastal areas are in the high heat generation rate, which is basically higher than 2.1 $\mu W/m^2$. In particular, large areas of granite in Guangdong Province and southwestern Jiangxi and southern Fujian have exposed, and the heat generation rate is more than 2.8 $\mu W/m^2$. Such a large area of high heat generation rate is rare in the global continent. Radioactive heat generation, as the most important source of heat in the crust, is of great significance to the occurrence of hot dry rock resources (Gao et al. 2014).

The southeast coastal areas, especially in Huizhou and Conghua, Guangdong Province, have good prospects for exploration and development of hot dry rock resources. According to the inversion results of the 5.5 km depth geothermal field in the southeastern coastal areas of China, it is basically consistent with the geothermal heat flow distribution map, the buried depth map of Curie interface and the burial depth profile of crust and mantle, indicating that the deep temperature of the southeast coastal area is affected by the tectonic setting. Fujian and Zhangzhouin Fujian, Yangjiang in Guangdong and Rucheng in Hunan have high deep temperatures, which can reach more than 250 °C. Considering the geothermal geological conditions and local needs of hot dry rock resources, the Xinzhou in Yangjiang and Leiqiong fault basins in Guangdong, the Huangsha Cave in Huizhou, Guangdong, and the Lingshui in Hainan are selected as the target areas for exploration of hot dry rock resources in the southeast coast.

2. Sedimentary basin type hot dry rock

The sedimentary basin type hot dry rock resource refers to the deep heat source reaching the shallow part through convection and conduction, and is blocked by the shallow sediment seal layer with small thermal conductivity. A type of dry hot rock resource is formed, where the seal layer of bedrock is thick, the surface geothermal gradient is large, and the increasing temperature is stable. The following focuses on the hot dry rocks of the sedimentary basins in the Guide Basin and the Gonghe Basin.

In the Mesozoic-Cenozoic fault basins in eastern Qinghai, the bases of Guide Basin and the Gonghe Basin are composed of the Triassic of Mesozoic and Indosinian granites. The basement of the Xining Basin consists of Paleozoic, Proterozoic metamorphic rocks and Caledonian granites. The deposits of the Mesozoic-to-Cenozoic are very thick, providing good formation and occurrence conditions for underground hot water and hot dry rock. The Gonghe-Guide Basin is located in complex part of the Kunlun-Qinling zonal tectonic belt and the Hexi tectonic belt. It is the Cenozoic fault basin of the Qin-Kun Pass. It is surrounded by the faulted fold uplifted mountain and is divided into two basins, Gonghe and Guide by the magmatic rock uplift belt of Waligong Mountain in the middle. The basin has a good conductive geothermal geological sedimentary setting. Its eastern and western sides are distributed with tectonic magma belt faulted geothermal high-temperature hot water, and the hot dry rocks are mainly distributed in the orogenic belts of the Qin-Kun-Combination. Controlled by strong tectonic activities, the mountains rise and form a Cenozoic fault basin, which creates a good geothermal geological background environment in the Gonghe-Guide Basin, and also deposits thick alluvial, river-lake and lake facies. It provides conditions for the formation of conductive hot dry rock resources in sedimentary basins. With the help of concealed fault convection and conduction, the deep heat source can be transmitted to the shallow part, and is subject to the overlying sedimentary seal layer, and the heat loss is significantly less. The dry heat rock mass is Indosinian granite. The heat source of this type of hot dry rock resource is high temperature melt mass, the thermal storage is the interbedded layer of the complete rock mass and the fractured rock mass, and the fault zone is distributed in some sections, and the thick alluvial deposit, river and lake facies and lake facies deposits and granite can be used as a thermal storage, the hidden fault provides a channel, the heat is transmitted in convection and conduction with good capping conditions (Qinghai et al. 1991).

3. Modern volcanic hot dry rock

China's Cenozoic volcanic activity is frequent, mainly distributed in two major areas: the first is distributed in the Qinghai-Tibet Plateau and its surrounding areas, which is part of the Mediterranean-Himalayan volcanic tectonic belt; the second is distributed along the north-north east plain and mountainous areas of eastern China, which is part of the western Pacific volcanic belt. The volcanoes of northeast area are the most densely populated in northeast-north China volcanic area. About 690 volcanic cones are distributed in Liaoning, Jilin, Heilongjiang and eastern Inner Mongolia in the area of about 1.5 million square kilometers. The Kunlun Mountain volcanic belt has about 12 volcanic groups within an extension of 1,000 km from east to west. The southeast coast and the Hainan Island volcanic group rise from the Weizhou Island in the Beibu Gulf in the west to the Datun in the northern end of Taiwan, and the southernmost point is the high-pointed stone volcano on the Xisha Islands. The Tengchong volcano group of Yunnan is suspended in the southwestern border. The relationship between the Cenozoic volcano and the hot dry rock resources is inseparable. Influenced by the bottom uncooled magma, the surface has obvious hydrothermal activity. Hot dry rocks of higher temperatures are usually obtained in shallower places.

4. Fault tectonic activity belt type hot dry rock

The strong tectonic activity of hot dry rock resources in China mainly refers to a type of hot dry rock resource due to the compression of the Eurasian plate and the Indian Ocean plate. It has been uplifted many times since the Cenozoic. The magma intrusion occurred in some areas, and the deep high-temperature convective heat source transfer through the large fault to shallow part. The uplift of the Qinghai-Tibet Plateau not only created a great change in the geological structure of the whole region, but also provided good conditions for the formation of hot dry rock resources. Influenced by tectonic activities, since the Quaternary, the Qinghai-Tibet Plateau has been strongly squeezed from north to south. With the constant changes in the geological stress of the Qinghai-Tibet Plateau, the early tectonic pattern dominated by east west distribution has been gradually destroyed. The series of strike-slip faults in the northwest, north-north-west and north-south extensional and torsional active faults have occurred in the active fault zone. The Naqu-Yangbajing-Duoqing fault active structure and the Chaquo-Duoqing tectonic belt are the most remarkable. The middle-high temperature geothermal occurring area in the Chaquo-Gudui-Cuona tectonic belt from the south to the north are Chuna, Gudui, Riduo, Voca, Songduo, and Chaquo. These areas can be used as the ideal prospect area for hot dry rock resources of strong tectonic activity belt type.

According to the evaluation data of the China Geological Survey, China's hot dry rock resources from 3 to 10 km are about 856 trillion tons of standard coal, which is of the same order of magnitude as the United States (Table 10.2). If calculated according to the lower limit of recoverable resources (2%), the available resource is about 3337 times of China's total national energy consumption in 2015. In particular, hot dry rock with a depth from 3.5 to 7.5 km has a high utilization value, the temperature is between 150 and 250 °C, and the resource is equivalent to 215 trillion tons of standard coal. It is the most potential strategic replacement energy source.

Table 10.2 Comparison of calculation results between China's dry hot rock resources and the United States

Country	No.	Formation depth for calculation (km)	Heat (10^{25} J)	Converted to standard coal (10^5 billion tons)
China	1	3.0–4.0	0.19	6.5
	2	4.0–5.0	0.25	8.4
	3	5.0–6.0	0.30	10.3
	4	6.0–7.0	0.36	12.2
	5	7.0–8.0	0.42	14.1
	6	8.0–9.0	0.47	16.1
	7	9.0–10.0	0.53	18.0
	3.0–10.0		2.52	85.6
US	3.0–10.0		1.67	57.2

From the temperature of hot dry rock geothermal resources, hot dry rock resources less than 75 °C account for 2% of total resources within 3–10 km depth; hot dry rock resources of 75–150 °C account for 43% of total resources; hot dry rock resources greater than 150 °C account for 55% of the total resources. From the regional distribution of hot dry rock geothermal resources, the hot dry rock resources in the southern part of the Qinghai-Tibet Plateau account for 20.5% of the total hot dry rock resources in the mainland of China, with the highest temperature; the hot day rock resources in the North China Basin and the Fenwei Graben account for The total resources are 8.6%; the hot dry rock resources in the Mesozoic magmatic activity areas in the southeastern coastal areas account for 8.2% of the total resources; the hot dry rock resources in the Songliao Basin in the Northeast account for 5.2% of the total resources; although the hot dry rock temperature in western Yunnan is higher, but the area is limited, accounting for 3.8% of the total resources.

10.2.2 Current Status of Development and Utilization of Hot Dry Rock

The United States, France, Germany, the United Kingdom, Japan, Australia and other countries started earlier. From the establishment of the world's first hot dry rock development project in the United States in 1970, 25 experimental EGS projects have been established worldwide (12 projects in European, 6 in the United States, 2 in Australia, and 2 in Japan, the cumulative power generation capacity is approximately 12 MW (Table 10.3).

The following is a brief introduction to the hot dry rock project of the United States, Australia, France, and Japan.

1. United States

In 1970, the United States chose to drill in the Valles Caldera area outside the ring structure fault in New Mexico, establishing the world's first enhanced geothermal system for the development of hot dry rock—the Fenton Hill project, the well bottom temperature of the Fenton Mountain hot dry rock is 200 °C and the well depth is about 2,000 m. In 1974, the pair well design (well GT-2 and well EE-1) was initially used. Due to the poor hydraulic connectivity to the well, one of the wells GT-2 was drilled deeper to 2,500 m to become the GT-2A well. After refracturing, the pair well system connecting with the EE-1 well was not established yet. Then a new well GT-2B well was drilled in the formation fracture between the two wells, and finally formed a good cycle pair well system with the well EE-1. The recirculation cycle produces a heat flow of 3 to 5 MW, experimentally driving a 60 kW generator. In 1978, the Fenton Hills project in the United States designed the EE-2 and EE-3 pair well systems, which are located in gneiss and schist at a depth of 3,000 m underground. For the first hydraulic fracturing, the EE-3 well failed to form a good hydrodynamic cycle with the well EE-2, and then drilled laterally at 2,830 m in the EE-3 well to form the EE-3A well based on the microseismic monitoring data. The well is 4,018 m deep

Table 10.3 Basic information of typical hot dry rock prcjects in major countries

Country	Project	Heat flow value (milliwatts/m^2)	Temperature (°C)	Depth (m)	Lithology	Well placement	Feature	Result
US	Fenton Hill	160	325	4391	Granite and ciorite	1 producer 1 injector		Pilot project, achieved valuable experience
	Desert Peak	128	135–204	2475	Metamorphic limestone	1 producer 2 injectors		Successful power generation in 2013, 1.7 MW
	The Geysers	168	400	3400	Metamorphic sandstone	2 producers 1 injector	Highest temp.	Successful grid-connected power generation, 5 MW added
	Newberry	160	315.6	3050	Marl			Under construction
	Raft River	–	130–200	1828	Granite			Under construction
Australia	Habanero	100	270	4911	Granite	Multiple producers, 1 injector		Started to produce in 2013, with capacity of 1 MW
UK	Rosemanowes	120	100	2600	Granite			Terminated

(continued)

Table 10.3 (continued)

Country	Project	Heat flow value (milliwatts/m^2)	Temperature (°C)	Depth (m)	Lithology	Well placement	Feature	Result
France	Soultz	176	210	5270	Granite	Multiple producers, 1 injector	Deepest	Started to produce in 2011, with capacity of 3 MW
Germany	Landau	100	160 (wellhead)	4200	Carbonate	1 producer, 1 injector		Power capacity of 2.9 MW, heating capacity of 3 MW
Japan	Hijiori	168	270	2200	Granite	Multiple producers, 1 injector		Terminated
	Ogachi	–	230	1100	Diorite	Multiple producers, 1 injector		Terminated

and enters the artificial fracturing system between the well EE-2 and the well EE-3. The cycle injection test between the well EE-2 and the well EE-3A showed that the water injection recovery rate was around 66%.

The Desert Peak geothermal field is located in the hot springs of northern Nevada and is a hot, hidden hydrothermal geothermal system. The geothermal activity in this area is mainly controlled by a series of faults, and the presence of these faults provides a good vertical channel for deep high temperature fluids. In order to increase the heat production of the Desert Peak geothermal field, the well 27-15 located at the edge of the geothermal field with low permeability were selected as target wells for hydraulic fracturing. Hydraulic fracturing execution began in September 2010, and the lithology of the fracturing target interval is dominated by rhyolite and mudstone. Hydraulic fracturing increases the injection capacity of the well by four times and provides better hydraulic connectivity to the geothermal field in the south, making it more commercially viable.

Geysers is the world's largest steam geothermal field, located more than 100 km north of San Francisco, California, with a generating capacity of 825 MW. Since the production in 1960, reservoir pressures have been declining. In 1980, the exploration well drilled a high-temperature metamorphic rock mass (280–400 °C) under the steam reservoir. The project plans to hydraulically fracture the deep low-permeability high-temperature rock mass to increase the thermal productivity of the upper steam reservoir. The project deepened the previously abandoned exploration wells P-32 and PS-31 to 3,000 m and 3,500 m respectively, and the two wells were about 500 m apart. In 2011, hydraulic fracturing tests were carried out on the two wells. The injection pressure was much lower than the fracturing pressure of the rock mass. It is hoped that the reservoir thermal contraction and shear failure will increase the reservoir permeability through low pressure cold water injection. The pressure response and monitoring data show that a new fracture system has been successfully established in the deep high temperature rock mass, and the cyclic injection test predicts a new reservoir heat extraction rate of about 5 MW.

The Raft River geothermal field is located at the junction of Idaho and Utah in the western United States, on the southern edge of the Raft River Canyon. The project began commercial power generation in 2007 and is operated by the US Geothermal Company with an installed capacity of 10.5–11.5 MW. In order to increase the thermal capacity and power generation of the geothermal field, the well RRG-9 about 1.6 km southwest of the geothermal field was hydraulically fractured to achieve better hydraulic connection with the existing geothermal field. Hydraulic fracturing has resulted in a significant increase in the injection capacity of the well RRG-9. Subsequent cyclic tests and tracer tests also showed connectivity to other wells in the geothermal field.

2. Australia

The Habanero project is located in the Cooper Basin in south-central Australia. Oil and gas exploration well data show that the 4,000 m well can reach a depth of 250 °C and the thermal reservoir lithology is granite. Drilling began in 2002 with a total of four development wells: Habanero-1, Habanero-2, Habanero-3, Habanero-4. The drilling depth is more than 4,000 m, the test formation temperature is about 240 °C, and the formation pressure is about 73 MPa. Through coring and logging interpretation analysis, the thermal reservoir is igneous rock and thermal alteration rock, the predicted porosity is only 0.3%, and the matrix has no permeability. The four wells are in the Habanero fault and there are a large number of natural fractures. The Habanero-1 well was drilled to a depth of 4,421 m and was hydraulically fractured using brine. The main fracture extends horizontally with a coverage area of approximately 3 km^2; the well Habanero-2 has fallen items at the bottom of the well after completion of the drill; The casing of well Habanero-3 is damaged. After the fracturing test, both the well Habanero-2 and the well Habanero-3 were abandoned. In 2012, the well Habanero-4 was drilled and the well Habanero-1 was used as the injection well to form one injector and one producer circulatory system. The flow rate reached 25 kg/s during the cyclic injection test and the production wellhead temperature was 210 °C. A two-well cycle injecting and producing (1 MW installed) power generation system with a distance of 705 m was built (Gallagher 1987).

3. France

Among the internationally enhanced geothermal system (EGS) demonstration projects, the most successful is the French Soultz project. The site of Soultz EGS project is located on the Rhineland graben and is part of the European Cenozoic Rift that extends from the Mediterranean to the North Sea coast. The sediment thickness in this area is about 1,400 m, the shallow geothermal flow value is 100–120 mW/m^2, and the geothermal gradient within 1,000 m is very high, reaching 100 °C/km (Genter et al. 1995).

The Soultz project began in the mid-1980s and can now be divided into three phases: the preliminary survey site selection phase; the exploration, drilling, and reservoir development phases; and the ground power plant construction phase. The second phase, which lasts longer, is the main phase of the project. The Soultz project has implemented five wells, which constitute two injection and production circulation systems in the shallow part (depth of 3.5 km, one injector and one producer comprising of well GPK1 and GPK2) and deep zone (depth of 5 km, one injector and two producers comprising of wells GPK2-GPK3-GPK4) (Roger 1999).

The Soultz project in France was initially tested as one injector and one producer. The well GPK1 was completed in 1992 with a depth of 3,590 m. According to the microseismic monitoring results of drilling and fracturing process in the well GPK1, the well GPK2 was located and designed. In 1995, the drilling was completed, the depth of the well was 3876 m, and the bottom of the well was 50 m away from the well GPK1. In the hydraulic fracturing process of well GPK2, the pressure response of well GPK1 is obvious, showing a good hydraulic relationship. The injection-production cycle test flow rate reached 25 kg/s (1.5 m^3/min), and the fluid recovery rate was close to 100%. In 2009, two injectors and two producers were adopted. When megawatt-scale power generation was carried out in 2011, in order to reduce induced earthquakes, two injectors and one producer were adopted.

4. Japan

The Hijiori project is Japan's first demonstration project for hot dry rock development. It is located on the southern edge of the Hijiori crater in Japan and can reach 225 °C at a depth of 1,500 m. The geothermal reservoir consists of two thermal reservoirs, shallow and deep. Four wells, SKG-2, HDR-1, HDR-2, and HDR-3, were drilled before and after. Natural fractures were found well developed at 1,800 and 2,200 m deep during drilling. The project's research was divided into two phases: the first phase began in 1985, and the hydraulic fracturing test was performed on the well SKG-2 in 1988, followed by the well SKG-2 as the injection well, the HDR-1 well and the HDR-2 well as production wells, water loss exceeds 70% during the cycle. In 1991, the injection-production cycle test was repeated, the water loss rate dropped to 22%, and the heat extraction rate reached 8.5 MW (Norio et al. 2008). The second phase began in 1992 with the well HDR-1 deepened to 2,205 m and hydraulic fracturing at the bottom of the well, then deepening the wells HDR-2 and HDR-3 to 2,303 m. In 1995, a cyclic injection test was conducted on deep heat storage, and the water loss rate was about 50%. A one-year cyclic injection and production test was started in 2000. Wells SKG-2 and HDR-1 were used as injection wells, HDR-2 and HDR-3 were used as production wells. The production temperature was about 172 °C, with total heat extraction rate of about 8 MW, driving a small generator set of 130 kW. Later, due to the continuous increase of water loss during the cyclic injection, and the production temperature of the well HDR-2 was reduced from 163 to 100 °C, the cyclic injection and production test was finally stopped (Hori et al. 1999).

The Ogachi project is located in Akita Prefecture, Japan, and the heat source is the Yamabushi volcano. The target reservoir lithology is granodiorite, which can reach a temperature of 230 °C at a depth of 1,000 m. The project began in 1990 with an injection well of OGC-1 with a depth of 1,000 m and bottom hole temperature of 230 °C. Two hydraulic fracturing were performed at the bottom of the well with a depth of 710 m. The production well OGC-2 has a depth of 900 m and a bottom hole temperature of 240 °C. The cyclic injection and production test showed that the water injection recovery rate was only 3%, and the recovery rate of the well OGC-2 after hydraulic fracturing was increased to 10%. After that, the well OGC-1 and the well OGC-2 were fractured again, and the recovery rate reached 25%. In order to improve the low recovery rate, combined with the previous hydraulic fracturing microseismic monitoring results, the newly drilled well OGC-3 touched the artificial fracture system. The new hydraulic circulation between the welOGC-3 l and the well OGC-1 has been greatly improved, and the recovery rate of the injected water has been greatly improved. Practice has shown that it is more effective to drill into an existing fracture system than to establish a fracture system by drilling a new well and followed by hydraulic fracturing (Kaieda 2015).

In summary, the development of early hot dry rock in the world is mainly concentrated on granite with high temperature. The current thermal storage stimulation technology is still difficult to achieve competitive commercial development, and then the research begins to extend to the edge or deep of the system of existing hydrothermal geothermal resource, and the hydrothermal heat reservoir is expanded to increase the power generation capacity of the existing hydrothermal geothermal field. Based on early research experience, the US Department of Energy has funded several EGS demonstration projects: Desert Peak in Nevada, Geysers in California, and Raft River in Idaho. The hydrothermal resources in these places have been developed for many years, but the production capacity has decreased slightly with time. There is an urgent need for the development of hot dry rock—hydraulic fracturing to increase production capacity (Walters 2013).

In short, from the depth point of view, 1000–3000 m of the upper 2,000–5,000 m are high temperature hydrothermal type; from the temperature point of view, the more successful cases are 180 to 200 °C or even higher, the highest has reached 400 °C; from the reservoir characteristics, the formation with natural fractures and relatively easy to form volumetric fractures is relatively easy to succeed; from the well construction process, mainly directional wells, open-hole completions, clean water, foam and other solid-free low-density drilling fluids are used; from the fracture parameters, water or salt water fluids, no proppant are used and the scale of fracturing is basically more than 2,000 m^3.

The development of hot dry rock is a worldwide problem. Unlike the shallow, medium and deep hydrothermal types, the development of hot dry rock is increasing with the increasing depth, temperature and complexity of deep strata. At present, there are 8 developed countries in the world involved in the development of hot dry rock, and only 4 projects in 3 countries have been commercialized.

10.2.3 Key Technologies

The engineering technology for hot dry rock development was first proposed by Los Alamos National Laboratory. The main practice is to form artificial fractures through hydraulic fracturing to communicate injection wells and production wells for the fluid to be continuously heated and produced in high temperature hot dry rock mass. The key technologies involved include well location identification, system design, efficient well construction, fracturing to stimulate formation, and stable operation of the system (Lu 2018).

1. Well location identification

Before the development of hot dry rock, the identification of area and well location should be carried out, and the target block with superior economic and technical indicators should be screened out (MIT-Led Report 2006). The preferred indicators include the temperature of the heat storage, the development of the fractures, the geothermal heat flow data, the buried depth of Curie interface, the acidic rock mass distribution, and thermal control structure characteristics. The main technical means is to optimize the target blocks of economic and technical indicators through geological, geophysical, geochemical, and remote sensing methods based on field experiments and research. The problems faced by the location identification of hot dry rock are burial depth of heat source, strong non-uniformity of geothermal field, complex main controlling factors of genetic mechanism and unclear availability of resources (Manuel 1999).

Hot dry rocks are generally buried deep under the surface of the earth. They cannot be directly surveyed and evaluated except for drilling work, but some parameters can be used to indirectly reflect underground thermal anomalies, such as geothermal flow values, buried Curie depths, acid rock mass distribution and thermal control construction features (Stober et al. 2012).

(1) Terrestrial heat flow value

The geodetic heat flow value, abbreviated as heat flow, is a phenomenon of heat transfer from the Earth's interior to the surface, which is the most important representation of geothermal field.

(2) Buried depth of Curie isothermal surface

In the earth's crust, when the ferromagnetic temperature rises to a certain value, the ferromagnetic susceptibility will drop sharply, and the ferromagnetic phase will be converted into a paramagnetic phase. This critical temperature is the Curie point, and the buried depth is Curie surface. The aeromagnetic data can be used to calculate the subsurface depth reaching the Curie point. The depth is relatively shallow, which means that the transmission and emission distance of deep heat source flows to the surface is short, conducive to the exploration and research of high temperature geothermal resources such as hot dry rock.

(3) Distribution of acidic rock

The geothermal heat sensed by the earth's surface is mainly composed of two parts, one part of the convective and conduction heat from the earth's core and the mantle up to the crust, referred to as the mantle heat flow; the heat flow generated by the radioactive elements present in the earth's crust, especially the isotope decay of U, Th, K is called the crustal heat flow. The higher crustal heat flow has an important influence on the temperature distribution of the shallow crust, which is conducive to the occurrence of hot dry rock resources. Acidic rock mass is rich in radioactive heat generating elements U, Th and K, and therefore is one of the important indicators for the occurrence of hot dry rock resources.

(4) Thermal control tectonic features

The neotectonic movement shapes the current geothermal resource background, and the active structure has important implications for the distribution of these resources. According to the heat flow tectonic regions, the intensity of the geothermal activity generally decreases as it moves away from the plate boundary. The plate boundary and its adjacent areas are the areas with the strongest tectonic activity, and have the necessary structural conditions and thermal background to generate intense hydrothermal activity and to inoculate high-temperature hydrothermal systems.

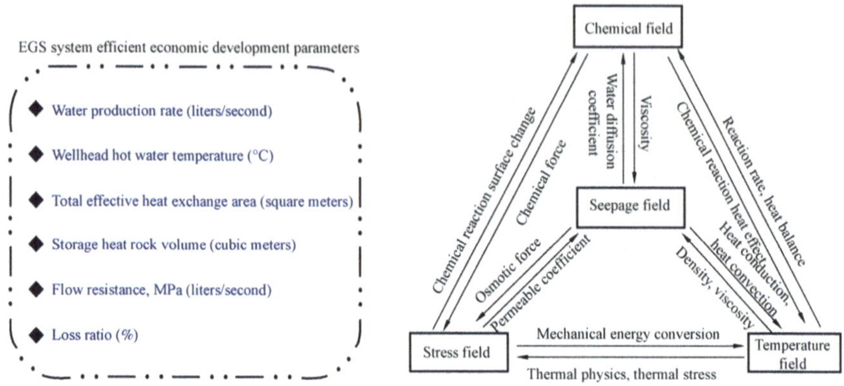

Fig. 10.3 EGS system design and adjustment basis

2. System design

After the site identification, we should design the hot dry rock development system, master the characteristics of thermal storage geological engineering, finely describe the heat storage, and design the heat storage parameters, well group, well pattern, well spacing, and injection and production system. The main technologies include geological modeling, seepage heat transfer simulation, multi-field coupling modeling and numerical solution, heat storage operation efficiency and life analysis. The difficulty lies in the development process of hot dry rock, which is a complex coupling process of temperature field, stress field, seepage field and chemical field in time and space. It involves many key development indicators and is difficult to coordinate (Fig. 10.3).

3. Efficient well construction

Deep drilling construction according to system design requirements, a reliable circulation heating channel is established. To rely on high-temperature and rock breaking bits and tools suitable for hard formations, high-temperature measuring instruments and wellbore fluid materials and supporting technologies will ensure smooth drilling in formations with ultra-high temperature, high strength, well developed fractures, complex structure and frequent thermal break out. At present, the drilling cost of hot dry rock is relatively high, mainly due to the wear of drilling bit, casing and high temperature resistant cement slurry (Warpinski et al. 2008). The statistics show that the drilling cost of a hot dry rock well with depth of 5,000 m is nearly CNY 100 million, and the drilling cost accounts for more than 50% of the total development cost.

4. Fracturing to stimulate formation

In the development of hot dry rock, fracturing to stimulate formation is currently the most difficult and the most critical technology which is restricting the development of hot dry rock. The purpose of fracturing is to establish a thermal reservoir of fractured development space with large volume and large heat transfer area. The difference between fracturing of hot dry rock and unconventional oil gas cracking is that oil and gas reservoirs require that the flow distance from oil and gas to fractures is the shortest, and the fractures are more complex. The hot dry rock needs to take into account both fracture complexity and fluid mobility. It requires both a large fracture area for heat exchange and a fluid flow resistance within the fractures within a certain range, and two conditions should be met for commercial exploitation.

There are many similarities between hydraulic fracturing of hot dry rock mass and hydraulic fracturing of oil and gas reservoirs. The purpose is to form hydraulic fractures in underground thermal storage/oil reservoirs through surface high pressure injection fluid to improve reservoir permeability and obtain maximum oil and gas productivity or heat. Through the analysis of technical parameters of hydraulic fracturing of hot dry rock abroad, it is found that the hydraulic fracturing scale of hot dry rock is roughly equivalent to the hydraulic fracturing scale of unconventional shale gas reservoir, and the fluid volume of single well reaches 30,000 m^3 or more. The purpose is to form a certain volume scale fracture network system in oil and gas reservoirs or thermal storage reservoirs by using clean water. The differences are the depth, temperature of the target zone of hydraulic fracturing and the injection rate of fracturing.

The goals of stimulating hot dry rock and unconventional oil and gas reservoirs are similar and different to some extent (Table 10.4). Unconventional oil and gas reservoir stimulation requires hydraulic fracturing to break down the reservoir to form a fracture network, shorten the distance of the oil flow from the rock matrix to the crack, and maximize the production rate of oil and gas. Fracturing stimulation of hot dry rock requires sufficient heat exchange space by hydraulic fracturing to form a certain fracture surface area for heat extraction/heat transfer. Therefore, it is necessary to consider the fracturing design by combining the different needs and understanding the corresponding fracturing mechanism.

Table 10.4 Analysis of the difference between geological and mechanical characteristics of hot dry rock reservoirs and unconventional reservoirs

Parameters	Hot dry rock	Shale	Tight oil
Density (g/cm^3)	2.64		
Porosity (%)	2	2–4	8–10
Permeability (mD)	0.01	<0.1	<0.1
Young's modulus (GPa)	69	36	
Poisson's ratio	0.16	0.239	
Shear modulus			
Compressive strength (MPa)		292	
Tensive strength (MPa)		2.5	

Note Hot dry rock data is the test result of 4,400 m rock sample in hot dry rock target area in northern Songliao basin

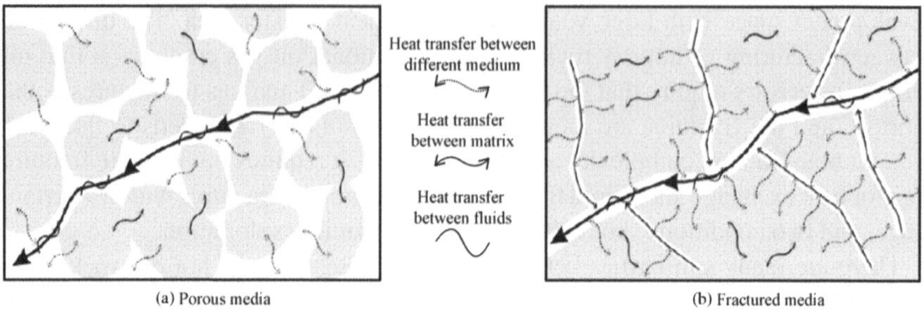

Fig. 10.4 Schematic diagram of heat conduction process of local thermal non-equilibrium model

The thermal energy of the hot dry rock power generation system is derived from the heating of the injected circulating fluid by the underground hot dry rock. The essence of heat is that the fluid flows in the fracture and extracts heat from the high temperature rock mass on both sides of the fracture. Driven by the temperature difference between the heat storage rock and the water flow in fracture, there is heat conduction inside the heat storage rock, and there is heat conduction and heat convection in the fracture wall and the fluid inside the fracture. According to the most basic heat transfer model, given the heat transfer coefficient, the larger the fluid contact area, the greater the heat transfer.

Hydraulic fracturing of thermal reservoirs creates new hydraulic fractures and alters the distribution and reconnection of the original natural fractures, which in turn affects the flow path of the fluid. The higher the complexity of artificial fracture during hydraulic fracturing, the more complex the fluid seepage path, which will increase the heat exchange area. Under the premise of ensuring that the fracturing has a certain stimulated volume, the longer the flow path of the fluid in the heat storage, the more heat exchange can occur between the fluid and the rock mass, and the higher the heat recovery temperature, the lower the risk of thermal breakthrough, and the longer the thermal reservoir life (Fig. 10.4).

Another commercial operating parameter of the hot dry rock power generation system requires that the fracture complexity be controlled within a certain range. This parameter is called the reservoir flow impedance. The flow resistance of the reservoir is related to the shape of the fracture formed after the fracturing. The more complicated the fracture morphology, the higher the frictional resistance formed, the greater the pressure difference between injection and production, and the higher the flow resistance. Therefore, hydraulic fracturing of hot dry rock power generation systems requires consideration of the complex relationship between the complexity of the fracture and the flow resistance of the reservoir.

The biggest difference between fracturing of hot dry rock and fracturing of oil and gas wells is the injection rate during fracturing (Table 10.5). This is because the heat extraction of the hot dry rock in the later stage requires that the flow rate of the fluid passing through the fracture surface should not be too fast, so that the fluid can fully exchange heat with the reservoir to achieve a certain heat exchange efficiency

Table 10.5 Comparison of fracturing processes of different resources

Parameters	Hot dry rock	Shale gas	Tight oil
Fracturing volume (m^3)	20,000–30,000	20,000–30,000	20,000–30,000
Injection rate (m^3/min)	<5	>10	>10
Depth (m)	>3500	<3500	<3000

from the flow into the production well while reducing the short circuit. The risk of circulation and thermal breakthrough requires that the fracture surface area formed by the rock mass is large enough and the fracture conductivity is relatively small.

Due to the influence of petrophysical parameters, there are fundamental differences in the breakdown mechanism of hot dry rock and unconventional oil and gas reservoirs during hydraulic fracturing. During the hydraulic fracturing of unconventional oil and gas reservoirs, the natural fractures slip and open with the increase of the net pressure of the fluid in the fracture, the hydraulic fractures extend to the surrounding, the natural fractures open and expand at the tip, the hydraulic fractures and natural fractures communicate with each other and eventually form a complex fracture network with continuous paths (Yusuke et al. 2013). Therefore, the unconventional oil and gas reservoir rock mass has both breakdown and extension and a certain degree of dislocation slip. The mechanism of the mixed breakdown makes the fracture network of the reservoir easier to establish, achieving the goal of increasing oil and gas production. The hot dry rock has high mechanical strength and low permeability. In a short time, the high-pressure fluid will directly penetrate into the natural fracture, and as the fluid pressure in the natural fracture increases, the natural fracture will slip and dilatate under the shearing action, and obtain a certain conductivity (White 1885).

From the hot dry rock power generation projects such as Fenton Hill in the United States and Soultz in France, it was found that before and after the hydraulic fracturing stimulation, the injected fluid flowed into the stratum from the natural fracture around the borehole wall, and there was no hydraulic fracture around the borehole wall, but the response of the microseismic monitoring is up to millions of square meters. Therefore, it is difficult for hot dry rock to form new hydraulic fractures under the existing operating parameters. The natural fractures existing in the rock mass itself are activated and displaced and slipped, and hydraulic shearing action occurs on the basis of existing natural fractures, and the stiffness of the fracture surface maintains the surface roughness of the crack, so that the fracture surface remains open after the pressure is released, forming a dilatancy. Compared with the idea that unconventional oil and gas reservoirs form a fracture network by opening and shearing together, the existing hydraulic fracturing process of hot dry rock may only have large-scale slip and dilatancy of natural fractures, although the response range from microseismic monitoring shows the slip and dilatancy is large, but the connectivity between the fractures may not be guaranteed, and sufficient flow heat exchange space cannot be formed, which cannot meet the parameters of the commercial operation of the hot dry rock power generation system (Warpinski et al. 2009).

5. System operation

In terms of system operation, it is necessary to maintain the operating life of the system for more than 20 years, and ensure that the temperature and flow rate of the produced fluid always meet the power generation requirements during the operation, ensuring the underground heat exchange efficiency and the smooth flow of the ground power generation system. System operation monitoring (trace), thermal storage dynamic simulation and dynamic optimization of operating parameters, optimization of power generation process, pipeline descaling and scale inhibition are required.

10.2.4 Outlook of Development

The exploration and development of China's hot dry rock is still in the exploration stage. In 2014, the land and resources system carried out geological exploration of hot dry rock resources in high heat flow areas such as Qinghai, Tibet, Sichuan, Fujian, Guangdong, Hunan, Songliao Basin and Hainan, and preliminary drilling of hot dry rocks in Guide and Gonghe of Qinghai, Lijin of Shandong, Huizhou of Guangdong and Kangding of Sichuan has been carried out. Most of the wells are about 1,000 m deep and the temperature is 100–120 °C (Table 10.6) (Gan et al. 2015).

Table 10.6 Statistics of hot dry rock survey in China

Region	Well name	Wellbore depth (m)	Bottom temperature (°C)	Description
Qiabuqia, Gonghe Basin	QR1	969	70	Geothermal well
	DR1	1455	87	Unutilized
	DR2	1853	100	Unutilized
	DR3	2927	183	Hospital heating (hot water)
	DR4	3102	181	Medium temperature hot water test power station
	GR1	3700	180	Unfinished
	GR2	3000	182	Plugging well
Zhacanggou, Guide Basin	R2	1709.56	97	Plugging-monitoring
	R3	2701.2	106.7	Plugging-monitoring
	ZR1	3050.68	151.34	Plugging-monitoring
	ZR2	2000	>120	Under construction
Zhangzhou, Fujian	HDR1	4000	109	Plugging-monitoring

In April 2014, after two years of drilling, the Qinghai Hydrogeology and Engineering Geological Environmental Geological Survey Institute finally drilled 183 °C hot dry rock at a depth of 2,927 m in the Gonghe Basin of Hainan Tibetan Autonomous Prefecture in Qinghai Province. In June 2014, hot dry rock above 168 °C was successfully drilled at 2,735 m. On October 6, the ZKD23 well in Gonghe County, Qinghai Province, drilled 2,889 m and the bottom hole temperature reached 181 °C. Up to now, seven exploration wells have been constructed in Chabucha, the Gonghe Basin in Qinghai, and four of them have reached the standard of hot dry rock. Four exploration wells were constructed in Zhacanggou of the Guide Basin, of which two reached hot dry rock standard; On May 21, 2015, China Geological Survey organized and implemented China's first deep well for hot dry rock scientific drilling and exploration, located in Qingquan Forest Farm in Dongpu Township, Longhai City, Zhangzhou, Fujian Province, with a drilling depth of 4,000 m. In May 2017, the GR1 hot dry rock exploration wellbore completed in southeast Chabucha Town, Gonghe County, reached a new high temperature, and the bottom temperature of 3,700 m reached 180 °C, achieving a major breakthrough in China's hot dry rock exploration (Xue and Gan 2013).

From the perspective of resources, hot dry rock is a potential strategic replacement energy source, which has attracted the attention of many countries. From a technical point of view, the development of hot dry rock in the world is still in its infancy, and it needs to continue to tackle the problem; from the perspective of geothermal resources itself, hot dry rock is the future of geothermal energy, mainly because conventional geothermal resources are mainly based on heat and are affected by geographical restrictions. Thermoelectric conversion is its most advanced form of utilization, but China's high-temperature geothermal power generation potential (7.12 GW) is relatively limited in resources compared to other new energy sources such as wind, light, hydropower, etc., and the received attention and policy support to continuously improve technology reduce the cost slowly. According to the Chinese Academy of Sciences, China's hot dry rock power generation potential is expected to exceed 700 GW, far exceeding the potential of the hydro power limit of 400 GW. Therefore, if geothermal energy plays an important role in the field of new energy, it must rely on hot dry rock (Zhao 1995).

References

Bai, Jiaqi, Mai Lin, and Mei-ling Yang. 2006. Geothermal resources and crustal thermal structure of the Qinghai-Tibet Plateau. *Journal of Geomechanics* 12: 354–362.

Gallagher, K. 1987. Thermal conductivity of sedimentary and basement rocks from the Eromanga and Cooper Basins, South Australia. *Exploration Geophysics* 18: 381–392.

Gan, Haonan, Guiling Wang, et al. 2015. Research on the occurrence types and genetic models of hot dry rock resources in China. *Science & Technology Review* 33 (19): 22–27.

Gao, Wanli, Zongxiu Wang, et al. 2014. Zircon U-Pb geochronological, geochemistry and tectonic implication of Indosinian Granite from southeastern Zhejiang, South China. *Acta Geological Sinica* 88 (6): 1055–1067.

Genter, A., H. Traineau, C. Dezayes, et al. 1995. Fracture analysis and reservoir characterization of the granitic basement in the HDR Soultz project (France). *Geothermal Science and Technology* 4 (3): 189–214.

Hori, Y., K. Kitano, H. Kaieda, et al. 1999. Present status of the Ogachi HDR project, Japan and future plans. *Geothermics* 28 (4/5): 637–645.

Kaieda, H. 2015. Multiple reservoir creation and evaluation in the Ogachi and Hijiori HDR projects, Japan. In *Proceedings world geothermal congress*, 19–25. Melbourne, Australia.

Manuel, N. 1999. The dependence of permeability on effective stress from flow tests at hot dry rock reservoirs at Rosemanowes (Cornwall) and Fenton Hill (New Mexico). *Geothermics* 28 (3): 315–340.

Massachusetts Institute of Technology. 2006. The future of geothermal energy: impact of enhanced geothermal systems (EGS) on the United States in the 21st century. London: Cambridge Press.

MIT-Led Report. 2006. The future of geothermal energy: impact of enhanced geothermal system (EGS) on the United States in 21st century. MIT-Led interdisciplinary panel, 372.

Norio, Y., M. Isao, S. Hajime, et al. 2008. Temperature-dependent scale precipitation in the Hijiori Hot Dry Rock system, Japan. *Geothermics* 37 (1): 1–18.

Qinghai Bureau of Geology and Mineral Resources. 1991. *Qinghai Province Regional Geology*. Beijing: Geological Publishing House.

Roger, P. 1999. The Rosemanowes HDR project 1983–1991. *Geothermics* 28 (4/5): 603–615.

Shyi-Min, Lu. 2018. A global review of enhanced geothermal system (EGS). *Renewable and Sustainable Energy Reviews* 81 (2): 2902–2921.

Stober, I., and K. Bucher. 2012. Enhanced-geothermal-systems (EGS), hot-dry-rock systeme (HDR), deep-heat-mining (DHM). Springer Geology, 163–176.

Walters, M. 2013. Demonstration of an enhanced geothermal system at the northwest geysers geothermal field. CA: U.S. Department of Energy Geothermal Technologies Office 2013 Peer Review.

Wang, Shaoyong. 2015. Some thoughts on new energy development by oil companies. *Sino-Global Energy* 20 (8): 18–22.

Wang, Jiyang. 2017. Advancement and developmental trend in the geothermics of oil fields in China. *Earth Science Frontiers* 24 (3): 2–10.

Wang, Guiling, Zhiming Liu, Wenjing Yan, et al. 2011. Assessment of geothermal resource potential in China. In *Geothermal energy development and utilization and low carbon economy seminar—summary of the 14th session of the 13th china association for science and technology annual meeting*, 14–25.

Warpinski, N.R., M.J. Mayerhofer, M.C. Vincent, et al. 2008. Stimulating unconventional reservoirs: maximizing network growth while optimizing fracture conductivity. SPE 114173.

Warpinski, N.R., M.J. Mayerhofer, M.C. Vincent, et al. 2009. Stimulating unconventional reservoirs: maximizing network growth while optimizing fracture conductivity. *Journal of Canadian Petroleum Technology* 48 (10): 39–51.

Weibin Ma. 2015. The present situation of geothermal energy exploitation and utilization and its development trend in China. *Proceedings of the Chinese Academy of Sciences* 31 (2): 199–204.

White, I.C. 1885. The geology of natural gas. *Science* 125: 521–522.

Xue, Jianqiu, and Bin Gan. 2013. Geological-geophysical characteristics of enhanced geothermal systems (hot dry rocks) in Gonghe-Guide Basin. *Geophysical and Geochemical Exploration* 37 (1): 35–41.

Yan, Wenjing, Liu Zhiming, Wang Yuli, et al. 2013. The assessment of geothermal resources potential of China. *Geology in China* 40 (1): 312–321.

Yusuke, M., A. Hiroshi, N. Hiroaki, et al. 2013. Characteristics of large-magnitude microseismic events recorded during and after stimulation of a geothermal reservoir at Basel, Switzerland. *Geothermics* 45 (45): 1–17.

Zhao, Ping. 1995. Characteristics of heat production distribution in SE China. *Acta Petrological Sinica* 11 (3): 292–305.

Chapter 11
Nuclear Energy

Nuclear energy is the energy released when the structure of the nucleus changes. Nuclear energy release includes nuclear fission and nuclear fusion. The raw material used for nuclear fission can release the equivalent of 30 tons of coal in 1 g of uranium, and 560 tons of raw material deuterium used in nuclear fusion can provide the world with one year of energy. The reserves of deuterium in the ocean can be used by humans for billions of years, and it is an inexhaustible new energy source. The current utilization of nuclear energy is mainly based on nuclear fission, and the development of nuclear energy will gradually move from nuclear fission to nuclear fusion (Sun et al. 2016).

11.1 Current Status of Nuclear Energy Utilization

11.1.1 Status of Uranium Resources

The distribution of uranium resources in the world is extremely uneven, mainly in Australia, Canada, Kazakhstan, Niger, Namibia, Russia, South Africa and other countries (Zhang 2017). The world's uranium resources amounted to 9,546,900 tons, of which the production cost of 80–130 US dollars/kg of resources totals 3,454,400 tons, 40–80 US dollars/kg of resources is 1,223,600 tons, less than 40 US dollars/kg of the resources with low production cost is 478,500 tons, and 130 to 260 US dollars/kg with high production cost is 4.384 million tons (Lei and Yan 2017).

The estimation of China's uranium resources is quite different. According to the data of the International Atomic Energy Agency, China has identified 68,000 tons of uranium resources, and the amount of uranium resources to be identified is 0.6 million tons, a total of 74,000 tons. The results of the "National Uranium Resources Potential Evaluation Project" of the China Geological Survey revealed that the identified uranium resources in China are distributed in 23 provinces, municipalities and

C. Zou, *New Energy*, https://doi.org/10.1007/978-981-15-2728-9_11

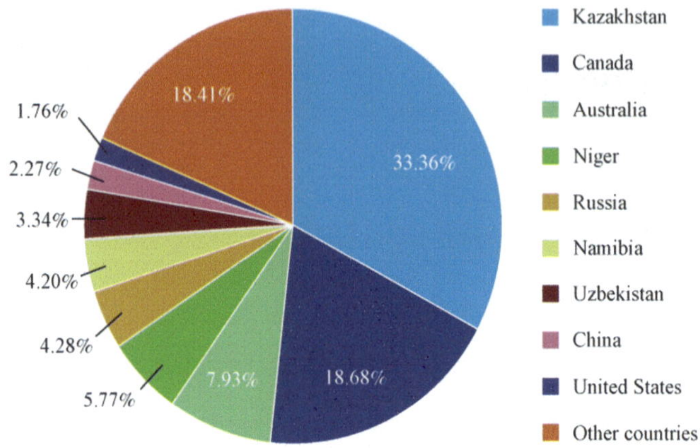

Fig. 11.1 Proportion of world's major uranium producing countries in 2015 as a percentage of world production

autonomous regions. The types of deposits are diverse and the metallogenic conditions are complex, with sandstone and granite types as the major (Sun 2016). China's uranium mine prediction area is expected to have a potential of more than 2 million tons of uranium resources (Pan 2012).

From 2007 to 2015, the world's uranium production showed an overall upward trend, up 46.5% year-over-year. In 2015, the world's uranium output was 71,343 tons, a year-over-year increase of 7.9%. It is expected that uranium production will continue to increase in the future. The world's uranium production is mainly concentrated in Kazakhstan, Canada, Australia, Niger, Russia, Namibia and Uzbekistan (Fig. 11.1). In recent years, Kazakhstan has occupied an extremely important position in the international uranium production market, using low-cost leaching mining technology (ISL) to develop sandstone-type uranium mines.

World uranium production as a whole was on the rise from 2007 to 2015, rising 46.5% year-over-year. World uranium production in 2015 was 71,343 tons, an increase of 7.9% year-over-year. Uranium production is expected to continue to show an increasing trend in the coming period. World uranium production is mainly concentrated in Kazakhstan, Canada, Australia, Niger, Russia, Namibia and Uzbekistan (Fig. 11.1). In recent years, Kazakhstan has occupied an extremely important position in the international uranium production market, using low-cost in-site leaching (ISL) mining technology to develop sandstone-type uranium deposits.

11.1.2 Development History of Nuclear Power Technology

A nuclear power plant is a power plant that uses electrical energy released by nuclear fission or nuclear fusion to generate electricity (Guan et al. 2012). Nuclear power

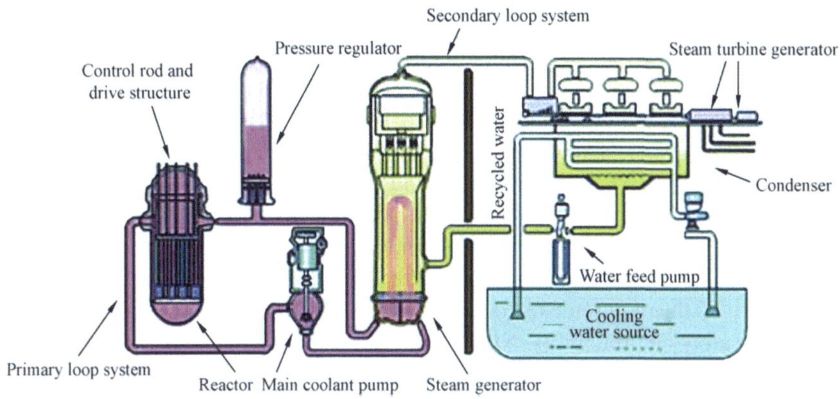

Fig. 11.2 Working principle of nuclear power plant (according to China Nuclear Energy Industry Association and Merchants Securities)

plants currently in commercial operation use nuclear fission reactions to generate electricity (Fig. 11.2). Nuclear power plants are generally divided into two parts, nuclear islands that use nuclear fission to produce steam (including reactor units and primary loop systems) and conventional islands that use steam to generate electricity (including turbine generator systems). The fuel used is generally radioactive uranium and thorium heavy metals. If a nuclear reactor is removed, the difference between the nuclear power plant and the thermal power plant is small except for the heat source that generates the steam. The development and construction of nuclear power plants began in the 1950s and has grown to the fourth generation of nuclear power plants since then.

At present, the third-generation reactor technology with passive safety technology is becoming more and more mature, and engineering projects such as AP1000, HPR1000, EPR and APR1400 have started construction or are close to completion; the fourth-generation reactor technology, including the development of accelerator-driven subcritical systems, etc. has been advanced in an orderly manner.

The first generation of nuclear power plants: In 1954, the Soviet Union built an experimental nuclear power plant with a power generation of 5 MW. In 1957, the United States built a Shipping Port prototype nuclear power plant with a power generation capacity of 90,000 kW. These achievements demonstrate the technical feasibility of using nuclear power to generate electricity. Internationally, the above experimental prototype nuclear power unit is called the first-generation nuclear power unit.

The second generation of nuclear power plants: In the late 1960s, on the basis of experimental and prototype nuclear power plants, nuclear power plants such as pressurized water reactors, boiling water reactors, heavy water reactors, and graphite water cooled reactors with a generating capacity of 300,000 kW were successively built. While further proving the feasibility of nuclear power generation technology, the economics of nuclear power can also be proved. At present, most of the more than 400 nuclear power units operating in the world are built during this period, and are customarily called second-generation nuclear power units.

The third-generation nuclear power plants: In the 1990s, in order to eliminate the negative impact of the accidents at the Three-Miles Island and Chernobyl nuclear power plants, the world's nuclear power industry concentrated its efforts on studying and solving problems of the prevention and mitigation of serious accidents. The United States and Europe have successively introduced the URD document of the Advanced Light Water Reactor User Requirements Document and the EUR document of European User Requirements for Light Water Reactor Nuclear Power Plants have further clarified the requirements for preventing and mitigating serious accidents and improving safety and reliability. Nuclear power plants that meet URD files or EUR files are often referred to internationally as third-generation nuclear power plants.

The third-generation reactors are advanced reactors, including light water reactors (ABWR, AP600/AP1000, IRIS, System 80, EPR, SWR1000, ESBWR, VVER-91), heavy water reactors (CANDU, ACR, AHWR), and high temperature gas cooled reactors (PBMR, GTMHR), fast neutron reactor (BN-800, BN-600, ALMR, Super-PRISM) and other reactor types.

The AP1000 is the third generation of pressurized water reactor designed by Westinghouse in the United States in accordance with the concept of passive nuclear safety. The CAP1400 is the third generation of pressurized water reactor developed by China on the basis of AP1000. The CAP1400 demonstration project was carried out in an orderly manner at Shidaowan, Rongcheng City, Shandong Province. HPR1000, Hualong No.1 is also the third-generation pressurized water reactor developed by China. The first demonstration project of Hualong No. 1—China Nuclear Industry Group Fuqing Nuclear Power Unit No. 5 is under construction. The APR1400 is the third generation of pressurized water reactor studied in Korea. The first unit of APR1400 (No. 3, Shingori Nuclear Power Station, Korea) was put into commercial operation in December 2016; the second unit (Shingori No. 4) is planned to be is put into commercial operation in September 2018. The four APR1400 units under construction in the United Arab Emirates are expected to be put into operation in 2020; in addition, the three nuclear power plants planned to be built in Mooreside, UK are planned to adopt the APR1400 design.

The fourth-generation nuclear power plants: As early as 1999, nuclear scientists proposed the concept of the fourth-generation nuclear power reactor (Gen-IV). The nuclear power developed countries such as the United States, France, Japan, and the United Kingdom established the Gen-IV Forum (GIF) International Forum in 2000. The overall goal is to provide the market with a fourth-generation nuclear energy system capable of addressing nuclear energy economics, safety, waste disposal and nuclear non-proliferation issues by around 2030. After extensive research and certification, the following six types of reactors have been selected as the focus of the fourth-generation reactor: sodium cooled fast reactor (SFR), lead alloy cooled fast reactor (LFR), gas cooled fast reactor (GFR), very high temperature reactor (VHTR), supercritical cooled water reactor (SCWR) and molten salt reactor (MSR).

The Accelerator Driven Subcritical System (ADS) is a fourth-generation nuclear energy system recognized by the International Atomic Energy Agency (IAEA) and is a subcritical fast reactor based on lead-bismuth alloy cooling. In December 2016, China's first lead-based nuclear reactor zero-powered device, Qimingxing II, achieved its first criticality.

11.1.3 Installation Capacity of World Nuclear Power

By the end of 2017, there were 30 countries using nuclear power generation in the world, with a total of 445 nuclear power generating units in operation, with a total net installed capacity of approximately 389.8 GW; 15 countries are constructing a total of 58 nuclear power units with a total installed capacity of approximately 62.7 GW (Table 11.1). Among them, there are 99 nuclear power units in operation in the United States with a total net installed capacity of 99.65 GW; and 2 units under construction with a total installed capacity of 2,500 MW. There are 35 nuclear power plants in operation in Russia with a total net installed capacity of 26.87 GW and 7 units under construction with a total installed capacity of 5,904 MW. There are 58 operating units in France with a total net installed capacity of 63.13 GW; one unit under construction with a total installed capacity of 1,750 MW. There are 15 operating units in the UK with a total net installed capacity of 8883 MW. There are 40 operating units in Japan with a total net installed capacity of 37.71 GW, but only 4 units are in operation, with a total net installed capacity of 3,352 MW; and 2 units under construction with a total installed capacity of 2,756 MW. There are 24 operating units in Korea with a total net installed capacity of 22.5 GW; 4 units under construction with a total installed capacity of 5,600 MW. There are 22 operating units in India with a total net installed capacity of 6,219 MW; 6 units under construction with a total installed capacity of 4,350 MW. China has 38 commercial nuclear power plants with an installed capacity of 34,647 MW; 20 units under construction with a total installed capacity of 21,516 MW. In 2017, the cumulative power generation of nuclear power was 247.469 billion kWh, an increase of 17.6% over 2016.

11.1.4 Challenges and Prospects of Nuclear Fission

The nuclear fission resources on the earth are mineral resources, and their reserves are limited. After the release of nuclear energy, they cannot recover their own nuclear fuel in a short period of time, which is a non-renewable energy source (Tang 2016). Although nuclear fission resources have a relatively long-life cycle, scarcity of resources has made it impossible to become the main body of world energy. Nuclear fusion can be regarded as a renewable energy to a certain extent, because the fusion raw material, that is, the isotope of hydrogen, is abundant in the ocean. However, its acquisition is more difficult than hydrogen, and the application technology is not yet mature, so it does not have industrial conditions in a short time.

Table 11.1 Status of nuclear power in countries/regions of the world

Country/Region	Units in operation		Units under construction		Units planned		Units proposed	
	Quantity (counts)	Net capacity (MW)	Quantity (counts)	Net capacity (MW)	Quantity (counts)	Net capacity (MW)	Quantity (counts)	Net capacity (MW)
United States	99	99,647	2	2500	14	3100	21	30,000
France	58	63,130	1	1750	0	0	0	0
Japan	40	37,712	2	2756	9	12,947	3	4145
China	38	34,647	20	21546	39	46,100	143	164,000
Russia	35	26,865	7	5904	26	28,390	22	21,000
Korea	24	22,505	4	5600	1	1400	6	8800
Canada	19	13,553	0	0	2	1500	0	0
Ukraine	15	13,107	0	0	2	1900	11	120,00
Germany	7	9444	0	0	0	0	0	0
United Kingdom	15	8883	0	0	11	15,600	2	2300
Sweden	8	8376	0	0	0	0	0	0
Spain	7	7121	0	0	0	0	0	0
India	22	6219	6	4350	19	17,250	46	52,000
Belgium	7	5943	0	0	0	0	0	0
Czech Republic	6	3904	0	0	2	2400	1	1200
Switzerland	5	3333	0	0	0	0	3	4000
Finland	4	2764	1	1720	1	1250	0	0
Bulgaria	2	1926	0	0	0	0	1	1200
Brazil	2	1896	1	1405	0	0	4	4000
Hungary	4	1889	0	0	2	2400	0	0
South Africa	2	1830	0	0	0	0	8	9600
Slovakia	4	1816	2	942	0	0	1	1200
Argentina	3	1627	1	27	2	1950	2	1300
Mexico	2	1600	0	0	0	0	3	3000
Pakistan	5	1355	2	2322	0	0	0	0
Romania	2	1310	0	0	2	1440	0	0
Iran	1	915	0	0	4	2200	7	6300
Slovenia	1	696	0	0	0	0	1	1000
Netherlands	1	485	0	0	0	0	0	0
Armenia	1	376	0	0	1	1060	0	0
United Arab Emirates	0	0	4	5600	0	0	10	14,400
Belarus	0	0	2	2388	0	0	2	2400

(continued)

Table 11.1 (continued)

Country/Region	Units in operation		Units under construction		Units planned		Units proposed	
	Quantity (counts)	Net capacity (MW)	Quantity (counts)	Net capacity (MW)	Quantity (counts)	Net capacity (MW)	Quantity (counts)	Net capacity (MW)
Bangladesh	0	0	1	1200	1	1200	0	0
Poland	0	0	0	0	6	6000	0	0
Turkey	0	0	0	0	4	4800	8	9500
Vietnam	0	0	0	0	4	4000	6	7100
Egypt	0	0	0	0	2	2400	2	2400
Jordan	0	0	0	0	2	2000	0	0
Indonesia	0	0	0	0	1	30	4	4000
Saudi Arabia	0	0	0	0	0	0	16	17,000
Thailand	0	0	0	0	0	0	5	5000
Chile	0	0	0	0	0	0	4	4400
Lithuania	0	0	0	0	0	0	2	2700
Malaysia	0	0	0	0	0	0	2	2000
Kazakhstan	0	0	0	0	0	0	3	1800
Israel	0	0	0	0	0	0	1	1200
Korea	0	0	0	0	0	0	1	950
Global total	445	389,801	58	62,710	157	16,2117	351	401,895

Nuclear radiation is very harmful to human health. To develop nuclear energy, we must first consider safety issues. Nuclear waste is a nuclear ash that remains after the nuclear material is burned in a nuclear reactor (atomic furnace). It is extremely radioactive and has a half-life of thousands of years, tens of thousands of years or even hundreds of thousands of years. After hundreds of thousands of years, these nuclear wastes can also harm humans and the environment. So how to handle nuclear waste safely and permanently is a major issue (Qiu and Yu 2017).

There are two prerequisites for safe and permanent disposal of nuclear waste: the first is to safely and permanently enclose nuclear waste in a container and to ensure that it does not leak for tens of thousands of years. For this purpose, it has been envisaged to enclose nuclear waste in a ceramic container or in a thick glass container. Scientific experiments have shown that these containers are ideal for storing nuclear waste within 100 years, but after 100 years, the container will not withstand the violent bombardment of radiation and burst, and then the radiation will be released into the surrounding environment. The consequences could be disastrous. The Royal Academy of Sciences has discovered that a new type of crystal can withstand the strong attacks of radiation and use it to produce containers for the storage of nuclear waste for greater safety. However, finding a substance that can withstand radiation from radiation for tens of thousands of years remains the direction of scientists' efforts.

The second is to look for a safe, permanent place to store nuclear waste. The site requires a particularly stable physical environment that is permanently protected from water and air and can withstand the impact of earthquakes, volcanoes, and explosions. Experiments have shown that granite, salt bed and clay layers can effectively ensure that nuclear waste containers are not damaged for hundreds of years. But after a few hundred years, these storage locations will be destroyed or not is unpredictable. The best way is to dig a hundred-meter-deep tunnel in such a stable place to store nuclear waste. After the science is developed in the future, look for a better way to deal with these "human killers".

International agencies such as the International Atomic Energy Agency (IAEA) and the International Energy Agency (IEA) announced their nuclear power development forecasts for the global future energy market in the second half of 2017, indicating that global nuclear power installed capacity will continue to grow in the future.

The IAEA has predicted both high and low values for future global nuclear power development scenarios: in high-value scenarios, global nuclear power installed capacity will increase from 392 GW at the end of 2016 to 554 GW in 2030, 717 GWin 2040 and 874 GW in 2050, the share of nuclear power generation in total global power generation in 2050 will increase from the current 11–13.7%; in the low-value scenario, the global installed capacity of nuclear power will decline to 345 GW in 2030, 332 GW in 2040 from 392 GW at the end of the 2016, and then gradually recovered to the level at the end of 2016. The share of nuclear power in 2050 will drop from the current 11% to about 6%.

The IEA pointed out that if the UN Sustainable Development Goals are to be achieved, nuclear power will need to provide at least 15% of the world's electricity by 2040. In the sustainable development scenario, the total global power generation will increase from 24,765 TWh in 2016 to 35,981 TWh in 2040, while nuclear power generation will increase from 2,611 to 5,345 TWh, and nuclear power share from 11% increased to 15%; in the new nuclear policy scenario, the total installed capacity of nuclear power worldwide will increase from 413 GW in 2016 to nearly 520 GW in 2040.

11.2 Development Prospect of Nuclear Fusion

The ultimate goal of using nuclear energy is to achieve controlled nuclear fusion (Xu et al. 2012). Existing reactors rely on heavy nuclear fission to release energy, such as uranium and plutonium, while fusion reactions release energy from lighter nuclei, such as isotopes of hydrogen deuterium and tritium polymerization to form helium, and the reaction product is ruthenium free from radioactive contamination.

11.2.1 Overview of Nuclear Fusion

Nuclear fusion, when light nuclei (such as deuterium and tritium) combine into heavier nuclei (such as helium), emits enormous energy. This fusion reaction between the nuclei is also called a thermonuclear reaction. Light nuclei involved in nuclear reactions, such as such as hydrogen (protium), deuterium, tritium, lithium etc., obtain the necessary kinetic energy from thermal motion to cause fusion reaction, which can generate a large amount of thermal energy in an instant, but it cannot be utilized at present. If the thermonuclear reaction can be produced and controlled in a controlled area according to people's intentions, the controlled thermonuclear reaction can be realized. The controlled thermonuclear reaction is the basis of the fusion reactor (Piao and Zhao 2017). Once a fusion reactor is successful, it may provide the cleanest and most inexhaustible source of energy to humans.

The fuels used for nuclear fusion energy are deuterium (D) and tritium. Deuterium is found in large quantities in seawater. About 6,500 hydrogen atoms in seawater has a deuterium atom, and the total amount of deuterium in seawater is about 45 trillion tons. The fusion energy released by the complete fusion of deuterium contained in each liter of seawater is equivalent to the energy of 300 L of gasoline fuel. In the foreseeable period of human survival on Earth, the deuterium in the sea are enough to meet the energy needs of humans for billions of years to come. In this sense, the fusion fuel on the earth can meet all future human needs for energy. The development of fusion energy will solve human energy needs "once and for all". After more than 60 years of unremitting efforts by scientists, significant progress has been made in the control of fusion energy.

The study of nuclear fusion began in the 1950s with magnetically constrained Tokamak achieving world-renowned achievements and is close to the reactor area. In the early 1950s, Soviet physicists Tam and Sakharov proposed that the polar magnetic field generated by large current induction in a ring plasma combined with a strong hoop (longitudinal) magnetic field could realize equilibrium shape of plasma. The Moscow Kurchatov Institute conducted the experimental study under the leadership of Azimovich. They placed a number of coils on the outer surface of the annular ceramic vacuum chamber. The charged capacitor group discharges the coils in series. A toroidal magnetic field is formed in the vacuum chamber, and the plasma current generated by the discharge of the transformer circuit itself induces a polar magnetic field, and the two magnetic fields synthesize a magnetic field having a spiral characteristic to have a good restraining effect on the plasma. They called this device Tokamak.

According to the triple product of temperature, density and time, the development history of Tokamak can be divided into five stages:

(1) The first stage: from 1950 to 1960, the triple production of Tokamak fusion device is in the range of 10^{18-19} (second · thousand electron volts)/m^3. As mentioned earlier, deuterium and tritium must be heated to a very high temperature and enter the plasma state before the fusion reaction occurs. The first fusion experiments using plasma under magnetic field constraints were published in 1951. Later, the United Kingdom,

the Soviet Union, the United States, etc. conducted research on fusion reactors, recognizing that it had unexpected difficulties. After 1958, the United States, the Soviet Union, China and other countries designed and experimented with various forms of magnetic field constraints. However, the amount of plasma leaked from the magnetic field is larger than the theoretical value, and the analysis of its cause is very difficult.

The year of 1958 was a major turning point in nuclear fusion research. The high expectations in the early stage of the study were not realized. The high temperature plasma of various devices generally showed instability, poor constrained performance, and plasma temperature is also far from the requirements of controlled nuclear fusion. On the one hand, strict secrecy between countries and mutual blockade of intelligence is an important reason hindering the development of nuclear fusion research. On the other hand, the focus at that time was to find specific ways to achieve nuclear fusion, and lack systematic research on the basic properties of high temperature plasma. Scientists have focused their research from rushing to build a fusion reactor to fundamental research on high-temperature plasma systems.

(2) The second stage: from 1960 to 1970, the fusion triple product of the Tokamak fusion device was around 10^{19} (second · kiloelectron volts)/m^3, represented by the Soviet T-3. In 1961, the Soviet Union's H. S. Yoffe and the Japanese Tihiro Ohkawa, who lived in the United States in 1963, used the so-called "minimum magnetic field" method to do the experiment of constraining the plasma, approaching the theoretical value (of leakage), and a glimmer of lightappeared.

In 1968, the controlled nuclear fusion research had a major turning point once again. The International Atomic Energy Commission held the second international conference on plasma physics and controlled nuclear fusion research in Novosibirsk, Soviet Union. At the meeting, Azimovich announced the latest experimental results obtained on the T-3 Tokamak, namely an electron temperature of 1 keV, an ion temperature of 0.5 keV, $n\tau$ (n is plasma density, t is the constraint time) reached 10^{18} (second · kiloelectron volts)/m^3. This pointed the way for nuclear fusion research, and then the Tokamak whirlwind was blown up in the world.

(3) The third stage: from 1970 to 1980, the fusion triple product of the Tokamak device was between 10^{19} and 10^{20} (seconds · keV)/m^3. Since the late 1970s, nuclear fusion research has made many important advances both in theory and in experimentation, and the Tokamak plasma parameters have been greatly improved. In order to repeat the T-3 experiment results as soon as possible, the Princeton Laboratory converted the original experimental device Star-C to the ST Tokamak; the American Oak Ridge National Laboratory built Ormak; France von Knai·O·Rhodes Institute built the TFR Tokamak; the British Karam Lab built Cleo; the Japan Atomic Energy Institute built the JFT-II Tokamak; and the Federal German Max Planck Institute built the Pulsator Tokamak. A few years later, China began research on Tokamak, and the small Tokamak CT-6 unit was put into operation in 1975. In 1978, a neutral injection heating method was used on the PLT Tokamak to raise the temperature of the plasma to 7.6 keV.

(4) The fourth stage: from 1980 to 1990, the fusion triple product of the Tokamak device was between 10^{20} and 10^{21} (second · keV)/m^3, represented by four large Tokamaks. TFTR in the United States, JT-60 in Japan, JET in Europe, and T-15 in the Soviet Union. The first three devices were built and put into operation in the early 1980s, and significant progress has been made. In 1986, TFTR's super-emission discharge set a world record for ion temperature, up to 20 keV, and maximum fusion output power greater than 10 MW, exceeding the requirements for fusion ignition. Japan's JT-60 also achieved the ratio of output power over input power, 1.25 and energy gains. Particularly exciting event is the first successful implementation of the controlled DT fusion reaction in Europe on November 9, 1991, at a temperature of 200 million °C for 2 s and a power of 1.7 MW. At present, TFTR and JT-6 have been stopped. The only one running is the European Joint Ring JET. It is one of the largest conventional Tokamaks ever built in the world. It has been ahead of the United States, the former Soviet Union and Japan, and has become a milestone in fusion research of magnetically constrained nuclear fusion.

The Southwest Institute of Physics of China's Nuclear Industry built the "China Circulator No. 1" nuclear fusion Tokamak device in Leshan City, Sichuan Province in September 1984 and put it into use. It is the first medium-sized Tokamak device in China. In May 1995, the "China Circulator New No. 1" was completed and improved in Leshan City, and its various parameters have been greatly improved, indicating that the experimental means for China to study controlled nuclear fusion has reached world-class level and will make a contribution to the search for new energy causes for mankind.

(5) The fifth stage: from 1990 to the present, the development goal is to make the fusion triple product of the Tokamak device reach 10^{21} (second · keV)/m^3 or more, basically achieving the first goal of the controllable nuclear fusion reaction, repre sented by China's EAST Superconducting Tokamak and International ITER Project. Under the current technology, the plasma temperature has reached 2×10^8–4×10^8 °C, the highest fusion output power exceeds 16 MW, and the ratio of fusion output power over input power is 1.25. The overall constraint performance has been greatly improved. However, the above progress can only be maintained for a few seconds, because the conventional Tokamak's magnet can only be pulsed, otherwise it will be burned, so achieving steady state operation has become the frontier research field of Tokamak. In this context, China timely proposed the construction of the advanced full-superconducting Tokamak program. After several years of demonstration and expert review, the EAST super-superconducting Tokamak nuclear fusion experimental device was used as the national "Ninth Five-Year Plan" major scientific project in 1997, approved by the Central Science and Technology Leading Group.

At present, the countries participating in the research and development of magnetically constrained nuclear fusion and inertial constrained fusion technology mainly include the United Kingdom, Russia, Japan, China, the United States and France. The types and parameters of devices developed by them are shown in Table 11.2.

Table 11.2 World-famous large laser devices

Laser device	Country and laboratory	Output energy (kJ)	Number of beams	Year built
GEKKO-XII	ILE, Osaka University, Japan	15	12	1983
PHEBUS	Rimmer Labs, France	10	2	
VULCAN	Rutherford Laboratory, UK	2	8	
HELEN	British Atomic Weapons Center	1	1	
NOVA	Livermore Lawrence National Laboratory	40	10	1984
OMEGA	LLE, University of Rochester, USA	30	60	
Beamlet	Livermore Lawrence National Laboratory	6.4	1	1994
Shenguang-3	China Academy of Engineering Physics	60	60	2020
NIF	Livermore Lawrence National Laboratory	1800	192	2010
MLF	French Nuclear Weapons Institute	1800	240	Unknown
100-TW	British Weapons Center	>100	32	Unknown
LEFX	Osaka University, Japan	>100		Not yet approved

11.2.2 Prospect of Nuclear Fusion Development Abroad

1. International plan

The International Thermonuclear Experimental Reactor (ITER) program is currently an international science engineering project after the world's space station. The program will integrate the major scientific and technological achievements of controlled magnetic constraint nuclear fusion in the world, and build a fusion experimental reactor that can realize large-scale fusion reactions for the first time. It will solve a large

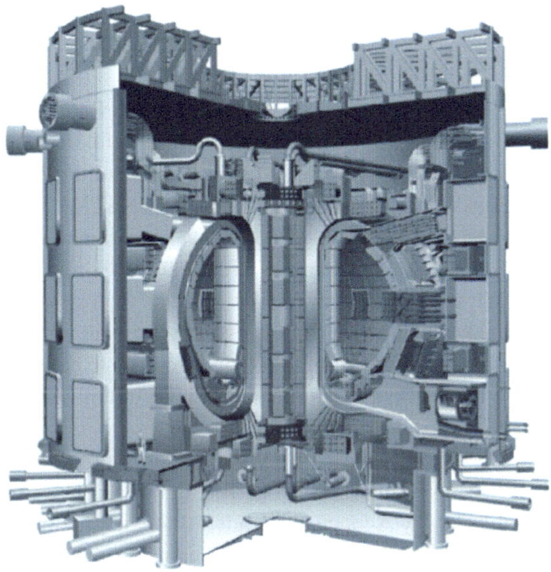

Fig. 11.3 Section view of the ITER device

number of technical problems and become a practical study of human controlled nuclear fusion research. A key step is to verify the scientific and technical feasibility of the peaceful use of fusion energy and to lay the scientific and technical foundation for the commercialization of fusion energy.

In November 2006, the European Union, the United States, Russia, Japan, South Korea, India and China signed an agreement in Paris to start the construction of the International Superconducting Magnetic Constraint International Thermonuclear Experimental Reactor (ITER). The project cost US$12 billion and will be built in France before 2020, the designed fusion power output is 500–700 MW, and the plasma discharge pulse is 500–1,000 s. ITER is a donut-shaped Tokamak (Fig. 11.3). When completed, it will be the world's largest Tokamak. Tokamak is known as the most likely way to first commercialize fusion energy. The construction of the ITER experimental reactor based on the Tokamak route indicates that the international fusion energy research has entered the research stage of the experimental reactor from basic research.

The ITER program is carried out in three phases: the first phase is the experimental reactor construction phase, from 2007 to 2020; the second phase is the thermonuclear fusion operation experimental phase, which lasts for 20 years, during which the performance of the nuclear fusion fuel, the reliability of the materials used in the experimental reactor, the developability of the nuclear fusion reactor, etc., will be verified for the scientific and technical certification of large-scale commercial development of fusion energy; the third stage is the experimental reactor disassembly phase, which will last for 5 years. After the end of the experimental phase, the participants will also carry out demonstration reactor construction to prepare for the final realization of commercial reactor development.

2. The United State

1) Sphere Tokamak Project

In 2016, physicists at the Princeton Plasma Laboratory (PPPL) of the U.S. Department of Energy published a paper on nuclear fusion, announcing their plans to develop a new generation of nuclear fusion equipment. PPPL completed the National Sphere Tokamak Experimental Upgrade Project (NSTX-U), which cost about US$94 million (about CNY 627 million) and is now operational.

The shape of the spherical Tokamak is like an apple with a core, while the traditional Tokamak is more like a donut than a spherical Tokamak. The spherical Tokamak can generate high-pressure plasma in a relatively weak, low-cost magnetic field. This special ability will help scientists to carry out a new generation of nuclear fusion experiments, complementing the International Thermonuclear Experimental Reactor (ITER).

"The reason why we want to develop a spherical Tokamak is to reduce the cost of nuclear fusion when using Tokamak," the newly appointed director of the UK Atomic Energy Agency, the leader of the magnetic constrain fusion research project at the Karam Science Center, Ian Chapman said. The size of the hole in the center of Tokamak is the key to the problem. In a spherical Tokamak, this hole is only half the size of a conventional Tokamak, so a relatively weaker magnetic field can be used to control the plasma. Once the holes are reduced, they can also be compatible with the system used to producetritium. In the next generation of Tokamaks, tritium will react with deuterium to produce nuclear fusion. In pilot power plants, researchers hope to replace the copper magnets in nuclear fusion devices with superconducting magnets. Superconducting magnets are much more efficient than copper magnets but require a thicker shield to protect them. However, high-temperature superconductors have recently made some progress, or can greatly reduce the thickness of superconducting magnets, thereby reducing the space occupied, and can also greatly reduce the size and cost of the machine.

The researchers also described a device called the "central beam injection device" that starts and maintains the plasma flow without the need for high-temperature coils in the Tokamak in their paper. The central beam injection device injects high-speed moving neutral atoms into the plasma and optimizes the magnetic field that constrains and controls the high temperature plasma.

The researchers said that the PPPL's Tokamak device is more powerful after the upgrade, and that Karam's Mega Ampere Spherical Tokamak (MAST) equipment upgrade is about to be completed, which will lead us one step further towards commercial nuclear fusion power plants. PPPL Director, Stewart Prager pointed out that the NSTX and MAST devices "will further advance the development of physics, enhance our understanding of high temperature plasmas, and build the scientific foundation for the development of nuclear fusion."

Some devices also face many physics challenges. For example, when ultra-high temperature plasma particles are exposed to strong electromagnetic fields, there are sharp fluctuations that must be controlled by these devices. In addition, they must carefully control the relationship between the plasma particles and the surrounding walls to prevent the plasma from becoming too dense or contaminated, thereby preventing nuclear fusion. Researchers at PPPL, Karam Labs, and other laboratories are working hard to find ways to address these challenges in order to better develop a new generation of nuclear fusion devices.

2) National Ignition Facility (NIF)

NIF is the abbreviation of National Ignition Facility, which is the national ignition Facility of the United States. Located at the Lawrence National Laboratory in Livermore, California, USA, there are 850 scientists and engineers, and about 100 physicists are designing experiments there. The building, which accommodates NIF, is 215 m long and 120 m wide, about the size of the Roman Colosseum and is currently the largest and most complex laser optical system in the world. The experimental process of NIF is: firstly, the external laser is enhanced by 10,000 times, then a laser beam is separated into 48 laser beams, and then enhanced, and further separated into 192 laser beams, the total energy is increased to 3,000 trillion times of the original energy, and then refocused on the diameter of 3 mm deuterium tritium pellets, a high temperature of 1×10^8 °C is generated, and the pressure exceeds 100 billion atmospheres, thereby triggering nuclear fusion. Each laser emits pulsed ultraviolet light that lasts for about three billionths of a second and contains 1.8 million joules of energy—more than 500 times the amount of electricity produced by all power plants in the United States. When these pulses impinge on the target reaction chamber, they produce X-rays that are concentrated on a plastic enclosure that is filled with heavy hydrogen fuel at the center of the chamber. X-rays will heat the fuel to 1×10^8 °C and apply enough pressure to release more than 15 times the input energy.

The program was launched in 1994 and officially started construction in 1997. It was completed in 2009 and the ignition test began in 2010. On July 5, 2012, NIF successfully merged 192 laser beams into a single pulse, generating 1.8 MJ of energy and 500 trillion watts of peak power, equivalent to more than 1,000 times of the national electricity consumption in the United States at any given time, it has become the most powerful laser pulse in human history. In the field of fusion energy research, the National Ignition Facilityaims to be the first facility to break through the balance point. Breaking the equilibrium point means that the energy produced is greater than the energy required to start it, the energy gain, which is the goal that nuclear fusion workers have been dreaming for more than half a century. This realization of the energy surplus means that NIF is one step closer to achieving this goal.

The problem with NIF is that its laser can only be fired every few hours, and the US Mercury laser solution is already planned. It is not necessarily larger than NIF and its goal is to emit pulses 10 times per second.

3. Germany

The Wendelstein 7-X unit in Greifswald, in the northeastern part of Germany, began operations in December 2015. According to the Max Planck Institute for Plasma Physics, the device was the world's largest controlled nuclear fusion device of Stellarator. In March 2016, the first round of experiments was successfully carried out to produce helium and hydrogen plasma. At the same time, after more than 2,000 experiments, the duration of the hydrogen plasma pulse was also from the initial 0.5–6 s.

The Stellarator is an imitation of a star and is a controlled nuclear fusion device. By imitating the nuclear fusion reaction inside the star, the Stellarator constrained the hydrogen isotopes deuterium and tritium of the plasma, and heated to about 1×10^8 °C for nuclear fusion to obtain continuous energy.

According to the introduction of researchers, with a 4 MW microwave heating device, the temperature inside the plasma reactor rises rapidly, the electron temperature reaches 1×10^8 °C, and the ion temperature rises to $1{,}000 \times 10^4$ °C. The results of the first round of experiments exceeded expectations and they were very satisfied with this. In 2016, the plasma reactor was being upgraded and completed in mid-2017. The Wendelstein 7-X was able to withstand higher temperatures and maintain plasma pulses for 10 s. According to the plan, after several upgrades, the plasma pulse can be achieved for about 30 min in about 4 years.

The Wendelstein7-X was built by the Max Planck Institute for Plasma Physics in Germany. The project invested more than 1 billion euros and the assembly of the equipment took 9 years until 2014.

4. Japan

The Nuclear Fusion Research and Development Division of the Japan Atomic Energy Research and Development Agency (Rokkaisho-mura Nuclear Fusion Research Institute), located in Rokkaisho-mura, Aomori Prefecture, Japanhas set up ITER's International Fusion Energy Research Center. When the ITER is put into operation, the ITER Remote Experiment Center with the same function as the ITER's control room will be set up. According to Ushigusa Kenkichi, director of the Rokkaisho-mura Nuclear Fusion Institute, the operation of ITER, which is being built in Kadarash, France, will be transmitted to the center in real time.

At present, the Rokkaisho Nuclear Fusion Research Institute is developing the generation and recovery technology and management technology of tritium, and verifying the radioactive activation of the reactor materials through experiments. In addition, in order to prepare tritium, the institute is also developing a technology for extracting lithium from seawater.

11.2.3 Prospects of China's Nuclear Fusion Development

1. Fully superconducting nuclear fusion experimental device (EAST)

EAST's full name Experimental Advanced Superconducting Tokamak is the world's first fully superconducting nuclear fusion experimental device independently designed and built by Chinese scientists (Fig. 11.4). The device passed the national acceptance in 2007. EAST is an experimental reactor. The main purpose is to study the experimental feasibility of plasma steady-state constraints, and to accumulate

Fig. 11.4 Superconducting nuclear fusion experimental device—"Keda Torus eXperiment (KTX)"

experience and data for the demonstration reactor construction for future power generation.

To stably use nuclear fusion, two conditions must be met: one is to instantaneously heat the tritium or deuterium plasma to 1×10^8 °C; the other is to lastfor at least 1,000 s to achieve a sustained reaction. This is the longest duration plasma discharge of the international Tokamak experimental device at an electron temperature of 5,000 $\times 10^4$ °C, achieving a span of 60 s to hundreds of seconds. The steady state operation mode will provide an important reference for ITER and future reactors.

The Science Engineering Management Committee believes that some previous fusion experiments lasted more than 100 s, but they were like riding a horse and it was difficult to control the unstable plasma. The experiment at EAST is more like a dressage show in which the plasma in an annular chamber shielded by a very strong electromagnetic field is controlled in an efficient and stable state, H-mode (high-constraint mode).

But at the same time, some experts believe that the continuous discharge of hundreds of seconds may not be as significant. The standard for whether nuclear fusion can be achieved is Lauson criterion, that is, the product of density, temperature, and constraint time is greater than a fixed value, and when the other two cannot reach a certain value, it does not make much sense to talk about the constraint time alone.

2. Shengguang-III

As the ignition device of the controlled nuclear fusion reactor, China Shenguang series laser ignition device has always been the focus of attention of the Chinese people. Recently, Shenguang-III, the first 100,000-J high-power laser device for high-energy density physics and inertial constraint fusion research, has been basically built in the Hefei Institute of Material Science, Chinese Academy of Sciences. The

Shenguang series has developed from Shenguang I and Shenguang II to Shenguang III, and the device has been completed and put into use.

In the past few decades, the Hefei Institute of Material Science and the Shanghai Institute of Optics and Fine Mechanics of the Chinese Academy of Sciences have continuously impacted the world's advanced level, and gradually established an independent and inertial fusion research system to build Shenguang I, II and III laser driving devices.

As the largest laser device in Asia and the second largest laser device in the world, Shenguang III has a total of 48 lasers with a total output of 180,000 J and a peak power of 60 trillion watts. In February 2015, the Shenguang IIImajor unit achieved an energy output of 7,500 J at a fundamental frequency and 2,850 J at a triple frequency. The main performance indicators of the laser have reached the design requirements, which indicates that the Shenguang III mainframe has been basically completed. China has become the second country to carry out multi-beam laser inertial constraint fusion experiments after the US national ignition facility.

3. Julong-I (PTS)

On December 27, 2014, the Julong Ifacility construction project of the Hefei Institute of Material Science, Chinese Academy of Sciences passed the national acceptance. As the first multi-channel parallel ultra-high power pulsed laser device in China, it uses ultra-high power pulse device to drive the cylindrical wire array load, which is vaporized and pinched to the shaft (i.e., Z-pinch), and the technical index reaches the international advanced level of the same kind.

At the first exhibition of military and civilian integration development of national defense science and technology industry held in Beijing in 2015, Julong-Iwas unveiled as the ideal energy source for the future ("the sun in the bottle"). Julong-I has an output current of 8 million to 10 million amperes, a current pulse rise time of less than one millionth of a second, and an instantaneous power of more than 20 MW, equivalent to twice the global average power generation.

As of May 2017, Julong-Ifacility carried out nearly 170 experiments and achieved a number of physical experiment results reaching the international advanced level. Its successful development marks China as one of the few countries that independently masters the development technology of several tens of trillion-watt ultra-high power pulse accelerators.

4. CFETR

China is developing a new program, the China Fusion Engineering Test Reactor (CFETR), to bridge the gap between ITER and future nuclear fusion power plants.

The conceptual design of CFETR was completed in 2015 and entered engineering design in 2016. According to Chinese scientists' assumptions, CFETR is completed in two phases: the first phase adopts ITER-like technology with the goal of stable operation; the second phase focuses on independent innovation with the goal of demonstrating nuclear fusion power generation.

CFETR will provide the world with the key capabilities to develop and test key elements needed for future commercial power plants, such as nuclear fusion technology, tritium production, self-sustainment, and fusion ion long-term stability. CFETR is expected to achieve a Q greater than 25 at a later stage, i.e., 25 times of each energy consumed, and ITER's goal is a Q greater than 10, which will provide China with a strong foundation for the frontier of fusion energy development.

CFETR will follow the goal of China's "two hundred years" and plan to complete the "China Fusion Dream" in three steps: from the first stage to 2021, CFETR will start construction; the second stage will be in 2035. It is planned to build a fusion engineering test reactor and start a large-scale scientific experiment. From the third stage to 2050, the fusion engineering test reactor experiment is successful, and the construction of the fusion commercial demonstration reactor begins.

5. Problems and challenges in the development of nuclear fusion

For the method of inertial constraint nuclear fusion, the laser heating method still has many drawbacks (Chen and Fan 2011). For example, China's Shenguang III still needs to tackle the problem of how to prevent fuel burn-through. The current plan is to hit all the energy of hundreds of laser heads into a very small nuclear fuel-filled target in a very short time and create a tiny fusion that will instantly complete the fusion process and release a lot of energy. The process is equivalent to the production of a tiny micro-hydrogen bomb explosion again and again, releasing the energy of the target core fuel without the impact of the explosion on the instrument (Chen and Jiang 2013). The Tokamak program needs to address key issues such as self-sustaining combustion and steady-state operation, and there are four difficulties (Xu et al. 2010):

(1) Physical theory. Although the motion of the plasma can be completely described by Maxwell's equations, even without quantum mechanics, because of the large number of particles involved, it will encounter essential difficulties. This is called "More is different". Just as in fluid mechanics, although the basic equation is the Navier-Stokes equation, the turbulence generated by it is a problem that has not been solved by physics for hundreds of years. Plasma also produces plasma turbulence because the presence of an external magnetic field is even more complicated than fluid turbulence. Physically, there is no way to find a concise model from the first principles to predict plasma behavior. What can be done now is to build some more suitable models, as well as numerical simulation techniques, like fluid turbulence studies.
(2) Physical experiments. Tokamak's high-temperature, high-density plasma is very unstable. If a probe is inserted into the plasma center, it will cause instability and the entire plasma will fall apart. For this reason, the means of experimental observations are very limited.
(3) Difficulties in engineering (Su 2016). If the ignition conditions for fusion are to be achieved, the project will produce a sufficiently strong magnetic field (approximately 10 T) in a sufficiently large volume. Nowadays, the maximum stable magnetic field that humans can achieve is about 10 T. The magnet that

produces such a large magnetic field requires a large current, and the large current will generate heat, and the material will be burned after the heat is generated. The largest Tokamak project ITER is being built, using a superconducting coil method to solve the problem of heat generation, but the coil needs to maintain superconductivity, which requires extremely low temperature, through liquid helium immersion. It is difficult to make the magnetic container itself reach −269 °C while maintaining the temperature in the core at a temperature of hundreds of millions of Celsiusdegrees.

(4) Economics. Large super-conducting magnets, the capital cost will be immeasurable. The largest Tokamak project ITER is currently being collaborated by 7 countries and is continuing to explore.

References

Chen, Guoyun, and Duping Fan. 2011. Characteristics of nuclear power and forecast of prospects. *Electric Power Technology and Environmental Protection* 27 (5): 48–50.

Chen, Wenjun, and Shenyao Jiang. 2013. Feasibility study on development of small nuclear power reactors in China. *Nuclear Power Engineering* No. 2 (134).

Guan, Genzhi, Xiaoqiong Zuo, and Jianping Jia. 2012. Nuclear power technology. *Hydropower and New Energy* (1): 7–9.

Lei, Fang, and Siqing Yan. 2017. The development of the world nuclear energy industry in the post-Fukushima era. *Old Liberated Area Built* 4: 37–40.

Pan, Ziqiang. 2012. Resolutely continue to develop nuclear power is an important way to solve the sustainable development of China's energy. *Science and Technology Review* 30 (31): 3.

Piao, Xuan, and Zhongxiu Zhao. 2017. The historical experience of world energy substitution and economic development is an important inspiration for China to restart nuclear power projects. *Modern Management Science* 2: 3–5.

Qiu, Weilin, and Wen Yu. 2017. Research on the status quo and countermeasures of China's nuclear energy industry in the new era. *Market Analysis*: 184–185.

Su, Gang. 2016. The "three steps development strategy" of China nuclear power science and technology. *Science & Technology Review* 34 (15): 33–41.

Sun, Xiaobing. 2016. The role of nuclear power in China's medium and long-term energy supply system. *Southern Energy Construction* 3: 6–15.

Sun, Guanglan, Longfang Duan, and Chunying Dong. 2016. Utilization of nuclear energy under energy crisis. *Journal of North China Institute of Aerospace Engineering* 26 (5): 3–6.

Tang, Yang. 2016. Advantages and development prospects of nuclear power generation. Water conservancy and electricity.

Xu, Buchao, Yanfei Zhang, and Ming Hua. 2010. Analysis on development modes and routes of nuclear energy in the context of low carbon economy. *Resource Science* 32 (11): 2186–2191.

Xu, Buchao, Wenbo Yan, and Ming Hua. 2012. A study on modes and routes of nuclear energy development under dual restraints from public security and low-carbon economic growth. *Sino-Global Energy* 47: 38–42.

Zhang, Yuxuan. 2017. Prospects for the development of new energy. *Hot Spotlight*: 028–029.

Chapter 12
New Energy Resources—Wind, Light and Tides

12.1 Solar Energy

12.1.1 Introduction

Solar energy refers to the energy carried by the sun. It is generally measured by the total amount of radiation that the sun shines on the ground, including the sum of the direct and diffuse scattered radiation of the sun. Solar energy is the most abundant energy resource on the earth. The sun radiates to the surface of the earth for one and a half hours of solar energy, enough to solve the world's energy consumption throughout the year. Thanks to the continuous advancement of technology, solar energy has an amazing potential to power human daily life. The use of solar power mainly includes two types of technologies: photovoltaic (PV) and concentrated solar power (CSP) (Xin, 2015).

Photovoltaic power generation is the most common and is usually used in the form of panels. When the sun illuminates the solar panel, photons from sunlight are absorbed by the cells in the panel, creating an electric field throughout the layer and causing power flow. The photoelectric direct conversion method utilizes the photoelectric effect to directly convert solar radiant energy into electrical energy. The basic device for photo-electrical conversion is a solar cell. A solar cell is a device that converts solar energy directly into electrical energy due to the photovoltaic effect. It is a semiconductor photodiode. When the sun shines on the photodiode, the photodiode turns the solar light into electrical energy and generates electricity.

Photovoltaic materials and instruments convert sunlight into electricity. A simple PV instrument is called a battery. A single PV cell is usually small and typically produces 1–2 W of electricity. In order to increase the power output of PV cells, they are connected in series to form a larger unit called a module or panel. Modules can be used alone or several connected to form an array. One or more arrays are connected into the grid as part of the overall PV system. Because of this structure of the module, the PV system can be built into a power supply system that meets the power requirements of different scales.

C. Zou, *New Energy*, https://doi.org/10.1007/978-981-15-2728-9_12

Concentrated solar power, also known as CSP, is mainly used in large power plants. Usually, mirrors or lenses are used to converge large areas of sunlight into a relatively small collection area by optical principle, so that the solar energy is concentrated, and the light collecting area on the generator is exposed to sunlight and the temperature rises. Solar energy is converted into heat, which is driven by a heat engine (usually a steam turbine engine) to generate electricity. Concentrated solar power generation is a light-heat-dynamic-electrical conversion method. Generally, solar collectors convert the absorbed heat energy into steam, and then drive the steam turbine to generate electricity. The former process is a light-to-heat conversion process; the latter process is a conversion process of heat-to-dynamic conversion to electricity (London: BP 2016).

12.1.2 Current Status of Solar Power

Since the birth of photothermal power technology in the 1950s, the global solar photothermal power generation industry has flourished. The tremendous growth of the global solar industry is paving the way for a cleaner, sustainable energy future. In the past few years, solar power costs have dropped significantly, making more homes and businesses affordable. At present, the proportion of solar power generation in the global power generation structure is still low, but its development trend is very strong (London: Shell Group of Companies 2017).

At present, China's total installed capacity is taking the lead in the global solar photothermal power generation market, the installed capacity of emerging markets has begun to release, and the entire industry is booming globally. In 2017, solar energy dominated the global investment in new energy power generation. The world's solar installed capacity increased by 98 GW, far exceeding the net increase in renewable energy, fossil energy and other power generation technologies such as nuclear energy. The main driving force for the sharp rise in solar investment in the past two years is China. China added 53 GW of installed capacity, accounting for more than half of the world's new installed capacity, and its investment increased by 58% year-over-year.

1. Development status of global solar photovoltaic power generation

(1) Global installed capacity of photovoltaic power generation continues to rise

From 2000 to 2016, the cumulative installed capacity of photovoltaic power generation in the world increased rapidly (Fig. 12.1). The cumulative installed capacity of photovoltaic power generation in the world increased from 1.3 GW in 2000 to 306.5 GW in 2016, with a compound annual growth rate of 40.7%. The newly added installed capacity increased from 0.3 GW in 2000 to 76.6 GW in 2016, with a compound annual growth rate of 41.4%. In 2016, the annual installed capacity of global photovoltaic power generation increased by 51.4% compared with 2015. China's newly installed capacity was 34.54 GW, ranking first; the United States added new installed capacity of 14.8 GW, ranking second; Japan added new installed capacity

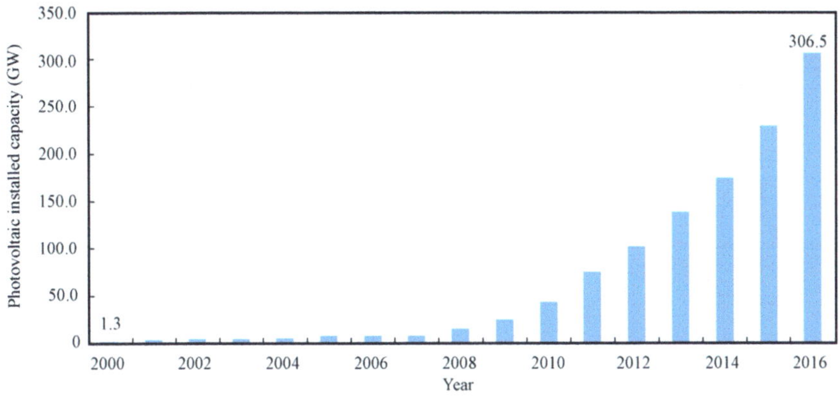

Fig. 12.1 Development status of global photovoltaic installed capacity

of photovoltaics is 10.5 GW, ranking third. Since 2011, the installed growth rate of the United States, China and other regions and countries in the Asia-Pacific region has surpassed the traditional European PV market and dominated.

At present, the United States is the country with the highest penetration rate of photovoltaic power generation in the world. Photovoltaic power generation has spread to the ordinary American households. In some of the larger states, middle-income and working-class families are increasing their investment in rooftop solar systems, and this trend is growing.

In 2016, the newly added solar PV installed capacity in the United States reached a record 14.8 GW, doubled from 2015. In 2017, due to the continuous revolution of shale gas in the United States and the rapid development of shale oil, the installed capacity of new photovoltaic power generation has declined, but still reached 12.5 GW.

In 2016, Japan's new photovoltaic installed capacity was 10,500 MW, down 2.18% from 10,800 MW in 2015. In 2017, Japan's new PV installed capacity was 6.8 MW.

From 2015 to 2017, the newly installed capacity of the European distributed PV market was 4.7 GW, 6.5 GW, and 7.1 GW, respectively. It is expected that the installed capacity of the distributed PV market will reach 10.4 GW in 2020.

(2) The cost of global photovoltaic power generation system has dropped significantly.

With the continuous advancement of technology, the global photovoltaic industry has developed rapidly and the cost of products has been declining (Fig. 12.2). The photovoltaic industry has entered a virtuous cycle of development. As a major component of photovoltaic systems, PV modules have continued to decline in production costs in recent years. According to the current development trend, the parity price of PV in most countries is expected to be realized by 2020, and photovoltaic power generation will become one of the major energy supply methods in the world (BP, BP World Energy Statistical Yearbook 2017).

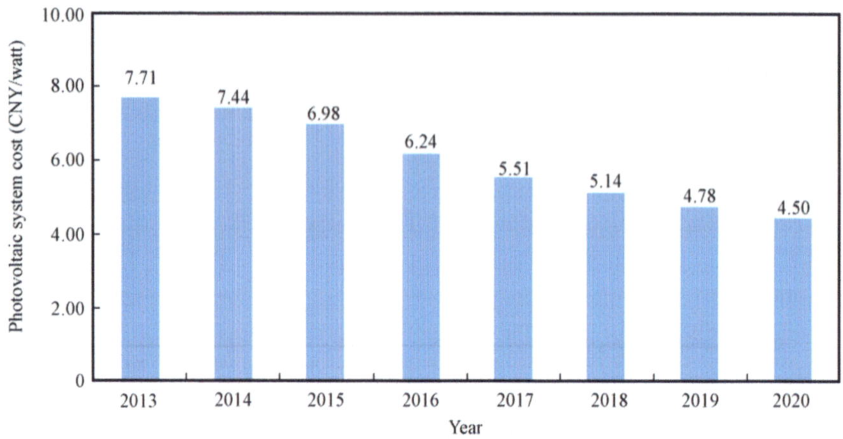

Fig. 12.2 Global PV system cost trend chart (2013–2020)

2. Development status of solar photovoltaic power generation in China

(1) China's photovoltaic power generation installed capacity is growing rapidly, with installed capacity of more than 130 GW

According to the National Energy Administration, from 2013 to 2015, the newly installed capacity of photovoltaic power generation in China was between 10 and 15 GW. The new installed capacity increased rapidly from 2016 to 2017. The installed capacity in 2016 exceeded 34 GW (Beijing: National Development and Reform Commission 2016). In 2017, the new installed capacity exceeded 53 GW (Beijing: National Development and Reform Commission 2017). As of the end of December 2017, the cumulative installed capacity of photovoltaic power generation in the country reached 130.25 GW, of which photovoltaic power station was 100.59 GW and distributed PV was 29.66 GW (Fig. 12.3).

According to statistics from the National Energy Administration, the newly installed capacity of photovoltaic power generation in China was 53.06 GW in 2017, an increase of 18.52 GW over the same period of the previous year, with a growth rate of 53.62%. The annual new installed capacity has once again set a new record. Among them, photovoltaic power plants were 33.62 GW, an increase of 11% year-over-year; distributed PV was 19.44 GW, an increase of 3.7 times. The growth rate of photovoltaic power plants dropped in 2017 (the PV power plant increased by 121% in 2016), which was mainly constrained by multiple factors such as arrears for photovoltaic power generation subsidies, limited land resources and indicators, and explosive growth of distributed photovoltaics. It is expected that the growth of photovoltaic power plants will gradually slow down in the next few years, and distributed photovoltaics will still maintain a high growth rate (Beijing: National Energy Administration 2017).

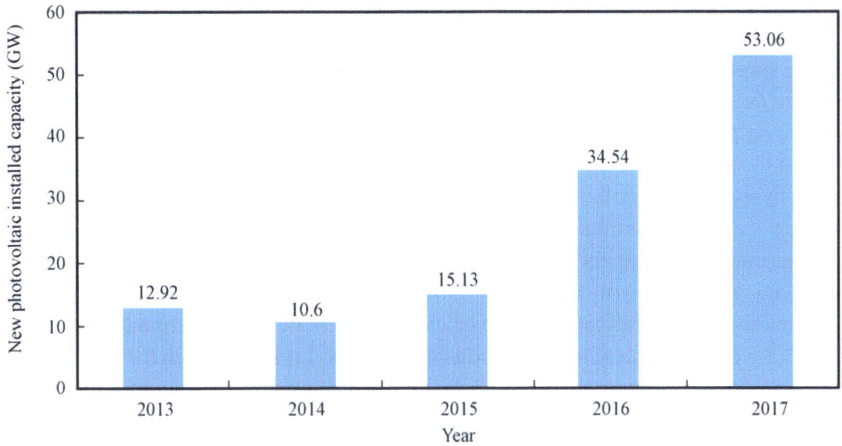

Fig. 12.3 Statistics on new installed capacity of China's PV

(2) Healthy development of China's photovoltaic industry chain

The upstream industry of the photovoltaic solar industry is silicon raw materials and polysilicon industry, and the downstream industry is solar power industry. China's solar cells, photovoltaic systems, solar thermal power generation systems and solar building integration system technologies have reached the world's leading level. The key technologies of large-scale grid-connected photovoltaic systems rank among the world's advanced level, and the development of photovoltaic micro-grid technology is basically synchronized with the international advancement. The megawatt tower solar thermal power station with completely independent intellectual property rights has made China the fourth country in the world after Spain, the United States and Germany to independently design and integrate large-scale solar thermal power stations.

In 2017, affected by the accelerated expansion of the domestic PV distributed market and the rapid rise of foreign emerging markets, China's PV industry continued to develop healthily, the industry scale grew steadily, the technical level improved significantly, production costs decreased significantly, corporate profits continued to improve and foreign trade maintained stability (State Grid Energy Research Institute 2017).

i. The industrial scale has grown steadily

In 2017, China's polysilicon production was 242,000 tons, up 24.7% year-over-year; silicon wafer production was 87 GW, up 34.3% year-over-year; cell wafer production was 68 GW, up 33.3% year-over-year; module production was 76 GW, up 31.7% year-over-year. The production scale of all links in the industrial chain accounts for more than 50% of the global total, and continues to be the world's first.

ii. The ability to innovate is constantly improving

P-type single crystal and polycrystalline cell technology continued to improve. The average conversion efficiency of conventional production lines reached 20.5% and 18.8%, respectively. The advanced production lines using passivated emitter rear contact technology (PERC) and black silicon technology reached 21.3% and 19.2% respectively. The polysilicon production process has been further optimized, and the industry's average combined power consumption has dropped to below 70 kWh/kg.

The innovation capability of photovoltaic cell technology has been greatly enhanced, creating a world record for the conversion efficiency of new battery technologies such as crystalline silicon. The establishment of an internationally competitive photovoltaic power generation industry chain has broken through the blockade of polysilicon production technology. Polysilicon production has accounted for about 40% of global production, and PV module production has reached 70% of global production. Technological advancement and expansion of production scale have caused PV module prices to fall by more than 60% during the "Twelfth Five-Year Plan" period, significantly increasing the economics of photovoltaic power generation.

iii. Production costs have dropped significantly

Driven by technological advancement and production automation and intelligent transformation, China's leading companies' polysilicon production costs fell to CNY 60,000 per ton, component production costs fell below CNY 2 per watt, photovoltaic power system investment costs fell to around CNY 5 per watt, the cost of power generation dropped to CNY 0.5–0.7 per kWh.

iv. The company's profits continue to improve

Benefiting from the expansion of the market scale, the company's shipments have increased significantly. At the same time, due to the advancement of technology and technology, the production cost has decreased. The profitability of Chinese PV companies has improved significantly. The highest gross profit margins of the upstream silicon materials, silicon wafers, raw and auxiliary materials downstream inverters, and power stations, etc. were 45.8%, 37.34%, 21.8%, 33.54% and 50% respectively.

v. Foreign trade remained stable

From January to November 2017, China's PV products exports totaled US$13.11 billion, up 1.4% year-over-year; polysilicon imports were 144,000 tons, up 17.3% year-over-year. Affected by the continued expansion of the global PV market, China's PV product exports have grown rapidly, but product export prices have continued to decline. Emerging markets such as Mexico, Brazil, and India have continued to expand, with exports to India ranking first.

12.1.3 Challenges Faced by Solar Energy and Solutions in China

1. Problems facing the photovoltaic industry

(1) Unbalanced industrial pattern

China's PV industry has an odd industrial structure of "big in the middle and small at both ends". The production pattern of raw materials relying on imports and finished products relying on exports depends too much on the international market. At the same time, due to the lack of deep understanding of technology, the production capacity is a simple copy. Production management does not adapt to the characteristics of the industry. Once the international market changes, Chinese companies will be greatly affected and lack the initiative to deal with the crisis.

From the perspective of the distribution of enterprises in the industrial chain, China's PV companies are mainly concentrated in the middle and lower streams of the industrial chain, and there are fewer enterprises in the production of polysilicon, silicon ingot casting and photovoltaic systems. The competition is much intense in the downstream, mainly due to the low production investment of downstream component manufacturing products, short construction period, low technology and capital threshold, and the closest to the market, attracting a large number of enterprises that do not have production technology and conditions to enter the photovoltaic industry. This kind of competition in the non-core technology market is not good for promoting the healthy development of the industry.

(2) Lack of core technology.

In the early stage of the photovoltaic industry, the replication and expansion of simple production scales were the mainstay. Most of the technology and production equipment were mainly imported from abroad. The arbitrariness of PV industry investment leads to the lack of core technology and the serious shortage of professional talents, which makes China's PV industry lack self-renewal ability, resulting in a large amount of equipment that has been invested can not run due to inefficiency and high production costs, forming a relatively surplus and ineffective capacity.

(3) Facing the dual risk of relying on subsidies and foreign markets

Although the annual growth rate of China's solar cell production far exceeds the world average, China's PV market has lagged far behind the development of the world PV market. Compared with China's PV manufacturing industry, it has already ranked first in the world as early as 2007, and the growth of China's PV installed capacity has been relatively slow. Insufficient domestic demand has made China's huge photovoltaic manufacturing industry rely on subsidies and relies heavily on external markets, facing double risks.

(4) Increased trade friction is not conducive to expanding the export of photovoltaic products.

According to data released by the Ministry of Commerce of China on January 21, 2016, in recent years, China's PV products have encountered anti-dumping and anti-subsidy investigations in major export markets such as the United States, with a total amount of US$25.3 billion.

Taking the United States to significantly increase the double anti-tax rate of China's photovoltaic products as an example, on July 8, 2015, the US Department of Commerce officially announced the final results of the 2012 Sino-US solar double-anti-tax rate review. According to a survey of imported silicon solar cells imported from China from May 25, 2012 to November 30, 2013, the final judgment drastically increased the double anti-tax rate of Chinese batteries.

2. Photovoltaic industry development initiatives

After 10 years of accumulation and nearly 5 years of leaping development, China's PV industry has reached a historic turning point: First, under the pressure of air pollution and emission reduction targets, the photovoltaic industry has met the unprecedented development opportunities; Second, it is gradually moving from substitution to replacement in the future, facing competition from other types of energy, and the market's consumption pressure will become more severe.

In accordance with the principles of technological progress, cost reduction, market expansion, and improvement of the system, we will promote the large-scale application and cost reduction of photovoltaic power generation, promote the industrialization of solar thermal power generation, and continue to promote the application of solar thermal energy in urban and rural areas.

(1) Comprehensively promoting distributed photovoltaic and "photovoltaic plus" integrated utilization projects.

Continue to support the large-scale promotion of rooftop photovoltaic power generation systems in power-concentrated areas such as industrial parks and economic development zones that have been completed and qualified; and actively encourage nearby cities to build photovoltaic power station projects in central and eastern cities and industrial areas with high power load and good industrial and commercial base in accordance with the principle of close utilization; combined with the comprehensive utilization of land, relying on agricultural planting, fish farming, forestry cultivation, etc., innovate various "photovoltaic" comprehensive utilization business models according to local conditions, and promote the organic integration of photovoltaics and other industries; In the possible areas such as the central and eastern regions, the "1 kW PV for all" demonstration project will be carried out to build a photovoltaic town and a photovoltaic new village.

(2) Orderly promotion of large-scale photovoltaic power station construction.

In the central and western regions with good resource conditions, access to grid conditions, and strong consumptive ability, under the premise of effectively solving the problem of abandoned light power, the photovoltaic power station construction will be promoted in an orderly manner. Actively support the implementation of photovoltaic "leader" program in the central and eastern regions by combining with environmental governance and land reuse requirements, promote advanced photovoltaic technology and product applications, and accelerate survival of the fittest in the market and the rapid decline of photovoltaic feed-in electricity prices and inferior prices. In areas with abundant hydropower resources, regulation capabilities of hydropower can be used to carry out hydropower, photovoltaic power complementarity or joint delivery demonstrations.

(3) Promoting the construction of solar thermal power generation demonstration project according to local conditions.

In accordance with the overall planning and step-by-step implementation, the solar thermal power generation industry will be actively promoted. The early development of solar thermal power generation is mainly based on demonstration. Through the construction of the first batch of solar thermal power generation demonstration projects, it promotes technological progress and large-scale development, drives localization of equipment, and gradually cultivates industrial integration capabilities. In accordance with the development principles of the first demonstration and promotion, timely summarize the experience of demonstration project construction, expand the market scale of thermal power generation projects, promote the construction of solar thermal power generation bases with good resource conditions in the western region, with conditions for consumption and ecological conditions, give full play to the peaking shaving effect of solar thermal power generation, and achieve complementary operation with wind power and photovoltaic power. The demonstrative operation mechanism of coupling coal power and solar thermal power generation was tried. Improve the technical level and system design capability of solar thermal power generation equipment, enhance system integration capability and industrial supporting capacity, and form China's independent solar thermal power generation technology and industrial system.

(4) Vigorously promoting the diversified development of solar thermal utilization

We will continue to expand the application of solar thermal utilization in urban and rural areas, actively promote the development of solar heating and cooling technologies, realize the scale utilization of solar energy hot water, heating, and cooling systems, and promote the complementary application of solar energy and other energy sources. We will continue to popularize solar water heating systems in urban civil construction and rural areas, accelerate the application of solar heating and cooling systems in the construction field, and expand the application scale of solar thermal utilization technology in industrial and agricultural production.

12.2 Wind Energy

12.2.1 Introduction

Wind Energy is the kinetic energy produced by air flow and a form of conversion of solar energy. Due to the solar radiation, the parts of the earth's surface are unevenly heated, causing the pressure distribution in the atmosphere to be unbalanced. Under the action of the horizontal pressure gradient, the air moves in the horizontal direction to form the wind. The total reserves of wind energy resources are very large, and the energy that technology can develop in a year is about 53 trillion kWh. Wind energy is a renewable, clean energy source with large reserves and wide distribution, but its energy density is low (only 1/800 of hydropower) and it is unstable. Under certain technical conditions, wind energy can be exploited as an important energy source. Wind energy utilization is a comprehensive engineering technology that converts the kinetic energy of wind into mechanical energy, electrical energy and thermal energy through wind turbines.

The history of human use of wind energy can be traced back to BC, but over the past thousands of years, wind energy technology has been slow to develop and has not attracted enough attention. Since the world oil crisis in 1973, under the double pressure of conventional energy emergency and global ecological environment deterioration, wind energy as a kind of new energy has only come back to great development. Wind energy as a pollution-free and renewable energy has great potential for development, especially for coastal islands, inaccessible remote mountainous areas, sparsely populated grassland pastures, as well as the rural, boarder areas away from the grid and the recent internal power grid is still difficult to reach, as a reliable way to solve production and living energy, is of great significance.

As an efficient and clean new energy source, wind energy has also received increasing attention. The United States began implementing the federal wind energy program as early as 1974. Its main contents are: assessing the country's wind energy resources; studying social and environmental issues in wind energy development; improving wind turbine performance and reducing cost; mainly researching wind turbines of less than 100 kW for agriculture and other users, and megawatt wind turbines designed for power companies and industrial users. In the 1980s, the United States successfully developed six wind turbines of 100, 200, 2,000, 2,500, 6,200, and 7,200 kW. At present, the United States has become the world's largest installed capacity of wind turbines, with more than 20,000 MW, and is growing at a rate of 10% per year.

At present, the world's largest new wind turbine has been built and operated in Hawaii. Its wind turbine blade diameter is 97.5 m and weighs 144 tons. The adjustment of the windward angle of the wind turbine and the operation of the unit are controlled by computer, and the annual power generation reaches 10 million kWh. From 2009 to 2013, the annual average wind power generation capacity is 7 GW, accounting for 4.5% of the end customer's electricity demand. From 2014 to 2020, it plans to account for 7% of the end customer's electricity demand. Sweden, the

Netherlands, the United Kingdom, Denmark, Germany, Japan, and Spain have also developed corresponding wind power plans based on their respective conditions. For example, in 2016, the new installed capacity of Swedish wind power was 493 MW, and the cumulative installed capacity reached 6,520 MW.

Denmark built the Jutland wind power station in 1978 with an installed capacity of 2,000 kW, a sweeping diameter of 54 m for three blades and a height of 58 m for concrete towers. In 1980, Germany built a wind power station at the mouth of the Elbe, with an installed capacity of 3,000 kW. British Isles are on the verge of the ocean, and the wind energy is very rich. The government also attaches great importance to wind energy development. In the first quarter of 2018, the UK's wind power generation exceeded the electricity produced by its eight nuclear power plants. It is also the first time that British wind power has surpassed nuclear power in one quarter. British wind power accounted for 18.8% of total electricity generation in the quarter, second only to natural gas.

In October 1991, Japan's largest wind power station in Aomori Prefecture, the Shizutetsu Strait, was put into operation, and five wind turbines could supply electricity to 700 households. For safety reasons, Japan began to phase off its nuclear power plants after the accident at the Fukushima nuclear power plant, and it was completely shut down in September 2013, resulting in a huge power gap in Japan. In support of the renewable energy power generation business, Japan followed the European countries such as Germany and Spain and launched the renewable energy fixed price acquisition system (Feed in Tariff, FIT) in July 2012. According to the latest statistics released by the Ministry of Economy, Trade and Industry, solar energy in 2016 has been able to supply 4.3% of electricity to Japan and wind power provide 1.5%.

China is located in the southeast of the Asian continent, on the west coast of the Pacific Ocean, and the monsoon is strong. The monsoon is a basic feature of China's climate. For example, the winter monsoon is 6 months in North China, 7 months in the Northeast, and the southeast monsoon is in the eastern half of China. According to the National Meteorological Administration, the total reserves of wind resources in the country are 1.6 billion kilowatts per year, and the recent development is about 160 million kW. The wind energy reserves in Inner Mongolia, Qinghai, Heilongjiang and Gansu are among the highest in China, and the days with annual average wind speed more than 3 m/s are more than 200.

The large-scale development of wind turbines in China began in the 1950s. In the late 1950s, wind turbines were awning-type windmills of various wood structures. In 1959, there were more than 200,000 wooden windmills in Jiangsu Province alone; wind power water pumps were mainly developed by the mid-1960s; after the mid-1970s, wind energy development and utilization was included in the "Sixth Five-Year" national key projects, and it has developed rapidly; after entering the mid-1980s, China has introduced a bunch of medium and large wind turbines from Denmark, Belgium, Sweden, the United States, and Germany, and established eight demonstration wind farms in Xinjiang, Inner Mongolia, and islands in Shandong, Zhejiang, Fujian and Guangdong. In 1992, the installed capacity reached 8 MW. Among them, the installed capacity of wind farms in Dabancheng, Xinjiang has reached 3,300 kW, making it the largest wind farm in the country. By the end of 1990, the irrigated area of wind pumping in the country had reached 25.8 thousand mu. In 1997, 100,000 kW of wind power were added. In 2017, China's new installed capacity was 19.5 GW, ranking first in the world in terms of new installed capacity. As of the end of 2017, China's wind power installed capacity reached 188.23 GW, accounting for 34.88% of the world. At present, China has developed more than 100 different types and capacities of wind turbines, and initially formed the wind turbine industry. Compared with developed countries, the development and utilization of wind energy in China is relatively falling behind and has not yet formed a scale. In recent years, China's wind power industry has ranked first in the world for many years, replacing the United States as the world's number one of wind power. Wind power surpassed nuclear power and became the third largest power source in China after thermal power and hydropower.

12.2.2 Current Status of Utilization

1. Development status of global wind power market

(1) North America

The market demand for wind turbine models with more than 3 MW has increased, making it a watershed for the grid connection capacity of the North American market in 2017. Vestas' 3 MW models (V117 and V126) and Nordex's AW3000 models are both popular, with new installed capacity accounting for 91% of total grid connections over 3 MW in North America; GE and Nordex become the main machine supplier of industrial and commercial power contractors. Among them, 56% of wind turbine units of industrial and commercial power contractors are provided by GE, and 2.X-116 models are used. In 2017, Nordex's new grid-connected market share exceeded 10% for the first time, a new high after 7% in 2015 (BP, BP World Energy Statistical Yearbook 2017).

(2) European Union

The 2017 Global Wind Turbine Commercial Market Share Report pointed out that the German and British offshore wind projects have become a powerful engine for the market. SGRE's SWT-6.0-154 wind turbines supplied to large offshore wind power projects in the UK and Germany are successfully connected to the grid. In the onshore wind power market, the demand for SWT 3 MW direct-drive wind turbines and G114 gearbox-driven wind turbines has increased, prompting the growth of SGRE's market share; Enercon has achieved double gains in the German and French markets with its excellent performance in low wind speed wind turbines, but has performed poorly in other European markets (Austria, Italy, Luxembourg and Portugal), leading to its drop of market share in 2017. Due to increased competition in the UK's onshore wind power market, Vestas' new grid-connected capacity declined by 5% year-over-year in 2017. The V105-3. X model has a flat performance, and Vestas has not added new grid connection to the Polish market. Dongfang Electric Company completed the grid connection of the Blaiken Phase IV wind power project in Sweden, using the DF2.5 MW-110 wind turbine model.

(3) Asia Pacific

SGRE and Suzlon have a good relationship with developers in the Indian market, and the market share accounts for more than 50% of the entire Asia Pacific (excluding China) market; the diversity of Vestas products can meet the diverse market needs of the region, the wind turbine platforms used by developers in Australia, India, Mongolia, South Korea, Thailand and Vietnam ranged from 1.8 to 3.45 MW; Hitachi's market share doubled year-over-year. Japan's local developers Eurus and Eco Powe's large-scale projects use Hitachi 2 MW models, which have been connected to the grid, becoming the main driving force; Goldwind's market share in the region increased by 4%, including Pakistan's 149 MW grid-connected capacity and the grid capacity of Australia's 175 MW White Rock project benefited from the better relationship between China and the Pakistani government and the open Australian wind power market.

(4) India

In 2017, the Indian government launched a three-year plan for renewable energy development, announcing that it will build more than 100 GW of solar and wind power projects over the next three years. Among them, by the end of fiscal year 2020, India plans to add 30 GW of wind power projects and build a wind and solar hybrid power generation project.

According to the latest statistics released by the Indian Wind Turbine Manufacturing Association (IWTMA), in the fiscal year from April 2017 to March 2018 (hereinafter referred to as "FY17"), the installed capacity of new wind power in India fell to the lowest level in 5 years. The country's total installed wind power capacity was 1,762 MW, a sharp drop of nearly 70% compared to the historical highs of the previous fiscal year.

The fiscal year of 2017 was a tough year for the Indian wind power industry. The industry has encountered a series of problems such as major policy adjustments and declining power demand, resulting in a sharp drop in new installed capacity. Since June 2016, the Indian government has stopped regulators to set on-grid tariffs, and wind power developers obtained projects through bidding. Since then, there have been few wind power projects in India for tendering, which has seriously affected the increase in wind power installed capacity.

The Indian government's adjustment of the previously set power generation based incentive (GBI) has also affected the industry's development of new projects. Originally based on the current GBI, enterprises who use wind power to generate can receive a certain subsidy per unit of electricity. At the beginning of the 2017 fiscal year, the GBI was suddenly stopped. Since then, in the middle of the 2017 fiscal year, some subsidies for wind power companies have been granted. The changing policies have confusing the entire industry and have led to a serious downturn in new installed wind power.

The Indian government actually prefers solar energy in the field of renewable energy, which has also affected the development of wind power to some extent. As early as the beginning of 2010, the Indian government proposed the JNNSM plan to increase India's solar power capacity to 20 GW by 2022. In June 2017, India officially approved the target plan for expanding solar power installations, which will increase the solar power generation target by 5 times based on the 2010 plan and achieve 100 GW of solar power generation by 2022. The Indian government has also provided a number of support policies for solar projects, such as preferential approval of land use, provision of supporting infrastructure, and tax exemptions. All of these initiatives have occupied the support that wind power has received from the Indian government. The Indian government only proposed that it plans to add 30 GW of wind power installed capacity by the end of 2020.

2. Status of world wind power development

Due to the booming industry, the Danish authoritative wind energy consultancy (MAKE) estimates that wind power generation capacity will grow by more than 65 GW per year from 2018 to 2027, with a compound annual growth rate (CAGR) of 4%. From 2017 to 2020, incentives and cross-market bidding mechanisms are the main drivers of growth, with annual production capacity expected to increase by more than 30%. From 2023 to 2027, the continued development of offshore wind power and global emerging markets will lead to a second capacity growth of more than 30%.

In North America, the US wind industry avoided a complete disaster through tax reform negotiations at the end of 2017, but the new policy has affected the project's plans and may have caused a growth bubble from 2019 to 2020. It is expected that annual productivity will drop significantly after 2021.

In 2018, the entire bidding schedule for Latin America (expected to be auctioned in Brazil, Mexico, Argentina, Colombia, and Peru) will result in a compound annual growth rate of more than 14% in the past 10 years. The return of the Brazilian wind power auction in December 2017, coupled with the 2018 bid, will help restore the market that has recently suffered political and economic turmoil.

Wind energy emerging markets in the Middle East and Asia will gradually evolve over the next 10 years. The power contract planned for South Africa and the 1.2 GW wind power contract in Saudi Arabia provide the basis for the market, and the growth in capacity in Iran and Egypt has also contributed to the expected compound annual growth rate of 26%. In the next 10 years, some emerging onshore wind power markets will increase the capacity of other parts of Asia to become the second mainland wind power market after China. The improved regulatory environment combined with new turbine technology creates opportunities for Thailand, Pakistan, Vietnam and other markets, with an overall capacity increase of over 9 GW.

(1) Offshore wind energy resources

Globally, all wind power markets rely on the growth of offshore wind power. In Northern Europe, offshore wind power accounts for 50% of new capacity, most of which are in the UK. Denmark (recently), Sweden and Ireland (long-term) are also committed to the development of offshore wind power. In Eastern Europe, the average production capacity in the region is expected to exceed 2 GW from 2024 to 2027.

The first offshore wind energy project began outside the Danish coastline in 1991. Since then, offshore wind energy facilities in the commercial sector have been operating in shallow waters around the world, mostly in Europe. Later, some wind energy projects were also formed in the United States. At the same time, the continuous development of new turbine and basic technology research has enabled wind energy projects to be implemented in deeper sea areas beyond the coastline. At the same time, the regulations of international wind energy resource management and database construction, as well as the establishment of rules and regulations, are also important factors.

The history of human utilization of wind energy resources has exceeded 2,000 years. For example, windmills are often used by farmers and herders to pump and grind grains. In modern times, wind energy is mainly used to generate electricity through wind turbines.

All wind turbines operate in the same way, that is, the wind blows the airfoil blades to drive the turbine blades to rotate. The fan blades are connected to the drive shaft to generate electricity from the generator. The latest wind turbines are highly sophisticated technologies that include engineering and mechanical engineering innovations to maximize the efficiency of generating electricity.

Offshore wind turbines have been used by many countries to extract energy from strong ocean winds. In the United States, 50% of the population lives in coastal counties, including areas along the coastline and coastal watersheds. Energy consumption and demand in these regions are generally high, but onshore renewable energy is limited. Abundant offshore wind resources make it possible to supply renewable energy to these large coastal cities in the United States, such as New York, Boston and Los Angeles.

Offshore winds are stronger and more uniform than onshore winds. The potential energy generated by wind is proportional to the cube of the wind speed. Every additional mile/hour of wind speed will generate a lot of electricity. For example, for the same wind turbine, at 16 mph, it produces 50% more electricity than at 14 mph. This is why developers are more willing to develop offshore wind energy.

The wind speeds of the South Atlantic coast and the Gulf of Mexico are generally lower than those of the Pacific Ocean. However, the current development of shallow waters in the Atlantic is more attractive and economical. Hawaii is rated as the most promising area because it has approximately 17% of the nation's wind resources.

Many countries, including the United States, have abundant coastal wind resources. The first wind farm in the United States, the Block Island wind farm, began operations in December 2016. In addition, more projects in the United States are in the planning stage, mostly in the northeastern and central parts of the Atlantic. Some projects are also considered to be built in the Great Lakes, the Gulf of Mexico and the Pacific.

Offshore wind facilities in the commercial sector are very similar to onshore wind facilities. However, offshore wind turbines require some improvements over land, including corrosion protection, and can withstand harsh environments such as strong typhoons and even ice. About 90% of wind resources are in deep water, and existing technologies cannot be applied to the area. Engineers are developing new technologies, including basic innovations and floating wind turbines, which can transform wind energy into deep seas and harsher environments.

(2) Offshore wind energy technology

The construction and design of offshore wind facilities depends on the actual situation of the construction site, especially in the deep water area. The geological conditions of the seabed and the wave conditions need to be considered. In shallow water areas, single piles are a more viable option. A steel pillar is inserted into the seabed to support the tower and the engine room. The engine compartment is used to enclose the transmission, generator, fan hub and remaining electronics components. Once the turbine is working, the yaw system connected to the wind sensor maximizes the power generated by making the engine compartment face the wind.

Technical improvements and substantial system upgrades for offshore turbines are required to adapt to the marine environment. These modifications include strengthening the tower to handle loading forces from waves or ice streams, pressurizing the engine compartment to protect critical electrical components from corrosive waves, and adding brightly colored channel platforms for safe navigation and access maintenance. Offshore turbines are often equipped with extensive corrosion protection and have climate control systems, advanced exterior paint and built-in service cranes.

To minimize routine maintenance costs, offshore turbines can have an automatic lubrication system to lubricate bearings and blades as well as heating and cooling systems to keep gear oil temperatures within specified limits. Lightning protection systems help minimize the risk of damage to some areas of the sea where lightning strikes frequently occur. Turbines and towers are usually painted in light grey or grayish white to help them blend into the sky and reduce the visual impact on the shore. The lower part of the support tower can be painted in bright colours to improve the safety of navigation of the passing through vessel.

In order to take advantage of more stable wind power, offshore wind turbines are larger than onshore turbines and have an enhanced version of the power generation capability. Offshore turbines typically have tower heights greater than 200 ft and rotor diameters of 250–430 ft. At the tip of the blade, the maximum height of the structure can easily reach 500 ft.

While the towers, turbines and blades of offshore turbines are often similar to onshore turbines, the substructure and the underlying system are significantly different. The most common type of substructure is a single pile—a large steel tube up to 20 ft in diameter. Single piles are typically used for water depths of 15–100 ft. The pile is driven into the seabed 80–100 ft below the mud line to ensure structural stability. A protruding transition piece is placed above the waterline and the flat flange is used to secure the tower frame. In shallow sea environments with a strong seabed base, a gravity-based system can be used to avoid the use of large piling hammers. The tripod and jacket base systems have been deployed in areas where the water depth exceeds the actual limit of the single pile.

(3) Transmission of wind power generation

All of the power generated by the wind turbine needs to be transmitted to shore and connected to the grid. Each turbine is connected to an Electric Service Platform (ESP) via a cable. The ESP is usually located somewhere within the turbine array and is a common collection point for all wind turbines and substations. In addition, ESP can be equipped with central service facilities, including helicopter landing pads, communication stations, crew quarters and emergency equipment.

After collecting electricity from the wind turbine, power is transmitted from the ESP operating high voltage cable to the onshore substation where it is collected into the grid. Cables for these projects are usually buried underneath the seabed, they are not damaged by anchors or fishing gear and can be avoided from exposure to the marine environment. This type of cable is expensive and is the main cost of the developer. The number of cables used depends on many factors, including the distance from shore of the project, the spacing between turbines, the need to route wires in certain directions due to the presence of obstacles, and other factors.

3. Status of development of China's wind power industry

(1) Wind energy resources are abundant, and the difference of regional distribution is obvious

China has a vast territory, a long coastline and abundant wind energy resources. According to the National Meteorological Administration's estimate in 2016, the national wind energy density is 100 W/m^2, and the total wind energy resources are about 160,000 MW, especially in the southeastern coastal and nearby islands, Inner Mongolia and Hexi Corridor, Northeast, Northwest, North China and Qinghai-Tibet Plateau. In some areas, the annual hours of wind at a speed of more than 3 m/s is about 4,000 or so. In some areas, the annual average wind speed can reach 6–7m/s or even higher, which has great development and utilization value.

The relevant experts of the National Meteorological Administration use the three-level zoning indicator system to divide China's wind energy resources.

The first level of zoning indicators: mainly consider the size of the effective wind energy density and the number of hours accumulated in the year. The areas where the annual average effective wind energy density is more than 200 W/m^2, and the accumulated hours with wind speed of 3–20 m/s is greater than 5,000 h are classified as rich wind energy zones, designated as "I"; the areas where the annual average effective wind energy density is 150–200 W/m^2 and the annual accumulated hours with wind speed of 3–20 m/s are 3,000–5,000 h, are classified as a relatively rich wind energy zones, indicated by "II"; the areas where the annual average effective wind energy density is 50–150 W/m^2 and the annual accumulated hours with wind speed of 3–20 m/s are 2,000–3,000 h, are classified as available wind energy zones, indicated by "III"; the areas where the annual average effective wind energy density is below 50 W/m^2 and the annual accumulated hours with wind speed of 3–20 m/s are less than 2,000 h, are classified as poor wind energy zones, indicated by "IV". Following the Roman numerals representing the four categories, an English letter is added to indicate each geographical area.

The second level of zoning indicators: mainly consider the distribution of wind energy density and the number of hours of effective wind energy in the four seasons. Using the wind speed data of 4 times of daily observations from 1961 to 1970, the annual changing curve of the effective wind speed hours of 483 stations with a wind speed of not less than 3 m/s is firstly formed. Then, the trends are consistently grouped together as a zone. Then add the accumulated hours of the effective wind speeds in each season and arrange them in order of magnitude. Spring refers to March to May, summer refers to June to August, autumn refers to September to November, and winter refers to December, January, and February. The spring, summer, autumn and winter seasons are represented by 1, 2, 3 and 4 respectively. If the spring has the most hours of effective wind speed (including effective wind energy), and the winter has more, it is represented by "14"; if the autumn has the most, and the summer has more, then "32" is used; the rest is the same.

The third-level zoning indicator: the maximum design wind speed of the wind turbine is generally taken from the local maximum wind speed. At this wind speed, the wind turbine is required to withstand the pressure applied to the plane perpendicular to the wind, so that the wind turbine remains stable and safe without tilting or being damaged. Since the life of the wind turbine is generally 20–30 years, for safety, the maximum wind speed value in 30 years is taken as the maximum design wind speed. According to the provisions of the Chinese Building Structure Code, the average maximum wind speed of an open site with 10 meters in height from the ground, once in 30 years, and self-recording for 10 min are used as the calculation standard, and the maximum wind speed of more than 700 meteorological stations and stations in the country have been counted for once in 30 years. According to the wind speed, the country is divided into 4 levels: the wind speed is above 35 m/s (instantaneous wind speed is 50–60 m/s), the maximum design wind speed is extra strong, called the extra strong pressure type; the wind speed is 30–35 m/s (instantaneous wind speed is 40–50 m/s), which is a strong design wind speed, called strong pressure type; wind speed is 25–30 m/s (instantaneous wind speed is 30–40 m/s), is the medium maximum design wind speed, called medium pressure type. The wind speed is below 25 m/s, which is the weakest maximum design wind speed, called weak pressure type. The four levels are indicated by the letters a, b, c, and d, respectively. According to the above principles, the national wind energy resources can be divided into 4 large regions and 30 sub regions.

Zone I (rich wind energy zone): I A34a—southeast coast, China Taiwan island and South China Sea islands, autumn and winter strong pressure type; I A21b—southern Hainan Island, summer and spring strong pressure type; I A14b—Shandong, Liaodong coastal, spring and winter strong pressure type; I B12b—West of Inner Mongolia and Xilin Gol League, spring and summer strong pressure type; I B14b—Yinshan of Inner Mongolia to the north of Daxing'anling, spring and strong pressure type; I C13bc—Songhua River downstream, spring and autumn strong-medium pressure type.

Zone II (relatively rich wind energy zone): II D34b—southeast coast (20–50 km from the coast), autumn and winter strong pressure type; II D14a—eastern Hainan

Island, spring and winter strong pressure type; II D14b—Bohai coast, spring and winter strong pressure type; II D34a—eastern Taiwan, China, autumn and winter strong pressure type; II E13b—northeast plain, spring and autumn strong pressure type in; II E14b—Inner Mongolia, spring and winter strong pressure type; II E12b—Hexi Corridor and its adjacent area, spring and summer strong pressure type; II E21b—northern Xinjiang, summer and spring strong pressure type; II F12b—Qinghai-Tibet Plateau, spring and summer strong pressure type.

Zone III (wind energy available zone): III G43b—Fujian coastal area (50–100 km from the coast) and Guangdong coastal area, winter and autumn strong pressure type; III G14a—Guangxi coastal and Leizhou peninsula, spring and winter strong pressure type; III H13b—Xiaoxin'anling, spring and autumn strong pressure type; III I12c—Liaohe River Basin and North Jiangsu, spring and summer medium pressure type; III I14c—Yellow River, Yangtze River middle and lower reaches, spring and winter medium pressure type; III I31c—Hunan, Hubei and Jiangxi, autumn and spring medium pressure type; III I12c—part of Northwest five provinces and eastern and southern areas of Qinghai-Tibet, spring and summer medium pressure type; III I14c—southwestern Sichuan and northern Yungui, spring and winter mediumpressure type.

Zone IV (wind energy weak zone): IV J12d—Sichuan, southern Gansu, Shaanxi, western Hubei, western Hunan, and northern Guizhou. spring and summer weak pressure type; IV Jl4d—north to Nanling Moutain, spring and winter weak pressure type; IV J43d—south to Nanling Mountain, winter and autumn weak pressure type; IV J14d—southern area of Yunnan and Guizhou, winter and spring weak pressure type; IV K14d—Yarlung Zangbo River valley, spring and winter weak pressure type; IV K12c—Changdu area, spring and summer medium pressure type; IV L12c—western Tarim Basin, spring and summer medium pressure type.

(2) The wind power industry tends to be rational after rapid development

The upstream of the wind power industry is a supplier of wind turbine components. Due to the wide span of parts and components required for wind turbine production and the difficulty of manufacturing, it is impossible for the whole machine manufacturer to manufacture all of them. Therefore, the technology, process level and production capacity of component suppliers have a great impact on the industry.

The downstream of the wind power industry is mainly domestic wind power investors and large power groups. These wind power investors and large power groups often invest in wind power and invest in other power sources such as thermal power, hydropower, etc. The investment direction and adjustment of investment focus directly affect the market demand of the wind power industry.

From 2013 to 2015, China's wind power installed capacity grew rapidly. In 2015, the newly installed capacity reached 30.75 million kW, a record high. After the "rushing installation tide" in 2015, the new scale of China's wind power installed capacity continued to decline in 2016 and 2017. In 2017, the newly added wind power installed capacity was 19.66 million kW, a year-over-year decrease of 3.71 million kW. The development of the wind power industry tends to be rational.

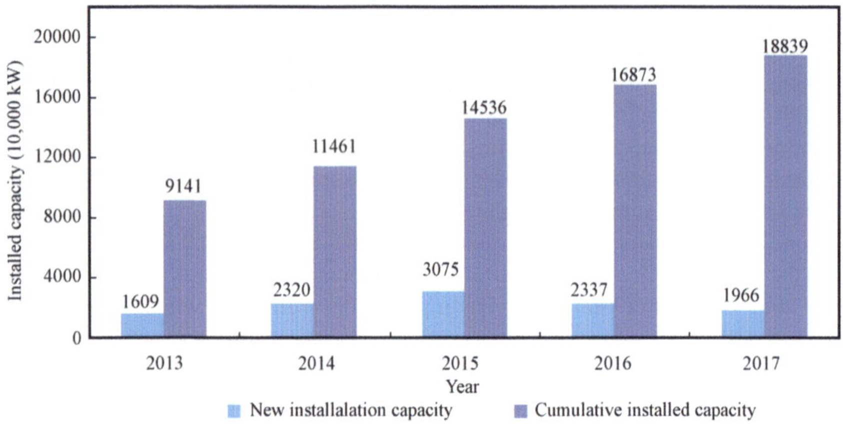

Fig. 12.4 Statistics on new installed capacity of wind power in China

(3) Wind power installed capacity ranks first in the world, abandoning wind power is common

As of the end of 2017, the cumulative installed capacity of wind power in China reached 188.39 million kW, accounting for 35% of the total installed capacity of wind power in the world, far exceeding the United States ranked second in the world (US installed capacity of 890.77 million kW) (Fig. 12.4).

In 2017, the regions with the highest average utilization hours of wind power in the country were Fujian (2,756 h), Yunnan (2,484 h), Sichuan (2,353 h) and Shanghai (2,337 h). In 2017, the area with a wind abandonment rate of more than 10% was in Gansu (33% abandonment rate, 9.2 billion kWh of abandoned wind power), Xinjiang (29% abandonment rate, 13.3 billion kWh of abandoned wind power), and Jilin (21% abandonment rate, 2.3 billion kWh of abandoned wind power), Inner Mongolia (15% abandonment rate, 9.5 billion kWh of abandoned wind power) and Heilongjiang (14% abandonment rate, and 1.8 billion kWh of abandoned wind power) (Beijing: National Energy Administration 2017).

12.2.3 Development Prospect and Challengers of Wind Energy in China

1. Development prospect of wind energy in China

(1) Operation and maintenance market will develop rapidly

With the maturity of China's wind power industry and the cumulative growth of out-of-warranty wind turbines, the wind power operation and maintenance service market has truly turned into an inflection point in 2016. In particular, the wind power machine giant that took the lead in development has more stock advantages in. The first batch of wind turbines entering the post-operation and maintenance market are from Sinovel Wind Group, Goldwind Science and Technology, China Mingyang Wind Power Group and so on. Sinovel Wind Power has more than 8,000 wind turbines with a capacity of 1.5 MW, and the current market volume has reached CNY 300 million. Sinovel Wind Power will have 6,000 wind turbines out of warranty. The operating life of a typical wind turbine is 20–25 years. After 15 years of operation, its economy will be greatly reduced. At this time, a large number of wind turbines are also facing the problems of upgrading and replacement, which will become another big demand point of the post-operation and maintenance service market (State Grid Energy Research Institute 2017).

When an industry develops to a certain stage, the post-service market will be a new driving force for industrial development. In recent years, China has also vigorously promoted the development of the service industry. With the gradual maturity of wind turbine manufacturing technology, the value-added space for operation and maintenance services will gradually expand, even surpassing industrial manufacturing.

(2) Operating mode

According to the life cycle theory, China's wind power operation and maintenance service is still in the growth stage, and the outbreak of wind power operation and maintenance market will optimize the wind power operation and maintenance service business model, update the operation and maintenance service content, and promote the industry to mature. According to the current development situation of wind power operation and maintenance industry, China's wind power operation and maintenance connotation will have a new direction of development.

i. The full lifecycle service concept becomes a future trend

In the early days of wind power industry development, wind farm developers often only focused on the quality and price of wind turbines, while ignoring the importance of operation and maintenance services. With the accumulation of experience in wind farm development, more and more wind farm developers realize that good operation and maintenance services are an important factor in maintaining the benefits of wind farms.

In foreign countries, wind farm developers have begun to consider the maintenance of the entire operating life of wind turbines from the moment they choose which wind turbine brand to use. Generally, wind power companies with wind power turbine manufactures have signed "2 + 8" or "2 + 10" wind power operation and maintenance service agreement. That is to say, after the first two years of the warranty period, the operation and maintenance services will continue to be provided by the whole machine manufactures to ensure the stability and power generation efficiency of the wind turbine.

In recent years, wind farm developers have also taken the strength of operation and maintenance services as an important evaluation standard when selecting wind turbine brands. Based on this, many whole machine manufactures have put forward the service concept of the whole life cycle, that is, to provide the owner with the whole process maintenance service from the early development site selection of the wind farm to the post-construction operation and maintenance, and then to the owner's wind farm development benefits. Sinovel Wind Power, Goldwind Sicene and Technology, United Power and many other wind farm machine manufacturers have proposed this service concept, which has led to the operation and maintenance services to be systematic and professional.

ii. Hierarchical structure of high, medium and low-end service markets

High-, medium-, and low-end operation and maintenance service market layering is another development trend in the future, which is related to the characteristics of the operation and maintenance service itself, and also related to the current development pattern of China's operation and maintenance enterprises.

As far as the operation and maintenance service itself is concerned, there are high, medium and low-end service forms and profit margins. Wind power operation and maintenance generally includes regular maintenance, daily operation and maintenance work, replacement of large components and overhaul of specific components. Among them, the regular maintenance has low requirements on the quality of operation and maintenance personnel, and the operation process is simple. It is also the highest part of wind power operation and maintenance homogenization, and the profit before tax does not exceed 5%.

Daily operation and maintenance work includes inspection and fault handling. The requirements for operation and maintenance personnel are high in technology and experience, and the operation process is also complicated. Therefore, the market competition is relatively low, the profit margin is large, and the profit before tax is between 15 and 100%. The replacement of large parts and the repair of specific parts are the most profitable parts of the operation and maintenance service, and the profit before tax is more than 30%. This part of the operation and maintenance service has high requirements for financial support, technical strength and engineering experience. The entry threshold is high. At present, only professional companies and strong machine manufacturers and developers have this capability, which is a high-end operation and maintenance service.

Due to the hierarchical structure of wind power operation and maintenance services, a high, medium and low-end market structure has been formed. At present, the main positioning of wind power manufactures is in high-end operation and maintenance services. For example, Sinovel Wind Power invested CNY 230 million for the technical upgrading of some wind power projects, achieving the results of 70% reduction in the downtime rate and more than 97% utilization rate of the unit, demonstrating its strength in this field. Sinovel also said that it will focus on high-end operation and maintenance market. Some wind farm developers also have high-end operation and maintenance strength. For example, China Datang Corporation Renewable Energy has established a subsidiary of Datang New Energy Experimental Research Institute, which can undertake all wind power operation and maintenance services of wind farms including technical transformation services.

The medium-end operation and maintenance service is mainly completed by the wind farm independently, or the wind farm invite a third-party operation and maintenance service company with strong strength to perform the work. Third-party operation and maintenance companies have more cost advantages in the field of low-end operation and maintenance services, so they are mainly completed by third-party operation and maintenance companies.

iii. Intelligent operation and maintenance to achieve value-added benefits

With the widespread application of emerging IT technologies such as big data and cloud computing, the application of emerging Internet technologies to improve wind turbine operation stability and wind farm power generation efficiency has become a new trend in the wind power industry. Smart operation and maintenance will become an important part of wind power operation and maintenance services.

"China Manufacturing 2025" released in 2015 proposed that smart manufacturing is an important direction for China's manufacturing industry. In the wind power industry, various attempts have been made in intelligent manufacturing, and intelligent operation and maintenance is an important aspect. In recent years, Sinovel has proposed the technology strategy of "Internet plus Smart Wind Energy Cloud PlatformTM". Its core connotation is the intelligent wind farm and smart wind turbine to realize the intelligent cloud manufacturing of the turbine enterprise and the intelligent management and operation and maintenance of the wind farm.

In terms of intelligent operation and maintenance, the combination of predictive diagnosis and expert diagnosis improves operational efficiency and reduces failure rate. Envision Energy also launched a smart wind energy management cloud platform to realize remote monitoring of wind turbines, wind towers and booster stations, thereby improving maintenance efficiency and reducing operation and maintenance costs. Many third-party operation and maintenance companies are also actively engaged in the layout of this field. AVIC Hi-Tech Intelligent Measurement and Control Co., Ltd. has launched an oil online monitoring system that can realize remote online detection of wind turbine gearbox fluids, thus enabling remote intelligent operation of wind turbines. The situation has improved the efficiency of operation and maintenance.

The advantage of intelligent operation and maintenance lies in the integration of big data, the life cycle operation evaluation of wind turbines, the prediction of wind turbine operation status, and timely maintenance to prevent failures, which helps to achieve the transformation of maintenance from failure operation and maintenance to planned operation and maintenance. More long-term significance is that smart operation and maintenance will also promote the construction of the energy Internet.

However, the current operation and maintenance of wind power in China is in a state of "noncooperation", which results in the inability to fully share the operational and maintenance data of various wind turbine manufacturers, which affects the integration and optimization of big data. Recently, this state has been broken. It is expected that as the technical communication and experience sharing between the whole machine manufactures, the whole machine supplier and the developers and owners gradually increase, this barrier will be broken and the interconnecting information sharing platform will be established in the future to maximize operational and maintenance benefits.

2. Challenges in future development

The development of China's wind power industry is characterized by the development of onshore wind power first, especially the establishment of onshore wind farms in areas such as the "Three North" (Northeast, Northwest, and North China) and then the development of more technically difficult offshore wind power. And simultaneously develop wind power in low wind speed regions with relatively few wind resources and relatively high technical difficulty. Correspondingly, it also formed the characteristics of first development of onshore operation and maintenance and late development of offshore operation and maintenance.

Compared with onshore operation and maintenance, offshore operation and maintenance has many difficulties due to weather and offshore conditions, high construction cost, short operation time, difficult transportation, and high risk factor. It has higher requirements for operation and maintenance service enterprises. In recent years, China's offshore wind power development has been slow, and the offshore operation and maintenance market is still in the process of being formed.

For the time being, Sinovel is the earliest and most mature enterprise in the development of offshore operation and maintenance services. In China's first offshore wind power demonstration project, the Shanghai Donghai Bridge 102 MW wind farm project, all 34 wind turbines were provided by Sinovel Wind Power and were out of warranty in 2010. It is understood that Sinovel has adopted measures such as intensive training, operation and maintenance mode innovation, enhanced data analysis and application, offshore large parts easy maintenance system, material management, intelligent operation and maintenance, etc., to achieve stable operation of wind turbines.

Developers who deployed offshore wind farms earlier also have strong offshore operations and maintenance capabilities. For example, before half year of the wind turbine warranty, Longyuan Power will send a team and the whole machine developer to do the wind turbine operation and maintenance work, learn the core operation and maintenance technology, and the specialized operation and maintenance company will be responsible for the main operation and maintenance of the wind turbine after its warranty, and secondary operations and maintenance services are undertaken by third-party operations and maintenance companies. Longyuan Power has also built a special operation and maintenance ship to save transportation costs.

Despite the high threshold for offshore operation and maintenance, the development of offshore wind power in China will form a sub-division market of the onshore operation and maintenance and offshore operation and maintenance.

12.3 Tidal Energy

In the bay or the tidal estuary, it can be seen that the sea or river has two fluctuations every day. In the morning, it is called low tide, and at night it is called high tide. As a natural phenomenon, tides provide convenience for human navigation, fishing and salt drying. This phenomenon is mainly caused by the tidal force of the moon and the sun and the effect of the earth's rotation. During the high tide, a large amount of seawater surged and had great kinetic energy. At the same time, the water level gradually increased and the kinetic energy was converted into potential energy. At the time of the ebb, the sea water rushed back, the water level declined, and the potential energy was turned into kinetic energy. The kinetic energy and potential energy of seawater in motion are collectively referred to as tidal energy.

The tide is a kind of renewable, inexhaustible, no need to mine and transport, clean and pollution-free energy with large reserves. The construction of tidal power stations does not require resettling, does not flood land, and has no environmental pollution problems. It can also be combined with tidal power generation to develop comprehensive utilization projects such as reclamation, aquaculture and marine chemicals. The main utilization of tidal energy is tidal power generation. The tidal power generation is similar to the principle of ordinary hydropower. Through the reservoir, the seawater is stored in the reservoir during the high tide, and it is stored in the form of potential energy. Then, when the tide is low, the seawater is released. The drop between the high tide and low tide makes the turbine rotate and drives the generator to produce electricity. The difference of seawater from river water is that the drop of accumulated seawater is small, while the flow rate is large and and the flow is intermittent. Therefore, the turbine structure of tidal power generation should have the characteristics of low head and large flow. The dam, gate and workshop are built in the conditional bay or the tidal estuary, and the reservoir is formed. A certain tidal drop (i.e. working head) is formed between the water level of the reservoir and the tidal level of the outer sea, so that the hydroelectric generating unit can be driven to generate electricity.

12.3.1 Current Status of Utilization

Tide Energy is the energy of periodic fluctuations of seawater. The water level difference is expressed as potential energy, and the velocity of the tide is expressed as kinetic energy. Both of these energies are available and are a renewable energy source. Because the tides are the most regular in various movements of seawater, and they rise to the shore, so they are also recognized and utilized by people. In the utilization of various ocean energies, the utilization of tidal energy is the most mature.

At the beginning of the 20th century, some countries in Europe and the United States began to study tidal power generation. In 1913, Germany established the first tidal power station on the North Sea coast. The world's first commercially available tidal power station was the Rance Tidal Power Plant, which was built at the Lens estuary in St. Malo Bay, Lens, France. It was built in 1960 and completed in 1966. The maximum tidal range at the Lens estuary is 13.4 m and the average tidal range is 8 m. A 750 m-long dam spans the Lens River. The dam is a road bridge for traffic vehicles. Ship locks, sluice gates and generator rooms are installed under the dam. There are 24 two-way turbine generators installed in the equipment room of the Lens tidal power station, which can generate electricity from high tide and low tide, with a installed capacity of 240 MW.

The first tidal power station in North America is the Annapolis Royal Generating Station, located at the entrance to the Bay of Fundy in Annapolis, Nova Scotia, Canada. The installed capacity is 20 MW and was completed in 1984. The first tidal power generator of the Race Rocks Tidal Power Demonstration Project was installed in September 2006 at the Race Rocks on Vancouver Island, Canada.

The MeyGen Tidal Power Station in the Scottish waters of the UK is the world's largest tidal power project with a designed installed capacity of 398 MW and was commissioned in November 2016. In August 2017, the project had two generators with a single-month tidal power generation of 700 MWh, breaking the world's highest monthly tidal power generation record.

The tidal power station in Sihwaho, South Korea, is designed and equipped with 10 generators with a combined generating capacity of 254 MW. In 2004, the construction of the Sihwaho tidal power station started. In April 2010, 6 generators entered the phased trial operation. In August 2011, 10 generators of the Sihwaho tidal power station were put into operation, with an annual power generation of 552 million kWh. China's coastline is tortuous, with a total length of about 18,000 km. There are more than 8,000 islands along the coast, which are rich in tidal energy resources. According to estimates, the theoretical reserves of China's tidal energy reach 110 million kW, the total installed capacity can be 21.79 million kW, and the annual power generation can reach 62.4 billion kWh. Among them, Zhejiang and Fujian have the largest reserves, accounting for 80% of the whole country.

China began to use tidal energy as early as the 1950s, and it is one of the countries in the world that used tidal energy earlier. In the late 1950s, China built more than 40 small tidal power stations or hydro force stations on the southeast coast. Due to the lack of scientific research and formal survey design, improper selection of sites, poor equipment, seawater corrosion and other reasons, after a period of operation, they were closed or abandoned. In the late 1970s and 1980s, the state built a number of larger tidal power stations, including tidal power stations such as Jiangxia, Xingfuyang, Baishakou and Haishan. The total installed capacity was nearly 6,000 kW, but now there are only two tidal power stations in Jiangsha and Haishan in operation.

China's largest tidal power station is the Jiangxia Tidal Power Station located in Jiangxia Port, the northern end of Yueqing Bay, Wenling City, Zhejiang Province. It is also the first two-way tidal power station in China. It has installed 6 tidal generating units with an installed capacity of 3.0 MW. In May 1980, the first unit was put into operation to generate electricity. The current annual generating capacity is about 10 million kWh, and the voltage supplied to Wenzhou Power Grid is at 35 kV. The

Jiangxia Tidal Power Station was the largest tidal power station in Asia at that time, and its installed capacity was second only to the French Lens tidal power station and the Canadian Annapolis tidal power station, ranking third in the world.

12.3.2 Main Utilization Techniques and Features

The tidal energy utilization technology can be divided into tidal range utilization technology and tidal current utilization technology.

1. Tidal range utilization technology

The potential energy generated by the difference in water level between low tide and high tide is called tidal energy. The tidal range can exist in large water bodies flowing into complex areas or bays and estuaries. The tidal range can be affected by the astronomical cycle such as the moon, the sun and the earth's gravity. It is not affected by weather conditions and the changes of its day, month or year can be accurately predicted.

The utilization of tidal energy is mainly used for power generation. Since the establishment of the world's first tidal power station in Germany in 1913, the development of tidal power generation technology has gone through a hundred years. France, South Korea and the United Kingdom have successively built large-scale tidal power stations.

Most conventional tidal systems use a cross-flow turbine, which is equivalent to a hydro turbine installed in a dam (a river power plant). The current mature technologies for tidal power generation include one-way low tide power generation, one-way high tide power generation, and two-way power generation.

One-way low tide power generation: At high tide, once the tide reaches the highest level, the gate or sluice is closed to fill the reservoir; at low tide, the water in the reservoir is released by the turbine and used to generate electricity. With a single cycle, only four hours a day generate electricity. Annapolis in Canada is a low tide power plant.

One-way high tide power generation: When the tide is high, the gate is kept closed to isolate the reservoir at the lowest water level; when the tide rises high, the water from the sea side flows into the reservoir through the turbine to generate electricity. The shortcomings of this cycle are small capacity, low power generation, and because the water level in the reservoir is kept low for a long time, it is likely to cause damage to the ecology. South Korea's Sihwaho is a high tide power plant.

Two-way power generation: The inflowing and outflowing tides are generating power by turbines. This cycle generates electricity twice a day for four hours. Two-way power generation requires reversible turbines. Lens project in France is a two-way power plant with a cross-flow turbine that optimizes pumping capacity.

2. Tidal current utilization technology

The tidal current is the kinetic energy that tidal water possesses when it moves horizontally, also known as ocean current energy. Currently, commercial power plants have not been built in the world, but equipment that uses tidal energy to generate electricity is under development, and some of them have been tested in full-scale in British waters.

Most demonstration projects for tidal power use horizontal axis turbines, which can be divided into three main types.

(1) Horizontal axis and vertical axial cross-flow turbine

Horizontal and vertical axis tidal turbines currently use blades that are either parallel to the direction of water flow or perpendicular to the direction of water flow. Turbines are similar to those used in wind turbines, but because of the higher density of water, the blades are smaller and rotate more slowly than wind turbines. In addition, they must withstand greater force and motion than wind turbines.

Most designs use blades that are connected to the center shaft and are connected to the generator shaft through a gearbox. The central open turbines have different designs, the blades are mounted inside and the center is open and placed on the shaft in the static tube. As the water flows through the shaft, it rotates and generates electricity. The blades of the horizontal or vertical turbine can also be enclosed within the conduit. The latter is known as a closed, conductive or closed turbine. Due to the closure, ocean currents are concentrated and streamlined to increase flow and power output from the turbine.

(2) Reciprocating device

The reciprocating device is also known as a hydrofoil. The blades of the hydrofoil, like the shape of an airplane wing, move up and down as the current flows on either side of the blade. The up and down motion of the hydrofoil is then converted into a rotation to drive the shaft, or connected to the piston to support the hydraulic system to generate electricity. The advantage of the reciprocating motion device is that the length of the blade is not limited by the depth of the water, however, it also requires a complicated control system to properly push the blade.

(3) Supporting structure

All utilization technologies of tidal currents require supporting structures to keep technology in place and withstand the harsh conditions of the ocean. The choice of foundation depends on the location of the tidal current utilization technology, water depth, subsea structure and supporting construction vessels and offshore drilling units. The support structure consists of three types: the first type is a gravity structure, which includes a large amount of concrete and steel, and the power generation equipment is connected to the seabed; the second type is a stacked structure, which drills or fixes one or more piles on the seabed; The third category is the so-called floating foundation, which is connected to the seabed by hard wires or cords or chains.

(4) Array pattern deployment

Single generators have limited capabilities, so multi-row arrays of tidal turbines need to be built to capture the full potential of the tidal current. The configuration in which they are placed is a key factor in determining potential production and output.

At present, the tide utilization technology is at the testing stage. Canada, China, France, Ireland, Japan, Korea, Spain, the United Kingdom, and the United States have tested the utilization technology of tide energy.

In 2016, the 1 MW generating unit of the offshore power generation platform was built and generated in the southern part of Xiushan Island, Daishan County, Zhoushan, China, marking the successful completion of the world's first megawatt-scale tide energy power station. After the unit is stably generating electricity, the annual power generation will reach 6 million kWh.

12.3.3 *Prospects for Development and Existing Challenges*

1. Good development prospect of tide energy

The world's tidal resources are considerable. Experts predict that global tidal energy resources will be 3 TW, with technology available resources of approximately 1 terawatt. The shape of the coast determines the fluctuation of the tidal range, and the level of the tidal range determines the abundance of tidal energy. Argentina, Australia, Canada, Chile, China, Colombia, France, Japan, Russia, South Korea, Spain, the United Kingdom and the United States have high tidal ranges and abundant tidal energy resources.

The utilization of tidal energy requires a water flow velocity of at least 1.5–2 m/s. The abundance of tidal energy also depends on the shape of the coast. It is estimated that the minimum available tidal resources in Europe is 12,000 MW. At present, most of the world's surveys of tidal energy are insufficient, and the understanding of tidal energy resources is still unclear.

Since the late 1960s, Canada, China, and France have begun to commercialize tidal energy, most recently in South Korea. With regard to tidal range, the initial installation-related costs are high, however, they have good long-term return characteristics. Many installations in the 1960s and 1970s are still without problem and are still running.

Because the cost of each site varies widely, there is little economic data available. The two main cost factors are: the size (length and height) of the tidal gate determines the cost of capital; the height difference between the high tide and the low tide determines the power generation intensity. Based on sources of network data, estimates for the installation of the largest and oldest lagoon installation in the Lance have shown that the cost per kWh is between 0.04 Euros and 0.09–0.12 Euros. The South China Lake Power Plant is the world's largest installed tidal range, with an estimated cost of approximately US$300 million and a power generation cost of US$0.024/kWh (Wyre Energy Ltd. 2013).

Construction costs do not necessarily need to be included in power generation. In the case of the Lens project, the building also has a highway function that can reduce the travel distance of 60,000 cars by 30 km per day. The Sihwaho tidal gate is built on top of the existing dam.

In addition to costs at the early stage, other major costs may be internal control, monitoring and ecological management of the reservoir. A study conducted by the Norwegian Ministry of Highway Management in 2012 showed that the cost of tidal range and tidal current utilization technologies can be reduced by 40% when buildings are integrated into the design of new infrastructure (such as coastal defense facilities, water quality measurements or roads). In addition, the integration of coastal protection facilities and bridges with tidal energy installations can significantly reduce equipment maintenance and operating costs.

The development of commercial arrays of tidal current utilization technologies is still in the demonstration stage. Therefore, the cost of leveling power generation is 0.25–0.47 Euro/kWh, and the lower range estimates are based on high capacity factors and low capital cost. The highest tidal technology costs are associated with installed capacity (35%), buildings (15%), and maintenance and operations (15%), with installed costs varying widely from place to place.

By 2020, the capacity factor of the array will increase from approximately 25% to 40%, and the availability factor will increase from 70 to 90%. By 2020, if the deployment is in a 200 MW sequence, the leveling power generation cost is 0.21–0.25 Euros/kWh. These estimates are similar to those of the Carbon Trust, which estimates that the cost of current technology in 2020 is 0.17–0.23 Euros/kWh. Deployment of high or low quality resource areas may increase to 0.16–0.30 Euros/kWh. Assuming the scale will be upgraded to 2–4 GW by 2030, the leveling cost of tide energy may fall below 0.2 Euros/kWh.

2. Technical barriers still exist

The potential of tidal energy is great, especially in certain places. In the past few years, the successful demonstration of full-scale tidal current technology has been supported by the government, as well as technology investment and project development of private capital in these areas.

The most important driving force for tidal range and tidal currents is that two types of technologies can generate renewable energy in areas close to cities, without negative environmental impacts on natural landscapes. On the positive side, tidal range installations have the least impact on the landscape, zero emissions and no noise, while contributing to or as part of the dyke facility or sluice, but there are still many obstacles to overcome.

(1) Technical barriers

The challenge for tidal range utilization technology is to increase turbine efficiency. For tidal current utilization technologies, basic technologies already exist, due to inadequate experience with materials, and the ability to work and install buildings in harsh environments, as well as lack of information and understanding in performance, life cycle, power plant technical operation and maintenance, so technical challenges are rising.

To become a true alternative to conventional energy, tidal current technology needs to increase its focus on technical risks such as design, construction, installation and operation. Costs need to fall by at least 50%, which is comparable to offshore wind power generation costs. In addition, the introduction of knowledge and experience in other industries is important, such as offshore oil and gas equipment and offshore wind farms, including risk assessment, environmental impact assessment and engineering standards.

More extensive research on materials and methods, new accessories and robust testing of the entire functional prototype require the establishment of these new technologies mentioned above. For tidal current utilization technology, the cost of seabed fixing, maintenance and installation needs to be reduced. In addition, more experience is required in array arrangement.

(2) Ecological issues

The potential of traditional tidal range utilization technologies to close rivers with dams or reservoirs is limited due to ecological constraints. Experience with artificially closed buildings has shown that managing artificial tidal basins is costly and requires careful monitoring and planning. The Canadian power plant deserves attention, and from the beginning of the plant operation, there is a discussion about the impact on fish and marine life and how to reduce the footprint with sufficient evidence. This information is currently valuable because ecological issues are an important requirement and condition for obtaining installation permits in protected waters.

The dams and tidal gates, which were usually built in the 1950s and 1970s, are reopened, which may have great ecological benefits for the water bodies located beneath them, as these increasing gradients contribute to the ecology of the waters and increase the oxygen content; Tide energy technology can also be used as a tool for water management while generating electricity. A more innovative tidal range utilization technique, namely the incomplete closure of the reservoir, is currently under development.

The challenges of tidal current utilization technology are various, and the ecological impact is smaller than tidal range utilization technology, but environmental regulators lack corresponding expertise or environmental risk assessment tools. Limited baseline data on biodiversity in seawater has led to increased costs for collecting evidence and deploying monitoring.

(3) Lack of industrial cohesion

The development of tidal current utilization technology has been linked to small and micro enterprises, many of which are by-products of university projects. As a result, there is a lack of cohesion in the industry, with many different designs and a large number of small-scale producers. Large turbine manufacturers such as ABB, Alstom, Andritz, Siemens and Voith Hydro have entered the emerging industry by intervening during the start-up phase. New interest is creating the necessary conditions to expand existing full-scale demonstration turbines into the array.

Tide energy also requires investment and R&D to develop and deploy viable, scalable business technologies and facilities, better understand environmental impacts and benefits, and achieve the entry of market. The orientation of most new projects is to help commercialize the technology, promote the access to research facilities, or support the establishment of new demonstration sites at sea. The tide energy utilization technology requires a support chain similar to the offshore wind power and oil and gas industry. It is expected that the participation of large and multidisciplinary industries will promote synergies that will generate economies of scale and reduce costs.

(4) Lack of new financing mechanisms

The cost of most projects using tidal current technology is provided by government funds or technology developers themselves. Australia, Canada, France, Ireland, South Korea and the United Kingdom have already had active policies to support the research and demonstration of tidal current utilization technology. Some countries have initiated more aggressive ocean energy policies.

Tidal current utilization technologies that have been tested on a full scale are particularly critical to the demand for new financing mechanisms and require scale deployment through market-driven incentives. Possible ways to attract investment: high-cost commercial installations can be made more attractive by providing tax returns to investors, attracting end users, or by fixing prices. In addition, appropriate mechanisms for risk sharing or reducing insurance risks also reduce the total cost of the project.

(5) Insufficient grid facilities

For the utilization of tidal current technology, the connection to the land grid may also be the problem. Some coastal countries, such as Portugal, the Netherlands, Norway, the southwest of the United Kingdom and some parts of Spain, have high-pressure transmission lines close to the waterfront, but many countries with tidal energy resources lack sufficient power transmission capacity to provide grid access for the power supply, and a large number of offshore test centers have not yet established grid connections.

Similar problems have been found for offshore wind energy. The European Commission, along with industry and member states, is supporting the development of offshore grid integration facilities. To transmission offshore wind energy to customers is proposed mainly through the North Sea International Offshore Power Grid. The possibility of offshore wind farm growth was considered and options for building an offshore grid in Europe were defined.

Port facilities are very important for further development. Marine systems are costly to install, operate, and maintain, and are more expensive if operated in high turbulence and variable waters. In order to reduce operating and maintenance time and costs, it is considered a good choice to remove transformer replacements from tidal energy offshore arrays and maintain them at safer and more accessible port facilities. Along with other ancillary services, appropriate space and port facilities are needed, and it is necessary to consider planning infrastructure management for coastal areas where tidal energy represents a true alternative energy source.

(6) How to integrate with economic and social functions.

Considering that the tidal range utilization technology is relatively new, most projects and projects pay special attention to the equipment itself and the technology of direct infrastructure. The combination of larger systems and other factors, such as shipping, recreation, water protection and ecological impact, can not only reduce installation costs, but reduce other types of costs and increase social acceptance. The Norwegian Highway Authority conducted some demonstrations.

For tidal current utilization technology, there are many hybrid system plans that combine floating offshore wind energy with tidal current utilization technology. In most cases, the tide current utilization technology cannot be well matched with the offshore wind farm, requiring strong tides which will increase the installation cost of the offshore wind farm. Therefore, the development of tidal energy requires technological innovation, cost reduction and consideration of local conditions.

References

BP. 2016. BP sustainability report 2016.
BP. 2017. BP world energy statistical yearbook.
National Development and Reform Commission. 2016. The 13th five-year plan for power development.
National Development and Reform Commission. 2017. The 13th five-year plan for power development.
National Energy Administration. 2017. Guidance on the implementation of the "13th Five-Year Plan" for renewable energy development.
Xin, Peiyu. 2015. *Comprehensive evaluation and application prospects of solar power generation technology*, 5–21. Beijing: North China Electric Power University.
Shell Group of Companies. 2017. World energy model.
State Grid Energy Research Institute. 2017. *China new energy power generation analysis report*, 1–35. Beijing: China Electric Power Press.

The manufacturer's authorised representative in the EU is Springer Nature Customer Service Centre GmbH, Europaplatz 3, 69115 Heidelberg, Germany. If you have any concerns regarding our products, please contact ProductSafety@springernature.com

Printed and bound by CPI Group (UK) Ltd, Croydon, CR0 4YY
25/09/2026
02222886-0005